*Alkaloids: Chemical
and Biological
Perspectives*

ALKALOIDS: CHEMICAL AND BIOLOGICAL PERSPECTIVES

Volume Four

Edited by

S. WILLIAM PELLETIER
Institute for Natural Products Research

and

The Department of Chemistry
University of Georgia, Athens

A Wiley-Interscience Publication

JOHN WILEY & SONS

New York Chichester Brisbane Toronto Singapore

Library of Congress has Cataloged this Serial publication as follows:
(Revised for volume 4)
Main entry under title:

Alkaloids: chemical and biological perspectives.

"A Wiley-Interscience publication."
Includes bibliographies and indexes.
1. Alkaloids.   I. Pelletier, S. W., 1924–

[DNLM: 1. Alkaloids.   QD 421 A4156]
QD421.A56   1983   574.19′242   82-11071
ISBN 0-471-89301-3 (v. 4)

Printed in the United States of America

10 9 8 7 6 5 4 3 2 1

Dedicated to
the memory of

Robert Burns Woodward

(1917–1979)

Considered by many as the greatest organic chemist of modern times, he was a master of synthesis and structure elucidation, innovator in the field of theoretical organic chemistry, and mentor to over 400 graduate and post-doctoral students. Woodward made monumental contributions to the field of alkaloid chemistry, including the total synthesis of quinine (1944), strychnine (1954), reserpine (1956), and colchicine (1963); the structure elucidation of strychnine (1948), cevine (1954), tetrodotoxin (1964), and of the antibiotic alkaloids penicillin (1945), terramycin (1952), aureomycin (1952), gliotoxin (1958), and streptonigrin (1963).

Contributors

J. W. Daly, Laboratory of Bioorganic Chemistry, National Institute of Arthritis, Diabetes, and Digestive and Kidney Diseases, National Institutes of Health, Bethesda, Maryland

W. Fenical, Institute of Marine Resources, Scripps Institution of Oceanography, La Jolla, California

R. F. Keeler, United States Department of Agriculture ARS, Poisonous Plant Research Laboratory, Logan, Utah

T. F. Spande, Laboratory of Bioorganic Chemistry, National Institute of Arthritis, Diabetes, and Digestive and Kidney Diseases, National Institutes of Health, Bethesda, Maryland

P. G. Waterman, Phytochemistry Research Laboratory, School of Pharmaceutical Sciences, University of Strathclyde, Glasgow, Scotland

Preface

Certain brightly colored frogs native to the rain forests of western Columbia, Costa Rica, and Panama elaborate an array of intensely poisonous alkaloids that have been used as poisons for arrows and blow darts for centuries. These and other amphibian alkaloids are remarkable from the standpoint of structure and biological activity. Over 200 alkaloids of various structural types have been isolated from amphibians. In Chapter 1 the chemistry, pharmacology, and biology of these amphibian alkaloids are reviewed in a monumental treatise by John W. Daly and Thomas F. Spande.

Chapter 2 provides the first comprehensive survey of marine alkaloids, compounds occurring in both marine plants and animals, but possessing structures different from the usual alkaloid structures found in terrestrial flora and fauna.

Chapter 3 treats Diels-Alder type dimeric alkaloids of the Rutaceae family, including biogenesis and synthesis. This interesting class of alkaloids has been discovered within the past few years and the review demonstrates the power of high-field NMR techniques in solving stereochemical problems.

Today evidence is substantial that congenital defects in mammals can be induced by dietary constituents. No longer can we be comforted by the idea that the developing mammalian embryo is protected *in utero* from passage of harmful compounds through the placenta. Compounds inducing congenital defects (teratogens) have only recently been identified, and in Chapter 4 of this volume Dr. Richard Keeler presents an exciting review of the biological and chemical aspects of the teratology of steroidal alkaloids, including species, structural and configurational specificity for teratogenicity, and the biochemical mechanism of action of steroidal alkaloid teratogens.

Each chapter has been reviewed by an authority in the field. Indexes for both subjects and organisms are provided.

I thank the authors for their work in preparing chapters for this volume and James L. Smith, Editor in the Wiley-Interscience Division, who has been a source of advice, help, and encouragement.

S. WILLIAM PELLETIER

Athens, Georgia
March 1986

Contents of Previous Volumes

VOLUME 1

1. The Nature and Definition of an Alkaloid 1
   S. William Pelletier
2. Arthropod Alkaloids: Distribution, Functions,
   and Chemistry 33
   Tappey H. Jones and Murray S. Blum
3. Biosynthesis and Metabolism of the Tobacco Alkaloids 85
   Edward Leete
4. The Toxicology and Pharmacology of Diterpenoid Alkaloids 153
   M. H. Benn and John M. Jacyno
5. A Chemotaxonomic Investigation of the Plant Families of
   Apocynaceae, Loganiaceae, and Rubiaceae by their Indole
   Alkaloid Content 211
   *M. Volkan Kisakürek, Anthony J. M. Leeuwenberg, and
   Manfred Hesse*

VOLUME 2

1. Some Uses of X-Ray Diffraction in Alkaloid Chemistry 1
   Janet Finer-Moore, Edward Arnold, and Jon Clardy
2. The Imidazole Alkaloids 49
   Richard K. Hill
3. Quinolizidine Alkaloids of the Leguminosae: Structural Types,
   Analyses, Chemotaxonomy, and Biological Properties 105
   A. Douglas Kinghorn and Manual F. Balandrin
4. Chemistry and Pharmacology of the Maytansinoid
   Alkaloids 149
   Cecil R. Smith, Jr. and Richard G. Powell
5. ^{13}C and Proton NMR Shift Assignments and Physical Constants of
   C_{19}-Diterpenoid Alkaloids 205
   *S. William Pelletier, Naresh V. Mody, Balawant S. Joshi,
   and Lee C. Schramm*

VOLUME 3

1. The Pyridine and Piperidine Alkaloids: Chemistry
    and Pharmacology   1
    Gabor B. Fodor and Brenda Colasanti
2. The Indolosesquiterpene Alkaloids of the Annonaceae 91
    Peter G. Waterman
3. Cyclopeptide Alkaloids 113
    Madeleine M. Joullie and Ruth F. Nutt
4. *Cannabis* Alkaloids 169
    Mahmoud A. El Sohly
5. Synthesis of *Lycopodium* Alkaloids 185
    Todd A. Blumenkopf and Clayton H. Heathcock
6. The Synthesis of Indolizidine and Quinolizidine Alkaloids of
    Tylophora, Cryptocarya, Ipomoea, Elaeocarpus and Related Species
    241
    R. B. Herbert
7. Recent Advances in the Total Synthesis of Pentacyclic
    Aspidosperma Alkaloids 275
    Larry E. Overman and Michael Sworin

Contents

1. AMPHIBIAN ALKALOIDS: CHEMISTRY, PHARMACOLOGY, AND BIOLOGY 1
  John W. Daly and Thomas F. Spande

2. MARINE ALKALOIDS AND RELATED COMPOUNDS 275
  William Fenical

3. THE DIMERIC ALKALOIDS OF THE RUTACEAE DERIVED BY DIELS-ALDER ADDITION 331
  Peter G. Waterman

4. TERATOLOGY OF STEROIDAL ALKALOIDS 389
  Richard F. Keeler

Subject Index 427

Organism Index 439

*Alkaloids: Chemical
and Biological
Perspectives*

Chapter One

Amphibian Alkaloids: Chemistry, Pharmacology, and Biology

John W. Daly and Thomas F. Spande
Laboratory of Bioorganic Chemistry
National Institute of Arthritis, Diabetes,
and Digestive and Kidney Diseases
National Institutes of Health
Bethesda, Maryland 20892

CONTENTS

1.	INTRODUCTION	4
2.	BATRACHOTOXINS	5
	2.1. Structures	5
	2.2. Syntheses	14
	2.3. Biological Activity	22
	2.3.1. Toxicity	22
	2.3.2. Pharmacological Activity	23
	2.3.3. Structure and Activity	28
	2.3.4. Binding of a Radioactive Batrachotoxin Analog	30
	2.3.5. Summary	35
	2.4. Addendum	35
3.	HISTRIONICOTOXINS	36
	3.1. Structures	36
	3.2. Syntheses	45
	3.2.1. Spiro Intermediates	47
	3.2.2. Cycloaddition Reactions	60
	3.2.3. An Intramolecular Ene Reaction	62
	3.2.4. Acyliminium Ion Cyclization	63
	3.2.5. Intramolecular Mannich Reactions	69
	3.2.6. Intramolecular Michael Additions	70
	3.2.7. Stereoselective *cis*-Enyne Syntheses	77
	3.2.8. Histrionicotoxin	80
	3.2.9. Chemical Modifications of Natural Histrionicotoxins	82

3.3. Biological Activity   83
   3.3.1. Toxicity and Pharmacological Activities   83
   3.3.2. Acetylcholine Receptor Channel Complex   84
   3.3.3. Binding of Radioactive Perhydrohistrionicotoxin   88
   3.3.4. Sodium Channels   91
   3.3.5. Potassium Channels   91
   3.3.6. Summary   92
3.4. Addendum   92

4. GEPHYROTOXINS   95
4.1. Structures   95
4.2. Syntheses   98
   4.2.1. Stereoselective Hydrogenation   98
   4.2.2. Diels-Alder Additions   100
   4.2.3. Sigmatropic Rearrangement   104
   4.2.4. Diels-Alder/Michael Addition   106
   4.2.5. Acyliminium Ion Cyclization   106
   4.2.6. The Synthesis of Gephran   110
4.3. Biological Activity   110
   4.3.1. Toxicity   110
   4.3.2. Pharmacological Activity   111

5. INDOLIZIDINES (BICYCLIC GEPHYROTOXINS)   112
5.1. Structures   112
5.2. Syntheses   117
   5.2.1. Synthesis of (5E,9E)-3-*n*-Butyl-5-*n*-propylindolizidine
       (Indolizidine **223AB**)   118
   5.2.2. Synthesis of Other Stereoisomers of 3-Butyl-5-propyl-
       indolizidine   119
   5.2.3. Enantioselective Synthesis of (−)-Indolizidine **223AB**   123
5.3. Biological Activity   123
   5.3.1. Toxicology   123
   5.3.2. Pharmacology   124
5.4. Addendum   124

6. PUMILIOTOXIN-A CLASS   124
6.1. Structures   124
6.2. Syntheses   135
   6.2.1. Pumiliotoxin **251D**   136
   6.2.2. Pumiliotoxin A   137
   6.2.3. Pumiliotoxin B   138
   6.2.4. Allopumiliotoxins   140
6.3. Biological Activity   142
   6.3.1. Toxicity   142
   6.3.2. Pharmacological Activity   142
   6.3.3. Summary   145
6.4. Addendum   145

7. DECAHYDROQUINOLINES (PUMILIOTOXIN-C CLASS)   146
7.1. Structures   146

7.2.   Syntheses   151
    7.2.1.   Intramolecular Diels-Alder Reactions   151
    7.2.2.   Intermolecular Diels-Alder Reactions   154
    7.2.3.   Tetrahydroindanones   156
    7.2.4.   Intramolecular Enamine Cyclizations   157
    7.2.5.   [3,3]Sigmatropic Rearrangement   160
    7.2.6.   Unsuccessful Routes   161
7.3.   Biological Activity   162
    7.3.1.   Toxicity   162
    7.3.2.   Pharmacological Activity   162
7.4.   Addendum   163

8. SAMANDARINES   163
8.1.   Structures   163
8.2.   Syntheses   170
    8.2.1.   Samandarone and the Hara-Oka Alkaloid   170
    8.2.2.   The Structure of Neosamane   175
    8.2.3.   Cycloneosamandione and Cycloneosamandaridine   176
    8.2.4.   Samanine   179
    8.2.5.   Interconversions of Samandarine Alkaloids   180
8.3.   Biological Activity   182

9. TETRODOTOXINS   182
9.1.   Structures   182
    9.1.1.   Tetrodotoxin   182
    9.1.2.   Chiriquitoxin   186
    9.1.3.   Zetekitoxin (Atelopidtoxin)   187
9.2.   The Synthesis and Chemistry of Tetrodotoxin   188
9.3.   Biological Activity   192
    9.3.1.   Tetrodotoxin   192
    9.3.2.   Chiriquitoxin   194
    9.3.3.   Zetekitoxin   194

10. OTHER ALKALOIDS   194
10.1.   Piperidine Alkaloids   194
    10.1.1.   Piperidines   195
    10.1.2.   Pyridyl piperidines   195
    10.1.3.   Piperidine-Based Alkaloids   196
10.2.   Pyrrolidine Alkaloids   202
10.3.   Indole Alkaloids   203
    10.3.1.   Dihydroindoles   203
    10.3.2.   Biogenic Amines   203
    10.3.3.   Dehydrobufotenine   204
    10.3.4.   Tetrahydrocarbolines   205
10.4.   Imidazole Alkaloids   206
    10.4.1.   Spinceamines   208
10.5.   Amidine Alkaloids   208
10.6.   Other Alkaloids   209

11. BIOLOGICAL SIGNIFICANCE, FORMATION, AND
     DISTRIBUTION OF AMPHIBIAN ALKALOIDS   210
     11.1. Biological Role   210
     11.2. Biosynthesis   221
     11.3. Distribution and Taxonomic Significance   224
            11.3.1. Batrachotoxins   224
            11.3.2. Other Dendrobatid Alkaloids   224
            11.3.3. Samandarines   254
            11.3.4. Tetrodotoxins   254
            11.3.5. Conclusions   254
     REFERENCES   255

1. INTRODUCTION*

During the past centuries the plant kingdom has been the source of more than 5000 alkaloids. The structure elucidation and synthesis of these diverse compounds has had a profound impact on the development of modern natural product and synthetic organic chemistry. Many of these plant alkaloids have unique biological activities, and their use as research tools has contributed to the development of modern pharmacology. The biological activities of some are undoubtedly beneficial to the plant in serving as a deterrent to herbivores. It is now recognized that alkaloids—defined as cyclic compounds containing nitrogen in a negative oxidation state and with a limited distribution in nature [1]—occur not only in the plant kingdom but also in both invertebrates and vertebrates of the animal kingdom. Amphibians have proven to be a particularly rich source and over 200 unique alkaloids have been characterized from this class of vertebrates. Structure elucidation of many of these diverse amphibian alkaloids often has been a challenge, because of the small amounts available to the chemist. The amphibian alkaloids have proven to be remarkable in terms of both structures and biological activity. Most are unique to amphibians and include batrachotoxins, his-

* This chapter is dedicated to Dr. Takashi Tokuyama of Osaka City University. His contributions to the chemistry of dendrobatid alkaloids have been central to the field, and his spirit of enthusiasm and dedication to science are an inspiration to all who have worked with him. The following brief and partial listing attests to the significance of the contributions of Dr. Tokuyama to this area of natural products chemistry. Between his preparation of a crystalline derivative of batrachotoxinin A in 1967 and his recent discovery and establishment of new classes of pumiliotoxins (indolizidines, decahydroquinolines, and amidines) lay years of major, often solitary, accomplishments by this modest investigator including the partial synthesis of batrachotoxin, the isolation and structural characterization of nearly a dozen histrionicotoxins, the difficult crystallization of gephyrotoxin, the detection and identification of indole and pyridyl alkaloids in amphibians, the establishment of the configuration of the side chains of pumiliotoxins A and B and the configuration of hydroxyls in their allopumiliotoxin congeners, and finally the isolation and structural definition of the 4-hydroxybatrachotoxins. It is to be hoped that the present review will bring the wider recognition so richly deserved by Takashi Tokuyama.

trionicotoxins, gephyrotoxins, indolizidines, the pumiliotoxin-A class, sa-mandarines, tetrodotoxins, decahydroquinolines, piperidine-based alka-loids, piperidines, pyrrolidines, indole alkaloids (such as calycanthine, chimonanthine, trypargine, and dehydrobufotenine), imidazole alkaloids (such as spinceamine), and amidine alkaloids. The majority of these alkaloids have been isolated from neotropical frogs of the family Dendrobatidae [2]. The unique structures and remarkable biological activity of the amphibian alkaloids provided the impetus for syntheses in many laboratories. A number of the amphibian alkaloids are now established as invaluable pharmacolog-ical tools for the investigation of physiological functions.

Alkaloids represent only one manifestation of the remarkably diverse sec-ondary metabolites that have been elaborated by amphibians. Most am-phibians, other than the dendrobatid frogs, elaborate peptides, biogenic amines, or bufodienolides. Such secondary metabolites appear to be stored in cutaneous glands of amphibians for release in chemical defense against predators.

The present review attempts to summarize current knowledge of the prop-erties, synthesis,† pharmacological activity, biological distribution, and sig-nificance of over 200 alkaloids from amphibians.

2. BATRACHOTOXINS

2.1. Structures

A unique source of poisons for arrows and blow darts are the skin secretions from certain brightly colored frogs native to the rain forests of western Col-ombia. The Noanamá and Emberá Indians of this region presumably used

---

† Abbreviations used in this chapter are as follows: Ac, CH_3CO; AIBN, azobisisobutyronitrile; Am, n-C_5H_{11}; aq., aqueous; Boc, t-butyloxycarbonyl; Bz, $C_6H_5CH_2$; cat., catalytic amount; Cbz, $C_6H_5CH_2OCO$; d, day; DABCO, 1,4-diazabicyclo[2.2.2]octane; DBU, 1,8-diazabicy-clo[5.4.0]undec-7-ene; DCC, N,N'-dicyclohexylcarbodiimide; DDQ, 2,3-dichloro-5,6-dicyano-1,4-benzoquinone; DMAP, 4-dimethylaminopyridine; DME, 1,2-dimethoxyethane; DMF, N,N-dimethylformamide; DMSO, dimethylsulfoxide; EGTA, ethylene glycol bis(β-aminoethyl ether)·N,N,N',N'-tetraacetic acid; eq., equivalents; HMPT, hexamethylphosphorous triamide; hr, hour; MEM, methoxymethylene; Ms, methanesulfonyl; NBS, N-bromosuccinimide; PCC, pyridinium chlorochromate; PPA, polyphosphoric acid; Py, pyridine; Red Al®, bis(2-methox-yethoxy)aluminium hydride (Aldrich Chemical Co.); TBS, tertiary-butyldimethylsilyl; TFA, CF_3CO_2H; THF, tetrahydrofuran; THP, tetrahydropyranyl; TIPS, triisopropylsilyl; TMS, tri-methylsilyl; Tr, triphenylcarbinyl; Ts, p-toluenesulfonyl; φ, C_6H_5; r.t., room temperature; Δ, generally reflux temperature; Lawesson's reagent,

$$p\text{---}CH_3OC_6H_4\text{---}P\underset{\underset{S}{\overset{\diagdown}{\diagup}}}{\overset{\overset{S}{\overset{\parallel}{\diagup}}\diagdown}{}}P\text{---}p\text{---}C_6H_4OCH_3$$

these frogs to poison blow darts even in pre-Colombian times: The first account of dart envenomation with poison frogs appeared in 1825 [3]. Secretions from a single frog are purported to have been sufficient for envenomation of at least 20 blow darts [4–7, see 8 for review of early literature]. The three species of frogs that are used to poison blow darts occur only in western Colombia, where the practice of poisoning blow darts with frog secretions still persists today. The poison-dart frog (*Phyllobates bicolor*) from the headwaters of the Río San Juan is called "neará" by the Indians, while lower in the same drainage, the poison-dart frog (*P. aurotaenia*) is called "kokoi." Envenomation of darts with the most toxic species (*P. terribilis*) from the Río Saija drainage is done simply by drawing the tip of the dart across the back of a living frog, while the other two species are impaled in order to elicit a copious flow of skin secretions [9]. There is no documentation of the use of the other neotropical frogs for the envenomation of blow darts or arrows.

Early studies demonstrated that the poison—obtained from the blow darts rather than from the frogs themselves—was highly active and probably owed its toxic effects to actions on nerve and muscle [5,10–13]. The chemical nature of the active principle was undefined although a chemist, J. Aronhson, had concluded in the late 1800s that it was an alcohol-soluble alkaloid [cited in 8]. Later studies did not appear to support this conclusion [11] but it was ultimately shown to be correct. The use of frogs as a source of poison for blow darts by Indians of western Colombia was brought to the attention of a chemist, B. Witkop, in the early 1960s by an animal collector, M. Latham [14]. The initial studies by Märki and Witkop [8] established that the active principle from the poison dart frog was an extremely toxic alkaloid. This provided the starting point and the impetus for some two decades of research on the skin alkaloids from this family of neotropical frogs, the Dendrobatidae, by Daly, Witkop, and colleagues at the National Institutes of Health.

The initial studies by Märki and Witkop in the early 1960s [8] on poison dart frogs established a pattern for subsequent studies. The frogs (*Phyllobates aurotaenia*) were obtained by Märki and M. Latham in the field in western Colombia, sacrificed and the skins placed in methanol. Preliminary field evaluation of extracts established that at least 90% of the toxic principles were in the skin and could be efficiently extracted into methanol. Methanol extracts were evaporated *in vacuo* at a field station and the residues transported to the National Institutes of Health for isolation of active principles, since the active principles appeared more stable in the dry residue than in the crude methanol extract. A bioassay was also established during these initial studies. It was found that the time of death of white mice was related to the amount of crude extract injected subcutaneously [8]. Only about 200 ng of pure toxin was needed for a single bioassay. Ultimately, a total of a "few hundred" micrograms of apparently pure toxin was isolated from extracts of 330 frog skins. The parent alkaloid of the first class of dendrobatid alkaloid was named batrachotoxin from the Greek *batrachos*, meaning frog. The initial studies [8,15] were reported to be with extracts from *P. bicolor*. This was due to an error in identification, and the extracts were actually from the species *P. aurotaenia* [see 9]. Subsequent studies have demonstrated the presence of batrachotoxins in *P. bicolor* [9]. The presence of batrachotoxins remains unique in nature to frogs of the genus *Phyllobates*, and indeed has been used as one character for the definition of the true poison dart frogs as a monophyletic genus, *Phyllobates* [9]. This genus, *Phyllobates*, contains five species ranging from Costa Rica through Panama and western Colombia. The distribution of batrachotoxins and other alkaloids in frogs of the genus *Phyllobates* is discussed in Section 11.

The most satisfactory isolation procedure developed by Märki and Witkop [8] for batrachotoxins was as follows: (1) extraction of minced skins twice with 70% aqueous methanol, followed by evaporation *in vacuo*; (2) trituration of the residue with isotonic NaCl, adjustment to pH 2 with HCl, followed by extraction of lipids into chloroform; (3) adjustment to pH 8.5 with aqueous ammonia, followed by extraction of toxic principles into chloroform; (4) purification of the active principles by thin-layer chromatography on silica gel with chloroform–methanol (6:1) after removal of solvent *in vacuo*. Two iodoplatinate-positive alkaloid spots were detected; the higher R_f material accounted for virtually all of the toxicity. In view of subsequent studies this higher R_f material was, undoubtedly, a mixture of two very similar compounds, batrachotoxin and homobatrachotoxin, while the lower R_f material was, undoubtedly, the much less toxic batrachotoxinin A. Little chemical characterization was possible. The material had basic properties, an apparent carbonyl absorption at about 1690 cm^{-1} and "end absorption" in the ultraviolet spectrum.

In order to accomplish a structure elucidation, larger numbers of frogs were clearly necessary. Fortunately, the frog occurs over a vast area in the Pacific drainage of Colombia and is fairly common in relatively accessible regions of the Río San Juan drainage. From extracts of 2400 frogs a total of 30 mg of "batrachotoxin," 15 mg of batrachotoxinin A, 6 mg of a compound referred to as batrachotoxinin B, and 1 mg of a compound referred to as batrachotoxinin C were isolated [15]. The "batrachotoxin" undoubtedly still represented a mixture of two very similar compounds, batrachotoxin and homobatrachotoxin. The batrachotoxinin B and C may have represented artifacts formed by oxidation during isolation or may have represented hydroxybatrachotoxins (see below). The extraction, partition, and thin-layer isolation procedures used in this study [15] were similar to those employed by Märki and Witkop [8]. Extractions and partitions were carried out at room temperatures and large losses were incurred during preparative thin-layer chromatography. The major compounds were partially characterized both by physical and chemical methods [15]. Mass spectral "element maps" for batrachotoxin ("kokoi venom") had been determined and presented earlier by Biemann [16]. The "batrachotoxin" fraction appeared to have a molecular ion in its mass spectrum corresponding to $C_{24}H_{33}NO_4$ (*m/z* 399) [15].

This apparent molecular ion was subsequently found to represent a pyrolysis fragment. Batrachotoxinin A afforded a true molecular ion corresponding to $C_{24}H_{35}NO_5$ (m/z 417). The two minor alkaloids, batrachotoxinin B and C, which had low R_f values, appeared, based on mass spectra, to be isomers of batrachotoxinin A. The isolated alkaloids all afforded positive reactions with iodoplatinate. A key observation was that the "batrachotoxin" fraction gave a positive Ehrlich reaction, red with p-dimethylaminobenzaldehyde and blue with p-dimethylaminocinnamaldehyde. This extremely sensitive reaction suggested a pyrrole moiety. However, the apparent absence of an ultraviolet chromophore argued against a pyrrole. Furthermore, only one nitrogen was present in the apparent molecular ion of the batrachotoxins, and "the nitrogen" of batrachotoxin was basic in character and could be converted under mild conditions to a methiodide. These data suggested that the pyrrole moiety *per se* was absent and only formed under the acid conditions of the Ehrlich reaction. This proved to be an incorrect conclusion; the pure batrachotoxin and homobatrachotoxin do contain the requisite ultraviolet chromophore and *two* atoms rather than one of nitrogen. The infrared spectrum of the "batrachotoxin" fraction showed a carbonyl absorption at about 1690 cm^{-1}. Proton nuclear magnetic resonance spectra were reminiscent of those expected from steroids and showed a methyl resonance singlet at 0.85 ppm and an N-methyl resonance peak at 2.3–2.5 ppm for both the "batrachotoxin" fraction and for the batrachotoxinins.

Chemical properties of the "batrachotoxin" fraction were assessed on a microscale. In retrospect it appears that most reactions resulted in loss of the pyrrole entity, and hence in a product that did not afford a positive Ehrlich reaction. These chemical reactions included catalytic hydrogenation with palladium on charcoal, reduction with lithium aluminum hydride, treatment with acidic methanol, oxidation with manganese dioxide, treatment with acid, reaction with 2,4-dinitrophenylhydrazine, and exhaustive methylation with methyl iodide. An Ehrlich-positive methiodide could be obtained under milder conditions with methyl iodide. Acetylation of the "batrachotoxin" fraction with acetic anhydride and pyridine afforded two Ehrlich-positive O-acetyl derivatives. Reaction with methoxylamine afforded an Ehrlich-positive O-methyloxime. Reaction with sodium borohydride afforded an Ehrlich-positive dihydro derivative. This product apparently isomerizes to other dihydro compounds [see 17]. Autoxidation, a serious problem during isolation of batrachotoxin, led to Ehrlich-negative products.

It appeared at this point that isolation of batrachotoxins would best be carried out at low temperatures and that thin-layer chromatography should be replaced with column chromatography to avoid large losses by autoxidation. Efforts were concentrated on the preparation of a crystalline derivative from batrachotoxinin A, the most stable and plentiful alkaloid present in the extracts. Extracts from 5000 frogs finally afforded, after silica gel column chromatographies, the following five alkaloids: batrachotoxin (11 mg), homobatrachotoxin (16 mg), batrachotoxinin A (46 mg), and a previously unsuspected labile compound pseudobatrachotoxin (1 mg) [17,18]. The latter compound spontaneously converted to batrachotoxinin A during isolation. Pseudobatrachotoxin represented a significant portion of the alkaloids in the initial alkaloid fraction when care was taken to maintain low temperatures during preparation and storage of extracts and during fractionation. During the collection of this particular extract, a portable refrigerator was taken into the field, but during most field work, low temperature storage has not been available for skin extracts. In the early studies, batrachotoxin and homobatrachotoxin were not separated from each other and the "batrachotoxin" fraction of the earlier studies undoubtedly contained some pseudobatrachotoxin and/or its conversion product, batrachotoxinin A. Pseudobatrachotoxin has no ultraviolet chromophore and its presence or that of its product, batrachotoxinin A, probably contributed to the low ultraviolet extinction coefficients of early batrachotoxin fractions. Homobatrachotoxin, when first isolated, was thought to be an isomer of batrachotoxin since the apparent mass spectral parent ion was identical with the apparent parent ion of batrachotoxin: Because of this, it was initially termed isobatrachotoxin [17].

A crystalline derivative for X-ray analysis was now obtained by T. Tokuyama from batrachotoxinin A by acylation with *p*-bromobenzoic anhydride under Schotten-Baumann conditions [18]. The resultant *O-p*-bromobenzoate (Fig. 1, I, R = 1) was purified by column chromatography on silica gel and

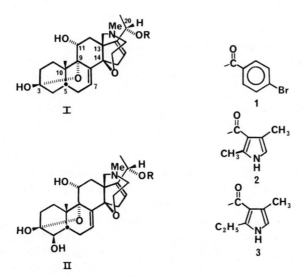

Figure 1. Structures of batrachotoxins. (I) Batrachotoxinin A (R = H) or batrachotoxinin A 20 α *p*-bromobenzoate (R = 1) or batrachotoxin (R = 2) or homobatrachotoxin (R = 3). (II) 4-Hydroxybatrachotoxin (R = 2) or 4-hydroxyhomobatrachotoxin (R = 3).

crystallized from acetone. X-ray diffraction analysis [18] afforded the structure of the batrachotoxinin A 20α-*p*-bromobenzoate: Batrachotoxinin A is 3α,9α-epoxy-14β,18β-(epoxyethano-*N*-methylamino)-5β-pregna-7, 16,diene-3β,11α,20α-triol (Fig. 1, I, R = H), and its structure provided the key to the structures of batrachotoxin and homobatrachotoxin. These and later X-ray crystallographic determinations of structures of dendrobatid frogs were done by I. Karle of the Naval Research Laboratories using the symbolic addition method developed by J. Karle. The structure of the labile pseudobatrachotoxin remains unknown due to a paucity of material for characterization. Pseudobatrachotoxin did not exhibit an Ehrlich reaction and appeared to exhibit only end absorption in the ultraviolet spectrum. The most plausible hypothesis for its structure is that of a highly labile ester, which hydrolyzes to batrachotoxinin A. Further X-ray analysis of batrachotoxinin A *p*-bromobenzoate [19,20] led to the absolute configuration of this steroidal alkaloid. The absolute configuration was identical with that of cholesterol at a number of points, namely carbons 3, 5, 9, 10, and 13. The configuration at carbon 14 was different from cholesterol and was instead reminiscent of the cardenolides. The configuration at 20 was **S**. An X-ray crystallographic analysis of a synthetic 20 **R**-epimer of 7,8-dihydrobatrachotoxinin A has appeared [21]. The configurations of these steroidal alkaloids are markedly constrained by the 3,9-oxygen bridge and the homomorpholine bridge at the C,D ring juncture.

A knowledge of the structure of batrachotoxinin A led to a reevaluation of the spectral properties of the much more toxic batrachotoxin and homobatrachotoxin. It soon became apparent that the ultraviolet absorption peaks (λ_{max} 234, 264 nm), the infrared absorption band at 1690 cm^{-1}, and additional nuclear magnetic resonance signals in the spectra of batrachotoxin and homobatrachotoxin ("isobatrachotoxin") were impossible to accommodate with an apparent molecular ion of $C_{24}H_{33}NO_4$ (*m/z* 399) for these two alkaloids. The molecular ion of batrachotoxinin A was $C_{24}H_{35}NO_5$ (*m/z* 417). Instead, batrachotoxin and homobatrachotoxin evidently contain another moiety that is responsible for the carbonyl absorption band, the ultraviolet chromophore, the positive Ehrlich reaction, a downfield singlet corresponding to one hydrogen in the nuclear magnetic resonance spectra at about 6.35 ppm, and resonance peaks corresponding to two methyl groups (singlets, 2.24 and 2.47 ppm) in batrachotoxin and to a methyl group (singlet, 2.24 ppm) and an ethyl group in homobatrachotoxin. Otherwise, the nuclear magnetic spectra of batrachotoxin and homobatrachotoxin contained most of the identifiable elements seen in the spectrum of batrachotoxinin A. However, the C(20)-hydrogen resonance of batrachotoxinin A, a quartet at about 4.6 ppm, was missing and, as in the 20-α-*p*-bromobenzoate of batrachotoxinin A, was replaced by a downfield quartet at about 5.9 ppm for both batrachotoxin and homobatrachotoxin. The most reasonable conclusion based on analysis of the spectral data was that batrachotoxin was a 20-α-dimethylpyrrole carboxylate and homobatrachotoxin a 20-α-ethylmethylpyrrole carboxylate of batrachotoxinin A. The ultraviolet spectra of batrachotoxin and homobatrachotoxin resembled those of pyrrole-3-carboxylates rather than pyrrole-2-carboxylates. Shifts of the resonances of methyl groups on model pyrrole-3-carboxylates and of the methyl or methylene resonances on the pyrrole moiety of batrachotoxin and homobatrachotoxin led to the conclusion that batrachotoxin contains a 2,4-dimethylpyrrole-3-carboxylate moiety and homobatrachotoxin a 2-ethyl-4-methylpyrrole-3-carboxylate moiety [17] (Fig. 1, I, R = 2 and 3). A 2-methyl substituent in model pyrrole-3-carboxylates is shifted downfield in $CDCl_3$ relative to its position in C_6D_6 by 0.07–0.08 ppm, a 5-methyl substituent is shifted downfield by 0.26–0.30 ppm, and a 4-methyl substituent is shifted upfield by 0.18–0.24 ppm. The shifts for the two pyrrole methyl groups of batrachotoxin were downfield by 0.10 and upfield by 0.11 ppm, while the shift of the methylene hydrogens of the pyrrole ethyl group of homobatrachotoxin was downfield 0.02 ppm and of the pyrrole methyl group upfield by 0.12 ppm. The methylene hydrogens of the pyrrole ethyl group in ethyl 2-ethyl-4-methylpyrrole-3-carboxylate were shifted downfield by 0.04 ppm. The ultraviolet absorption spectrum of ethyl 2,4-dimethylpyrrole-3-carboxylate was virtually identical with those of batrachotoxin and homobatrachotoxin. The visible absorption spectra of the Ehrlich reaction products, obtained from ethyl 2,4-dimethyl-pyrrole-3-carboxylate and from the two batrachotoxins, were also essentially identical.

The mass spectra of batrachotoxin and homobatrachotoxin could now be readily interpreted. Indeed, it even proved possible to detect the molecular ion at m/z 538 for batrachotoxin, but its intensity even under the most favorable conditions was less than 1% of the apparent molecular ion at m/z 399. Furthermore, the true molecular ion at m/z 538 could never be observed except transiently, probably because of pyrolytic elimination of the pyrrole carboxylate moiety. The pyrrole carboxylate moiety was responsible for major ions of $C_7H_9NO_2$ (m/z 139) and C_6H_9N, C_6H_8N (m/z 95, 94) in batrachotoxin and for major ions of $C_8H_{11}NO_2$ (m/z 153), $C_7H_8NO_2$ (m/z 138), $C_7H_{11}N$ (m/z 109), and C_6H_8N (m/z 94) in homobatrachotoxin [17].

Hydrolysis of the hindered pyrrole ester moiety of batrachotoxin proved possible with 2 N NaOH at 60° for 16 hr [17]. Under those conditions a partial hydrolysis to a compound identical with batrachotoxinin A was attained. Reaction of the mixed anhydride of ethyl chloroformate and 2,4-dimethylpyrrole-3-carboxylate with batrachotoxinin A under Schotten-Baumann conditions (Scheme 1) afforded a product identical in properties with batrachotoxin, thereby confirming the proposed structure.

In the early 1970s, a new species of *Phyllobates* was discovered in the Río Saija drainage in western Colombia by an anthropologist, B. Malkin. This frog, one of the three used by Colombian Indians to poison blow darts, contained about twentyfold higher levels of batrachotoxins than did the species *P. aurotaenia* and *P. bicolor*, some 150 km to the north, in the Río San Juan drainage [9]. The frog was named *P. terribilis* to refer to the extraor-

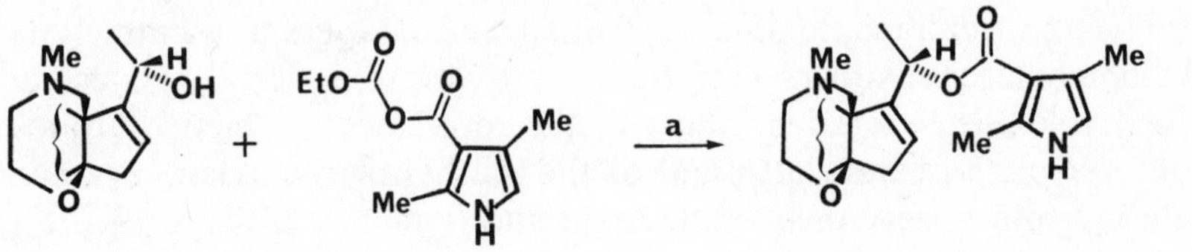

Scheme 1. Partial synthesis of batrachotoxin from batrachotoxinin A [17]. (*a*) $CHCl_3$/aq. NaOH.

dinary toxicity of the skin secretions and also to the fear evoked by poisoned blow darts: One frog of the species may contain up to 1.9 mg of batracho-toxin. Extracts from some 426 skins of *P. terribilis* yielded large amounts of batrachotoxin, homobatrachotoxin, batrachotoxinin A, and additional al-kaloids of the batrachotoxin class [22]. The alkaloid extract was fractionated by high-pressure liquid chromatography on a reversed phase silica gel col-umn to yield batrachotoxin (175 mg), homobatrachotoxin (113 mg), batrach-otoxinin A (153 mg), and two minor alkaloids (2 and 3 mg), which appeared to be monohydroxy derivatives of batrachotoxin and homobatrachotoxin, respectively. Analysis of the mass spectral fragmentation patterns indicated that the additional hydroxy group was in the A-ring. This was confirmed by analysis of proton and ^{13}C magnetic resonance spectra: The spectra were compatible only with a 4β-hydroxy group with carbon 4 appearing at 77.1 ppm (doublet) and the 4α-H (equatorial) appearing as a broad singlet. These two additional alkaloids were therefore 4β-hydroxybatrachotoxin and 4β-hydroxyhomobatrachotoxin (Figure 1, II, R = **2** and **3**). 3-O-Methylbatrach-otoxin (38 mg) and 3-O-methylhomobatrachotoxin (16 mg) were also isolated from extracts of *P. terribilis*. But these undoubtedly were formed as artifacts through acid-catalyzed reaction of batrachotoxins with methanol during preparation and fractionation. In earlier studies, small amounts of artifactual "dimers" derived from reaction of pyrrole rings of batrachotoxins with ace-tone (present in methanol) had been isolated (unpublished results). The for-mation of such a "dimer" from homobatrachotoxin and acetone and its proton magnetic resonance spectrum is reported in [22]. Extracts of *P. ter-ribilis* also yielded noranabasamine, chimonanthine, and calycanthine (see Sections 10.1, 10.3).

The properties of various batrachotoxins are presented in the following section in a format similar to that employed originally for "piperidine-based" dendrobatid alkaloids [23]. For the batrachotoxins, the following are in-cluded: (1) Trivial name, (2) empirical formula, (3) R_f value for thin-layer chromatography on silica gel with chloroform:methanol 9:1, (4) electron impact mass spectral ions (*m/z*) followed in parentheses with elemental com-position and the intensity relative to the base peak set equal to 100. Only the most diagnostic peaks are reported. An expanded version of this format including gas chromatographic data is used for histrionicotoxins (Section 3.1), gephyrotoxins (Section 4.1), indolizidines (Section 5.1), pumiliotoxin-

A class alkaloids (Section 6.1), decahydroquinolines (Section 7.1), piperidine-based dendrobatid alkaloids (Section 10.1), pyrrolidines (Section 10.2), amidines (Section 10.5), and certain other nondendrobatid alkaloids (Section 10.6). The batrachotoxin alkaloids emerge as pyrolysis products (m/z 399) at about 280° on temperature-programmed gas-chromatographic columns (1.5% OV-1, 150–280°) used for analysis of other dendrobatid alkaloids (see Section 3.1).

Batrachotoxin, $C_{31}H_{42}N_2O_6$, 0.45, m/z 538 ($\ll$1), 399 ($C_{24}H_{33}NO_4$, 3), 312 ($C_{20}H_{24}O_3$, 13), 294 ($C_{20}H_{22}O_2$, 10) 286 ($C_{18}H_{22}O_3$, 10), 184 ($C_{13}H_{12}O$, 30), 139 ($C_7H_9NO_2$, 65), 138 ($C_7H_8NO_2$, 24), 122 (C_7H_8NO, 20), 121 (C_7H_7NO, 13), 120 (C_7H_6NO, 10), 109 ($C_9H_{11}N$, 15), 95 (C_6H_9N, 100), 94 (C_6H_8N, 100), 88 ($C_4H_{10}NO$, 34), 71 (C_4H_9N, 26).

Homobatrachotoxin, $C_{32}H_{44}N_2O_6$, 0.50 m/z 552 ($\ll$1), 399 (6), 312 (25), 294 (20), 286 (22), 184 (60), 153 ($C_8H_{11}NO_2$, 90), 139 (22), 138 (100), 122 (12), 121 (15), 120 (26), 109 (60), 95 (23), 94 (94), 88 (72), 71 (28).

4-Hydroxybatrachotoxin, $C_{31}H_{42}N_2O_7$, 0.10, m/z 554 ($\ll$1), 415 (6), 386 (5), 328 (30), 310 (18), 184 (38), 139 (68), 84 (100).

4-Hydroxyhomobatrachotoxin, $C_{32}H_{44}N_2O_7$, 0.10, m/z 568 ($\ll$1), 415 (5), 386 (6), 328 (7), 310 (17), 184 (14), 153 (60), 139 (100), 84 (66).

Batrachotoxinin A, $C_{24}H_{35}NO_5$, 0.28, m/z 417 ($C_{24}H_{35}NO_5$, 2), 399 (11), 330 ($C_{20}H_{26}O_4$, 100), 312 (30), 202 (15), 184 (11), 158 ($C_{11}H_{10}O$, 14), 88 (60).

Pseudobatrachotoxin,—, 0.25, m/z 399 (6), 312 (17), 294 (15), 286 (6), 202 (3), 184 (48), 88 (100), 71 (45). Ions at m/z 342 ($C_{22}H_{32}NO_2$, 6) and 166 ($C_{10}H_{16}NO$, 12) were detected, but may have been due to an impurity. Pseudobatrachotoxin converts to batrachotoxinin A during storage of extracts at room temperature.

Further physical and spectral properties of the various batrachotoxins are summarized in Table 1. Mass spectra and their interpretation have been presented [17,22] and reviewed [24]. Proton magnetic resonance spectra for batrachotoxin, homobatrachotoxin, and batrachotoxinin A are depicted and assignments are discussed in a recent review [24]. A proton magnetic resonance spectrum of synthetic batrachotoxinin A is depicted in Imhof et al. [25]. Proton magnetic spectra for batrachotoxin, homobatrachotoxin, and batrachotoxinin A are also depicted in Tokuyama et al. [17]. Carbon-13 magnetic resonance data and assignments for the batrachotoxins and 4-hydroxy congeners are reported [22], as is a portion of the proton magnetic resonance spectrum of 4-β-hydroxybatrachotoxin. The effect of 3-O-methylation of (homo)batrachotoxin on the ^{13}C magnetic resonance peaks is also documented [22]. The pK_a of batrachotoxin had been estimated to be in the range of 7.1–8.0 based on partitions between buffers and organic solvents [8]. Recently, the pK_a of batrachotoxinin A was determined by titration and found to be 8.2 or slightly greater [26].

The toxicity of batrachotoxin and the sensitivity of the Ehrlich reaction provide two sensitive probes for occurrence of batrachotoxin and homobatrachotoxin in amphibian extracts. Positive identification of (homo)batra-

Table 1. Physical and Spectral Properties of the Batrachotoxins[a]

Property	Reference
Batrachotoxin	
Ultraviolet: λ_{max} 234 nm, ϵ 9800	[17]
262 nm, ϵ 5000	
Infrared: 1690 cm^{-1}	[17]
Optical rotation: (c 0.23, CH_3OH) $[\alpha]_{584}^{24}$ -5 to $-10°$	[15]
$[\alpha]_{300}^{24}$ $-260°$	
Homobatrachotoxin	
Ultraviolet: λ_{max} 233 nm, ϵ 8900	[17]
264 nm, ϵ 5000	
Infrared: 1690 cm^{-1}	[17]
Batrachotoxinin A	
mp 160–162° (synthetic)	[25]
Ultraviolet: end absorption	[17]
Optical rotation (synthetic): (c 0.45, CH_3OH) $[\alpha]_D^{20}$ $-42°$	[25]
Pseudobatrachotoxin	
Ultraviolet: end absorption	[17]

[a] Ultraviolet spectra are depicted in [17].

chotoxin has been made only with skin extracts from five species of dendrobatid frogs, now all grouped together in the monophyletic genus, *Phyllobates* [9]. The occurrence and levels of batrachotoxins in these frogs and other biological aspects of these skin toxins are presented in Section 11.

2.2. Syntheses

The structure of batrachotoxin was initially confirmed by a partial synthesis from batrachotoxinin A [17]. This proved feasible using a mixed anhydride prepared from 2,4-dimethylpyrrole-3-carboxylic acid and ethyl chloroformate (Scheme 1). A variety of analogs of batrachotoxin were prepared by Tokuyama in a similar manner. These compounds are documented in Section 2.3. A dihydrobatrachotoxin was prepared by sodium borohydride reduction (Scheme 2) and apparently is subject to allylic rearrangement to other dihydro products [17].

The elaboration of the steroidal portion of a molecule as complex as batrachotoxin represented a formidable challenge. This challenge was accepted by H. Wehrli and colleagues at Eidgenössischen Technische Hochschule, Zurich, and led over a period of two years to the partial synthesis of batrachotoxinin A and 7,8-dihydro and 7,8,15,16-tetrahydro analogs from steroid precursors [25,27–34]. The partial synthesis of batrachotoxinin A itself

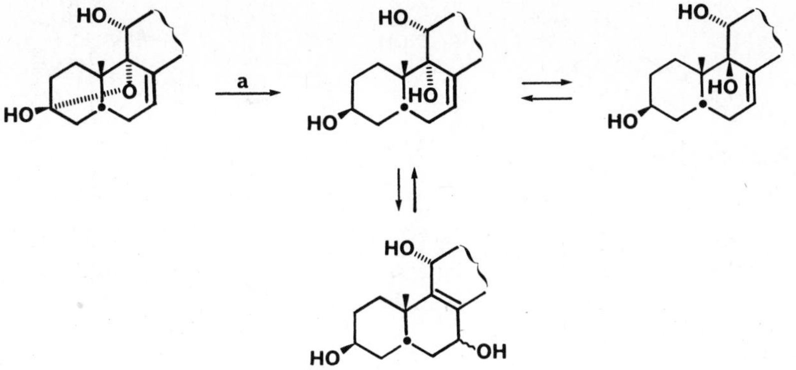

Scheme 2. Formation of dihydrobatrachotoxin and postulated conversions to lower R_f isomers [17]. (*a*) NaBH$_4$.

was ultimately accomplished in 36 steps from lactone **1** in 1972 [33] (Scheme 11). Lactone **1** was available in ten steps from 11α-hydroxyprogesterone, a microbiological oxidation product of progesterone. Twelve milligrams of batrachotoxinin A were produced in a synthesis that averaged an impressive 78% yield per step. This remarkable synthesis followed after a series of meticulously executed model studies. Two other laboratories [35–37] had also conducted model studies. The formidable synthetic problems involved and their solutions are discussed in the final publication of the Wehrli group [25] and in two recent reviews [24,38]. The model syntheses are briefly sketched in Schemes 3–10, and deal with the construction of the uncommon 3α,9α-oxide-bridged hemiketal (Schemes 6–10), the $\Delta^{7,8}$-olefinic linkage (Schemes 6,8), the required S-configuration at C(20) (Scheme 10), and the C-D ring-fusion homomorpholine propellane system (Schemes 4,5,7,9,10). Scheme 10 shows the most successful approach to the latter unprecedented structural feature. As applied to batrachotoxinin A (Scheme 11), the following steps were involved: Δ^{14-15}-unsaturation was produced by two cycles

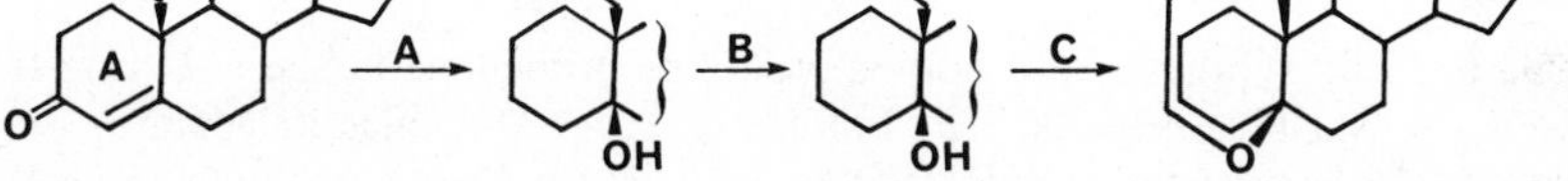

Scheme 3. The synthesis of a homomorpholine propellane at the A-B ring fusion [27]. A: (*a*) H$_2$O$_2$/aq. NaOH/MeOH-CH$_2$Cl$_2$(→4β,5β-oxide); (*b*) Ac$_2$O/Py (reacetylation); (*c*) N$_2$H$_4$·H$_2$O/EtOH-HOAc (→Δ^4-5β-ol); (*d*) H$_2$/Pt/EtOH. B: (*a*) NaHCO$_3$/aq. EtOH, Δ (→19-ol); (*b*) Ag$_2$CO$_3$-celite/φH, Δ (→19=O); (*c*) MeNH$_2$/φH, 120° (→19=NMe); (*d*) NaBH$_4$/aq. MeOH (→19-NHMe); (*e*) ClCH$_2$COCl/aq. NaOH/CHCl$_3$. C: (*a*) NaH/THF-φH (trace EtOH); (*b*) LiAlH$_4$/Et$_2$O.

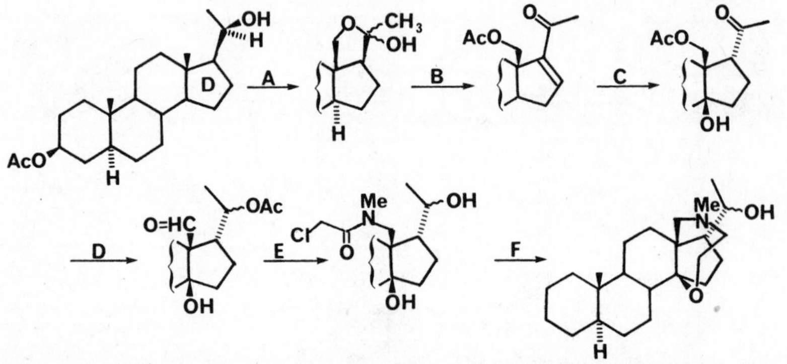

Scheme 4. The synthesis of a homomorpholine propellane at the C-D ring fusion [28]. A: (*a*) Pb(OAc)$_4$/CaCO$_3$/C$_6$H$_{12}$, Δ then I$_2$, hv ($\rightarrow$18-I); (*b*) CrO$_3$/Me$_2$CO/H$_2$SO$_4$ ($\rightarrow$20=O); (*c*) AgOAc/MeOH, Δ, chromatography on H$_2$O-treated SiO$_2$-gel ($\rightarrow$20-hemiacetal). B: (*a*) Ac$_2$O/Py, Δ ($\rightarrow$18-OAc); (*b*) Py·HBr·Br$_2$/CH$_2$Cl$_2$ ($\rightarrow$17-Br); (*c*) DMF, Δ. C: (*a*) NBS/CCl$_4$, AIBN, hv, Δ ($\rightarrow$15-Br); (*b*) NaI/Me$_2$CO, Δ ($\rightarrow$14,16-diene); (*c*) *p*-NO$_2\phi$CO$_3$H/CHCl$_3$ ($\rightarrow$14β,15β-oxide, Δ^{16}); (*d*) H$_2$/Pd-BaSO$_4$/EtOH. D: (*a*) LiAl(OBu^t)$_3$H/THF ($\rightarrow$20-ols); (*b*) Ac$_2$O/Py ($\rightarrow$20-acetates); (*c*) NaHCO$_3$/aq. MeOH, Δ ($\rightarrow$18-ol); (*d*) CrO$_3$/Me$_2$CO/H$_2$SO$_4$. E: (*a*) MeNH$_2$/ϕH, Δ ($\rightarrow$18=NMe); (*b*) Ac$_2$O/Py (reacetylation); (*c*) NaBH$_4$/aq. MeOH ($\rightarrow$18-NHMe); (*d*) ClCH$_2$COCl/ aq. NaOH/CHCl$_3$. F: (*a*) NaH/ϕH-THF (trace EtOH); (*b*) LiAlH$_4$/Et$_2$O-THF.

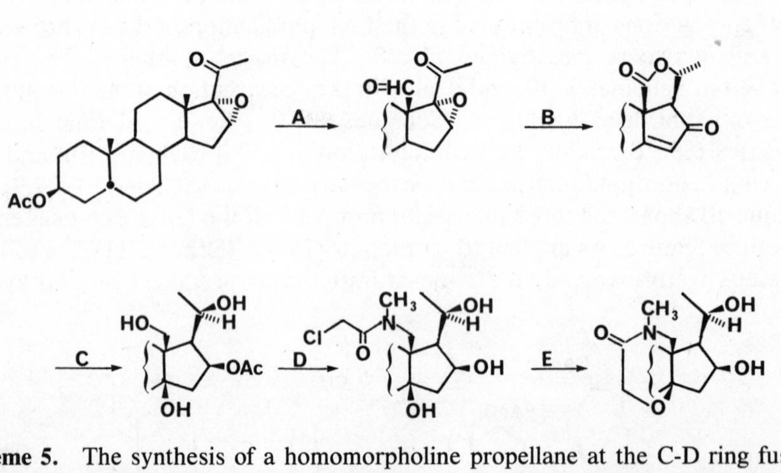

Scheme 5. The synthesis of a homomorpholine propellane at the C-D ring fusion [35]. A: (*a*) NaBH$_4$ ($\rightarrow$20-ols); (*b*) Pb(OAc)$_4$, I$_2$ ($\rightarrow$20-hemiacetalacetate); (*c*) hydrolysis ($\rightarrow$20-hemiacetal); (*d*) DMSO/Py·SO$_3$. B: (*a*) Cr(OAc)$_2$/Py ($\rightarrow$16-α-ol); (*b*) LiAl(OBu^t)$_3$H ($\rightarrow$20-ol); (*c*) CrO$_3$ ($\rightarrow$lactone, 16=O); (*d*) Br$_2$ ($\rightarrow$15-Br); (*e*) dehydrobromination. C: (*a*) LiAl(OBu^t)$_3$H ($\rightarrow\Delta^{14}$,16-β-ol); (*b*) CH$_3$CO$_3$H (14β,15β-oxide, 16-β-ol); (*c*) LiAlH$_4$ ($\rightarrow$14β,16β-diol); (*d*) Me$_2$CO/BF$_3$·Et$_2$O ($\rightarrow$14,18-acetonide); (*e*) Ac$_2$O ($\rightarrow$16β-OAc); (*f*) H$^+$. D: (*a*) DMSO/Py·SO$_3$ ($\rightarrow$18=O); (*b*) MeNH$_2$/EtOH, Δ ($\rightarrow$18-(NHMe)-O-20); (*c*) H$_2$/Pt/HOAc (18-NHMe); (*d*) ClCH$_2$COCl/THF. E: KOBu^t/ THF, $-20°$.

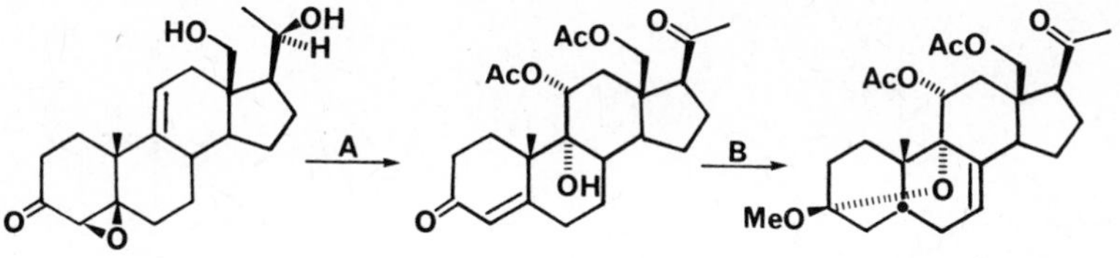

Scheme 6. A model synthesis of the ABC ring system of batrachotoxin [32]. A: (*a*) Ac$_2$O/Py (→18-OAc); (*b*) CrO$_3$/Me$_2$CO/H$_2$SO$_4$ (→20=O); (*c*) OsO$_4$/Py then H$_2$S/ dioxane − aq. NH$_4$Cl (→9α,11α-diol); (*d*) Ac$_2$O/Py (→11α-OAc); (*e*) Zn/HOAc, Δ (→Δ^4). B: (*a*) DDQ/dioxane-aq. HCl (→Δ^6); (*b*) *p*-NO$_2$ϕCO$_3$H/CHCl$_3$ (→6α,7α-oxide); (*c*) Pd-BaSO$_4$/cyclohexene-dioxane, Δ (→7α-ol); (*d*) H$_2$/Pd-C/MeOH (→5β-H); (*e*) HCl/aq. MeOH (→3-ketal); (*f*) POCl$_3$/Py (→Δ^7).

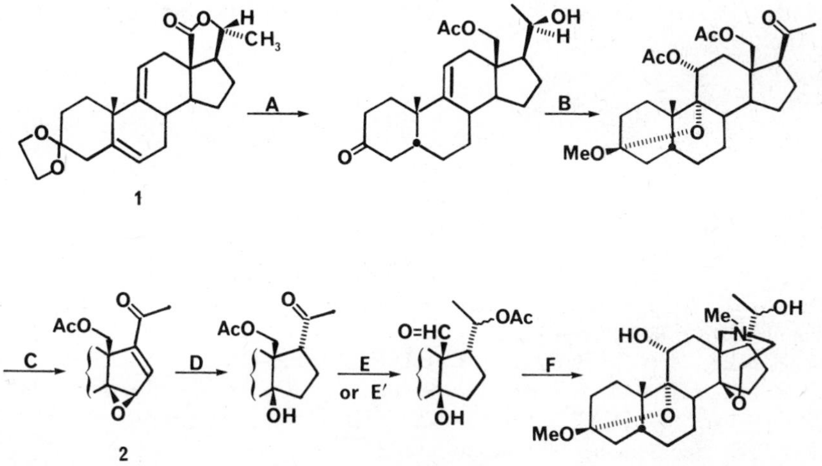

Scheme 7. The synthesis of the C-20 alcohol diastereomers of 7,8,16,17-tetrahydrobatrachotoxinin A [29,30]. A: (*a*) LiAlH$_4$/THF, Δ (→18,20-diol); (*b*) TsOH/aq. Me$_2$CO, Δ (→Δ^4,3=O); (*c*) H$_2$/Pd-C/KOH-aq. EtOH (→5β-H); (*d*) Ac$_2$O/Py. B: (*a*) CrO$_3$/Me$_2$CO/H$_2$SO$_4$ (→20=O); (*b*) OsO$_4$/Py then H$_2$S/dioxane-aq. NH$_4$Cl (→9α,11α-diol); (*c*) HCl/aq. MeOH (→3-ketal); (*d*) Ac$_2$O/Py. C: (*a*) NBS/CCl$_4$, AIBN, hv, Δ (→17-Br); (*b*) LiBr-Li$_2$CO$_3$/DMF, Δ (→Δ^{16}); (*c*) NBS/CCl$_4$, AIBN, hv, Δ (→15-Br); (*d*) LiBr-Li$_2$CO$_3$/DMF, Δ (→14,16-diene); (*e*) *p*-NO$_2$ϕCO$_3$H/CHCl$_3$-MeOH, Δ. D: H$_2$/Pd-C/EtOH; or preferably (*a*) Pd/BaSO$_4$/cyclohexene-MeOH, Δ (→Δ^{16},14β-ol) then (*b*) H$_2$/Pd-C/EtOH [30]. E: (*a*) NaBH$_4$/aq. MeOH (→20-ols); (*b*) Ac$_2$O/Py (→20-acetates); (*c*) NaHCO$_3$/aq. MeOH, Δ (→18-ol); (*d*) CrO$_3$/Me$_2$CO/H$_2$SO$_4$. E′ alternative procedure [30]: (*a*) NaOH/aq. dioxane (→18-ol); (*b*) CrO$_3$/CH$_2$Cl$_2$-Py (→18=O); (*c*) HCl/MeOH (→18-acetal); (*d*) H$_2$/Pd-C/EtOH; (*e*) NaBH$_4$/aq. MeOH (→20-ols); (*f*) Ac$_2$O/Py (→20-acetates); (*g*) TsOH/Me$_2$CO. F: (*a*) MeNH$_2$/ϕH, 120° then Ac$_2$O/Py (→18=NMe); (*b*) NaBH$_4$/aq. MeOH (→18-NHMe); (*c*) ClCH$_2$COCl/aq. NaOH/CHCl$_3$; (*d*) NaH/ϕH-THF (trace EtOH); (*e*) LiAlH$_4$/Et$_2$O-THF.

17

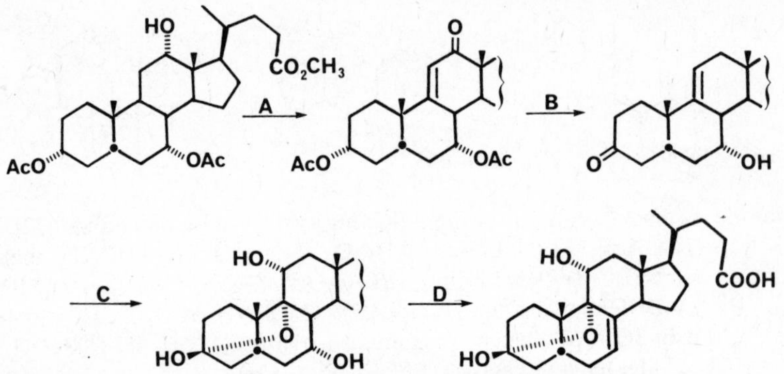

Scheme 8. A model synthesis of the ABC ring system of batrachotoxin [36,37]. A: (a) CrO_3/HOAc, Δ ($\rightarrow$12=O); (b) SeO_2/HOAc, Δ ($\rightarrow\Delta^{9-11}$, 12=O); B: (a) $(CH_2SH)_2$/ $CHCl_3$, HCl ($\rightarrow$12-dithiolone); (b) Raney Ni/EtOH; (c) K_2CO_3/aq. MeOH ($\rightarrow$3-ol); (d) CH_2N_2/Et_2O-dioxane (reesterification); (e) $Na_2Cr_2O_7$·$2H_2O$/HOAc ($\rightarrow$3=0); (f) NaOMe/MeOH ($\rightarrow$7-ol). C: OsO_4/Py then H_2S/dioxane $-$ aq. NH_4Cl. D: (a) HBr/ MeOH ($\rightarrow$3-ketal); (b) Ac_2O/Py, Δ ($\rightarrow$11-OAc); (c) $POCl_3$/Py ($\rightarrow\Delta^7$); (d) $HClO_4$/aq. HOAc ($\rightarrow$3-hemiketal); (e) NaOH/aq. EtOH (saponification).

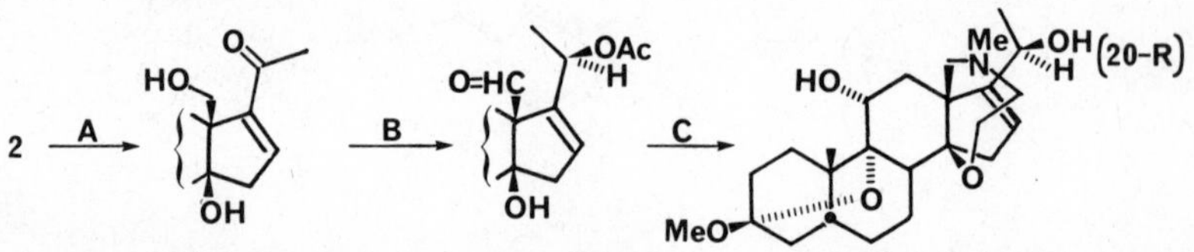

Scheme 9. The synthesis of 3-O-methyl-7,8-dihydro-20-epibatrachotoxinin A [34; see also 31]. A: (a) Pd-$BaSO_4$/cyclohexene-MeOH, Δ ($\rightarrow\Delta^{16}$, 14β-ol); (b) NaOH/aq. dioxane. B: (a) CrO_3/CH_2Cl_2-Py ($\rightarrow$18=O); (b) HCl/MeOH ($\rightarrow$18-acetal); (c) Al(i-Bu)$_2$H/ϕCH_3, $-78°$ ($\rightarrow$20(**R**)-ol); (d) Ac_2O/Py ($\rightarrow$20(**R**)-OAc); (e) TsOH/Me_2CO. C: (a) $MeNH_2$/ϕH, 4 Å sieves ($\rightarrow$18=NMe); (b) LiCNBH$_3$ ($\rightarrow$18-NHMe); (c) Cl-CH_2COCl/aq. NaOH/$CHCl_3$; (d) NaH/THF-ϕH (trace EtOH); (e) LiAlH$_4$/ Et_2O-THF. See Scheme 7 for structure of **2**.

(m,n) of allylic bromination-dehydrobromination followed by stereospecific β-epoxidation (i) and an unusual catalytic hydrogen transfer reaction (o), which cleaves the 14β,15β-epoxide regiospecifically to produce the 14 β-alcohol **3**. Another key step is the Goldman-Allbright oxidation (u) of the 14,18-diol with DMSO-Ac_2O to produce the C(18)-aldehyde, while simultaneously protecting the C(14)-tertiary alcohol as a hemithioketal ether **7**. Imine formation with methylamine, reduction, and acylation affords, after deblocking, the chloroacetamido alcohol **8**, which is ring-closed to the lactam ether **9**. Reduction with lithium aluminum hydride (a') yields the homomorpholine ring and cleaves both acetates. A final ketal hydrolysis produces batrachotoxinin A. The conversion to batrachotoxin by acylation with the mixed anhydride of 2,4-dimethylpyrrole-3-carboxylic acid had been demonstrated in 1969 by Tokuyama (Scheme 1) [17].

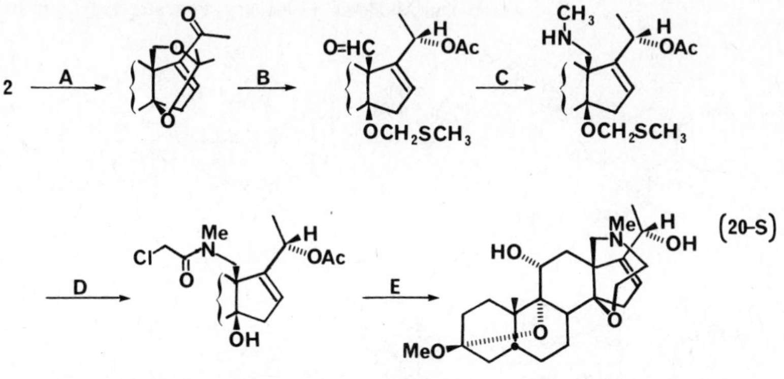

Scheme 10. The synthesis of 7,8-dihydrobatrachotoxinin A [34; see also 31]. A: (*a*) Pd-BaSO$_4$/cyclohexene-MeOH (→Δ^{16}, 14β-ol); (*b*) NaOH/aq. dioxane; (*c*) Me$_2$C(OMe)$_2$, TsOH. B: (*a*) NaBH$_4$/MeOH, −30° (→20(S)-ol); (*b*) Ac$_2$O/Py; (*c*) TsOH/MeOH (acetonide exchange); (*d*) DMSO/Ac$_2$O. C: (*a*) MeNH$_2$/φH, Δ (→18 = NMe); (*b*) NaBH$_4$/aq. MeOH. D: (*a*) ClCH$_2$COCl/aq. NaOH/CHCl$_3$; (*b*) HCl/ aq. MeOH. E: (*a*) NaH/THF-φH (trace EtOH); (*b*) LiAlH$_4$/Et$_2$O; (*c*) TsOH/aq. Me$_2$CO. See Scheme 7 for structure of 2.

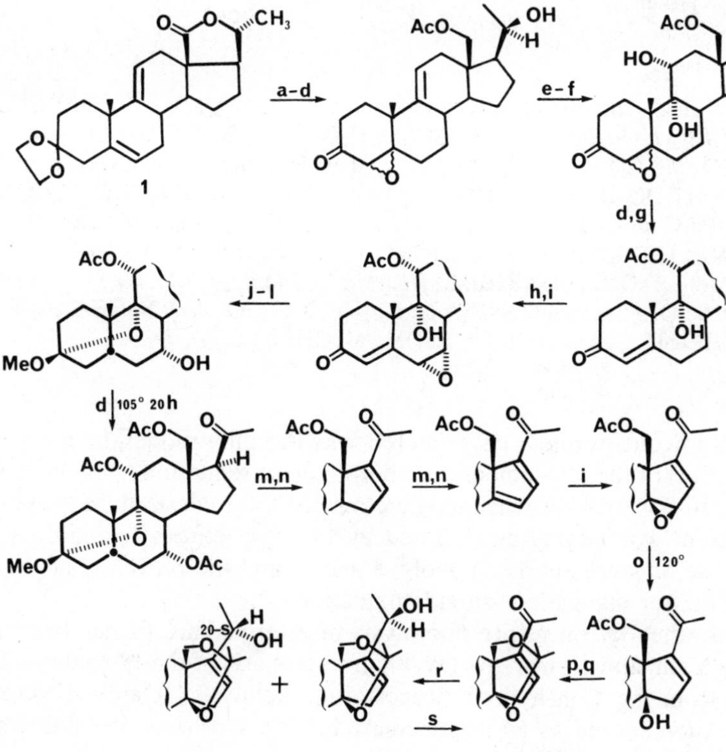

Scheme 11. (*Continued on next page*)

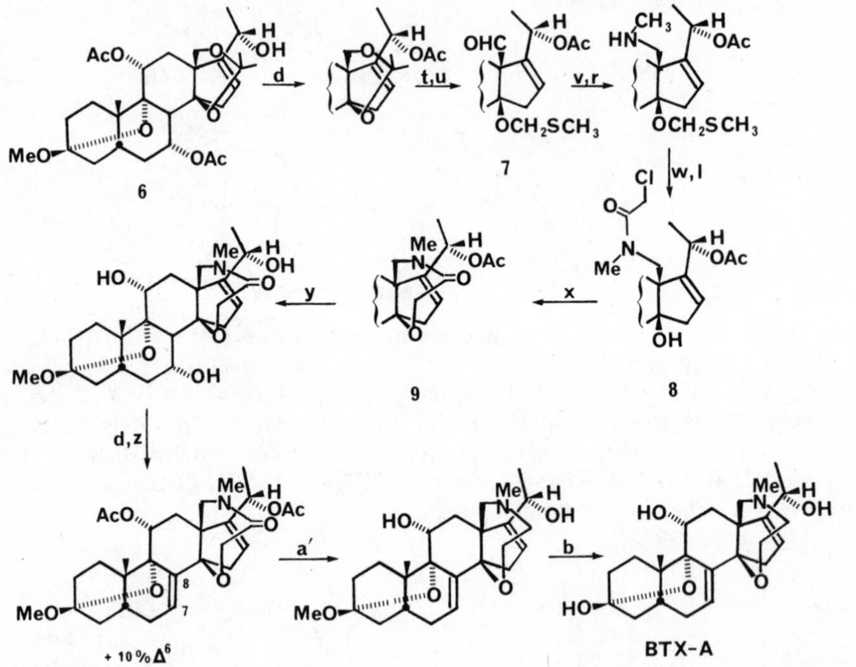

Scheme 11. The synthesis of batrachotoxinin A (BTX-A) [25,33]. (*a*) LiAlH$_4$/THF, (*a'*) LiAlH$_4$/Et$_2$O; (*b*) TsOH/aq. Me$_2$CO; (*c*) H$_2$O$_2$, KOH/MeOH-CH$_2$Cl$_2$; (*d*) Ac$_2$O/ Py; (*e*) CrO$_3$/Me$_2$CO/H$_2$SO$_4$; (*f*) OsO$_4$/Py, then H$_2$S/dioxane-aq. NH$_4$Cl; (*g*) Zn/ HOAc, Δ; (*h*) DDQ/dioxane/aq. HCl; (*i*) *p*-NO$_2\phi$CO$_3$H/MeOH-CHCl$_3$; (*j*) Cr(OAc)$_2$/ Py; (*k*) H$_2$/Pd-C/MeOH; (*l*) HCl/aq. MeOH; (*m*) NBS/CCl$_4$, AIBN, hv, Δ; (*n*) LiBr, Li$_2$CO$_3$/DMF, 130°; (*o*) Pd-BaSO$_4$/cyclohexene-MeOH; (*p*) NaHCO$_3$/aq. MeOH; (*q*) (Me)$_2$C(OMe)$_2$, TsOH; (*r*) NaBH$_4$/MeOH, −30°; (*s*) MnO$_2$-C/ϕH; (*t*) TsOH/MeOH; (*u*) DMSO/Ac$_2$O, 15 hr; (*v*) MeNH$_2$/ϕH, Δ; (*w*) ClCH$_2$COCl/aq. NaOH-CH$_2$Cl$_2$, 0°; (*x*) NaH/THF-ϕH, trace EtOH; (*y*) NaOMe/MeOH; (*z*) SOCl$_2$/Py.

One significant problem encountered was the only moderate stereospecificity (2.5:1) of the low temperature borohydride reduction (*r*) of the C(20)-ketone **4**. Bulky hydride reagents gave mainly the undesired 20-**R**-epimer. This problem was largely ameliorated by the easy chromatographic separation of the diastereomeric alcohols, **5** and **6**, and the recycling of the 20-**R**-epimer **5** after manganese dioxide oxidation (*s*).

Recently the Barton nitrite photolysis of intermediate **11** has been employed by Yelin and co-workers [39,40] in a synthesis of the 18-acetoxy Δ^{9,11} olefin **12** from the 11-α-hydroxyprogesterone ketal **10** (Scheme 12). Compound **12** then served as an intermediate for the synthesis of 7,8-dihydrobatrachotoxinin A by Wehrli's methodology. Levchenko and co-workers [41] reported in 1978 a novel new C,D-ring fused homomorpholine synthesis of possible application to batrachotoxinin A analogs (Scheme 13).

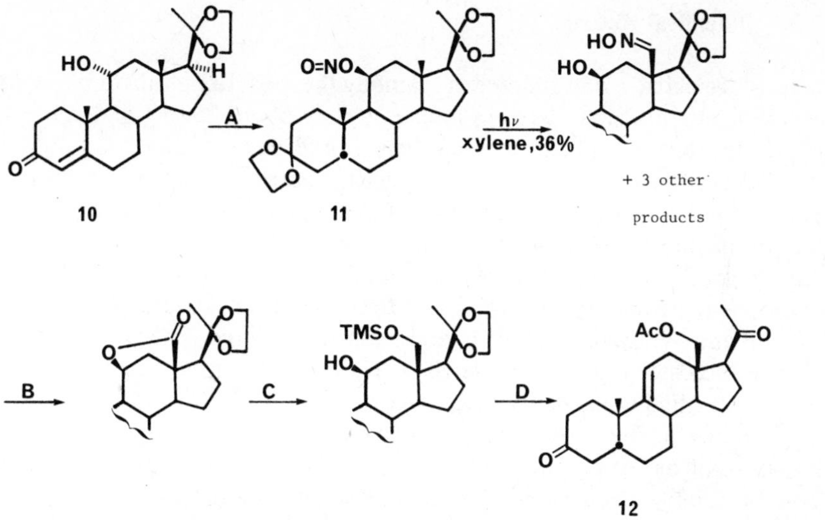

Scheme 12. The synthesis of a 7,8-dihydrobatrachotoxinin A intermediate [39,40].
A: (a) H_2/Pd/CaCO$_3$ (→5β-H); (b) H_2CrO_4 (→11=O); (c) (CH$_2$OH)$_2$, H$^+$ (3,20-ketals); (d) NaBH$_4$ (→11β-ol); (e) NOCl/Py. B: (a) HNO$_2$ (oxime hydrolysis); (b) H$^+$ (→18-hemiacetal); (c) H_2CrO_4 (→lactone, 20=O); (d) (CH$_2$OH)$_2$, H$^+$ (→20-ketal). C: (a) LiAlH$_4$ (→11,18-diol); (b) Me$_3$SiNEt$_2$. D: (a) SOCl$_2$/Py (→Δ^{9,11}); (b) KOH/MeOH (desilylation); (c) Ac$_2$O/Py; (d) H$^+$ (ketal hydrolysis).

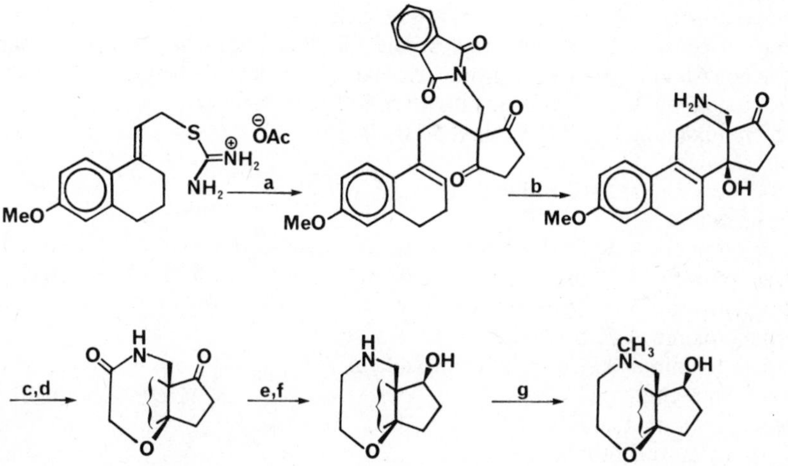

Scheme 13. The synthesis of a C, D ring fused homomorpholine batrachotoxinin A analog [41]. (a) 2-Phthalimidomethylenecyclopentan-1,3-dione; (b) H$^+$/MeOH; (c) ClCH$_2$COCl; (d) NaH/THF; (e) NaBH$_4$; (f) LiAlH$_4$; (g) H$_2$CO, HCO$_2$H.

2.3. Biological Activity

2.3.1. Toxicity. Batrachotoxin is among the most toxic substances known: A lethal subcutaneous dose in mouse is only about 200 ng and it has been estimated that in humans a lethal dose would be less than 200 µg [9]. Undoubtedly toxic effects on the heart leading to arrhythmias and cardiac arrest play a dominant role in the toxicity of this agent. However, the partial protection of mice against batrachotoxin afforded by anticonvulsants, such as diazepam [R. Cadenas Carrera, personal communication], suggests a central component to toxicity as well. The toxicity of batrachotoxin and various natural and synthetic analogs is documented in Table 2. Obviously, the nature of the ester function at the 20α-position is of critical importance. Thus the unesterified congener batrachotoxinin A is fully 500-fold less toxic than batrachotoxin. But it remains a very toxic compound, being still about one-half as toxic as strychnine. The 20α-4,5-dimethylpyrrole-3-carboxylate homolog is notable because of the marked reduction in toxicity compared to batrachotoxin. It appears that two alkyl substituents, flanking the carboxyl group of 20α-pyrrole-3-carboxylates, are necessary for high toxicity. The 20α-2,4,5-trimethylpyrrole-3-carboxylate is the most toxic of the various synthetic analogs. The 20α-pyrrole-2-carboxylate and 20α-*p*-bromobenzoate

Table 2. Toxicity of Batrachotoxin and Related Compounds [17,22,42,43]

Compound	LD$_{50}$, µg/kg[a]
Batrachotoxin	2
Homobatrachotoxin	3
Batrachotoxinin A	1000
Batrachotoxinin A 20-(2,5-dimethylpyrrole-3-carboxylate)	2.5
Batrachotoxinin A 20-(4,5-dimethylpyrrole-3-carboxylate)	260
Batrachotoxinin A 20-(2,4,5-trimethylpyrrole-3-carboxylate)	1
Batrachotoxinin A 20-(2,4-dimethyl-5-ethylpyrrole-3-carboxylate)	8
Batrachotoxinin A 20-(2,5-dimethyl-5-acetylpyrrole-3-carboxylate)	280
Batrachotoxinin A 20-(*N*,2,4,5-tetramethylpyrrole-3-carboxylate)	>1000
Batrachotoxinin A 20-(pyrrole-2-carboxylate)	>1000
Batrachotoxinin A 20-(*p*-bromobenzoate)	>1000
Batrachotoxinin A 20-benzoate	2
Batrachotoxinin A 20-(*N*-methylanthranilate)	15
Dihydrobatrachotoxin	250
Batrachotoxin methiodide	500
3-*O*-Methylbatrachotoxin	30
4-β-Hydroxybatrachotoxin	200

[a] Toxicity determined by subcutaneous injection into 20 g mice. Dihydrobatrachotoxin was formed by NaBH$_4$ reduction of the 3α,9α-oxide function.

esters of batrachotoxinin A show low toxicity. This might suggest that there is something unique about the pyrrole-3-carboxylates, but such a generalization has been proved incorrect: The 20α-benzoate ester of batrachotoxinin A is fully as active as batrachotoxin itself, indicating that the bulky *p*-bromo moiety of the 20α-*p*-bromobenzoate is responsible for its low activity.

The nature of the steroidal portion of the molecule also appears critical to the toxicity of batrachotoxin. Reduction with borohydride to yield dihydrobatrachotoxins or formation of a 3-*O*-methyl ether results in a marked decrease in toxicity of batrachotoxin (Table 2). The methiodide of batrachotoxin is also much less toxic. 4β-Hydroxybatrachotoxin has relatively low toxicity compared to batrachotoxin.

2.3.2. Pharmacological Activity.

Batrachotoxin depolarizes neurons and muscle cells via a specific interaction with sites on voltage-dependent sodium channels in plasma membranes. These actions were delineated in the early 1970s by E. X. Albuquerque and his colleagues [44–49]. Tetrodotoxin was shown to prevent batrachotoxin-elicited depolarization through blockade of sodium channels. The lack of effect of batrachotoxin on voltage-dependent calcium channels was also documented at that time [50]. Subsequent studies have revealed allosteric effects of a variety of agents, particularly polypeptide toxins and local anesthetics, on the biological actions of batrachotoxin (see below). The effects of batrachotoxin on cells with voltage-dependent sodium channels have been comprehensively reviewed [51]: Interaction of batrachotoxin with a site on sodium channels appears to prevent the normal physiological inactivation of the channel. A resultant massive influx of sodium ions through open channels then leads to membrane depolarization. Batrachotoxin causes marked ultrastructural damage to nerve and muscle. Undoubtedly, this damage is due to osmotic changes secondary to the massive influx of sodium.

Activation of Sodium Channels by Batrachotoxin. The action of batrachotoxin in nerve and muscle is time- and stimulus-dependent, suggesting that the binding or action of batrachotoxin requires a prior activation or opening of the channel [51]. Indeed, in some preparations, such as eel electroplax, batrachotoxin has no effect unless channels are opened by stimulation [52,53]. Electrical stimulation in frog nerve enhances the rate of activation of channels by batrachotoxin [54]. Batrachotoxin slows the rate of activation of sodium channels in neuroblastoma cells, markedly shifts the voltage at which channels spontaneously activate to very negative values, and eliminates both slow and fast inactivation of channels [55]. Voltage clamp and patch clamp experiments confirm that batrachotoxin-activated channels do not inactivate. In Ranvier nodes of frog nerves, inactivation of batrachotoxin-activated channels does not occur, but conductances decrease, perhaps because of depletion of perinodal sodium due to the massive influx of sodium [56]. Analysis of single channel currents in neuroblastoma cells by patch clamp techniques confirms that batrachotoxin decreases conductances of sodium channels, prolongs the open state by blocking inactivation, and increases the spontaneous opening of channels even in hyperpolarized cells [57]. Similar results have been obtained with patch clamp techniques on single muscle fibers [58]. The effects of batrachotoxin in many nerve and muscle preparations are relatively irreversible. This apparent irreversibility probably reflects a slow removal of the alkaloid from tissues because of lipid solubility and because only a small percentage of sodium channels (<5%) need to be activated to cause and maintain complete depolarization in most electrogenic membranes. The effects of batrachotoxin are reversible in neuroblastoma cells [59,60].

Batrachotoxin has been used to select for neuroblastoma cell clones resistant to the cytotoxic effects of this alkaloid [6]. All of the 15 resistant clones were deficient in sodium channels. Recently, batrachotoxin-sensitive sodium channels have been detected at low levels in fibroblast cells [62,63], suggesting that such channels may exist at low levels in a variety of cell types other than nerve and muscle.

Other Activators of Sodium Channels. Certain other alkaloids, such as veratridine and aconitine, and the diterpene grayanotoxin, appear to interact with the same site on the sodium channel as batrachotoxin, but these compounds are much less potent and efficacious [see 64]. All cause depolarization in similar manner, that is by preventing inactivation of voltage-dependent sodium channels in nerve and muscle. The effects of batrachotoxin with the sodium channel can be antagonized by the "partial agonists" aconitine and veratridine. However, the mode of binding of batrachotoxin may differ slightly from the mode of binding of the "partial agonists." Thus muscles of the poison dart frogs, which produce batrachotoxin, are virtually insensitive to its action, while still retaining some sensitivity to the action of veratridine and grayanotoxin [65,66]. Muscles of other dendrobatid frogs, such as *Dendrobates histrionicus*, are sensitive to batrachotoxin [cited in 51,66]. The potencies of batrachotoxin, veratridine, aconitine, and grayanotoxin as activators of sodium channels are markedly increased by certain polypeptide toxins and the "partial agonists" now become full agonists [see 64]. The apparent potencies of batrachotoxin-elicited effects in different nerve and muscle preparations extend over nearly a fiftyfold range, but it is not known whether this is due to actual changes in affinity for binding sites on sodium channels, density of sodium channels, or to pharmacokinetic factors.

Selective Blockers of Sodium Channels. Tetrodotoxin and saxitoxin, which block the voltage-dependent sodium channels, prevent and actually reverse batrachotoxin-elicited depolarizations [see 51]. The blockade by tetrodotoxin occurs at a different channel site than that at which batrachotoxin acts. Thus batrachotoxin does not block binding of radioactive tetrodotoxin to its site [67], nor does tetrodotoxin block binding of a radioactive batrachotoxin to its site [42]. Batrachotoxin does not cause depolarization in the absence of sodium ions.

After denervation, muscles develop sodium channels that are not sensitive to tetrodotoxin, but that do retain sensitivity to batrachotoxin [47]. During development of chick muscle and cardiac cells, tetrodotoxin and saxitoxin sensitivity develops more slowly than sensitivity to batrachotoxin or polypeptides [68–70]. Both tetrodotoxin-sensitive and -insensitive sodium channels exist in cultured muscle cells, but both can be activated by batrachotoxin [71,72].

Batrachotoxin causes a massive release of acetylcholine in neuromuscular preparations. This effect undoubtedly is due to depolarization of the presynaptic nerve terminal and, like the effects of batrachotoxin in muscle, can be prevented by tetrodotoxin [45,47,73]. Botulinus toxin also prevents the effects of batrachotoxin on neurotransmitter release [74].

Allosteric Interactions with Polypeptide Toxins. In addition to a tetrodotoxin-site and a batrachotoxin-site, the sodium channel contains a site that interacts with certain polypeptide toxins, including scorpion (*Leiurus*) toxin and anemone toxins [see 64,75]. There is an apparent cooperative interrelationship between binding of agents at the batrachotoxin site and binding of the polypeptide toxins at the peptide site. Thus the polypeptide toxins increase the apparent affinity of batrachotoxin for the batrachotoxin site. In the presence of the polypeptide toxin, the affinities of the "partial agonists" veratridine, aconitine, and grayanotoxin for the batrachotoxin site also increase, and these agents now appear to be full agonists with respect to activation of sodium channels. A converse cooperative interrelationship is also true, since batrachotoxin, veratridine, and aconitine increase the affinity of polypeptide toxins for the peptide site. Batrachotoxin enhances the photoaffinity labeling of the α- and β-subunits of brain sodium channels with derivatives of scorpion (*Leiurus*) toxin [75]. During development of chick muscle cells, the batrachotoxin-sensitive channels develop more slowly than channels that can be activated by batrachotoxin in the presence of a polypeptide toxin [68,69].

The activation of sodium channels by batrachotoxin also is enhanced by a toxin from the dinoflagellate *Ptychodiscus brevis* [76]. This toxin does not appear to act via the same polypeptide site at which scorpion (*Leiurus*) toxin and anemone toxins act. Crotamine, a basic polypeptide from a South American rattlesnake, potentiates the depolarizing action of batrachotoxin in rat diaphragm [77].

Batrachotoxin can enhance scorpion (*Leiurus*) toxin binding in synaptosomal membranes, indicating retention of allosteric interactions in the absence of membrane potentials and ion gradients [78]. Scorpion toxins from other genera (*Centruroides, Tityus*) have different effects on sodium channels and appear to interact with sites that are different from those at which scorpion (*Leiurus*) toxin binds. Batrachotoxin has no effect on binding of an iodinated scorpion (*Centruroides*) toxin to muscle membranes [79]. Tityus γ toxin, another scorpion (*Tityus*) toxin, appears to interact with the same site as scorpion (*Centruroides*) toxins. Its binding is not affected by batrachotoxin [80].

In neuroblastoma cells various pyrethroids markedly enhance batrachotoxin-elicited sodium flux [81,82]. The mechanism appears different from that of the polypeptide toxins, since such pyrethroids also enhance sodium flux elicited by polypeptide toxins.

Allosteric Interactions with Local Anesthetics. Local anesthetics have been shown to block the action of batrachotoxin in a variety of preparations including squid axon [83], rat diaphragm [84], eel electroplax [52], frog nerve [85–91], synaptosomes [92–96], and neuroblastoma cells [93,96–99]. Although the antagonism of the action of batrachotoxin by local anesthetics often appears competitive in nature, binding studies provide strong evidence that local anesthetics act allosterically to reduce the affinity of batrachotoxin for its binding site (see Section 2.3.4).

Altered Properties of Batrachotoxin-Activated Sodium Channels. Batrachotoxin changes the voltage-dependency of sodium channels so that the open form is now stable at resting membrane potentials. The batrachotoxin-activated channel shows differences from stimulus-activated channels not only in voltage-dependency, but in interactions with local anesthetics, in kinetics, and in pore size, apparently being larger than the stimulus-activated channel [85–92,100–123]. Studies on altered properties of batrachotoxin-activated channels in Ranvier node of frog nerve and in neuroblastoma cells has been pioneered by B. Khodorov and his colleagues [see 118–120 and references therein]. Batrachotoxin-activated channels are less selective for sodium ions, and passage through the channel of hydrogen ions and larger ions, such as ammonium, potassium, and calcium, can now be detected. The batrachotoxin-activated channel is more subject to blockade by hydrogen ions and can be blocked by calcium ions. The tetrodotoxin sensitivity is unaltered, although tetrodotoxin does slow the rate of activation. Scorpion toxin does not accelerate inactivation of batrachotoxin-activated channels, suggesting a different pathway for inactivation than that which pertains for normal sodium channels. The sensitivity of sodium channels to blockade by a variety of local anesthetics and other agents appears reduced after activation by batrachotoxin. These agents include procaine, trimecaine, lidocaine, quinidine, ethmozine, strychnine, ajmaline, yohimbine, phenobarbital, and certain quaternary derivatives of local anesthetics [85–91]. Most of these are cationic compounds and are probably interacting with anionic sites on the channel. Neutral compounds such as benzocaine and butanol are equally effective versus normal and batrachotoxin-activated channels. However, oenanthotoxin, a polyunsaturated heptadecan-1,14-diol, blocks normal but not batrachotoxin-activated channels [121]. The sensitivity to phenobarbital, another neutral compound at physiological pH values, is also reduced by batrachotoxin [122].

Cardiac and Ganglionic Actions. Batrachotoxin is an extremely potent cardiotonic agent [46,48,124–128]. Its actions lead ultimately to arrhythmias and cardiac arrest. The basis for the action of batrachotoxin in cardiac preparations is linked to activation of sodium channels and can be antagonized by tetrodotoxin. Unlike the cardiotonic cardiac glycosides, batrachotoxin has little effect on Na^+-K^+-ATPase, causing only a slight inhibition of the enzyme at 60 μM [129], which is a concentration many fold higher than that usually employed for depolarization of cells.

Batrachotoxin is a potent ganglionic blocking agent [48]. The blockade is presumably due to depolarization and resultant blockade of nerve conductance.

Blockade of Axonal Transport. Batrachotoxin is extremely potent in blocking axonal transport [130]. The effect is prevented by tetrodotoxin, and thus appears dependent on the activation of sodium channels and a resultant influx of sodium ions. Sodium ions have been shown to be essential for blockade of axonal transport in desheathed mammalian nerve [131]. External calcium ions can antagonize the blocking effect of batrachotoxin. However, for the mollusc *Aplysia californica*, it has been proposed that the inhibition of axonal transport by batrachotoxin is not due to interactions with sodium channels [132]. This interpretation of the molluscan data has been questioned [133].

Blockade of axonal transport by batrachotoxin reduces both retrograde transport of nerve growth factor and orthograde transport of choline acetylase [134], decreases slightly muscle membrane potential [135], decreases activity of muscle acetylcholine esterase [136], increases activity of glucose-6-phosphate dehydrogenase [137] and certain muscle lysosomal enzymes [138,139], increases Ca^{2+}-ATPase [140] and alters uptake of calcium in muscle sarcoplasmic reticulum [136,140,141], and elicits morphological changes followed by recovery over a period of 1–3 weeks [142–145]. Batrachotoxin inhibits saltatory movements in neuroblastoma cells [133]. This inhibitory effect is blocked by tetrodotoxin and by sodium-free media.

Calcium and Potassium Channels. Batrachotoxin has no effect on a calcium channel in crustacean muscle [50], or on potassium channels in molluscan nerve [146]. At very high concentrations of calcium ions, batrachotoxin, veratridine, and a dihydrograyanotoxin, but not aconitine, were reported to block calcium currents in neuroblastoma cells [147,148]. Half maximal blockade was elicited by 40 nM batrachotoxin. Polypeptide toxins had no effect on calcium currents, and the blockade was not reversed by washing. As yet, direct effects of batrachotoxin on calcium currents have not been seen in other systems. Batachotoxin has no effect on binding of the calcium channel antagonist nitrendipine to brain membranes [149]. Batrachotoxin had no effect on ATP-dependent Ca^{2+} uptake or on Na^+-Ca^{2+} exchange in resealed synaptosomes [150]. Veratridine did reduce Na^+-Ca^{2+} exchange.

Acetylcholine Receptor Channel Complexes. Batrachotoxin antagonizes the increase in conductance elicited by nicotinic agonists in striated neuromuscular preparations [151] and adrenal glands [152]. The mechanism involved in the inhibition of nicotinic receptor-controlled conductances by batrachotoxin remains unclear. Batrachotoxin causes depolarization by activating sodium channels in muscle [151] and adrenal glands [153,154]. The inhibitory effects of batrachotoxin on nicotinic responses are mimicked by aconitine and veratridine and can be observed only in the presence of tetrodotoxin, present to prevent depolarization due to activation of sodium channels by batrachotoxin.

Effects on Purified Sodium Channels. Batrachotoxin activates purified sodium channels isolated from electric eel electroplax [155], rat muscle [156–162], and rat brain [159], and reconstituted into phospholipid vesicles. Batrachotoxin also activates sodium channels of lobster nerve membranes [160–162] or brain membranes [163,164] reconstituted in phospholipid vesicles or bilayers. Batrachotoxin, presumably by activating sodium channels, antagonized ATP-dependent sodium uptake into inside-out vesicular preparations from skeletal muscle [165].

2.3.3. Structure and Activity. Correlations of structure and biological activity have been studied to a limited extent for analogs of batrachotoxin. Toxicity in white mice (Table 2), effects on nerve-striated muscle preparations [166], cardiac preparations [127], ATPase [129], and eel electroplax [cited in 51] show similar profiles with different analogs. The substitution pattern on the pyrrole moiety is important. The relative order of activity for batrachotoxin derivatives modified at the 20α-substituent as depolarizing agents in a neuromuscular preparation was as follows: batrachotoxin > homobatrachotoxin = 2,4,5-trimethylpyrrole carboxylate analog > 5-acetyl-2,4-dimethylpyrrole carboxylate analog > 4,5-dimethylpyrrole carboxylate isomer > p-bromobenzoate analog > batrachotoxinin A [166]. It thus appeared that replacement of a 20α-pyrrole-3-carboxylate with another function would result in marked reduction in activity. Certainly the 20α-p-bromobenzoate analog showed relatively low activity in neuromuscular preparations and the nonesterified congener, batrachotoxinin A, was even less active. Recently it was found that the 20α-benzoate analog is virtually as toxic and as potent in neuromuscular preparations as batrachotoxin [42]. Although the pyrrole-3-carboxylate has not proven to be essential to the high toxicity of batrachotoxin alkaloids, the effect of such a moiety on activity of other drugs has been probed. 2,4,5-Trimethylpyrrole carboxylates of codeine, ephedrine, jervine, scopoline, and methylreserpate were prepared [167], and the scopoline derivative showed modest analgesic activity.

Recently a fluorescent 20α-N-methylanthranilate analog of batrachotoxin was prepared and shown to be highly toxic in mice [43,168]. This analog is a potent depolarizing agent in frog muscle and a potent inhibitor of binding of a radioactive batrachotoxin analog to brain preparations (see Section 2.3.4). Scorpion (*Leiurus*) toxin causes a red shift of the fluorescence emission spectrum and an enhancement in fluorescence of batrachotoxin 20-α-(N-methylanthranilate) bound to sites on the sodium channel [43]. Such changes are consonant with a conformational change in the channel leading to a more hydrophilic environment at the batrachotoxin site. Fluorescence resonance energy transfer studies suggest that the distance separating the batrachotoxin-binding site and the scorpion toxin site is about 40 Å [43]. The N-methylanthranilate analog has been used as a fluorescent marker for sodium channels in mammalian nerves [168].

The steroid configuration of batrachotoxin has appeared quite important to its biological activity since reduction of the 3α,9α-hemiketal function with borohydride yields a relatively inactive dihydro derivative [166]. The quaternary methyl derivative of batrachotoxin shows low activity. However, such low activity might be due to lack of penetration of this positively charged compound.

Batrachotoxin is most effective in neuromuscular preparations at pH 9, at which pH much of the alkaloid would be in the unionized form [166]. It had been suggested that a membrane constituent ionized at pH values >6.0 is essential to the action of batrachotoxin [51]. Binding data support such a hypothesis (see Section 2.3.4). Early studies had indicated that sulfhydryl reagents could prevent the action of batrachotoxin [49], while later studies [26] suggest that a histidine residue also may be essential to binding of batrachotoxins. Inhibition of protein glycosylation by tunicamycin reduces batrachotoxin-elicited sodium uptake in chick muscle cells [169]. Batrachotoxin is more effective applied internally to squid axons than when present in the external media [146]. This result suggests that batrachotoxin acts at a site more accessible from the interior than from the exterior of cells.

A triad of oxygen atoms in batrachotoxin, veratridine, aconitine, and grayanotoxin have been proposed to be essential for activation of the sodium channel [170]. These oxygens are the 3α-OH, the 3α,9α-oxide, and the 11α-OH in batrachotoxin. A proximate hydrophobic moiety was also proposed as essential. This proposal has been extended by Kosower [171] to include complexation by the oxygen triad of an ε-ammonium ion of a channel lysine concerned with the anionic gating mechanism. Earlier very tentative models for possible interactions of batrachotoxin and other neurotoxins with sodium channels have been presented [172–175]. The recent elucidation of the primary sequence and a tentative tertiary structure of the voltage-dependent sodium channel from electric eel [176] should allow future modeling to be less speculative in nature.

2.3.4. Binding of a Radioactive Batrachotoxin Analog.

Batrachotoxinin A 20α-p-[³H]benzoate ([³H]BTX-B) has proven to be a satisfactory ligand for the investigation of batrachotoxin-binding sites in brain synaptosomes and microsacs [26,42,177]. The affinity constant for [³H]BTX-B is about 50 nM in the presence of scorpion (*Leiurus*) toxin and about 500 nM in the absence of the scorpion toxin. Binding is optimal in sodium-free media with tetrodotoxin present to prevent depolarization. Depolarization would reduce the effectiveness of the scorpion toxin. The density of binding sites for [³H]BTX-B is similar to that noted for tetrodotoxin, saxitoxin, and the scorpion toxin. The ability of scorpion (*Leiurus*) toxin to enhance [³H]BTX-B binding has been used as a biochemical assay for this polypeptide toxin [178]. Binding of [³H]BTX-B is antagonized by batrachotoxin and analogs and by aconitine, veratridine and grayanotoxin [42,43,177]. A variety of local anesthetics also antagonize the binding of [³H]BTX-B [95,179–181]. Their potencies as antagonists of [³H]BTX-B binding correlate well with potencies as local anesthetics, and ranges of IC$_{50}$ values of 1–2 μM for dibucaine and proparacaine have been reported, as have IC$_{50}$ values of 110 μM for procaine, 240 μM for lidocaine, and 910 μM for benzocaine. Other drugs that have ancilliary activity as local anesthetics inhibit binding of [³H]BTX-B [179,180,182–184]. These include anticonvulsants, antihistamines, antidepressants, adrenergic antagonists, calcium antagonists, phenothiazines, and butyrophenones. Certain plant alkaloids such as reserpine, cocaine, yohimbine, strychnine, and quinine, and certain amphibian alkaloids, such as the histrionicotoxins, pumiliotoxin C, and samandarine, are potent antagonists of [³H]batrachotoxin binding. Many of these alkaloids, including yohimbine and strychnine, antagonize pharmacological effects of batrachotoxin. Another indole alkaloid, epiervatamine, is a potent antagonist of the action of batrachotoxin [93].

Agents such as batrachotoxin, veratridine and aconitine that compete directly for the binding site of [³H]BTX-B have no effect on the rate of dissociation of [³H]BTX-B, while local anesthetics, although appearing to be competitive inhibitors of [³H]BTX-B binding, markedly increase the rate of dissociation of [³H]BTX-B [181,184], consonant with an indirect allosteric mechanism.

Recently, specific binding of 7,8-dihydrobatrachotoxinin A 20α-[³H]benzoate to rat brain synaptosomes and enhancement by various polypeptide toxins was reported [185]. Local anesthetics, oenanthotoxin, and a pyrethroid, cyanpermethrin, inhibited binding. Certain pyrethroids have, however, been reported to enhance binding of [³H]BTX-B in the presence of scorpion (*Leiurus*) toxin [186]. The binding of the [³H]7,8-dihydroBTX-B was stated to be decreased by phospholipase A, proteases, N-ethylmaleimide, dithioth-

Table 3. Effects of Batrachotoxin on Biological Systems

System	Species	Comments	References
Whole animal	Dog, cat, rabbit, mouse, frog, toad, sheep	High toxicity; cardiac effects have a major role	[5,8–13,18,48, 190,194]
		Tissue distribution	[51]
Neuromuscular preparation (striated)	Rat, frog, toad	Depolarization of striated muscle; slower depolarization of nerve axon; depolarization of nerve terminal rapid, resulting in an initial increase and then blockade of transmitter release	[8,10–13,42–45, 47,65,66,73,84, 128,151,166]
	Lobster, crayfish	No effect on voltage-dependent calcium channels	[50]
Cardiac preparations	Cat, rabbit, dog, guinea pig, toad, mouse	Initial increase of contractile force; depolarization; arrhythmias; cardiac arrest	[11–13,46,48, 124–128]
		Blockade of isoproterenol-elicited accumulation of cyclic AMP	[195]
Heart cells	Rat, chick	Increased sodium fluxes, blocked by tetrodotoxin, enhanced by polypeptide toxins; increased rate of beating followed by arrhythmias and arrest	[72,196,197]
Smooth muscle	Rabbit, guinea pig	Rhythmic contractures in ileum, blocked by tetrodotoxin; no effect in uterus.	[13,128]
Muscle cells	Chick, rat	Increased sodium fluxes, enhanced by polypeptide toxins	[68,69,71,72,82, 169,198–200]

31

Table 3. (*continued*)

System	Species	Comments	References
Glial cells	Mudpuppy	No effect on membrane potential	[201]
Fibroblasts	Human, hamster	Increased sodium fluxes; enhancement by polypeptide toxins	[62,63]
Nerve (axons)	Rat, rabbit, sheep, mudpuppy, frog, lobster, squid, mollusc, garfish	Depolarization (blocked by tetrodotoxin); batrachotoxin-activated channels show alterations in voltage-dependency and properties, including lack of inactivation, decreased ion selectivity, and decreased sensitivity to local anesthetics	[45,49,54,56,67, 83,85,87– 89,100, 101,103–112, 118–123,146, 194,201–203]
	Mouse	Localization of sodium channels with fluorescent analog	[168]
	Rat, cat, mouse, mollusc	Reduction in axonal transport and resultant distal effects on nerve and muscle; axonal degeneration and regeneration	[130–132,134– 145,194, 204–207]
Superior cervical ganglion	Rabbit	Depolarization and blockade of ganglionic transmission; lowering of tyrosine hydroxylase levels	[48,208]

Table 3. (*continued*)

System	Species	Comments	References
Adrenal medulla	Bovine	Increased sodium fluxes; release of catecholamines; inhibition of nicotinic response in presence of tetrodotoxin	[152–154]
Electroplax	Eel	Depolarization; no effect in unstimulated preparations	[52,53,209–211]
	Catfish	Depolarization	[212]
Skin	Frog	No effect on short circuit current	[51]
Neuroblastoma cells	Mouse	Increased sodium fluxes; no effect in cells that lack voltage-dependent sodium channels; cooperative interactions with polypeptide toxins, pyrethroids; inhibition of effects by local anesthetics, certain anticonvulsants, and certain antiarrhythmic agents; reduction in selectivity, resulting in calcium influx via sodium channels; cytotoxicity	[55,59,61,76,82, 90,92,93,96–99, 102,114– 117,147, 199,213–221]
		Blockade of calcium currents	[147,148]
Pituitary cells	Rat	Enhanced release of luteinizing hormone	[222]
Erythrocytes	Human	No effect on sodium permeability [cited in 51]	

Table 3. (*continued*)

System	Species	Comments	References
Brain	Rat, guinea pig, mouse	Depolarization—resulting in cyclic AMP and cyclic GMP formation in brain slices and microsacs	[223–231]
	Rat, guinea pig, mouse	Increased sodium fluxes; depolarization of microsacs and synaptosomes and of neurons in fetal brain cell cultures; inhibition of effects by local anesthetics, certain anticonvulsants; potentiation by polypeptide toxins	[93–96,150, 232–242]
	Rat, cat	Inhibition of uptake and increase in release of neurotransmitters	[243,244]
	Rat, mouse, guinea pig	Binding of radioactive batrachotoxin analogs; potentiation by polypeptides; allosteric antagonism by local anesthetics	[26,42,43,95, 177–186]
	Rat	Enhancement of photoaffinity labeling of sodium channels by derivatives of a polypeptide toxin	[43]
ATP levels	Rat, cat, mouse	Little effect on levels of ATP in nerve-muscle; reduces phosphocreatine; reduction of ATP in slices and atria	[129,130,195,229]

Table 3.   (*continued*)

System	Species	Comments	References
Na$^+$-K$^+$-ATPase	Eel	Slight inhibition at high concentrations	[129]
Ca^{2+}-ATPase	Rat	No inhibition	[150]
Cyclic AMP-phosphodiesterase	Guinea pig	Slight inhibition at high concentrations	[223]

reitol, glutathione, and lubrol [185]. It was also stated that photolysis of tritiated 7,8-dihydrobatrachotoxin A 20α-*o*-azidobenzoate with brain preparations led primarily to labelling of membrane lipids. Radiation inactivation studies indicate that the [^{3}H]BTX-B binding units have target sizes of 287,000 and 51,000 daltons [187].

The specific binding of [^{3}H]BTX-B is greatest at pH 8.5 and is markedly reduced at pH < 8 and pH > 9 [26]. It would appear that pH-dependence relates to two groups with pK_as of 7.7 and 8.8. The pK_a of batrachotoxinin A is about 8.2. Two interpretations are possible: Either the protonated form of batrachotoxin might be the most active form, accounting for reduction in binding at pH > 8.8; or both protonated and nonprotonated alkaloid might be active, and deprotonation of a pK_a 8.8 group in the channel proteins might be unfavorable to binding of batrachotoxin. In either case, protonation of a group with pK_a of 7.7, perhaps representing a histidine residue in the sodium channel, appears unfavorable, accounting for the reduction in binding at pH < 7.7. Photochemical or chemical treatments that should selectively modify histidine residues do reduce binding of [^{3}H]BTX-B.

[^{3}H]BTX-B has been used to localize the site of action of batrachotoxins in studies on axonal transport [135]. The tissue distribution of [^{3}H]batrachotoxin has been reported [51].

2.3.5. Summary.   Batrachotoxin has proven to be an invaluable tool for the mechanistic study of voltage-dependent sodium channels and for investigation of effects of depolarization and/or influx of sodium ions on physiological functions. A summary of such investigations is provided in Table 3. A number of reviews on toxicology and/or pharmacology of batrachotoxin are available [24,38,51,64,86,118–120,188–193].

2.4. Addendum

An intramolecular Diels-Alder reaction has been employed to prepare a model compound containing the ABC rings of batrachotoxin with its triad of oxygen atoms (3α,9α-oxide-bridged-3-ketal and 11α-hydroxyl) as well as additional functionality required for a total synthesis of batrachotoxin [P. Magnus, T. Leapheart, and C. Walker, *Chem. Commun.*, **1985**, 1105)].

3. HISTRIONICOTOXINS

3.1. Structures

A preliminary investigation of the alkaloids in extracts from a brightly colored, extremely variable species of dendrobatid frog, *Dendrobates histrionicus*, was carried out in the late 1960s. The results indicated the presence of simple, highly unsaturated, alkaloids with distinctive mass spectral properties, namely major fragment ions at m/z 218 ($C_{14}H_{20}NO$) and m/z 96 ($C_6H_{10}N$) [J. W. Daly, unpublished results]. The source of these extracts was a population of *D. histrionicus* sympatric with the poison dart frog *Phyllobates aurotaenia* in the Río San Juan drainage of western Colombia, and had been obtained during field work on the latter frog. Since the Río San Juan populations of *D. histrionicus* were not particularly abundant, further field work was carried out in collaboration with C. W. Myers with an extremely abundant population of the same species that was known to occur near the town of Guayacana in southwestern Colombia. Methanol extracts from some 400 specimens of the Guayacana population of *D. histrionicus* afforded, after partitions and silica gel column chromatography, two major alkaloids, histrionicotoxin (53 mg) and isodihydrohistrionicotoxin (93 mg) [245]. The names were derived from the species name of the frog, and the toxin designation was used (incorrectly, since the compounds are relatively nontoxic) because earlier isolated alkaloids, batrachotoxin (Section 2) and pumiliotoxin A and B (Section 6), were quite toxic. In addition to the two histrionicotoxins, an alkaloid fraction (40 mg) was obtained that contained one major and several minor alkaloids. One of these alkaloids ultimately proved to be gephyrotoxin (Section 4).

The hydrochloride and hydrobromide salts of histrionicotoxin and the hydrochloride salt of isodihydrohistrionicotoxin were obtained in crystalline form. X-ray crystallographic analyses revealed the structures and absolute configurations of histrionicotoxin, [2R,6R,7S,8S]-7-(1-*cis*-buten-3-ynyl)-2-(*cis*-2-penten-4-ynyl)-1-azaspiro[5.5]undecan-8-ol, and isodihydrohistrionicotoxin [245,246] (Fig. 2). These compounds were the first examples of an unusual class of spiropiperidine alkaloids with acetylene, allenic, or olefinic moieties in the side chains. Such histrionicotoxins occur in many dendrobatid frogs. Their mass spectra are diagnostic, exhibiting a fragment ion that corresponds to loss of the side chain at the 2-position and a major fragment ion at m/z 96 ($C_6H_{10}N$).

In a subsequent study, the alkaloids from methanol extracts from some 1100 skins of the same abundant population of *Dendrobates histrionicus* were subjected by Tokuyama to multiple column chromatographies on silica gel and Sephadex LH-20 [247]. Histrionicotoxin (226 mg), isodihydrohistrionicotoxin (320 mg), and four analogs were isolated. The mass spectra and nuclear magnetic spectra defined the structures of the analogs as neodihydrohistrionicotoxin (19 mg), tetrahydrohistrionicotoxin (2 mg), isotetrahy-

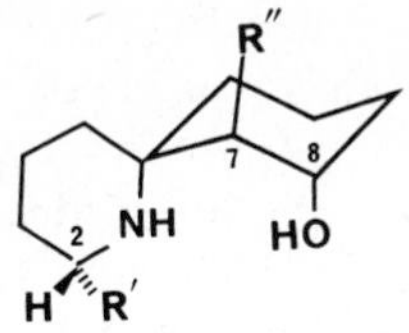

	R'		R''	
A	cis	$-CH_2CH=CHC\equiv CH$	cis	$-CH=CHC\equiv CH$	283A
B	cis	$-CH_2CH=CHCH=CH_2$	cis	$-CH=CHC\equiv CH$	285E
C	cis	$-CH_2CH=CHC\equiv CH$	cis	$-CH=CHCH=CH_2$	285B
D	cis	$-CH_2CH=CHCH=CH_2$	cis	$-CH=CHCH=CH_2$	287B
E		$-CH_2CH_2CH=C=CH_2$	cis	$-CH=CHC\equiv CH$	285A
F		$-CH_2CH_2CH=C=CH_2$	cis	$-CH=CHCH=CH_2$	287A
G		$-CH_2CH_2CH_2C\equiv CH$	cis	$-CH=CHC\equiv CH$	285C
H		$-CH_2CH_2CH_2C\equiv CH$	cis	$-CH=CHCH=CH_2$	287D
I		$-CH_2CH_2CH_2CH=CH_2$		$-CH_2CH_2CH=CH_2$	291A
J	cis	$-CH_2CH=CHC\equiv CH$	trans	$-CH=CHC\equiv CH$	283A
K		$-CH_2CH_2CH_2CH_2CH_3$		$-CH_2CH_2CH_2CH_3$	
L		$-CH_2CH=CH_2$	cis	$-CH=CHC\equiv CH$	259
M		$-CH_2CH=CH_2$		$-CH=CH_2$	235A

Figure 2. Structures of the histrionicotoxins. (*A*) Histrionicotoxin; (*B*) Dihydrohistrionicotoxin; (*C*) Neodihydrohistrionicotoxin; (*D*) Tetrahydrohistrionicotoxin; (*E*) Isodihydrohistrionicotoxin; (*F*) Isotetrahydrohistrionicotoxin; (*G*) Allodihydrohistrionicotoxin; (*H*) Allotetrahydrohistrionicotoxin; (*I*) Octahydrohistrionicotoxin; (*J*) Δ-17-*trans*-histrionicotoxin; (*K*) Perhydrohistrionicotoxin (not detected in nature); (*L*) Histrionicotoxin **259**; (*M*) Histrionicotoxin **235A**.

drohistrionicotoxin (6 mg), and octahydrohistrionicotoxin (9 mg) (Fig. 2). A fifth compound referred to as HTX-D (47 mg) was obviously not a histrionicotoxin, although it contained a five- carbon side chain (*cis*-CH₂CH=CHC≡CH) identical with one of the side chains of histrionicotoxin. The structure of HTX-D was later elucidated by X-ray crystallographic analysis and the tricyclic alkaloid was renamed gephyrotoxin (Section 4). A number of minor alkaloids were detected as mixtures in certain column fractions, but their structures were not elucidated at this time. Four additional histrionicotoxins were later isolated by Tokuyama and characterized from extracts of further large samples of skins of this population of *D. histrionicus* [248,249]. These were allodihydrohistrionicotoxin, allotetrahydrohistrionicotoxin, dihydrohistrionicotoxin, and Δ-17-*trans*-histrionicotoxin (Fig. 2). The former is a major alkaloid constituent, while the latter two are minor or trace constituents. All of these histrionicotoxins reduce to a common perhydro derivative, dodecahydrohistrionicotoxin ($C_{19}H_{37}NO$) (Fig. 2). This perhydro derivative has not as yet been detected in any extracts of dendrobatid frogs.

A histrionicotoxin **259** with the empirical formula $C_{17}H_{25}NO$ had been detected in various species of dendrobatid frogs, using a gas chromatographic mass spectral analytical protocol for alkaloid fractions, first introduced by Myers and Daly [250] in 1976. This protocol has now led to the identification of over 200 alkaloids from dendrobatid frogs (see Section 11). It was apparent from the initial studies that dendrobatid frogs would afford a large number of alkaloids and that a code system of nomenclature would be needed. A system was formalized [23] in 1978. Alkaloids were designated by molecular weight in boldface type with an added letter to identify alkaloids of the same nominal molecular weight. Histrionicotoxin **259** was finally isolated in sufficient quantities for a nuclear magnetic resonance spectrum from extracts of a Panamanian population of *Dendrobates auratus* [249]. Histrionicotoxin **259** (Fig. 2) represents the first member of this class of dendrobatid alkaloids shown to contain a three-carbon side chain on the piperidine ring rather than a five-carbon chain. Histrionicotoxin **235A**, with an empirical formula $C_{15}H_{25}NO$, has been identified by the gas chromatographic–mass spectral analytical protocol and appears to contain a three-carbon side chain on the piperidine ring and a two-carbon, rather than a four-carbon, side chain on the cyclohexanol ring. Its tentative structure is as shown in Fig. 2.

The protocol for the analysis of constituent alkaloids from skin extracts of dendrobatid frogs by gas chromatography–mass spectrometry [23,250] has been developed and refined over the years. Such a protocol now allows the characterization of alkaloid profiles in extracts from a single frog skin or less and has in several instances provided sufficient data for gross structural assignment of trace alkaloids. A current protocol for such analyses is as follows.

Frogs are skinned in the field, and skins stored in methanol (1 part skins to from 2 to 100 parts methanol), when possible at $-5°$, otherwise at ambient temperature. Skins are macerated twice with methanol to extract alkaloids. Skin samples with a total wet weight of 50–500 mg are macerated twice each time with 5–10 mL of methanol; skin samples from 1 to 4 g wet weight are macerated three times, each time with 10–20 mL of methanol. Larger skin samples are macerated twice, each time with about 2–4 volumes of methanol (weight/volume). The methanol extracts are then diluted with an equal volume of water. In some instances for larger samples, the methanol extract is first concentrated *in vacuo* to a smaller volume before dilution. The aqueous methanol is then extracted twice, each time with 2 volumes of chloroform. The basic chloroform-soluble alkaloids are then extracted three times from the combined chloroform layers, each time into a one-half volume of 0.1 N HCl. The combined 0.1 N HCl fractions are adjusted to pH 9 with 1 N aqueous ammonia followed by reextraction into chloroform, twice, each time with an equal volume of chloroform. The combined chloroform layers are dried over anhydrous sodium sulfate and then evaporated *in vacuo* to dryness. Methylene chloride may be used in place of chloroform during these partitions. Certain of the dendrobatid alkaloids have appreciable volatility and, therefore, evaporations *in vacuo* must be done carefully. The resulting alkaloid residue is dissolved in methanol so that 10 μL corresponds to 10 mg of the original wet weight of the skins. This alkaloid fraction contains primarily alkaloids, though traces of steroids and environmental artifacts, such as phthalates, sometimes are present as contaminants.

Such alkaloid fractions are analyzed by thin-layer chromatography, gas chromatography, and mass spectrometry [23,24]. The studies have been carried out in a manner designed to facilitate quantitative and qualitative comparisons of alkaloid profiles between populations and species of dendrobatid frogs and other amphibians.

Thin-layer chromatographic analyses are routinely carried out with alkaloid fractions equivalent to 10 mg wet weight skin on silica gel plates with chloroform:methanol (9:1). Detection is routinely performed with iodine vapor. In some instances preparative thin-layer chromatography on a microscale, followed by analysis by gas chromatography–mass spectrometry, has been used to determine R_f values of alkaloids.

Gas chromatographic analyses are routinely carried out with alkaloid fractions equivalent to 2 mg wet weight skin on a 1.5% OV-1 column. The column is programmed from 150° to 280° at 10° per min with a flow rate of 20–25 cm^3/min of nitrogen. Analysis with a flame ionization detector provides a quantitative profile of alkaloid components [see 23 for further details]. The second step in analysis is combined gas chromatography–chemical ionization mass spectrometry with nitrogen carrier gas and ammonia as the ionizing gas. Ammonia chemical ionization provides virtually exclusively the protonated parent ion of the various dendrobatid alkaloids. Computer-assisted analysis of the ammonia chemical ionization mass spectra obtained during the gas chromatographic run provides the number, elution sequence, and parent ions of the alkaloids corresponding to peaks in the flame ionization profiles. Often an apparently single flame ionization peak is found to represent two or more alkaloids. Capillary columns are proving useful for the separation of such mixtures. Repetition of chemical ionization mass spectral–gas chromatographic analysis using deuteroammonia reveals the number of exchangeable hydrogens (hydroxyl and secondary amine functions) in the deuterated parent ion [251]. The position of the exchangeable moiety is sometimes interpretable from the fragment ions that retain the deuterium. Recently, "pseudo" electron impact spectra of dendrobatid alkaloids have been determined using N_2-NO in a chemical ionization mode [251]. Such spectra, like electron-impact spectra, appear to be sufficiently detailed for characterization and identification of individual dendrobatid alkaloids. After chemical ionization determinations, the gas chromatography–mass spectral analysis is repeated in the electron impact mode. Such electron impact spectra are usually sufficiently detailed to allow characterization and identification of individual alkaloids.

A gas chromatographic profile is presented for the alkaloid fraction from one population of *Dendrobates histrionicus* [251] in Fig. 3 with various peaks identified as to alkaloids by the code designations, based on the nominal molecular weight, and by an added letter as necessary. Further examples of gas chromatographic profiles of alkaloid fractions from dendrobatid and nondendrobatid frogs are depicted in other publications [23,24,250–261]. A capillary gas chromatographic profile is depicted in [183].

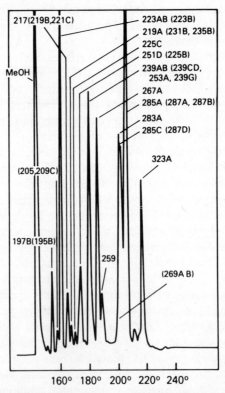

217(219B,221C)
223AB (223B)
219A (231B, 235B)
225C
251D (225B)
239AB (239CD, 253A, 239G)
267A
285A (287A, 287B)
283A
285C (287D)
323A
MeOH
(205,209C)
197B(195B)
259
(269A B)

160° 180° 200° 220° 240°

Figure 3. Gas chromatographic profile for alkaloids from a population of *Dendrobates histrionicus* (species/population 14A of Table 17). A sample of 2 μL of methanolic alkaloids equivalent in amount to 2 mg wet skin was injected at 150°C onto a 1.5% OV-1 column. After the maximum for the solvent peak was passed, the column was programmed at 10°C per min from the initial 150°C to 280°C. Emergent temperatures differ somewhat with different batches of column packing. A flame ionization detector was used. Alkaloids identified and characterized by combined gas chromatography–mass spectrometry are designated by their molecular weights and where necessary with an added code letter. Trace constituents are in parentheses (see Table 16 for complete listing of dendrobatid alkaloids).

Finally, empirical formulas for the parent ion or the protonated parent ion and fragment ions are determined for alkaloids by gas chromatography–high resolution mass spectrometry. Often this is done in a mixed chemical ionization–electron-impact mode on the protonated parent ion, since the parent ions of many dendrobatid alkaloids are vanishingly small in the electron-impact mode [251].

Additional data on the alkaloids can be obtained after perhydrogenation, acetylation, or phenylboronation. In such instances, alkaloid fractions corresponding to 100–300 μg wet weight skin are either reduced in methanol with 10% palladium-on-charcoal catalyst and 30 psi hydrogen gas overnight, acetylated in acetone with acetic anhydride and sodium acetate at room temperature overnight, or treated with phenylborane in acetone to react with vic-diol functions. The resultant derivatized alkaloid fraction is subjected to analysis by gas chromatography–mass spectrometry. The above isolation and analytical protocols have been used to detect and characterize over 200 dendrobatid alkaloids from some 35 species of *Dendrobates* (see Section 11).

Histrionicotoxins, as a class, are characterized by a major mass spectral fragment ion ($C_6H_{10}N$) at m/z 96. They have two exchangeable hydrogens and contain two unsaturated side chains. A significant mass spectral fragment results from loss of the side chain at the 2-position of the piperidine ring. Over a dozen histrionicotoxins have been detected in various species and populations of dendrobatid frogs by gas chromatographic–mass spectral analysis. All except four (**235A,239H,261,263C**) of the histrionicotoxins have been isolated in sufficient quantities for magnetic resonance spectroscopy and unambiguous determination of structure.

The properties of various natural histrionicotoxins are presented below in a format developed by Daly for dendrobatid alkaloids [23]. The entries are as follows:

1. A trivial name (if any) and boldface numerical designation based on molecular weight and, for alkaloids with the same nominal molecular weight, a letter in boldface. For certain very similar alkaloids, a prime and double prime designation has been used to indicate their close structural relationship.

2. An empirical formula based on high resolution mass spectrometry. Tentative empirical formulas that are based on analog and chemical and chromatographic properties and that have not been confirmed by high resolution mass spectrometry are in quotations.

3. An R_f value from thin-layer chromatography on silica gel with chloroform:methanol, 9:1.

4. An emergent temperature on a 1.5% OV-1 packed gas chromatographic column programmed from 150–280° at 10°/min.

5. The electron impact mass spectral ions (m/z) followed in parentheses with elemental compositions and the intensities relative to the base

peak set equal to 100; only the most diagnostic peaks and not all elemental compositions are reported.

6. The perhydrogenation derivative obtained with a palladium-on-carbon catalyst in methanol and 30 psi hydrogen (H_0 = no addition of hydrogen) and mass spectral data for the perhydro derivative.

7. The number of hydrogens exchangeable with deuteroammonia (0D, 1D, etc.—no exchangeable, one exchangeable, etc.).

8. Any other pertinent comments.

Most of the other dendrobatid alkaloids are tabulated in this format in appropriate sections of this review.

Histrionicotoxins

235A. $C_{15}H_{25}NO$, 0.36, 176°, m/z 235 (5), 234 (2), 218 (15), 194 ($C_{12}H_{20}NO$, 76), 176 ($C_{12}H_{18}N$, 25), 150 (8), 96 ($C_6H_{10}N$, 100). H_4 derivative, m/z 239, 196, 178, 96. 2D. Tentative structure: Fig. 2.

239H. "$C_{15}H_{29}NO$,"—, 182°, m/z 239 (7), 238 (4), 221 (6), 196 (35), 178 (10), 96 (100). H_0 derivative. 2D. A perhydro-**235A**.

259. $C_{17}H_{25}NO$, 0.36, 190°, m/z 259 (4), 242 (2), 218 ($C_{14}H_{20}NO$, 18), 200 (6), 96 ($C_6H_{10}N$, 100). H_8 derivative, m/z 267 (20), 250 (13), 224 (39), 196 (15), 268 (19), 152 (100), 96 (68). 2D. Structure: Fig. 2.

261. $C_{17}H_{27}NO$,—, 190°, m/z 261 (8), 220 (100), 204 (10), 96 (68). H_6 derivative. 2D. A dihydro-**259**.

263C. $C_{17}H_{29}NO$,—, 192°, m/z 263 (1), 222 (100), 204 (10), 96 (48). H_4 derivative. 2D. A tetrahydro-**259**.

283A. Histrionicotoxin, $C_{19}H_{25}NO$, 0.50, 210°, m/z 283 (9), 282 (2), 266 (5), 250 (2), 218 ($C_{14}H_{20}NO$, 48), 200 (27), 160 (22), 96 ($C_6H_{10}N$, 100). H_{12} derivative, 0.36, 214°, m/z 395 (12), 294 (2), 278 (13), 252 (18), 224 (73), 196 (27), 180 (100), 168 (39), 96 (68). 2D. Structure: Fig. 2.

283A′. Δ-17-*trans*-Histrionicotoxin, $C_{19}H_{25}NO$, 0.53, 210°, m/z 283 (26), 282 (12), 266 (11), 218 (60), 200 (21), 188 (11), 174 (14), 160 (40), 124 (14), 96 (100). H_{12} derivative. 2D. This isomer of histrionicotoxin was not detected by gas chromatographic analysis, since it emerges with histrionicotoxin, but was later isolated from extracts [249]. Because of the close relationship to histrionicotoxin, it is given a prime designation. Structure: Fig. 2.

285A. Isodihydrohistrionicotoxin, $C_{19}H_{27}NO$, 0.39, 215°, m/z 285 (7), 284 (2), 268 (8), 252 (12), 238 (3), 218 (6), 200 (9), 190 (4), 176 (24), 162 (18), 96 (200). H_{10} derivative. 2D. Structure: Fig. 2.

285B. Neodihydrohistrionicotoxin, $C_{19}H_{27}NO$, 0.46, 211°, m/z 285 (4), 284 (1), 268 (3), 250 (2), 220 (37), 202 (9), 260 (20), 96 (100). H_{10} derivative. 2D. Structure: Fig. 2.

285C. Allodihydrohistrionicotoxin, $C_{19}H_{27}NO$, 0.40, 211°, m/z 285 (4), 284 (1), 268 (2), 252 (2), 218 (5), 200 (3), 190 (4), 176 (15), 162 (17), 96 (100). H_{10} derivative. 2D. Structure: Fig. 2.

285E. Dihydrohistrionicotoxin, $C_{19}H_{27}NO$, 0.50, 212°, m/z 285 (13), 268 (10), 218 (100), 200 (84), 176 (13), 96 (78). H_{10} derivative. 2D. Structure: Fig. 2.

287A. Isotetrahydrohistrionicotoxin, $C_{19}H_{29}NO$, 0.42, 216°, m/z 287 (12), 286 (4), 270 (3), 220 (30), 202 (34), 176 (45), 162 (60), 148 (24), 96 (100). H_8 derivative. 2D. Structure: Fig. 2.

287B. Tetrahydrohistrionicotoxin, $C_{19}H_{29}NO$, 0.43, 213°, m/z 287 (13), 286 (2), 270 (2), 220 (43), 202 (18), 176 (6), 162 (4), 148 (4), 96 (100). H_8 derivative. 2D. Structure: Fig. 2.

287D. Allotetrahydrohistrionicotoxin, $C_{19}H_{29}NO$, 0.35, 215°, m/z 287 (14), 270 (8), 220 (24), 202 (36), 176 (38), 162 (49), 96 (100). H_8 derivative. 2D. Structure: Fig. 2.

291A. Octahydrohistrionicotoxin, $C_{19}H_{33}NO$, 0.35, 212°, m/z 291 (12), 290 (2), 274 (14), 250 (54), 222 (24), 194 (18), 192 (12), 178 (100), 96 (52). H_4 derivative. Structure: Fig. 2.

Further physical and spectral properties of histrionicotoxins are presented in Table 4. Mass spectra for histrionicotoxins and various congeners have been presented [245,247–249; for discussion see 247; for a detailed tabulation see 24]. Proton and ^{13}C magnetic resonance assignments for various histrionicotoxins have been presented [247–249] and reviewed [24]. Proton magnetic resonance spectra have been depicted for histrionicotoxin [249], Δ-17-*trans*-histrionicotoxin [249], isodihydrohistrionicotoxin [24], allodihydrohistrionicotoxin [248], histrionicotoxin **259** [249], and perhydrohistrionicotoxin [247]. The natural histrionicotoxins have all been levorotatory. The pK_a values for histrionicotoxin and perhydrohistrionicotoxin, as determined by titration, were in the range 9.0–9.3 [unpublished results].

The occurrence of histrionicotoxins in dendrobatid frogs and in various populations of *Dendrobates histrionicus* has been discussed as a biological

Table 4.   Physical and Chemical Properties of Histrionicotoxins [248,262, Unless Otherwise Noted]

Histrionicotoxin	
m.p. (HCl) 225–228°	
(Free base) 79–80°	[unpublished result]
Ultraviolet (C_2H_5OH)	λ_{max} 224 nm, ϵ 22,300
Infrared ($HCCl_3$)	2100 cm^{-1} (acetylene)
	1664 cm^{-1} (*cis*-olefin)
Optical rotation (HCl) (c 1.0, C_2H_5OH)	$[\alpha]_D^{25} = -96.3°$
Dihydrohistrionicotoxin	
Ultraviolet (C_2H_5OH)	λ_{max} 226 nm, ϵ 24,700
Infrafed ($HCCl_3$)	2100 cm^{-1} (acetylene)
	1670 cm^{-1} (diene)
Optical rotation (HCl) (c 1.0, C_2H_5OH)	$[\alpha]_D^{25} = -122°$ [222]

Table 4. (*continued*)

Isodihydrohistrionicotoxin

m.p. (HCl) 240–243°

Ultraviolet (C$_2$H$_5$OH) λ_{max} 225 nm, ϵ 8100

                                                     λ_{max} 235 nm, ϵ 7200

Infrared (HCCl$_3$)   2100 cm^{-1} (acetylene)

                                                     1598 cm^{-1} (allene)

Optical rotation (c 0.5, C$_2$H$_5$OH) $[\alpha]_D^{25} = -35.3°$

Neodihydrohistrionicotoxin

m.p. (HCl) 195–200°

Ultraviolet λ_{max} 224 nm, ϵ 17,300

Infrared (HCCl$_3$)   2100 cm^{-1} (acetylene)

                                                     1670 cm^{-1} (diene)

Optical rotation (HCl) (c 1.1, C$_2$H$_5$OH) $[\alpha]_D^{25} = -125.9°$

Allodihydrohistrionicotoxin

m.p. (HCl) 247–250°

Optical rotation (HCl) (c 1.2, C$_2$H$_5$OH) $[\alpha]_D^{25} = -43.4°$

Tetrahydrohistrionicotoxin

Ultraviolet (C$_2$H$_5$OH) λ_{max} 228 nm, ϵ 3900

Infrared (HCCl$_3$)   1670 cm^{-1} (diene)

Isotetrahydrohistrionicotoxin

Ultraviolet (C$_2$H$_5$OH) λ_{max} 228 nm, ϵ 19,200

Infrared (HCCl$_3$)   1950 cm^{-1} (allene)

                                                     1665 cm^{-1} (diene)

Octahydrohistrionicotoxin

Natural

  m.p. (HBr) 180–181°   [247]

  Ultraviolet (C$_2$H$_5$OH)   End absorption

  Synthetic racemate   [263]

  m.p. (HCl) 151–154°

Perhydrohistrionicotoxin

"Natural"

  m.p. (HCl) 184–186°

  Optical rotation (HCl) (c 1.0, C$_2$H$_5$OH, CHCl$_3$) $[\alpha]_D^{25} = -34.6, -36.2°$

Synthetic racemate

  m.p. (HCl) 159–161°   [264]

Synthetic (2**S**)

  Optical rotation (HCl) (c 1.0, C$_2$H$_5$OH, CHCl$_3$) $[\alpha]_D^{25} = -34.5, -36.0°$

Synthetic (2**R**)

  Optical rotation (HCl) (c 1.0, CHCl$_3$) $[\alpha]_D^{25} = +35.8°$

character [23,250; see also Section 11). Histrionicotoxins are fairly widely distributed in frogs of the genus *Dendrobates*, occurring in about half of the 35 species that have been examined. Often they occur at relatively high levels. Recently histrionicotoxin, allodihydrohistrionicotoxin, and isodihydrohistrionicotoxin were identified by gas chromatographic–mass spectral analysis in extracts of a Madagascan mantellid frog, *Mantella madagascariensis* [261].

3.2. Syntheses

The synthesis of histrionicotoxin itself (see Section 3.2.8) has only very recently been completed in Kishi's laboratory at Harvard [265], although a naturally occurring congener, octahydrohistrionicotoxin, was prepared by Kishi's laboratory more than ten years ago [263]. The majority of the synthetic work has been directed toward racemic perhydrohistrionicotoxin 1 (Scheme 14). (±)2-Depentylperhydrohistrionicotoxin 2 has also been synthesized, either as an intermediate to 1 or, more recently, for pharmacological studies [262]. Synthesis of histrionicotoxin analogs has been reviewed [24,38,266].

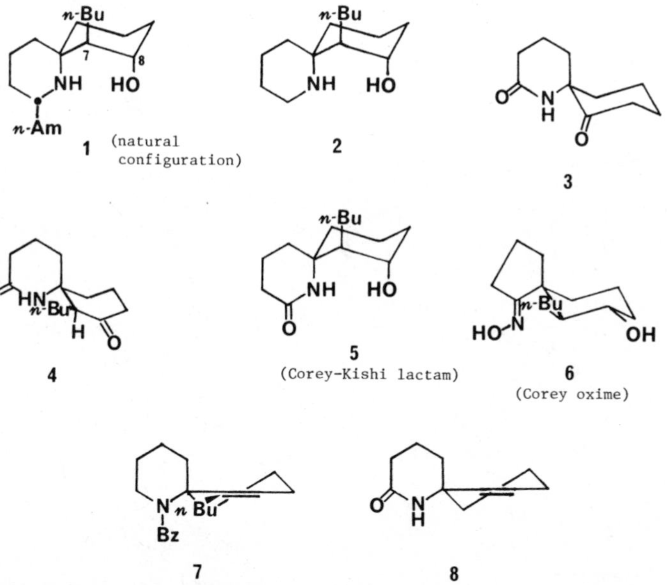

Scheme 14. Histrionicotoxin analogs and azaspirocyclic intermediates. It is assumed that, wherever an intramolecular N···HO or =N···HO bond is possible, the 8-axial configuration is preferred for 8β-hydroxyl groups (cf. **1,2,5**). Otherwise preferred conformations for 1-azaspiro[5.5] undecanes will have the 1-aza substituent in the equatorial (**3,4**) or quasi-equatorial (**7,8**) configuration.

Most early work concentrated on the expeditious construction of the 1-azaspirane system with subsequent elaboration of the two side chains and the 8-hydroxyl group. Consequently, identical intermediates were often approached independently by different groups. These intermediates are still sought as target or relay compounds in developing other routes.

Section 3.2.1 covers synthesis of 1 and 2 via intermediates 3, 6, and 7 (Scheme 14). These intermediates are responsible for three actual syntheses of 1 [264,267] and five preparations of 2 [267,269–273]. Other sections (3.2.3, 3.2.4, 3.2.6) cover four syntheses of 1 via 5 (two of these also via 4) and one synthesis of 2 (Section 3.2.6). Intermediates 2, 4, and 5 are synthetically so proximate to 1 that their preparations are considered equivalent to a synthesis of 1.

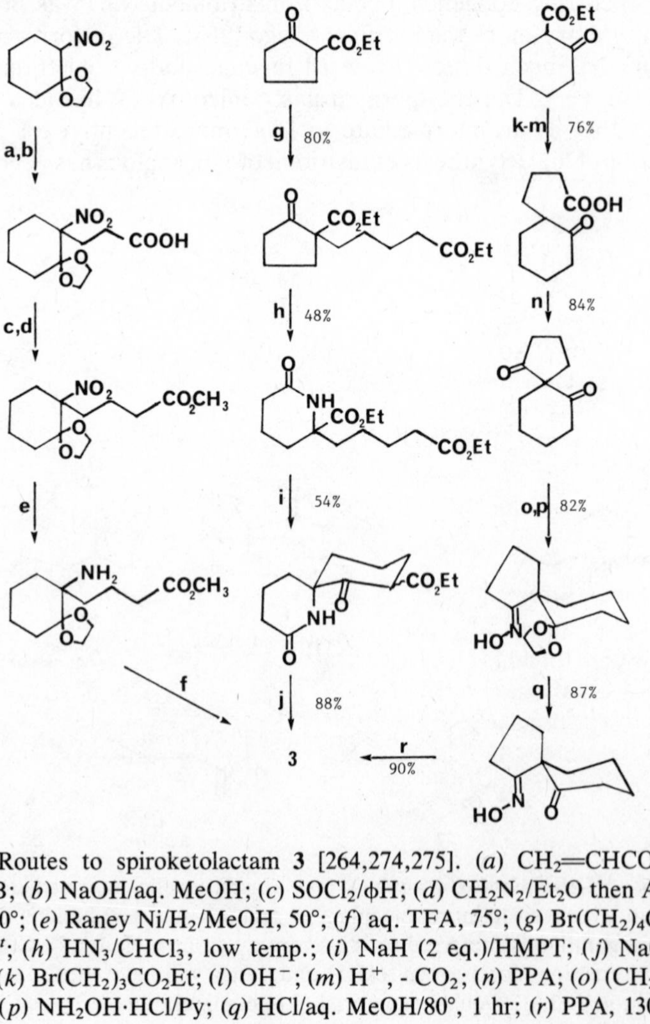

Scheme 15. Routes to spiroketolactam **3** [264,274,275]. (*a*) CH$_2$=CHCO$_2$Me/*t*-BuOH/triton B; (*b*) NaOH/aq. MeOH; (*c*) SOCl$_2$/ϕH; (*d*) CH$_2$N$_2$/Et$_2$O then AgBF$_4$/MeOH-Et$_3$N, 0°; (*e*) Raney Ni/H$_2$/MeOH, 50°; (*f*) aq. TFA, 75°; (*g*) Br(CH$_2$)$_4$CO$_2$Et/DMSO/KOBu^t; (*h*) HN$_3$/CHCl$_3$, low temp.; (*i*) NaH (2 eq.)/HMPT; (*j*) NaOH, Δ then HCl, Δ; (*k*) Br(CH$_2$)$_3$CO$_2$Et; (*l*) OH$^-$; (*m*) H$^+$, - CO$_2$; (*n*) PPA; (*o*) (CH$_2$OH)$_2$, Py·HCl, ϕH; (*p*) NH$_2$OH·HCl/Py; (*q*) HCl/aq. MeOH/80°, 1 hr; (*r*) PPA, 130° then r.t. 18 hr.

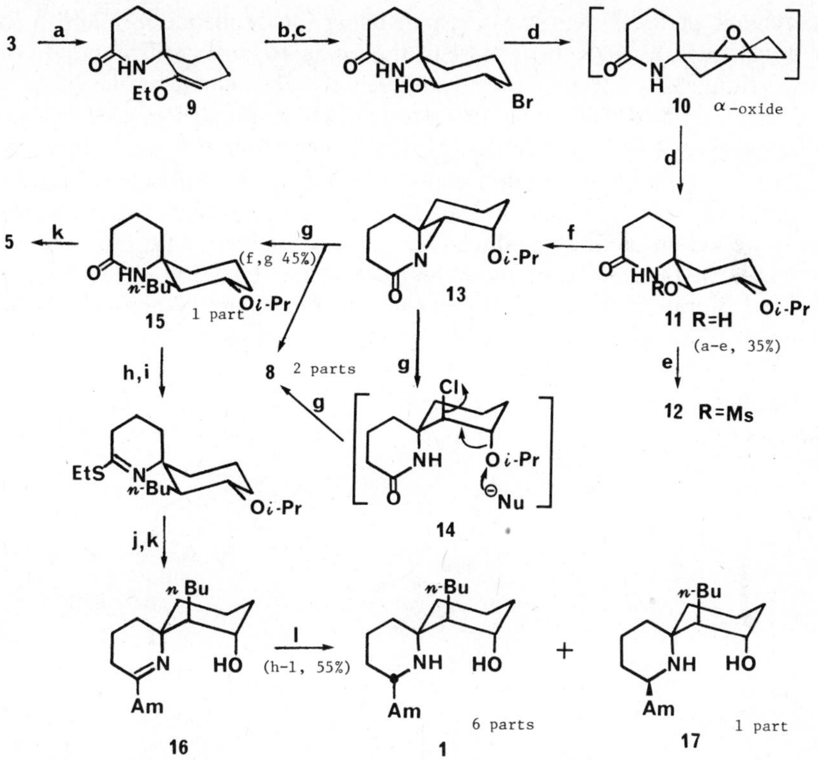

Scheme 16. Synthesis of ($\pm$)-perhydrohistrionicotoxin **1** from **3** via an aziridine intermediate [264]. (*a*) CH(OEt)$_3$, H$^+$ then Δ; (*b*) Br$_2$; (*c*) NaBH$_4$; (*d*) *i*-PrONa/*i*-PrOH; (*e*) MsCl/Py; (*f*) NaH/wet ϕH; (*g*) LiBu$_2$Cu/THF; (*h*) P$_2$S$_5$/ϕH, Δ; (*i*) Et$_3$O$^\oplus$BF$_4^\ominus$; (*j*) *n*-AmLi/C$_6$H$_{14}$-Et$_2$O, Al(*i*-Bu)$_2$H (cat.); (*k*) BBr$_3$/CH$_2$Cl$_2$; (*l*) AlH$_3$/C$_6$H$_{12}$.

3.2.1. Spiro Intermediates. *Synthesis via a Spiroketolactam.*

Three groups have prepared the azaspiroketone **3** as a potential intermediate for histrionicotoxin analogs. Kishi's group developed one such route [264], and successfully converted **3** to **1** (Schemes 16, 17) in two of the earliest syntheses of histrionicotoxin analogs.

Kishi's laboratory [264] prepared **3** in 60% overall yield using the alkylation, Arndt-Eistert homologization, reduction, and cyclization sequence shown in Scheme 15. Kissing and Witkop [274] applied the Schmidt reaction to an α,α-disubstituted cyclopentanone prepared by alkylation, then used the Dieckmann cyclization and decarboxylation to prepare **3**. Bond and co-workers [275] developed a seven-step route to **3** featuring a polyphosphoric acid mediated acylation, selective ketalization, and a Beckmann rearrangement.

Kishi's first approach to **1** (Scheme 16) [264] exploited the stereospecific opening of an acylaziridine **13** with lithium dibutylcuprate (*g*), followed by attachment of the 2-pentyl side chain using the sequence *h–k* and a final stereoselective alane reduction (*l*) of the imine **16** under critical conditions. The aziridine **13** is obtained by intramolecular displacement of mesylate from **12** derived from the ether alcohol product **11**, resulting from the action of sodium isopropoxide upon the α-oxide **10**, generated *in situ*. The intervention of substantial reductive β-elimination (*g*) during the aziridine cleavage is thought to arise via the chloro ether **14** produced from **13** by generated LiCl. The spirolactam olefin **8** produced has proved useless so far (see below) in syntheses of histrionicotoxins. When the alane reduction is conducted in tetrahydrofuran instead of cyclohexane, the epi-2-pentyl isomer **17** predominated. It is the sole product using $NaBH_4$ in methanol.

Kishi's second route (Scheme 17) [264] used a high-yield sequence (*a–f*) to introduce the butyl side chain and shift the ketone group of **3** to C(8). The lactam enol ether **9** after conversion to the α-phenylthioenone (*b*), was

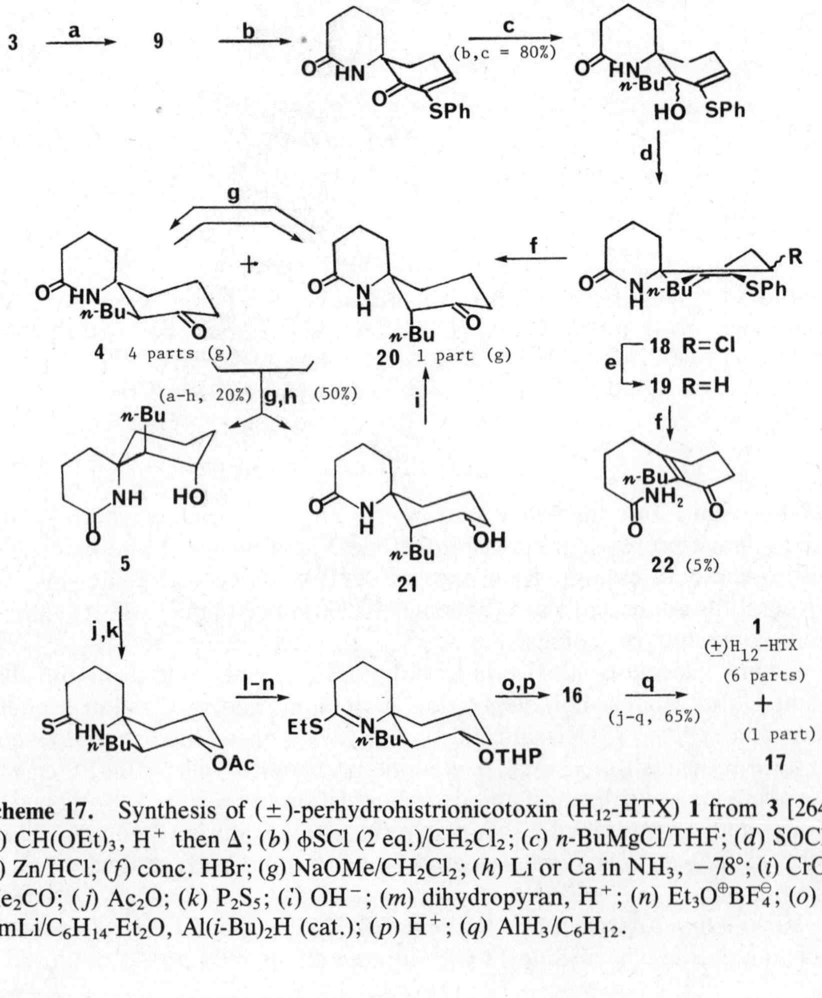

Scheme 17. Synthesis of (±)-perhydrohistrionicotoxin (H_{12}-HTX) **1** from **3** [264]. (*a*) $CH(OEt)_3$, H^+ then Δ; (*b*) φSCl (2 eq.)/CH_2Cl_2; (*c*) *n*-BuMgCl/THF; (*d*) $SOCl_2$; (*e*) Zn/HCl; (*f*) conc. HBr; (*g*) $NaOMe/CH_2Cl_2$; (*h*) Li or Ca in NH_3, −78°; (*i*) CrO_3/ Me_2CO; (*j*) Ac_2O; (*k*) P_2S_5; (*l*) OH^-; (*m*) dihydropyran, H^+; (*n*) $Et_3O^{\oplus}BF_4^{\ominus}$; (*o*) *n*-AmLi/$C_6H_{14}$-$Et_2O$, $Al(i$-$Bu)_2H$ (cat.); (*p*) H^+; (*q*) AlH_3/C_6H_{12}.

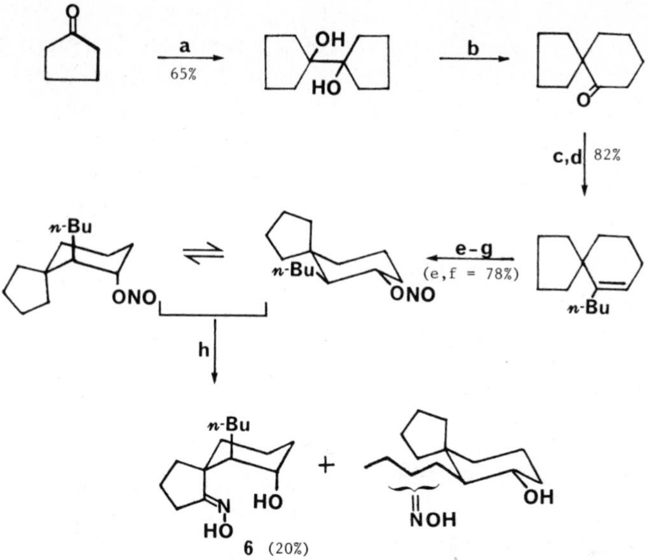

Scheme 18. Synthesis of the Corey oxime **6** by nitrite photolysis [267]. (a) Mg(Hg)-TiCl$_4$/THF, $-10°$; (b) H$^+$; (c) n-BuLi/hexane-Et$_2$O, Δ (3 cycles); (d) SOCl$_2$/Py (2 eq.)/pentane, $-78°$; (e) B$_2$H$_6$; (f) H$_2$O$_2$, OH$^\ominus$; (g) NOCl/Py/CH$_2$Cl$_2$, $<0°$; (h) hv (pyrex)/Py.

treated with butylmagnesium chloride (c) to provide the tertiary carbinol. Thionyl chloride treatment (d) produced the rearranged allylic chloride **18**, which was reductively dechlorinated with zinc (e) to yield **19**. Acid hydrolysis of the phenylthioenol ether group produced a 3:1 mixture of ketolactams **4** and **20** which could be equilibrated with base to increase the proportion of the desired stereoisomer **4**. A lithium or calcium in ammonia reduction provided the important intermediate **5**, also available by boron tribromide cleavage of **15** (Scheme 16). The minor alcohol diastereomers **21**, epimeric at C(7), could be recycled by Jones' oxidation and equilibration. The lactam alcohol **5** could be converted to perhydrohistrionicotoxin in 65% yield by a slight modification (j–q) [263] of Kishi's first route. A 13% overall yield of **1** resulted. A minor ring-cleaved by-product **22** in the HBr hydrolysis (f) of the phenyl thioenol ether **19** proved important in designing a third synthesis of **1** [263]. It was shown *not* to arise from **4** or **20** under the base equilibration conditions and most likely stems directly from **19**. More importantly, it can be recyclized to a mixture of **4** and **20** (see Section 3.2.6; Scheme 43).

Synthesis via a Spirooxime Alcohol. The oxime alcohol **6**, originally prepared by Corey, has been synthesized as a precursor to **1** by four different routes (Schemes 18–21). The first approach of Corey's group (Scheme 18)

[267] employed an eight-step sequence (20% overall yield) exploiting Barton's nitrite ester photolysis procedure. Their second preparation of **6** (Scheme 19) [269] utilized, in key steps, an oxime-assisted stereoselective hypobromous acid addition to the olefin **23** via **24** yielding mainly the *trans*-bromohydrin **25** and a novel stereo- and regiospecific alkylation of a bromooxime **26** via the nitroso olefin **27** formed *in situ*. Axial attack by the acetylide at C(7) from the least sterically hindered face (probable conformation **27**) produces in high yield the desired configuration in the acetylene dioxime **28**. The acetylenic side chain was reduced, the C(8) free oxime hydrolyzed, and the resulting keto oxime ether **29** converted to **6** by O-debenzylation and reduction.

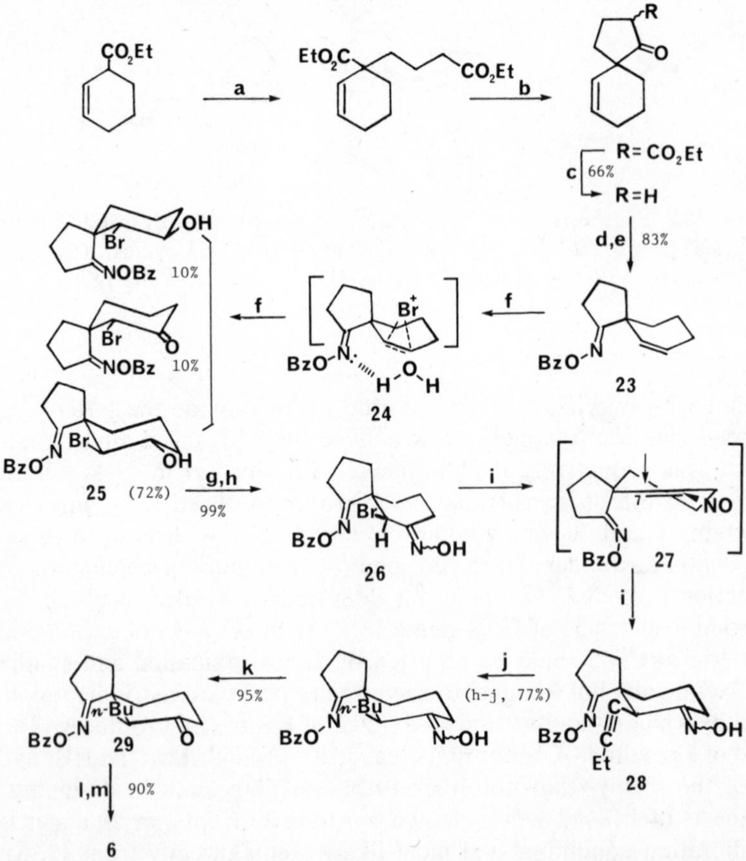

Scheme 19. Synthesis of the Corey oxime **6** by stereoselective hypobromous acid addition to an olefin [269]. (*a*) LiN(*i*-Pr)$_2$/Br(CH$_2$)$_3$CO$_2$Et/HMPT; (*b*) NaH; (*c*) H$^+$/aq. THF; (*d*) NH$_2$OH·HCl/Py/EtOH; (*e*) φCH$_2$Br/DME on K-salt; (*f*) NBS/wet DME, −20°; (*g*) CrO$_3$/Me$_2$CO; (*h*) NH$_2$OH·HCl/NaOAc/HOAc; (*i*) EtC≡CLi/THF, −75° → −10°; (*j*) H$_2$/Pd-C/EtOAc; (*k*) TiCl$_3$/aq. MeOH, pH 6; (*l*) H$_2$/Pd-C/EtOH; (*m*) Na/NH$_3$-THF-*i*-PrOH, −78°.

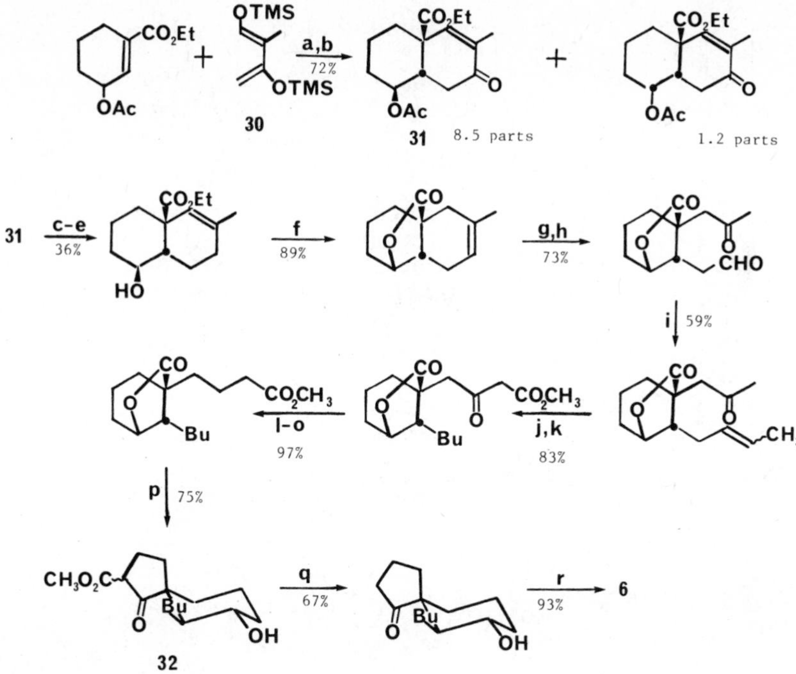

Scheme 20. Synthesis of the Corey oxime **6** by a Diels-Alder, Dieckmann reaction sequence [270,272]. (*a*) Mesitylene, 170°, 48 hr, then 190°, 72 hr; (*b*) 5% HCl; (*c*) (CH₂SH)₂, H⁺; (*d*) Raney Ni/THF; (*e*) KOH/aq. EtOH, 50°; (*f*) TsOH/φCH₃/Δ, 8 hr; (*g*) OsO₄/*N*-methylmorpholine *N*-oxide; (*h*) HIO₄/aq. THF; (*i*) EtP⊕φ₃Br⊖/ DMSO-THF, −50° → −40°; (*j*) H₂/Pt/MeOH; (*k*) methylmagnesium carbonate, DMF, Δ then CH₂N₂/Et₂O; (*l*) NaBH₄/MeOH, −40° → −30°; (*m*) MsCl/φH-Et₃N; (*n*) DBU, Et₃N, 4–7°; (*o*) H₂/Pt/MeOH; (*p*) KH/THF, 0–2°; (*q*) DABCO/xylene, Δ; (*r*) NH₂OH·HCl/Py.

Ibuka, Inubushi, and colleagues [270,272] employed the Diels-Alder reaction of a *bis*-trimethylsilyloxy butadiene derivative **30** to construct the key α,β-unsaturated ketone **31** having the requisite stereochemistry at the three chiral centers (Scheme 20). The keto group was removed and what was the α,β-double bond was then isomerized (*f*) to the thermodynamically more stable position. Subsequent steps of ring-cleavage (*g,h*), double homologization (*i–k*), and Dieckmann cyclization (*p*) provided the keto alcohol derivative **32**, which could be decarbomethoxylated and converted to the Corey oxime, **6**.

The second approach of Ibuka, Inubushi and colleagues [271,272] (Scheme 21) employed a stereospecific aluminium chloride–catalyzed addition of butyl cuprate to the enone **33**, where the bulky γ-substituent forces the reagent to add from the opposite face. Another key step is the intramolecular acyloin condensation (*l–n*) of **35** to the mixture **36** and **37**. A minor

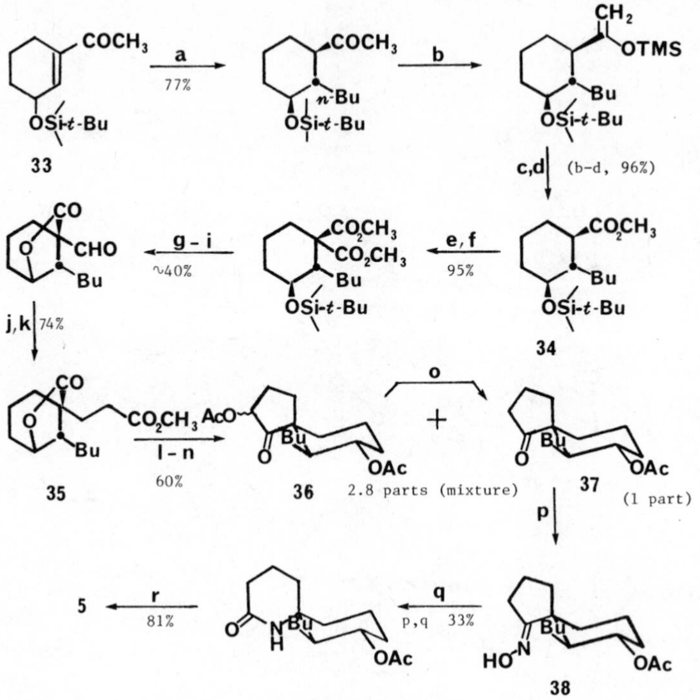

COCH₃   **a**   77%   COCH₃   **b**

OSi-t-Bu   **33**   n-Bu   OSi-t-Bu

CH₂ / OTMS   Bu   OSi-t-Bu   **c,d** (b–d, 96%)

CO / CHO   Bu   **35**   j,k 74%

g – i   ~40%   CO₂CH₃ / CO₂CH₃   Bu   OSi-t-Bu   **34**

e, f   95%   CO₂CH₃   Bu   OSi-t-Bu

CO₂CH₃   **l – n**   60%   **35**

AcO / Bu / OAc   **36** 2.8 parts (mixture)   **o**   +

Bu / OAc   **37** (1 part)   **p**

5   **r**   81%   O=N-Bu / OAc (H)   **q**   p,q 33%   HO-N / Bu / OAc   **38**

Scheme 21. Synthesis of the Corey-Kishi lactam **5** via the Corey oxime acetate **38** [271,272]. (*a*) *n*-BuCu-AlCl₃; (*b*) LiN(*i*-Pr)₂, −70° → −40° then TMS—Cl/Et₃N, −70° → −10°; (*c*) O₃, −70°; (*d*) CH₂N₂; (*e*) LiN(*i*-Pr)₂, CO₂; (*f*) CH₂N₂; (*g*) aq. HCl, 50°; (*h*) Al(*i*-Bu)₂H/φCH₃-hexane, −70°; (*i*) PCC/CH₂Cl₂; (*j*) MeO₂CCH₂P(=O)(OCH₂-)₂/φH-Et₂O, 0°; (*k*) H₂/Pt/MeOH; (*l*) Na/φCH₃/TMS—Cl; (*m*) 5% HCl, 0°; (*n*) Ac₂O/Py/DMAP/CHCl₃; (*o*) Zn/HOAc; (*p*) NH₂OH·HCl/Py; (*q*) TsCl/Py; (*r*) NaOMe/MeOH.

by-product diastereomeric mixture, isomeric with **36** and having the cyclopentyl carbonyl and acetoxyl groups interchanged, was also detected [272]. The major product, the acyloin diacetate diastereomeric mixture **36** can be converted to the keto acetate **37** by zinc/acetic acid reduction (*o*). A final oximation affords the O-acetate **38** of **6** preferred by Ibuka, Inubushi, and colleagues for subsequent conversion (*q,r*) to the Corey-Kishi lactam **5**.

Corey's conversion of **6** to perhydrohistrionicotoxin [267] involves six steps (Scheme 22), commencing with a regiospecific Beckmann rearrangement to **5** followed by reduction of the amide to 2-depentyl-perhydrohistrionicotoxin **2**, then steps of hydroxyl group protection, *N*-bromination, and hydrobromide elimination (*c–e*). *n*-Pentyllithium is added to the resulting silylated imine **39** under nonpolar conditions to give **41** and small amounts of the 2-epimer **40**, addition occurring preferentially from the least hindered face. A final fluoride-ion desilylation (*g*) provided

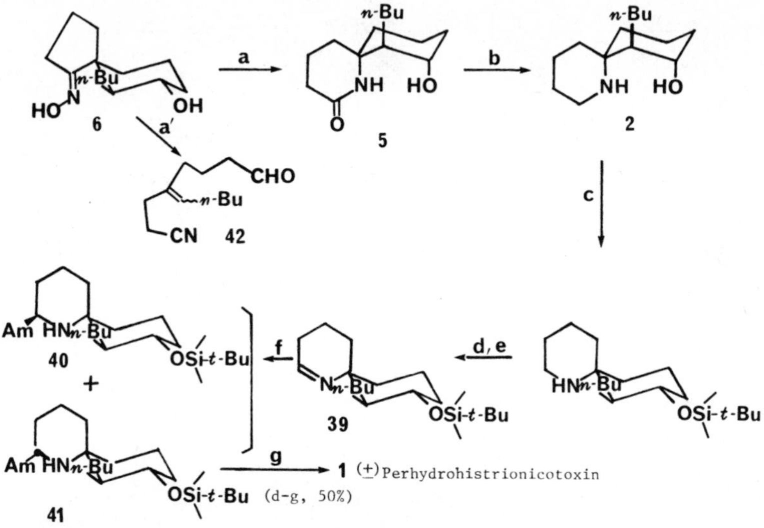

Scheme 22. Conversion of the Corey oxime **6** to ($\pm$)-perhydrohistrionicotoxin [267]. (a) TsCl/Py (2.4 eq.)/ϕH, 12 hr; (b) LiAlH$_4$/THF, Δ, 36 hr; (c) t-BuMe$_2$SiCl/THF, NaH (1.3 eq.), 2 hr; (d) NBS/THF, 0°, 1 hr; (e) KOBu^t/THF, $-40°$, 2 hr; (f) n-AmLi/C$_6$H$_{14}$, 2 hr; (g) n-Bu$_4$N$^\oplus$F$^\ominus$/THF, 25°.

the first synthesis of ($\pm$)-perhydrohistrionicotoxin. If step f is carried out on unsilylated **39**, the abnormal 2-epipentyl isomer predominates. Ibuka and colleagues [272] report that under Corey's Beckmann rearrangement conditions (a), significant amounts ($\sim$50% product) of the second-order Beckmann cleavage product **42** was produced and for that reason they prefer rearranging the O-acetate.

Ibuka, Inubushi and colleagues [271] have prepared the 7,8-epi isomer **47** of the Corey-Kishi lactam **5** by the steps of Scheme 23. Acid hydrolysis of the silyl ether ester **34**, reblocking with the less bulky tetrahydrofuranyl group and allyl bromide alkylation afforded **43**. The olefinic side chain was cleaved to an aldehyde (f), and converted to the α,β-unsaturated ester by the Wadsworth-Emmons reaction (g). Saturation of the side chain (h) and a Dieckmann cyclization (i) provided the spiro keto ester **44**. Decarboethyloxylation (j) and oximation (k) yielded the oxime **45**, epimeric to the Corey oxime **6** at both butyl and hydroxyl groups. Beckmann rearrangement, surprisingly, yielded the rearranged tosylate **46**, which was cleaved to **47** using sodium naphthalide. X-ray analysis indicated **46** as the conformation of the tosylate.

Syntheses via an Azaspiro Olefin. Three routes to the azaspiro olefin have been developed (Schemes 24,25). One route (Scheme 24), developed independently by two groups [273,276], features the palladium catalyzed

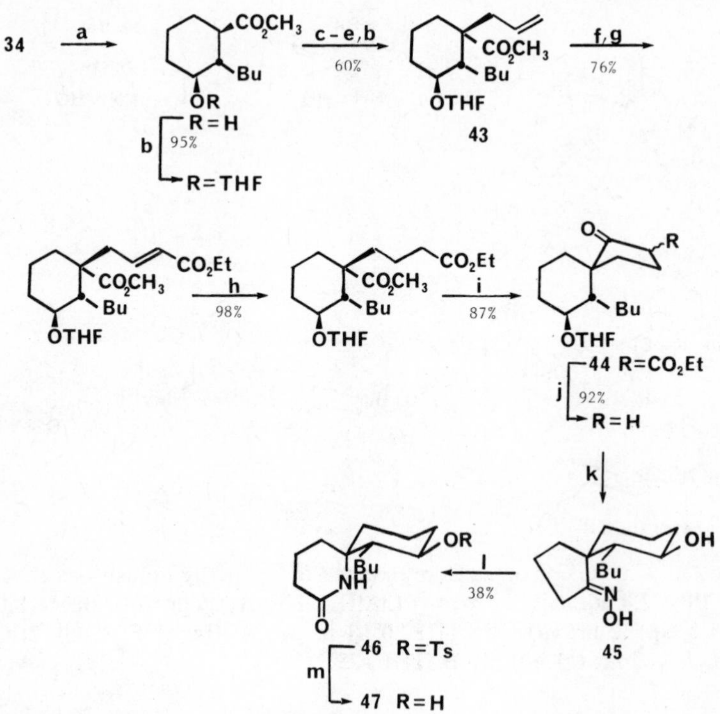

Scheme 23. Synthesis of the 7,8-epi isomer of the Corey-Kishi lactam [271,272]. (*a*) HCl/aq. THF-MeOH; (*b*) 2,3-dihydrofuran/Py·HOTs; (*c*) KN(TMS)$_2$/THF; (*d*) CH$_2$=CHCH$_2$Br; (*e*) 5% HCl; (*f*) OsO$_4$—NaIO$_4$, *N*-methylmorpholine *N*-oxide/aq. THF; (*g*) EtO$_2$CCH=PO(OEt)$_2$; (*h*) H$_2$/Pt; (*i*) KH; (*j*) LiCl-NaHCO$_3$/aq. DMSO, 150°; (*k*) NH$_2$OH·HCl/NaOAc/aq. MeOH, 170°, 36 hr; (*l*) TsCl/Py; (*m*) Na-naphthalene/THF.

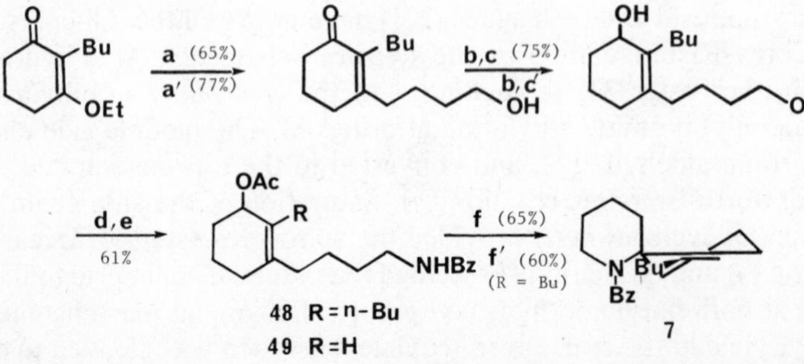

Scheme 24. Synthesis of the azaspirane intermediate **7** by cyclization with palladium acetate [273,276]. (*a*) ClMg(CH$_2$)$_4$OMgCl/THF, −40°; (*a'*) ClMg(CH$_2$)$_4$OTHP/THF, −78° then TsOH·Pyr/EtOH; (*b*) TsCl/Py, −10°; (*c*) Al(*i*-Bu)$_2$H/φCH$_3$, −50° [273]; (*c'*) NaBH$_4$—CeCl$_3$(H$_2$O)$_6$/MeOH [276]; (*d*) Ac$_2$O/Et$_3$O/DMAP/CH$_2$Cl$_2$; (*e*) φCH$_2$NH$_2$/NaI/DMSO; (*f*) Pd(Pφ$_3$)$_4$/Et$_3$N (1 eq.)/CH$_3$CN, 150°, sealed tube; (*f'*) Pd(Pφ$_3$)$_4$/Et$_3$N/CH$_3$CN, Δ, 72 hr.

54

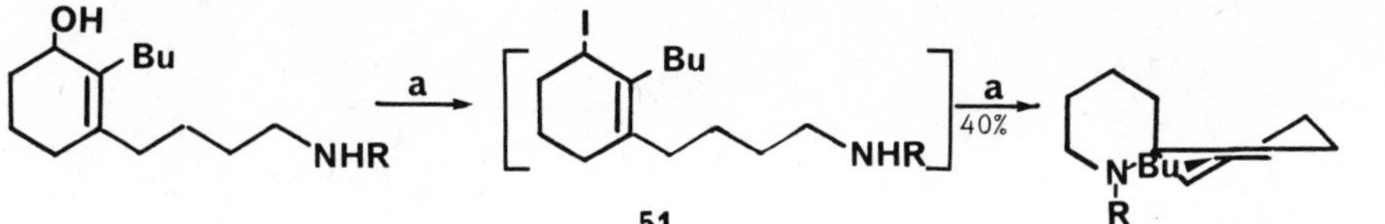

Scheme 25. Synthesis of intermediate **7** by cyclization with trimethylsilyl iodide [273]. (*a*) TMS—I (1.5 eq.)/Et$_3$N (1 eq.)/CH$_3$CN, $-20°$.

cyclization of the allylic acetate **48** to **7** in 60–65% yield. The allylic acetate **48** was prepared in one study by a low-temperature Al(*i*-Bu)$_2$H reduction (*c*) [273], while the other [276] used a modified borohydride reagent (*c'*). In contrast to the rather strenuous cyclization conditions (*f*,*f'*) required to produce **7**, Godleski's group [277] noted earlier the much more facile cyclization of the unsubstitued enone **49**, where quantitative cyclization occurs after 2 hr at 70°C.

A second pathway (Scheme 25) to **7** was developed by Godleski and colleagues [278], who generated the reactive allylic iodide **51** *in situ* by action of trimethylsilyliodide upon the allylic alcohol **50**. A facile S$_N$2′ displacement yields **7**. The analogs **50a** and **50b**, prepared by Al(*i*-Bu)$_2$H reduction of the appropriate enones, were also cyclized to **7a** and **7b**.

Another route to **7** (Scheme 26) was developed by Pearson and Ham [279], who discovered a high-yield entry into the azaspirocyclic system by treatment of the iron-carbonyl complex **52** with benzylamine. Unfortunately, butyl cuprate addition to the derived enone **53** proceeded preferentially from the β-face to give predominantly the 7-epibutyl isomer **54**, necessitating a lengthy sequence (*d–h*) to remove the abnormal chirality at C(7).

Holmes and colleagues [280] recently reported a synthesis of **7** using an ozonolysis-condensation sequence on intermediate **56** prepared conveniently by nitrosylchloride-addition to the octalin **55** (Scheme 27). The resulting azaspiroketoolefin **57** is converted in two steps to the aminoketone **58**, to which is added butyl lithium. The unfortunate tendency of this ketone to enolize necessitated the repetition of the addition step (*i*) to obtain even modest yields. The corresponding *N*-benzyl or *N*-benzyloxycarbonyl (Cbz) derivatives gave solely enolization or reduction even with organometallic reagents that normally minimize these reactions. Furthermore, the subsequent dehydration reaction unfortunately gave chiefly the exo olefin, evidently favored by relief of A$_{1,2}$-strain. The final benzylation proceeded in modest yield to a 1:4 mixture of **7** and its exo isomer **59**.

Duhamel and Kotera [268] had prepared earlier the *N*-methyl analog of **58** by the facile ring-contraction of a tetrahydroazepine ester with bromine and base (Scheme 28). The resulting spiroaldehyde ester is converted to the

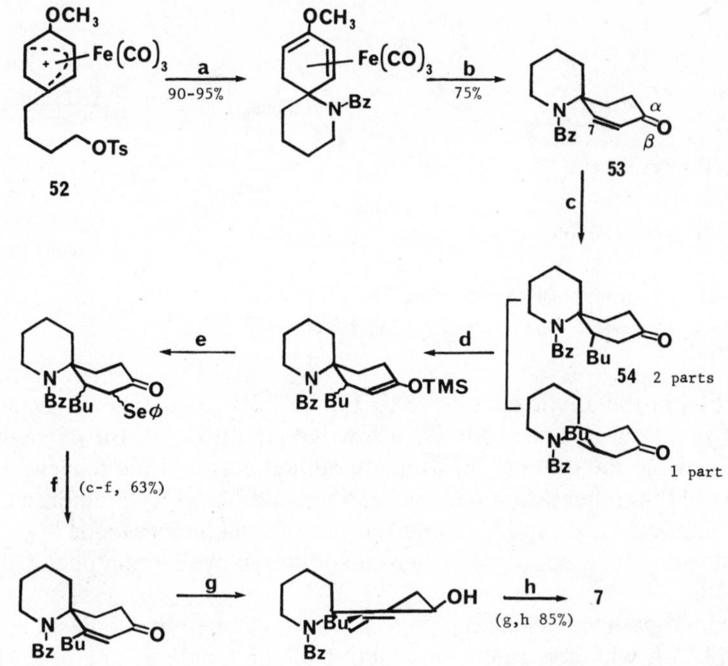

Scheme 26. Synthesis of intermediate **7** by cyclization of an iron-carbonyl complex [279]. (a) $\phi CH_2NH_2/CH_3CN$; (b) $Me_3NO/AcNMe_2$ then aq. H_2SO_4/THF; (c) Li-Bu_2Cu/THF; (d) TMS—$Cl/Et_3N/THF$, $-20°$ (trapping enolate from step c); (e) $\phi SeCl/THF$; (f) H_2O_2/aq. HOAc; (g) $NaBH_4/MeOH$; (h) $LiAlH_4-AlCl_3/Et_2O$, $0°$, 1 hr.

N-methylazaspiroketone by the Wittig reaction (c), hydrogenation (d), Dieckmann cyclization (e), and a chloride-ion mediated decarbomethoxylation.

Intermediate **7** has been converted to 2-depentylperhydrohistrionicotoxin (**2**) by strenuous oxidation of the organoborane adduct [273,279] with hydrogen peroxide (Scheme 29). With elevated temperatures and lengthy reaction periods, modest (35,40%) yields of diastereomeric alcohols **60** and **61** (ca. 2:1) could be obtained. Godleski and colleagues [273], after separating the alcohols, debenzylated the major diastereomer to **2** in 85% yield, whereas Pearson and Ham [279] used the three-step sequence (c–e) to obtain **2**. Godleski's group improved the yield of **2** by Swern oxidation (f) of the *N*-benzyl alcohol mixture **60** and **61** to ketones (**62,63**), which on treatment with base (g) gave a 13:1 equilibrium mixture in favor of the desired stereoisomer **62**, a much more favorable ratio than that (4:1) arising between **4** and **20** (Scheme 17). Lithium in ammonia effected both reduction of the ketone and debenzylation and provided **2** in 29% yield from the alcohol mixture.

Carrothers and Cumming [276] improved the borane oxidation procedure

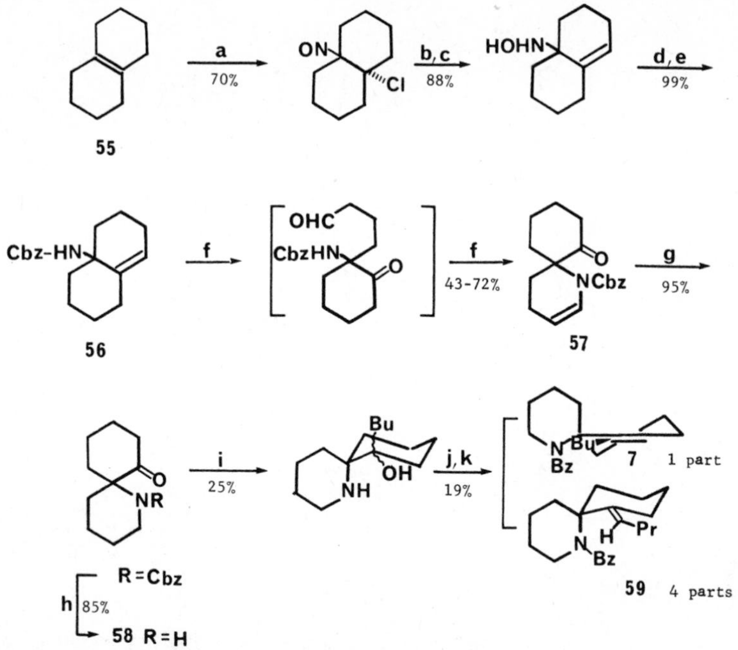

Scheme 27. Synthesis of intermediate **7** by an ozonolysis-condensation route [280]. (a) i-AmONO/HCl; (b) H$_2$/Pt; (c) NaOMe/MeOH; (d) Al(Hg)/aq. THF; (e) CbzCl; (f) O$_3$/CH$_2$Cl$_2$, −78° then Me$_2$S; (g) H$_2$/Pt/EtOH; (h) TMS—I/CH$_3$CN, 16 hr; (i) n-BuLi then MeOH (6 cycles); (j) KHSO$_4$, 170°; (k) φCH$_2$Br, KI, (i-Pr)$_2$NEt/CH$_3$CN.

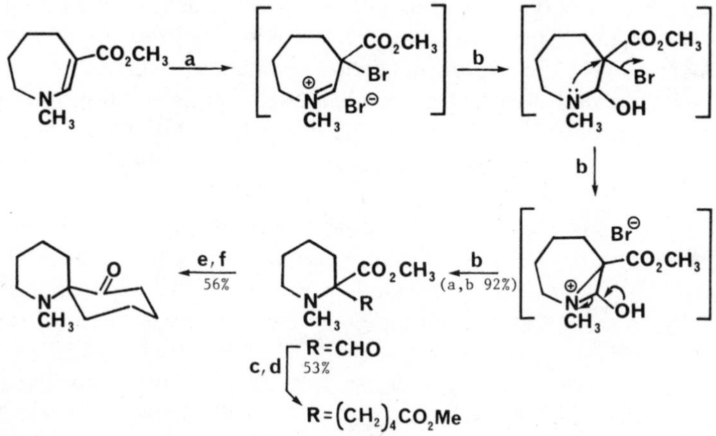

Scheme 28. The synthesis of an azaspiroketone by ring contraction of a tetrahydroazepine [268]. (a) Br$_2$/Et$_2$O, −70°; (b) aq. Et$_3$N; (c) φ$_3$P=CHCH=CHCO$_2$Me/φCH$_3$, Δ; (d) H$_2$/Pt/MeOH; (e) KH/dioxane, Δ; (f) KCl/DMSO, 120°.

57

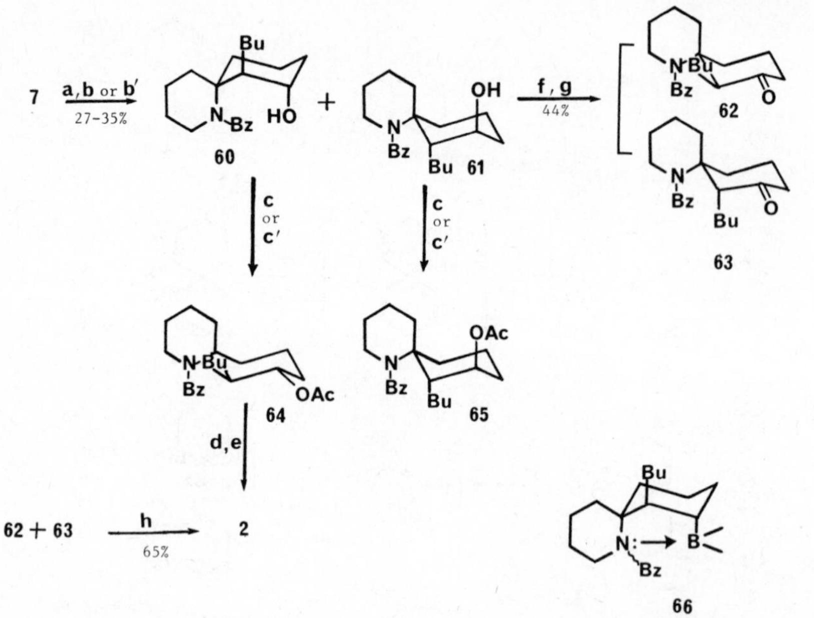

Scheme 29. Conversion of intermediate **7** to (±)-depentylperhydrohistrionicotoxin (**2**) [273,276,279]. (a) B$_2$H$_6$/THF [276,279] or BH$_3$·Me$_2$S/THF, 18 hr [273]; (b) NaOH, H$_2$O$_2$/THF, 50°, 5–6 hr [279] or NaOH, H$_2$O$_2$/DME, 85°, 10 hr [273]; (b') Me$_3$NO/diglyme, Δ, 18 hr; (c) AcCl/Py [279]; (c') Ac$_2$O/DMAP/CH$_2$Cl$_2$ [276]; (d) NaOH/MeOH; (e) H$_2$/Pd-C/EtOH; (f) DMSO-(COCl)$_2$/CH$_2$Cl$_2$, −78°; (g) NaOMe/CH$_2$Cl$_2$; (h) Li/MeOH (2 eq.)/NH$_3$, −78°.

by employing trimethylamine oxide in boiling diglyme (b') to prepare the diastereomeric alcohols **60** and **61** in a 9:1 ratio. These could be separated as their O-acetates. The major product **64** was hydrolyzed and debenzylated to **2**. Interestingly, the proton magnetic resonance spectrum of the major acetate **64** indicated clearly the *trans*-diequatorial orientation of the butyl and acetoxyl groups [276,279], while the minor O-acetate **65** showed a *trans*-diaxial orientation [279]. The latter conformation may also be that preferred by **47**.

All three groups ascribe the difficulty in borane oxidation to the stabilization of the borane adduct by intramolecular amine complexation. Pearson and Ham [279] have even presented proton magnetic resonance spectral evidence for a pair of stable borane complexes **66**, epimeric at nitrogen.

A Spirolactam-Olefin Intermediate.   Another intermediate that initially appeared to hold promise in approaches to perhydrohistrionicotoxin was the spirolactam olefin **8**. It was prepared by three different groups [274,275,281] and isolated as a reaction by-product by another [264] (Scheme 16). So far **8** has not proved to be convertible to any useful intermediate, although the

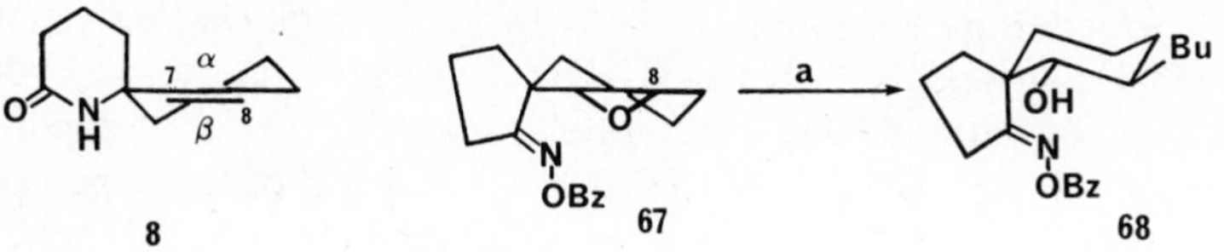

Scheme 30. An unsuccessful approach to perhydrohistrionicotoxin [270]. (*a*) *n*-Bu(*n*-PrCH=CH₂)CuLi/Et₂O-HMPT, −45°.

closely related hydroxamic acid derivative (**102**) has (Scheme 35). The β-oxide, arising as the major product on *m*-chloroperbenzoic acid oxidation of **8**, as well as the α-oxide (**10**) (Scheme 16) were opened with nucleophiles, such as butyl cuprate or sodium isopropoxide, only at the less-sterically congested 8-position (cf., **11**, Scheme 16) [264]. A related observation has been reported [269] in the reaction of the oxime epoxide **67** with a mixed cuprate reagent (*n*-Bu(C₃H₇CH=CH₂)CuLi) where **68** was the only product obtained (Scheme 30).

The Bamford-Stevens reaction of the keto lactam **3** has been used to prepare **8** in good yield [274,275]. Overman [281], in an unusual approach to the azaspirocyclic system, used a [3,3]-sigmatropic rearrangement of the hydroxamate ester **69** to the amido olefin **70** (Scheme 31).

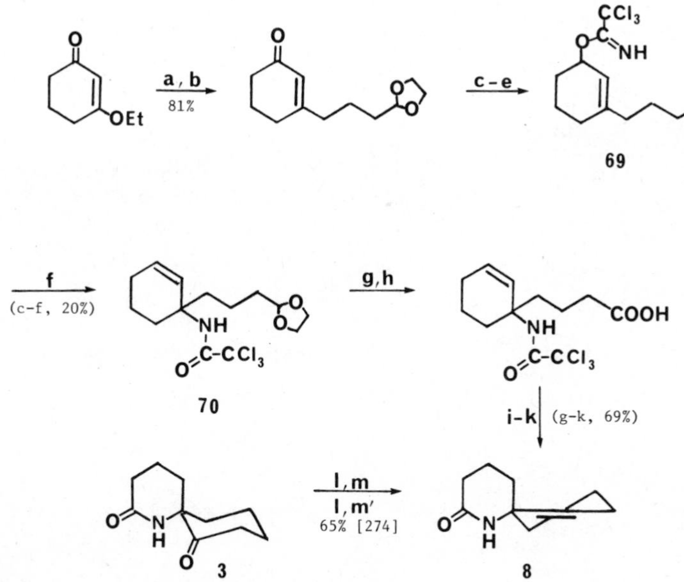

Scheme 31. Synthesis of lactam olefin **8** [274,275,281]. (*a*) ClMg(CH₂)₃CH(OCH₂)₂; (*b*) pH 1.5; (*c*) LiAlH₄/THF, −78°; (*d*) NaH/Et₂O; (*e*) Cl₃CCN; (*f*) hexane, 69°, 5 d; (*g*) (CO₂H)₂/aq. Me₂CO; (*h*) Ag₂O; (*i*) NaOH, 25°, 2 hr; (*j*) H₂SO₄/MeOH; (*k*) OH⁻; (*l*) TsNHNH₂; (*m*) MeLi [275]; (*m'*) BuLi [274].

3.2.2. Cycloaddition Reactions *[3 + 2] Thermal Azide-Olefin Cyclization.* Corey and Balanson [282] proposed a rather direct route to perhydrohistrionicotoxin by intramolecular 1,3-dipolar addition of a side chain azido group to the enone moiety of **71** (Scheme 32). Instead of the expected triazoline (**73**) they obtained, after 6 hr of reflux in xylene, a 2:1 mixture of ketoaziridines **75** and **76**. The triazoline, however, was detectable as an intermediate. Regrettably, the unnatural 2-pentyl epimer **75** predominated, as shown by X-ray analysis of the amino alcohol **78** (2,7-epiperhydrohistrionicotoxin), which results after aziridine ring cleavage (*b*), followed by reduction of the ketone (*c*). The minor ketoaziridine diastereomer **76** on similar treatment gave some 15% of perhydrohistrionicotoxin (**1**), but the epi-7-butyl isomer **79** was the major product, probably, as in the case of **78**, a consequence of preferential protonation on the α-face (via **81**) of the enolate arising from the aziridine cleavage. Attempts to cleave aziridine **75** under acidic conditions (zinc/acetic acid) gave instead the amino enone **21** (Scheme 17) by retro-Michael reaction of **80**.

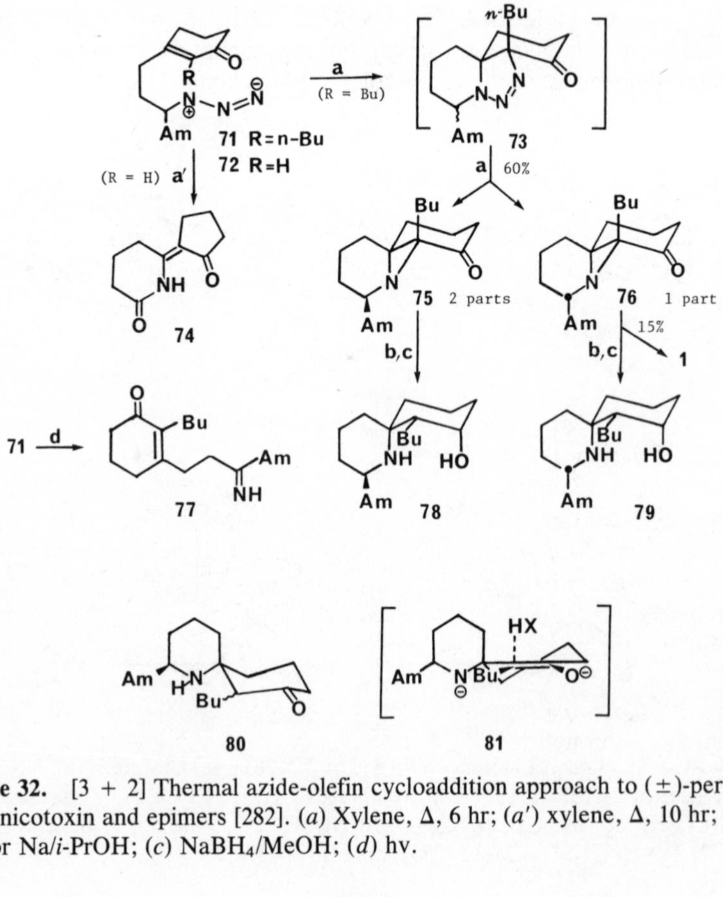

Scheme 32. [3 + 2] Thermal azide-olefin cycloaddition approach to (±)-perhydrohistrionicotoxin and epimers [282]. (*a*) Xylene, Δ, 6 hr; (*a'*) xylene, Δ, 10 hr; (*b*) Li/NH₃ or Na/*i*-PrOH; (*c*) NaBH₄/MeOH; (*d*) hv.

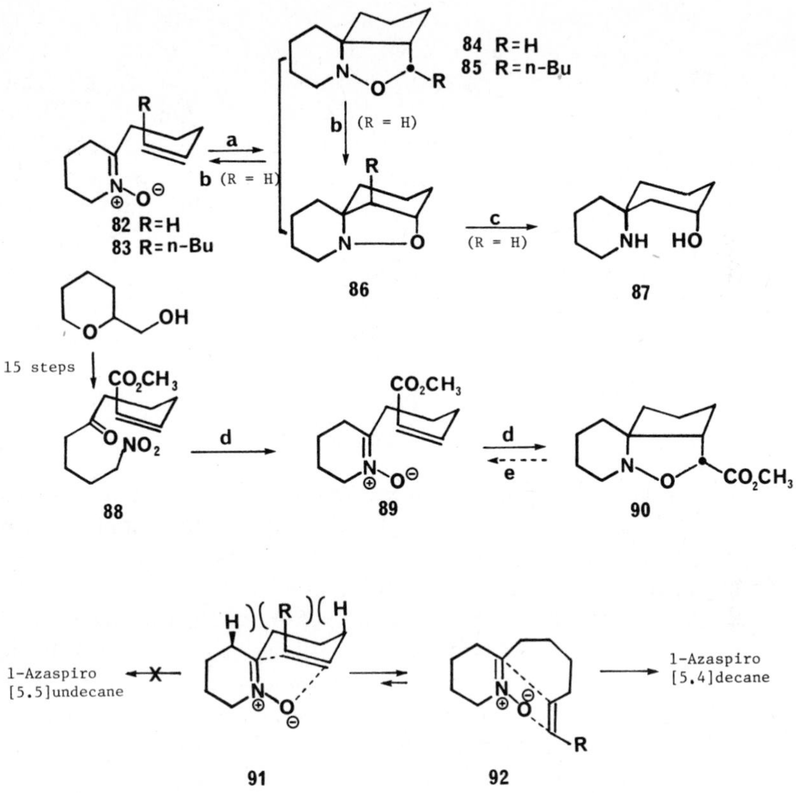

Scheme 33. [3 + 2] Thermal nitrone-olefin cycloaddition approach to histrionico-toxin analogs [283,284]. (*a*) r.t.; (*b*) ϕCH$_3$, 195°, 2 d, sealed tube; (*c*) H$_2$/Raney Ni/EtOH; (*d*) Zn/NH$_4$Cl; (*e*) xylene, Δ, 30 hr.

The azide **71** was also photolyzed, generating the reactive nitrene, with the hope that it would form the aziridine directly, possibly with a different stereochemical outcome. The imine **77**, resulting by hydrogen transfer, was, however, the sole product [282].

When the azido enone **72**, lacking the α-butyl substituent, was heated in xylene, no desbutylaziridines analogous to **75** and **76** were produced; instead the sole product in high yield was the ring-contracted vinylogous amide **74**, presumably arising by a diradical rearrangement [282].

[3 + 2] Thermal Nitrone-Olefin Cyclization. Another 1,3-dipolar cy-cloaddition approach to the azaspirane system of histrionicotoxins was de-veloped by Gössinger, Imhof, and Wehrli [283], who prepared the nitrone olefin **82** in three steps from commercial *N*-hydroxypiperidine (Scheme 33). The nitrone olefin **82** cyclized spontaneously to a 5:1 mixture of isoxazo-lidine cycloadducts **84** and **86**. By thermally reversing the cycloaddition,

Gössinger and colleagues were able to isomerize this mixture to the more stable spiro [5.5]undecane product **86**. Hydrogenolysis of the N—O bond produced **87**, the azaspirane parent of histrionicotoxin. When the disubstituted olefin **83** was subjected to the same procedure, the spiro[5.4]decane **85** was the only isolable product. All attempts at thermal isomerization failed.

Tufariello and Trybulski [284] had attempted an analogous nitrone olefin cycloaddition, hoping that the conjugating carbonyl function in **89** might favor the desired regiospecificity. The nitrone **89**, formed *in situ* by the hydroxylamine-ketone condensation after partial reduction of the nitro group (*d*) of intermediate **88**, gave spontaneously a single oxazolidine, which was shown to be the undesired spiro[5.4]decane product **90**. All thermal isomerization attempts proved fruitless. Both groups [283,284] have invoked unfavorable steric interactions to explain the destabilization of the transition state leading to the spiroundecane system (**91** vs. **92**).

[2 + 2] Photocycloaddition.   The intramolecular addition of a side chain acetylenic linkage to an enone under irradiation has been applied by Koft and Smith [285] in an ingenious but, as yet, impractical approach to perhydrohistrionicotoxin (Scheme 34). Irradiation of the acetylenic enone **93** led to the slow formation of the tricyclic cyclobutene ketone **94** along with some polymerization. The Wilkinson decarbonylation (*h*) of the reduced and ozonolyzed product **96** unfortunately failed, necessitating the abandonment of this rather direct route to the Ibuka-Inubushi compound **37** (see Scheme 21) and the use of a longer nonstereospecific route to its precursor **99**. The photochemical adduct **94** was oxidized to the diketoacid **97**, which on decarboxylation afforded a 10:1 mixture of diketones (**98,99**), with the abnormal 7-butyl isomer (**98**) predominating. Acid equilibration (*k*) reduces this ratio to a more useful 2.3:1. It is interesting to note that here, in contrast to ketolactam (Scheme 17) or ketoamine (Scheme 29) equilibrations, the unnatural 7-epibutyl isomer predominates. The minor isomer (**99**) was identical with the product resulting on saponification and oxidation (*l, m*) of the Ibuka-Inubushi compound **37**. No attempts to reduce **99** were reported.

3.2.3.   An Intramolecular Ene Reaction.   Keck and Yates [286] have generated the reactive acylnitroso compound **101** by a retro Diels-Alder reaction of **100** in refluxing toluene (Scheme 35). The acylnitroso compound **101** underwent an intramolecular ene reaction to afford the cyclic hydroxamic acid **102** in quantitative yield. Bromination with *N*-bromosuccinimide produced the bromooxaziridine **104** in high yield acquiring the requisite hydroxyl group configuration at C(8) by neighboring group participation (cf. **103**). The intermediate **104** proved labile toward ionic reactions and underwent facile reductive elimination and reformation of **102**. But a one-electron radical reaction with allyl tributyltin was devised [287], which gave the allyl substituted derivative **105** as a single diastereomer in 88% yield. Unfortunately, the product possessed the 7-epi configuration, necessitating the reoxidation

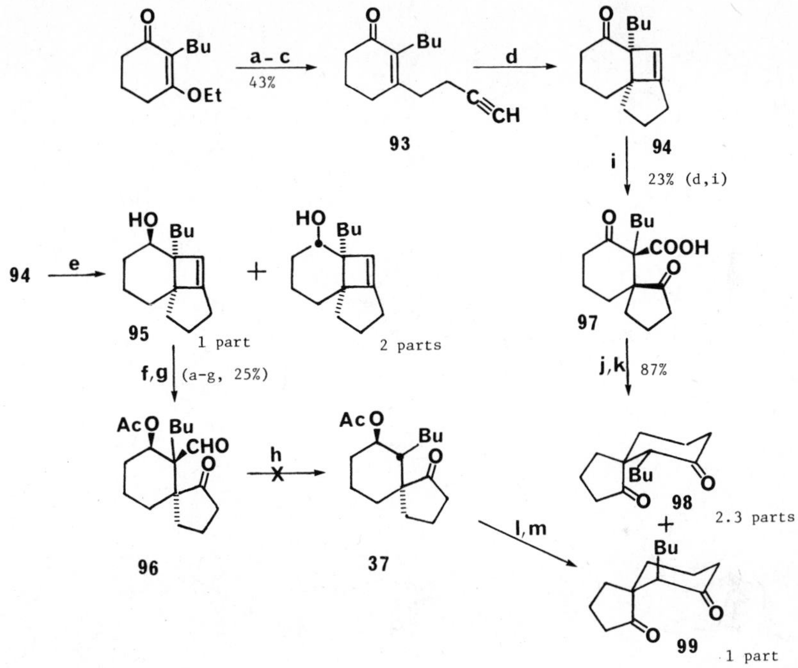

Scheme 34. [2 + 2] Photocycloaddition approach to the Ibuka-Inubushi compound (**37**) [285]. (*a*) TMSC≡C(CH$_2$)$_3$MgI/THF; (*b*) 10% HCl; (*c*) *n*-Bu$_4$N$^⊕$F$^⊖$/THF; (*d*) hv (U-glass), NaOAc/MeOH, 48–72 hr; (*e*) NaBH$_4$/MeOH; (*f*) O$_3$/φ$_3$P; (*g*) Ac$_2$O/Py/DMAP/CH$_2$Cl$_2$; (*h*) Wilkinson's catalyst, φCN, 190°, 4 hr; (*i*) RuO$_2$/NaIO$_4$; (*j*) φCH$_3$, Δ; (*k*) aq. HCl/THF, 48 hr; (*l*) K$_2$CO$_3$/MeOH; (*m*) PCC, NaOAc/CH$_2$Cl$_2$.

of the ensuing **106** and an equilibration step (*i*) after the side chain was converted (*d–g*) to the butyl group. Reduction of the resultant major ketolactam (**4**) produced the Corey-Kishi lactam **5**.

3.2.4. Acyliminium Ion Cyclization

The groups of Speckamp [288–290] and Evans [291,292] independently prepared the formate of the Corey-Kishi lactam **110** (Scheme 36) by application of acyliminium ion-induced azaspirocyclization methodology, developed earlier in Speckamp's laboratory. The desired 6,6-cyclization and proper configurations at both C(7) and C(8) result when the acylimininium ion **109** generated *in situ* in anhydrous formic acid interacts with the (E)-non-4-enyl side chain. Although reaction details and intermediates differed somewhat, the basic approach of the two groups is indicated in Scheme 36 and affords one of the most direct and convenient routes to perhydrohistrionicotoxin at present. Evans and colleagues [291,292] showed that the reaction (*a'*) of glutarimide magnesium iodide (**107**; R = MgI) with (E)-non-4-enyl magnesium chloride **108** (X = Cl) gave a mixture of cyclized carbinolamide **114** and the ketoamide **113** (66%), which

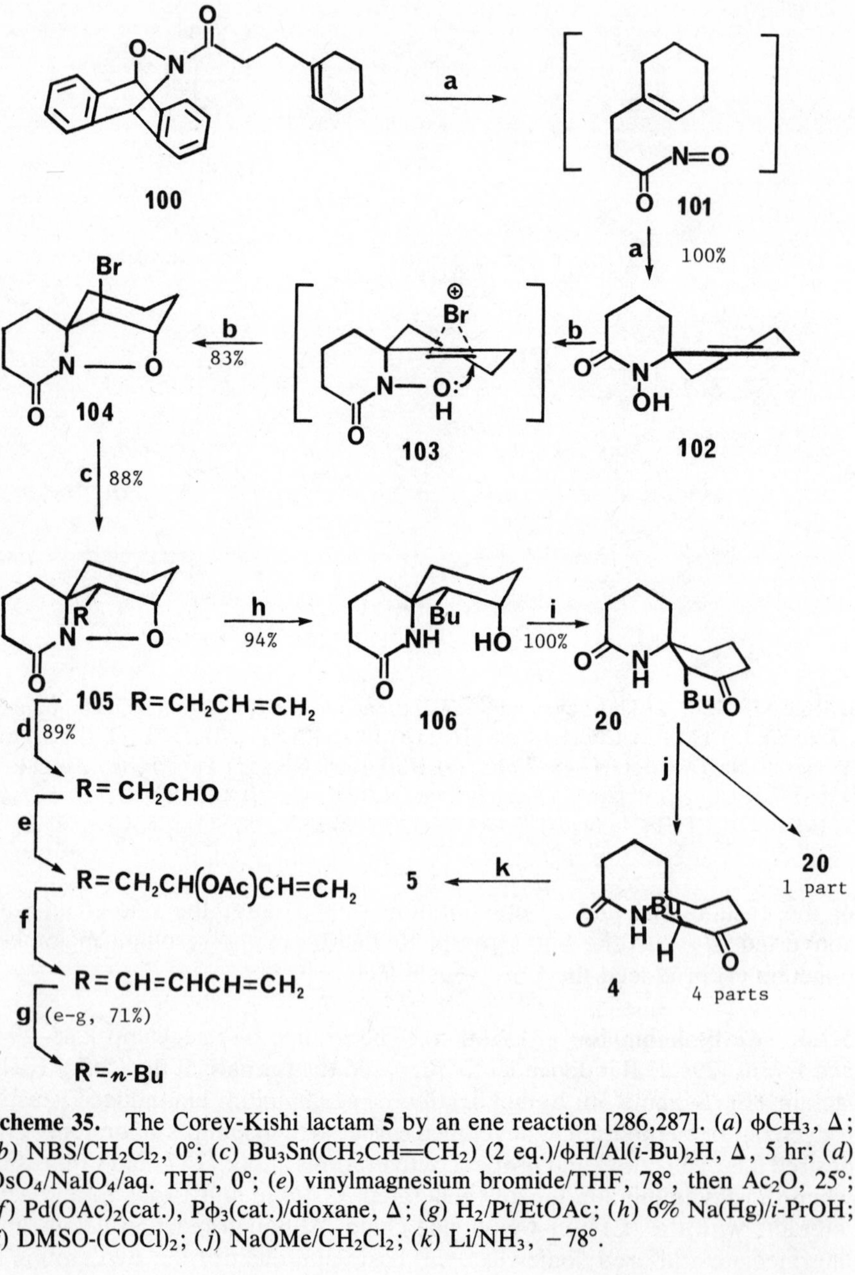

Scheme 35. The Corey-Kishi lactam **5** by an ene reaction [286,287]. (*a*) ϕCH_3, Δ; (*b*) NBS/CH_2Cl_2, 0°; (*c*) $Bu_3Sn(CH_2CH{=}CH_2)$ (2 eq.)/ϕH/Al(i-Bu)$_2$H, Δ, 5 hr; (*d*) OsO_4/NaIO$_4$/aq. THF, 0°; (*e*) vinylmagnesium bromide/THF, 78°, then Ac$_2$O, 25°; (*f*) Pd(OAc)$_2$(cat.), Pϕ_3(cat.)/dioxane, Δ; (*g*) H_2/Pt/EtOAc; (*h*) 6% Na(Hg)/i-PrOH; (*i*) DMSO-(COCl)$_2$; (*j*) NaOMe/CH_2Cl_2; (*k*) Li/NH$_3$, $-78°$.

64

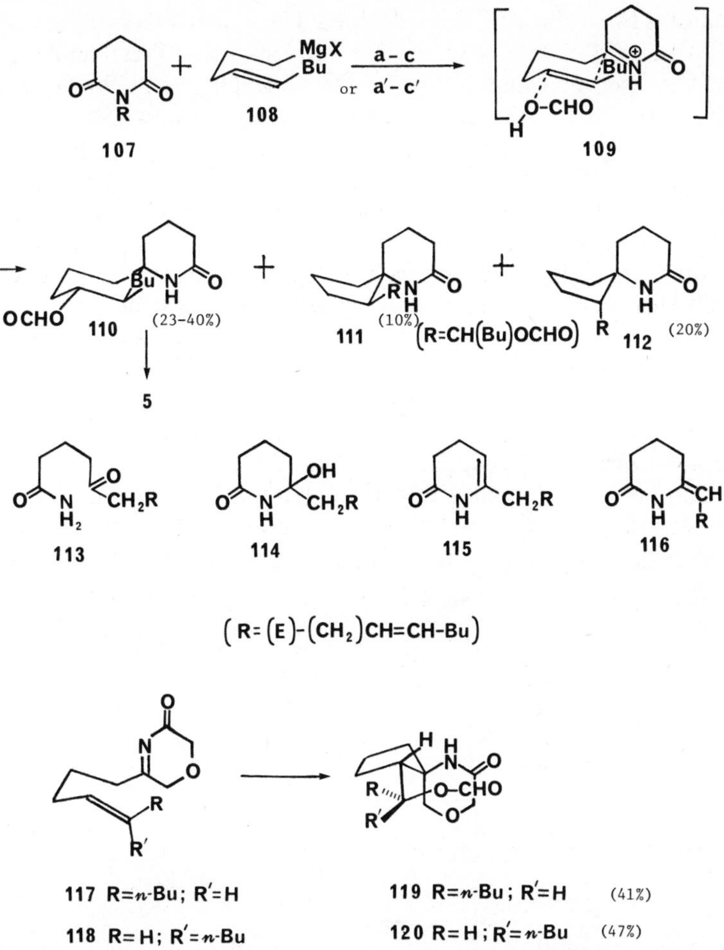

Scheme 36. The acyliminium ion pathway to the Corey-Kishi lactam 5 [288–292]. Conditions of Speckamp [288,290] (107 R = H; 108 X = Br). (a) THF, 30–35°, 20 hr, 2.2–5.9 eq. 108; (b) Evaporate; (c) HCO₂H, 44°, 8–14 d, 23–30% yield. Conditions of Evans [291,292] (107 R = MgI; 108 X = Cl). (a′) Et₂O, 0.5 hr, Δ, 1.2 eq. 108; (b′) DMF: φCH₃ (50:1), TsOH, Δ, 48 hr; (c′) HCO₂H, 25°, 32 hr, 26–40% yield.

could be cyclized under acidic conditions (b′ or TFA) to an 8–9:1 mixture of enamides 115 and 116. Carbinolamide 114 alone resulted when the Grignard reaction was conducted in methylene chloride. In ether, Shoemaker and Speckamp [290] observed rapid ring-opening to 113. Formolysis of the enamides 115 and 116 (c′) gave 110 in 40% yield along with two diastereomers of the spiro[5.4]decane system 111, 112 (30%), and 10% dimerized enamides, as well as recovered starting material (10%) [291,292]. Interestingly, Evans and colleagues [291,292] showed it to be unnecessary to convert the carbinolamide **114** to the enamides **115** and **116**, since **110** could be obtained directly from **114** in 33% overall yield on formolysis. Presumably the same intermediate pertains under the conditions employed in Speckamp's procedure. Only traces (0.5%) of [5.4]spiranes were reported [290] under Speckamp's conditions (*a–c'*). But surprisingly, [5.4]spiranes were indicated as the *sole* cyclized products on formolysis of the (E)- and (Z)-oxoanalogs **117** and **118** [290]. These diastereospecific cyclizations yielded **119** and **120**, respectively (Scheme 36).

The differences between the yields of the 5,4-spirane by-product in the two laboratories has not yet been explained, nor has the reason why the 5,4-spirane **119** (evidently via **117a**) (Scheme 37) is the sole product in formolysis of the 4-oxo analog (**117**) of **109** (Scheme 36). Structure (**120**) was determined by X-ray analysis [290].

The high degree of stereoselectivity in the acyliminium cyclizations leading to the 1-azaspiro[5.5] undecane system may be a result of a severe steric interaction between the butyl side chain and the 5-methylene group of **121**, thereby destabilizing that transition state relative to the one (**109**) leading to **110** [290,292] (Scheme 37). The formation of two spirodecanes **111** and **112**

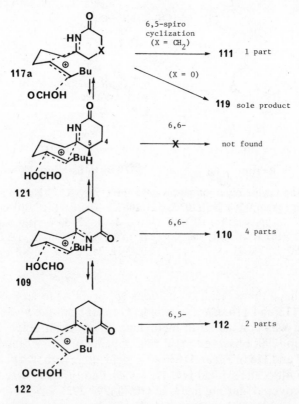

Scheme 37. Acyliminium ion cyclization intermediates.

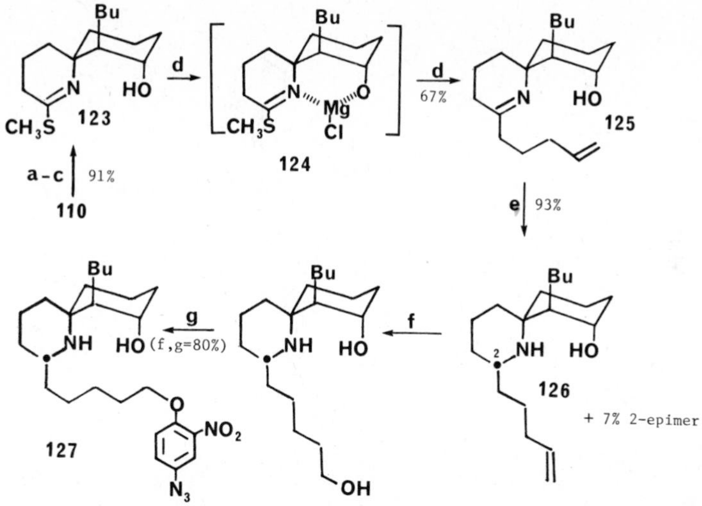

Scheme 38. Synthesis of a perhydrohistrionicotoxin analog (**127**) [292] suitable for photoaffinity labeling studies of biological targets. (*a*) P$_2$S$_5$/φH, Δ; (*b*) NaOH/aq. MeOH; (*c*) MeI/CH$_2$Cl$_2$, 18 hr; (*d*) MgCl$_2$/CH$_2$Cl$_2$ then CH$_2$=CH(CH$_2$)$_3$MgCl; (*e*) AlH$_3$/φCH$_3$, −70°; (*f*) BH$_3$·Me$_2$S/CH$_2$Cl$_2$ then NaOH/H$_2$O$_2$; (*g*) 4-fluoro-3-nitro-phenylazide, KOBu^t/THF.

[291,292] indicates that the 6,5 cyclization process (**117a, 122**) is less sensitive to this steric interaction.

Evans and colleagues [292] prepared a perhydrohistrionicotoxin analog (**127**), suitable for photoaffinity labeling of biological targets, from the lactam formate (**110**) by application of Kishi's synthetic sequence (Scheme 38). One important modification, namely the use of the magnesium chelate (**124**) of **123**, led to improved yields of imine **125** in the Grignard reaction with 4-pentenyl magnesium chloride. The alane reduction provided the decahydrohistrionicotoxin **126** with excellent stereospecificity. Hydroboration, hydrolysis, and alkylation with 3-nitro-4-fluorophenyl azide provided the aryl azide **127** for potential use in photoaffinity labeling studies.

Brossi's group [262,293] prepared the Corey-Kishi lactam **5** in overall yields of 20% by applying the Speckamp-Evans procedure on half-molar scale with only slight modifications. The Corey-Kishi lactam then was reduced with lithium aluminum hydride in tetrahydrofuran to 2-depentyl-perhydrohistrionicotoxin (**2**). X-ray analysis of the crystalline hydrobromide of **2** revealed a conformation similar to that of histrionicotoxin: that is, axial butyl and hydroxyl groups with an intramolecular N···H—O hydrogen bond [293]. Various derivatives (**128–132**) were also prepared [293] (Scheme 39).

The *N*-methyl alcohol **128** could be oxidized to the unstable 8-ketone **133**, which on attempted hydrochloride formation underwent some ring opening

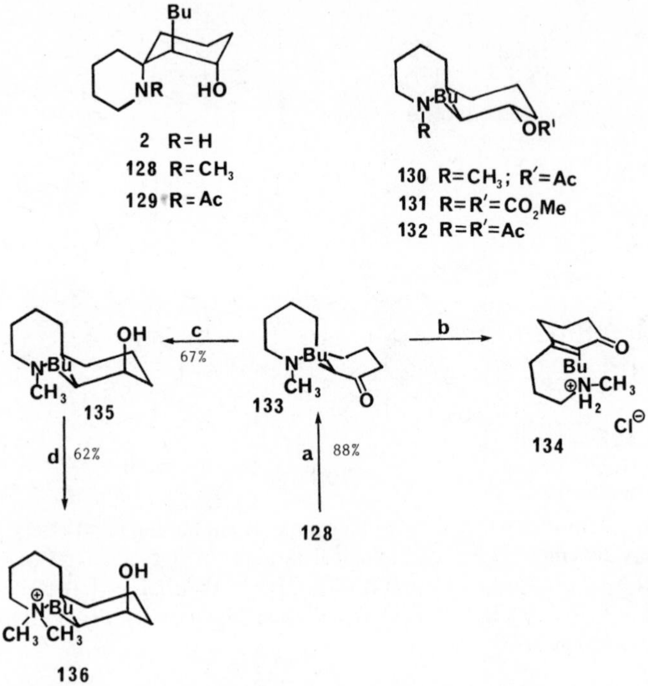

Scheme 39. Derivatives and transformations of 2-depentylperhydrohistrionicotoxin (2) [293]. (a) PCC/CH_2Cl_2; (b) $HCl/MeOH$; (c) $NaBH_4/MeOH$; (d) MeI/Me_2CO, 25°.

to **134** by a retro-Michael reaction. When reduced **133** gave the epi-8-alcohol **135**, showing no intramolecular hydrogen bond by infrared spectroscopy. Methylation of **135** proceeded readily at room temperature to yield the quaternary methyl ammonium salt **136**, in contrast to **128** (with its strong intramolecular hydrogen bond), which required nearly 2 weeks at 40–50°C for quaternization [293]. The O-acetate of **128** was more readily quaternized.

Takahashi and colleagues [262] resolved the Corey-Kishi lactam (**5**) (Scheme 40) by preparation of diastereomeric isocyanates, **137** and **138**, separation of these by high-pressure liquid chromatography, and hydrolysis to **5a** and **5b**. The resolved lactams were converted using Kishi's sequence (Scheme 17) to (+)- and (−)-perhydrohistrionicotoxin (**1a, 1b**) or were reduced directly with lithium aluminum hydride to (+)- and (−)-2-depentylperhydrohistrionicotoxin (**2a,2b**). (−)-Perhydrohistrionicotoxin (**1b**), derived from levorotatory **138**, had the same rotation as perhydrohistrionicotoxin, obtained by reduction of histrionicotoxin or dihydrohistrionicotoxin. There was only moderate stereoselectivity in the Kishi alane cyclohexane reduction step (Scheme 17, step q), and some 30% of the 2-epipentyl derivative **17** was obtained as a by-product [262].

3.2.5. Intramolecular Mannich Reactions.

Corey and colleagues [282,294] initiated two imaginative, but unsuccessful, approaches to perhydrohistrion-icotoxin, featuring the intramolecular alkylation of Mannich imine or imi-nium ion intermediates (Scheme 41). The first approach [282] envisaged an intramolecular alkylation of the keto imine **140**, generated *in situ* on acid hydrolysis of the dioxolone **139**, prepared in turn by a nine-step synthesis. Instead of the desired azaspiro ketone **143**, the quinolizidine **141** resulted, presumably by isomerization to the enamine, amine addition to the carbonyl group, and dehydration. An attempt was made to circumvent this outcome by methylating the intermediate **139**. Acidification would then generate the iminium ion intermediate **142**, as a more reactive substrate for intramolecular alkylation. Instead of the hoped for reaction product **144**, the hexahydro-quinoline derivative **145** was produced by another enamine condensation pathway.

The second approach [294] employed the pyrrolidine enamine **146**, thereby, preventing the carbonyl addition reactions, as well as increasing the nucleophilicity of the alkylating carbon. This intermediate was prepared in good yield by a ten-step synthesis from commercial cyclopent-1-enyl ace-tonitrile and was cyclized with acid (*d,e*), reduced (*f*), and converted to a cyclic urethane (*g*) with phosgene. X-ray analysis showed **148** to have the epi-configuration at *both* C(2) and C(7). It was converted in four steps as shown in Scheme 42 to 2,7-epiperhydrohistrionicotoxin (**78**) (11% overall) and to the dioxo analog (**149**) of this compound as well.

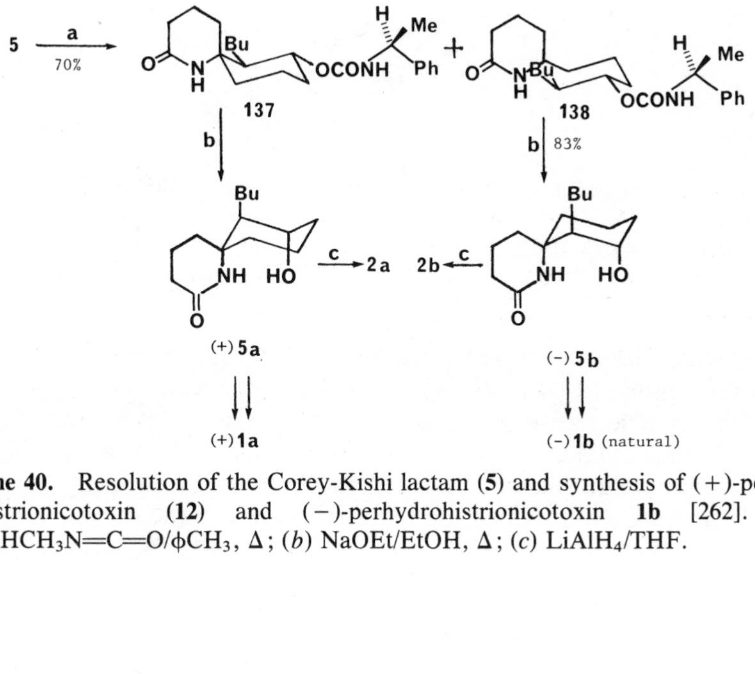

Scheme 40. Resolution of the Corey-Kishi lactam (**5**) and synthesis of (+)-perhy-drohistrionicotoxin (**12**) and (−)-perhydrohistrionicotoxin **1b** [262]. (*a*) (S)ϕCHCH₃N=C=O/ϕCH₃, Δ; (*b*) NaOEt/EtOH, Δ; (*c*) LiAlH₄/THF.

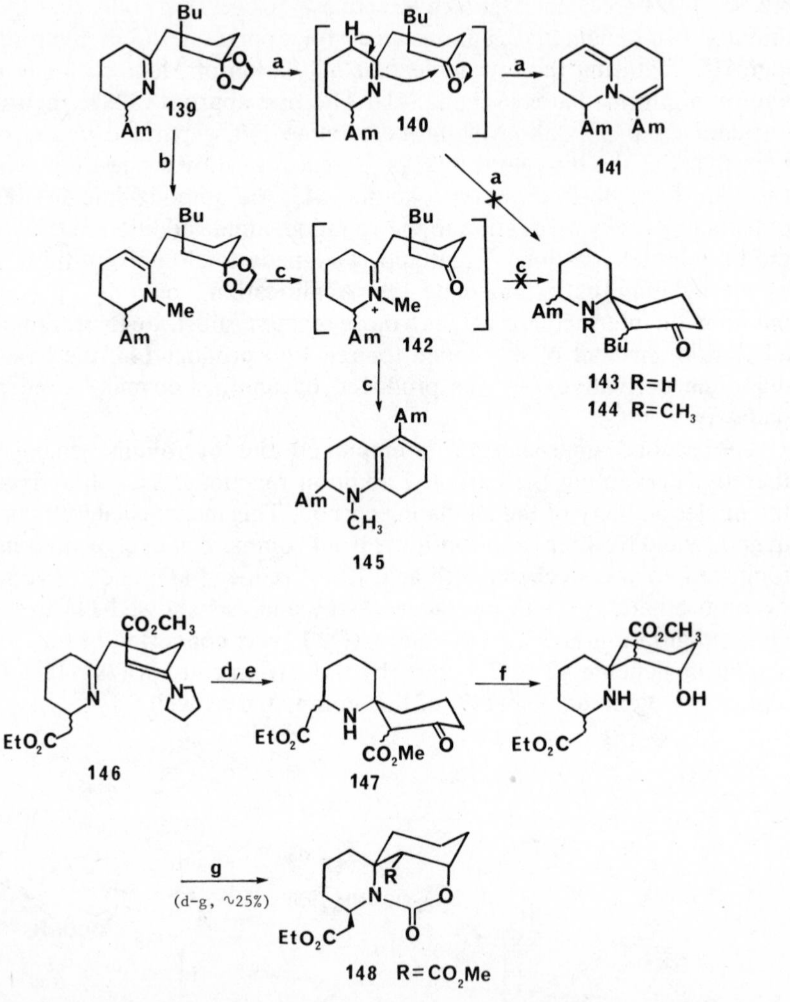

Scheme 41. Attempted synthesis of perhydrohistrionicotoxin via Mannich inter-mediates [282,294]. (*a*) Various acids; (*b*) MeOSO$_2$F/CH$_2$Cl$_2$, 0°; (*c*) H$_3$O$^\oplus$; (*d*) TsOH·H$_2$O (1.8 eq.)/CH$_2$Cl$_2$, 0°; (*e*) NaHCO$_3$; (*f*) NaBH$_4$/EtOH, −20°; (*g*) COCl$_2$/ CH$_2$Cl$_2$, 0° then Py, 20 hr.

3.2.6. Intramolecular Michael Additions. The observation by Kishi that the amido enone **22** (Scheme 17) cyclized to a 1:2 mixture of keto lactams **4** and **20** on acid treatment was the basis for a more direct route [263] to both **1** (Scheme 43) and to octahydrohistrionicotoxin (Scheme 44) in about 14% overall yield. Substituting 5-pentenyl cadmium for *n*-pentyl cadmium in (*a*) (Scheme 43) provided the required precursor **150** for the latter syn-thesis. The resulting butenyl ketolactam mixture **151**, **152** (Scheme 44) was

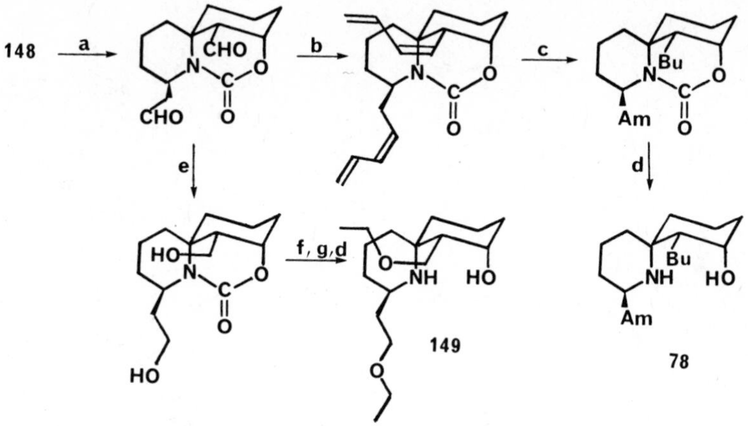

Scheme 42. Synthesis of 2,7-epiperhydrohistrionicotoxin **78** and a dioxo analog
(**149**) [294]. (*a*) Al(*i*-Bu)$_2$H (2.5 eq.)/CH$_2$Cl$_2$, $-78°$; (*b*) CH$_2$=CHCH$_2$P$^\oplus\phi_3$Br$^\ominus$ (5 eq.)/
DMSO anion in THF, 0°; (*c*) H$_2$/Pd-C/THF; (*d*) Li-MeNH$_2$, $-78°$; (*e*) LiBH$_4$/THF;
(*f*) KH; (*g*) EtI/HMPT.

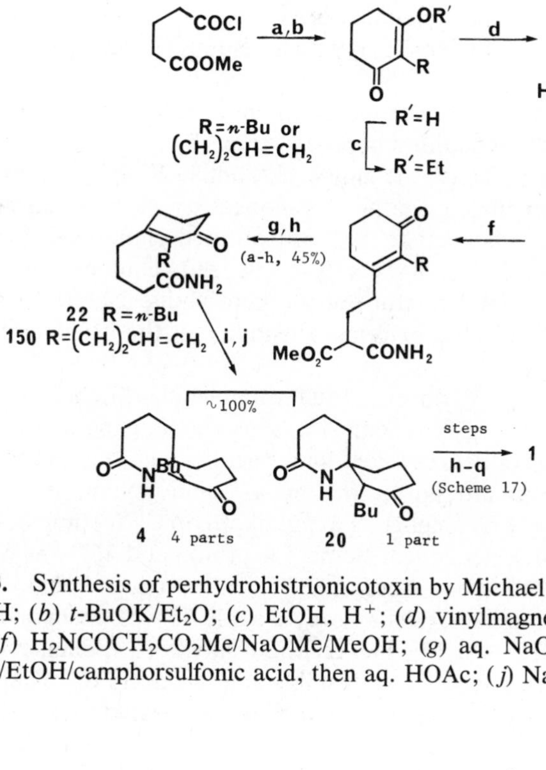

Scheme 43. Synthesis of perhydrohistrionicotoxin by Michael addition [263]. (*a*) *n*-
Am$_2$Cd/ϕH; (*b*) *t*-BuOK/Et$_2$O; (*c*) EtOH, H$^+$; (*d*) vinylmagnesium bromide/THF;
(*e*) H$^+$; (*f*) H$_2$NCOCH$_2$CO$_2$Me/NaOMe/MeOH; (*g*) aq. NaOH; (*h*) aq. HCl; (*i*)
HC(OEt)$_3$/EtOH/camphorsulfonic acid, then aq. HOAc; (*j*) NaOMe/CH$_2$Cl$_2$.

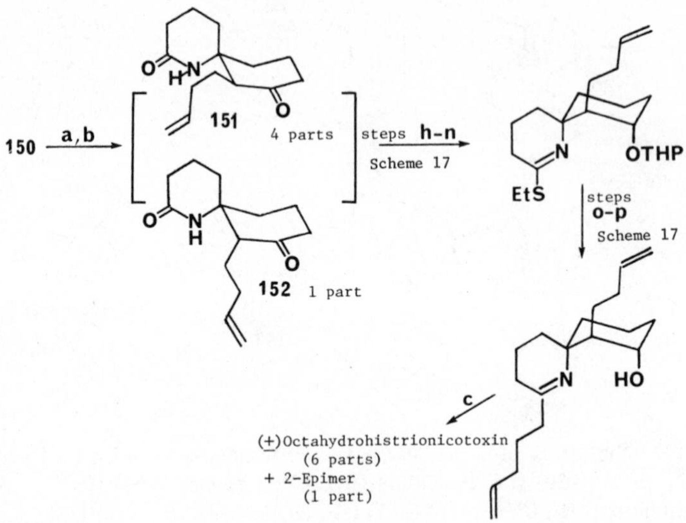

Scheme 44. Synthesis of (±)-octahydrohistrionicotoxin by Michael addition [263]. (*a*) HC(OEt)₃/EtOH/camphorsulfonic acid then aq. HOAc; (*b*) NaOMe/CH₂Cl₂; (*c*) AlH₃-C₆H₁₂.

carried through the same sequence, originally used by Kishi [264] for perhydrohistrionicotoxin (Scheme 17), with 5-pentenyl lithium replacing *n*-pentyl lithium in (*p*).

The Corey azide **71** served as a convenient precursor of the amino enone **153** in another Michael addition approach to perhydrohistrionicotoxin [282] (Scheme 45). Unfortunately, the amine **153**, unlike Kishi's carboxamide **22**, failed to undergo cyclization either spontaneously or under usual Michael conditions. On the other hand, the unsubstituted enone **154**, prepared by reduction of **72** with the Lindlar catalyst, cyclized spontaneously and quantitatively to an equimolar mixture of the keto amine 2-pentyl epimers **155** and **156**, demonstrating the extreme sensitivity of the Michael addition to steric effects.

Venit and Magnus [295] observed an intramolecular Michael addition with an even simpler enone **157** (Scheme 45). This enone, generated *in situ* from a Birch reduction product, cyclized under acidic conditions to the unstable amino ketone **158**. An attempt to trap this as the dioxolone gave instead the hexahydroquinoline **159**, analogous to pumiliotoxin C (Section 7), by a retro-Mannich → Mannich sequence. Some 5% protonated **157** was detected in equilibrium with **158** on exposure of **158** to trifluoroacetic acid.

The mesylate precursor **160** of Corey's azide **76** was examined by Corey and Balanson [282] as a Michael-reaction substrate (Scheme 45). When treated with ammonia at 60°C, an unstable amino ketone **162** resulted in 40% yield, presumably via the Michael adduct **161** generated *in situ*. The amino ketone **162** reverted to ring-opened **153** on standing at room temperature for 10 hr or immediately on exposure to acid. When **162** was reduced with borohydride immediately after formation, a number of amino alcohol stereoisomers resulted, among which was 5% of perhydrohistrionicotoxin.

A number of investigators have observed that the equilibrium position in Michael reactions of amino α-substitued enones favors the ring-opened form in acid or base. Godleski and Heacock [278] were able, however, to induce a Michael-type addition under *neutral* conditions by generating a reactive intermediate from the enone substrate **163** using trimethylsilyliodide (Scheme 46). The intermediate **164** or **165** then cyclizes by either an S$_N$2 or S$_N$2′ process. An intramolecular proton transfer (via **166**) presumably accounts for the 5:1 diastereoselectivity observed. Base equilibration (*e*) raises the ratio of amino ketone products **167** and **168** to an even more favorable

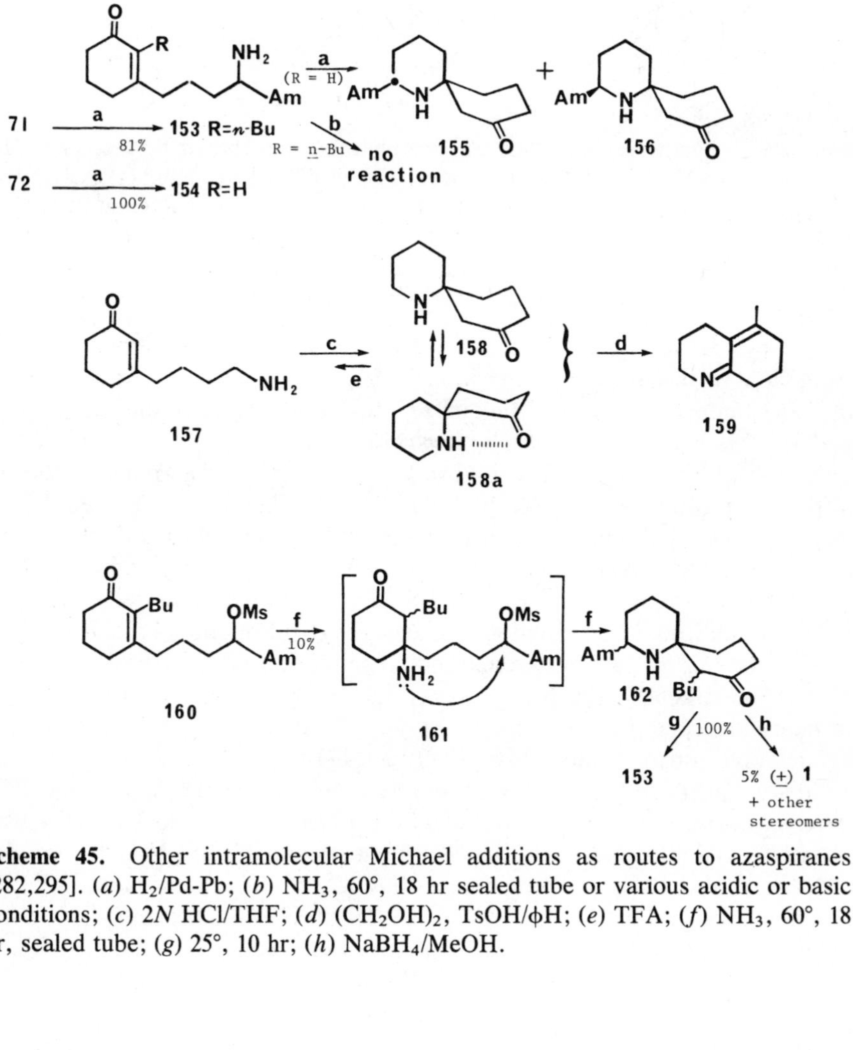

Scheme 45. Other intramolecular Michael additions as routes to azaspiranes [282,295]. (*a*) H$_2$/Pd-Pb; (*b*) NH$_3$, 60°, 18 hr sealed tube or various acidic or basic conditions; (*c*) 2*N* HCl/THF; (*d*) (CH$_2$OH)$_2$, TsOH/ϕH; (*e*) TFA; (*f*) NH$_3$, 60°, 18 hr, sealed tube; (*g*) 25°, 10 hr; (*h*) NaBH$_4$/MeOH.

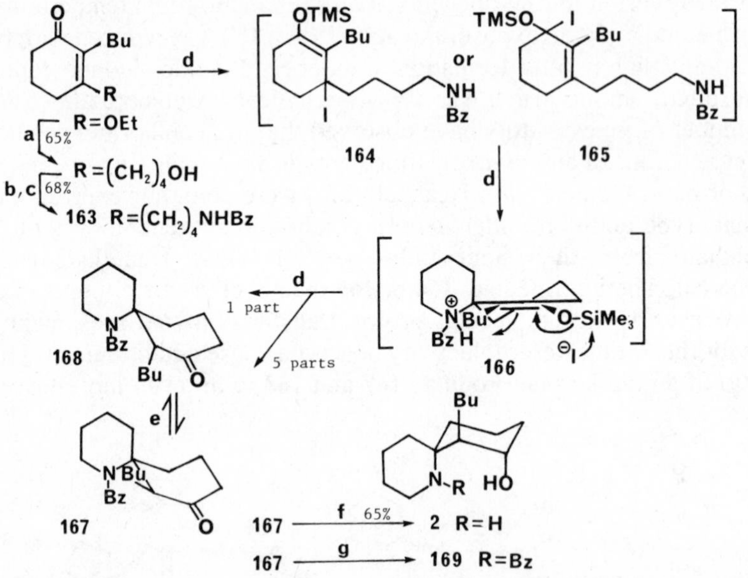

Scheme 46. Synthesis of ($\pm$)-2-depentylperhydrohistrionicotoxin by a modified Michael addition [278]. (*a*) ClMg(CH$_2$)$_4$OMgCl/THF; (*b*) TsCl/Py, $-10°$; (*c*) ϕCH$_2$NH$_2$/ DMSO, NaI (cat.), 25°, 18 hr; (*d*) TMS—I (2 eq.)/Et$_3$N (1 eq.), NaI (1 eq.)/CH$_3$CN, $-20°$, 12 hr; (*e*) NaOCH$_3$/CH$_2$Cl$_2$, 25°, 24 hr; (*f*) Li/NH$_3$/MeOH (2 eq.), $-78°$; (*g*) Li/NH$_3$, $-78°$.

13:1 ratio. Reductions afford either **2** or its *N*-benzyl derivative (**169**). In agreement with Corey's observations [282], Godleski and Heacock found that **167** undergoes a retro Michael reaction in dilute HCl and **163** fails to undergo Michael reaction with either acid or base.

Winterfeldt and colleagues [296] successfully achieved a Michael addition of the Corey amino enone **153** by ketalization (Scheme 47). They exploited an attractive "biomimetic" route to the amino enone based on the symmetrical triketone **170**, which by aldol condensation (*a,a'*) and amination (*b,b'*) generates precursors to both the histrionicotoxin and decahydroquinoline (pumiliotoxin C) (Section 7) classes of dendrobatid alkaloids. Others [23,24,38,297,298] have also remarked on similar possible biosynthetic pathways. The triketone **170** was prepared in a high-yield three-step sequence from commercial 1,7-dichloroheptan-4-one and allowed to cyclize under acidic conditions to a mixture of diketones **171** and **172**, which on reductive amination produced the easily separable amino enone **153** and imine **173**. After cyclization and *in situ* ketalization of **153**, the amino ketal **174** was isolated. On deblocking, it produced a small amount of retro Michael product **153** and the mixture of ketoamines **175** and **176**. The ketal is mainly one stereoisomer, indicating that epimerization at C(7) must occur during the ketal hydrolysis. Base-catalyzed equilibration (*e*) converts the ketoamine mixture cleanly to one stereoisomer (**175**), which on hydride reduction (*f*), gives rise to two isomeric alcohols **177** and **17**. The minor alcohol (**17**) forms a cyclic urethane with phosgene, consonant with the desired axial 8-hydroxy configuration. The proton magnetic resonance spectrum, however, indicated the 2-epipentyl configuration. The alcohol **17** was converted to the imine **178** as shown, providing a formal synthesis of perhydrohistrionicotoxin, since this imine has been converted to perhydrohistrionicotoxin by Corey's method.

Winterfeldt and colleagues [296] hypothesized that the abnormal 2-epi configuration in **174** arises because the pentyl group adopts the structurally more favorable quasi-equatorial configuration in the cyclization transition

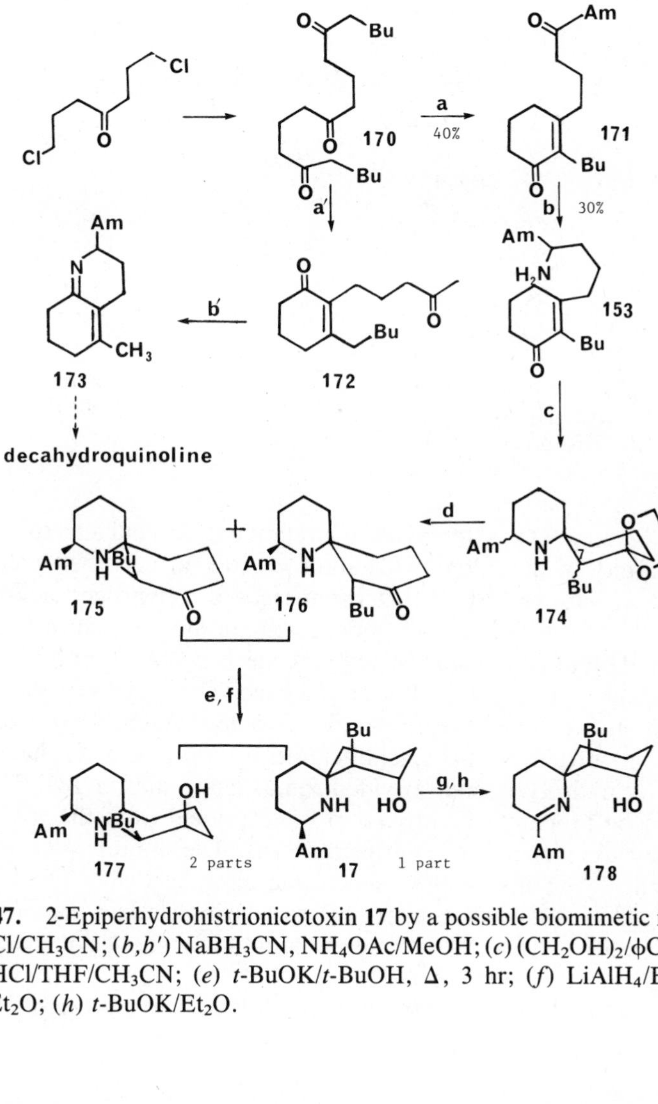

Scheme 47. 2-Epiperhydrohistrionicotoxin **17** by a possible biomimetic route [296]. (*a,a′*) HCl/CH$_3$CN; (*b,b′*) NaBH$_3$CN, NH$_4$OAc/MeOH; (*c*) (CH$_2$OH)$_2$/φCH$_3$, TsOH, Δ; (*d*) HCl/THF/CH$_3$CN; (*e*) *t*-BuOK/*t*-BuOH, Δ, 3 hr; (*f*) LiAlH$_4$/Et$_2$O; (*g*) *t*-BuOCl/Et$_2$O; (*h*) *t*-BuOK/Et$_2$O.

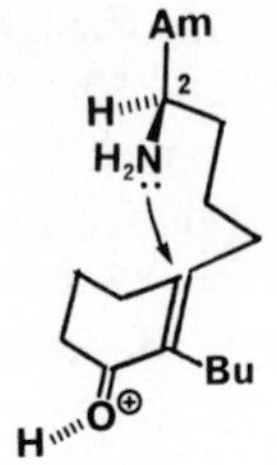

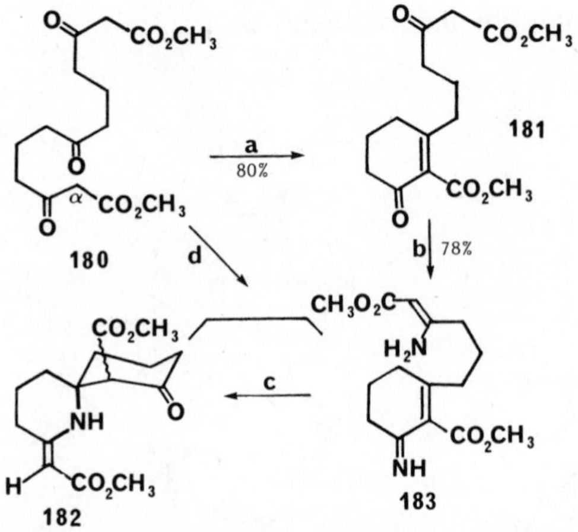

Scheme 48. Regiospecific azaspirocyclization by a possible biomimetic pathway [296]. (*a*) (*i*-Pr)$_2$NEt/HOAc/MeOH, 5°, 10 d; (*b*) NH$_3$/NH$_4$NO$_3$/MeOH, 0°, 5 hr; (*c*) HCl/THF; (*d*) NH$_3$-dioxane, 15 hr.

state **179** (Scheme 48). Attempting to circumvent this and also to restrict the regiospecificity of the initial cyclization (*a*) still further, they prepared the triketodiester **180**, reasoning that the increased α-hydrogen acidity would cause pathway *a* of Scheme 47 to predominate over *a'* and that an sp^2 hybridized carbon at C(2) would remove the steric problems encountered with an sp^3 carbon. This was found to be the case. The initial cyclization of **180** proceeded with complete regiospecificity to **181** which, when treated with ammonia (*b*) produced **183**. Acidification (*c*) induced a Michael addition followed by hydrolysis and provided the azaspiro diester **182**. The diester could also be prepared directly (*d,c*) in 88% yield from the triketodiester **180** without the isolation of intermediate **181**. These interconversions demonstrate the feasibility of this biomimetic approach, and it is hoped that methods will be found to convert **182** to histrionicotoxin analogs with the natural chirality at C(2) and C(7) (see Scheme 41 for studies on the analogous Corey-Ruden intermediate **148**).

Husson and colleagues [298] discovered a potentially useful azaspirane synthesis, while studying intramolecular displacement of cyanide by side chain enolates in alkylated 2-cyano Δ^3-piperidines, for example **184** (Scheme 49). The kinetic enolate failed to cyclize, while the thermodynamic enolate yielded a Dieckmann-Thorpe product **185**. Reduction of **184** and attempted acidic cyclization of **187** gave, however, the amino enone **189**, indicating the intermediacy of the azaspiro ketone **188**. Attempts to recyclize **189** under acidic conditions failed, but ketalization did produce the diastereomeric ketals (**190,191**) along with the rearranged noncyclized ketal **186**. The authors comment that the successful acid catalyzed Michael additions of Corey and Balanson [282] and Venit and Magnus [295] without ketalization may be a result of the cyclic forms being stabilized by an N—H⋯O=C hydrogen bond (e.g., **158a**, Scheme 45). In their approach, such a bond is impossible and the cyclized form is obtainable only by trapping. No rearranged quinoline derivatives, as obtained by Corey (e.g., **145**), could be detected.

3.2.7. Stereoselective *cis*-Enyne Syntheses. Several methods exist to construct the *cis*-enyne functionality found in side chains of histrionicotoxin and gephyrotoxin (Section 4). Three use aldehyde precursors, while others feature an epoxide, a vinyl chloride, or azaspiro cyclic phosphonium salt.

The approach of Corey and Ruden [299] (Scheme 50) uses the Wittig phosphorane from methylene chloride on the *cis*-α,β-unsaturated aldehyde **192** to produce the chlorodiene **193**. Dehydrochlorination and trapping the

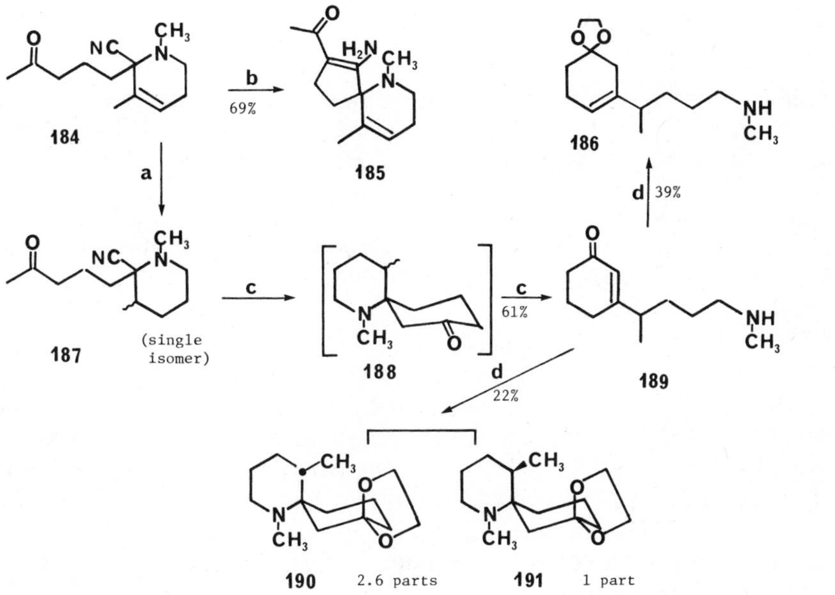

Scheme 49. A novel approach to azaspiranes via alkylated piperidines [298]. (*a*) H₂/Pd-C/EtOH; (*b*) *t*-BuOK/*t*-BuOH; (*c*) TsOH/φH, Δ; (*d*) TsOH/φH, (CH₂OH)₂.

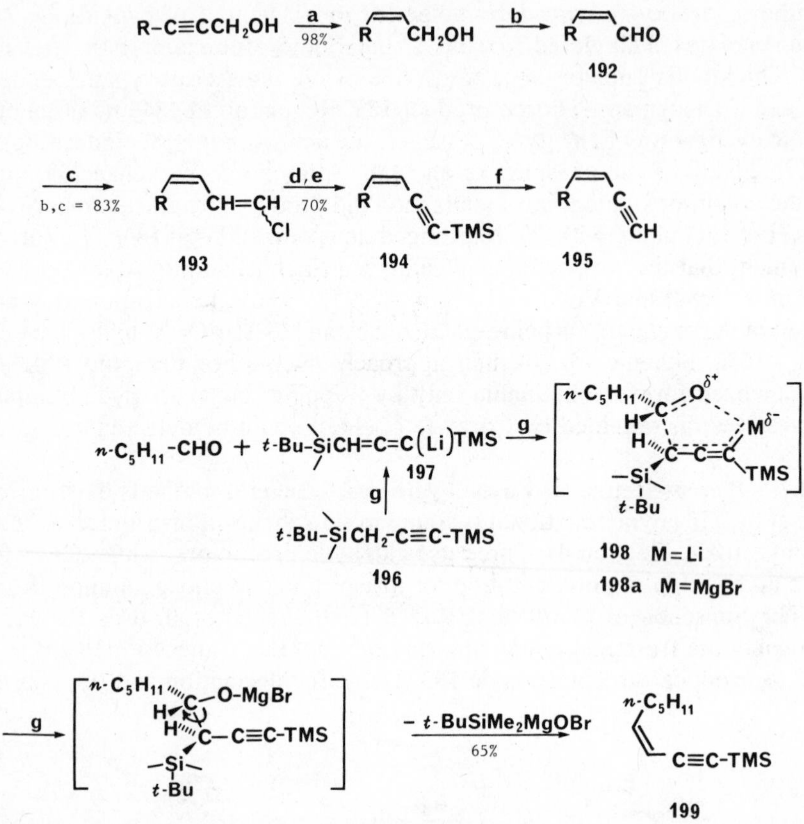

$$R-C{\equiv}CCH_2OH \xrightarrow[98\%]{a} R{\diagup}{=}{\diagdown}CH_2OH \xrightarrow{b} R{\diagup}{=}{\diagdown}CHO$$

192

$$\xrightarrow[b,c\,=\,83\%]{c} R{\diagup}{=}{\diagdown}CH{=}CH{\diagdown}_{Cl} \xrightarrow[70\%]{d,e} R{\diagup}{=}{\diagdown}C{\diagdown}_{C-TMS} \xrightarrow{f} R{\diagup}{=}{\diagdown}C{\diagdown}_{CH}$$

193   **194**   **195**

$$n\text{-}C_5H_{11}{-}CHO + t\text{-}Bu{-}\overset{|}{\underset{|}{Si}}CH{=}C{=}C(Li)TMS \xrightarrow{g}$$

197

$$\Big\uparrow g$$

$$t\text{-}Bu{-}\overset{|}{\underset{|}{Si}}CH_2{-}C{\equiv}C{-}TMS$$

196

$$\left[\begin{array}{c} n\text{-}C_5H_{11}{\cdots}\overset{O^{\delta+}}{\underset{}{C}} \\ H{\diagdown}{\cdots}C{-}C{\equiv}C \\ -Si{-} \quad TMS \\ t\text{-}Bu \end{array} M^{\delta-}\right]$$

198 M = Li

198a M = MgBr

$$\xrightarrow{g} \left[\begin{array}{c} n\text{-}C_5H_{11}{\cdots}\overset{O-MgBr}{\underset{}{C}} \\ H{\cdots}C{-}C{\equiv}C{-}TMS \\ -Si{-} \\ t\text{-}Bu \end{array}\right] \xrightarrow[65\%]{-\,t\text{-}BuSiMe_2MgOBr} \quad n\text{-}C_5H_{11}{\diagdown}{=}{\diagdown}_{C{\equiv}C-TMS}$$

199

Scheme 50. The synthesis of *cis*-enynes from aldehydes [299,300]. (*a*) H_2/Pd-CaCO$_3$/ϕH; (*b*) MnO$_2$/CH$_2$Cl$_2$; (*c*) ϕ_3P=CHCl/THF; (*d*) MeLi/THF, 12 hr; (*e*) TMS—Cl; (*f*) *n*-Bu$_4$N$^{\oplus}$F$^{\ominus}$/THF; (*g*) *t*-BuLi/THF-Et$_2$O-pentane, $-78°$.

acetylide with trimethylsilylchloride produces the trimethylsilyl (TMS)-blocked *cis*-enyne **194**. Conventional deblocking affords **195** in good overall yield.

Yamamoto and colleagues [300] introduced an easier, highly stereoselective route to *cis*-enynes based on the Petersen olefination reaction of aldehydes with the bissilylpropargyl reagent **196** (Scheme 50). The allenic anion **197**, formed with strong base, is thought to form a transition state complex **198**, where the bulky silyl group and the chelating metal ion favor the orientation shown. The addition of one equivalent of MgBr$_2$ was found to increase greatly the stereoselectivity, evidently by forming a more stable complex (**198a**). Rotation and *syn*-elimination produces the *cis*-enyne **199** with, in the case of *n*-hexanal, a (**Z/E**)ratio greater than 50.

Corey and Rücker [301] independently developed a similar approach with the reagent **200** having bulky triisopropylsilyl (TIPS) groups (Scheme 51).

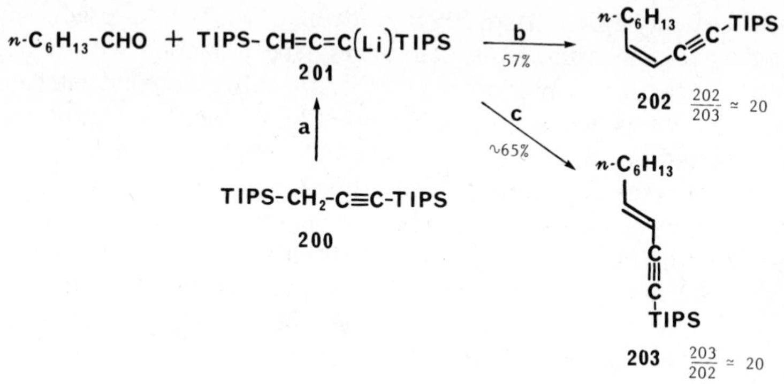

Scheme 51. Stereoselective *cis*- or *trans*-enyne syntheses [301]. (*a*) *n*-BuLi/THF, −20°; (*b*) −78° → r.t.; (*c*) *n*-BuLi/HMPT (5 eq.)/THF, −78°.

They observed a 20:1 stereoselectivity for the *cis*-enyne product **202** in the reaction of *n*-heptanal with the allenic anion **201** from **200**. Interestingly, inclusion of hexamethylphosphorous triamide in the reaction completely reverses the stereochemical outcome with the *trans*-enyne **203** now greatly predominating. Hexamethylphosphorous triamide was shown to reverse the stereoselectivity of the initial carbonyl addition step and not the elimination step.

Holmes and colleagues [302] used the oxidative coupling of acetylenes and the known resistance of trimethylsilyl (TMS) acetylenes to hydrogenation in a short, highly stereospecific enyne synthesis (Scheme 52). An epoxide substrate (**204**) is opened by *trans*-diaxial attack of lithium acetylide to

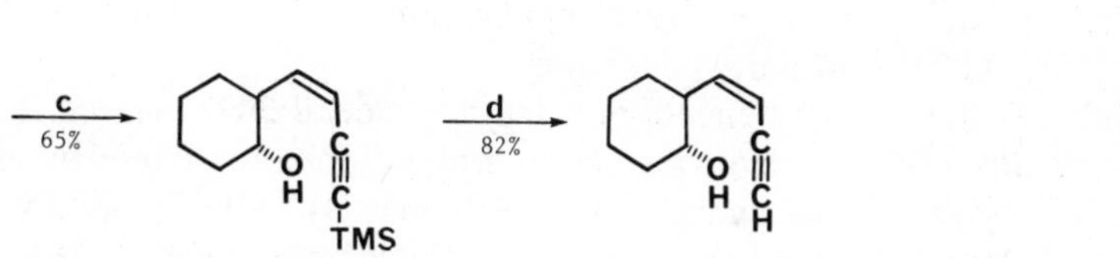

Scheme 52. A stereospecific *cis*-enyne synthesis from epoxides [302]. (*b*) LiC≡CH/(CH₂NH₂)₂; (*b*) HC≡C—TMS, Hay catalyst; (*c*) H₂/Pd-BaSO₄/quinoline/hexane/MeOH; (*d*) *n*-Bu₄N⊕F⊖/THF.

provide the acetylenic alcohol **205** intermediate, which after oxidative coupling undergoes the usual (Z)-specific acetylene reduction with the Lindlar catalyst. So far it has proved impossible to utilize epoxide intermediates successfully in this manner in a synthesis of histrionicotoxin and the alcohol function would be superfluous in synthesis of gephyrotoxin.

Negishi's group [303] has reported a direct, highly stereoselective (>97%) synthesis of *cis*-enynes that requires no protection of the acetylenic moiety and utilizes the commercially available lithium acetylide ethylenediamine complex (Scheme 53). This complex is converted to chlorozinc acetylide using zinc chloride in tetrahydrofuran, then treated with a (Z)-alkenyl iodide in the presence of a palladium catalyst to provide the *cis*-enyne. Alkenyl bromides can also be utilized.

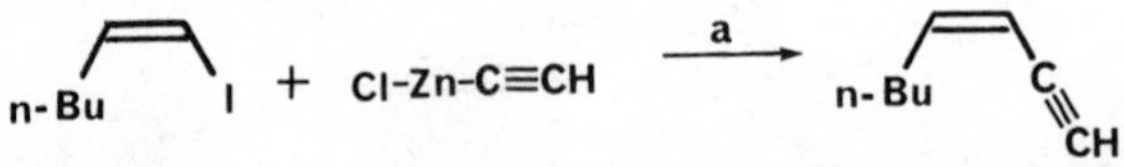

Scheme 53. A stereospecific *cis*-enyne synthesis from (Z)-alkenyl halides [303]. (*a*) THF-5% Pd(Pϕ_3)$_4$, 0–25°.

The general topic of *cis*- and *trans*-enyne syntheses has been reviewed in the thesis of Kishi's student, S. C. Carey [265]. Scheme 54 (Section 3.2.8) illustrates their newly developed application of the Wittig reaction in constructing the *cis*-enyne side chain at C(2) of histrionicotoxin.

3.2.8. Histrionicotoxin. The first approach to histrionicotoxin had to be abandoned near its completion by Kishi and Carey [265] when they discovered that the allylic alcohol **218** (Scheme 54) underwent a Grob cleavage during attempts to prepare the unsaturated aldehyde. Presumably, this occurred via intermediates such as **219**, produced in this instance by a Swern oxidation. Attempts to circumvent this dismal outcome, by appropriate functionalization of the nitrogen atom, proved fruitless.

An alternative approach, which ultimately proved successful, constructed the C(7) side chain *before* that at C(2). Their new *cis*-enyne synthesis, developed after extensive model studies [265] and applied in the preparation of **218**, was used to create the C(2) *cis*-enyne moiety. This step featured a Wittig reaction between an azaspirophosphonium salt and tertiary butyldimethylsilyl-(TBS)- propargylaldehyde.

Their synthesis commenced with the hemiacetal **207**, derived originally by Kishi and colleagues from the octahydrohistrionicotoxin intermediate **206** by steps (*b–g*) in 35% overall yield (Scheme 54). After reduction of the masked aldehyde (*h*) and conversion to a thiolactam diacetate **208**, the Eschenmoser episulfide contraction (*k*) was used to prepare the α-acetyl vinylogous urethane **209**, whereby the nitrogen atom is effectively deactivated toward the potentially disasterous Grob cleavage. Selective hydrolysis of the primary allylic acetate (*l*) and pyridinium chlorochromate oxidation (*m*) provided the (Z)-unsaturated aldehyde **210**, which was carried through the Corey-Ruden procedure and acyl cleavage (*o*), to yield the silylated *cis*-enyne **211**. The vinylogous urethane could be reduced with good stereoselectivity to a 1:1 mixture of C(2) epimers **212** and **213**. Reduction in a polar solvent, where no chelation of the reducing agent with the 8-hydroxyl group is likely, led solely to **213**, the product of attack from the sterically least-hindered face. Interestingly, another reduction procedure (*r*) provided the 2-ethyl imine **214** directly. Reduction of the ester side chain (*s*), acetylation (*t*), desilylation (*u*), and another preferential primary acetate hydrolysis (*v*) afforded the C(2) hydroxyethyl intermediate **215**. After conversion to the phosphonium salt **216** (*w–y*), the crucial Wittig reaction (*z*) proceeded in modest yield to the desired *cis*-enyne **217** with remarkably high (24:1) (Z)-selectivity. After desilylation and deacetylation (deacetylation occurs as a side reaction in step *z*), the first synthetic sample of histrionicotoxin was

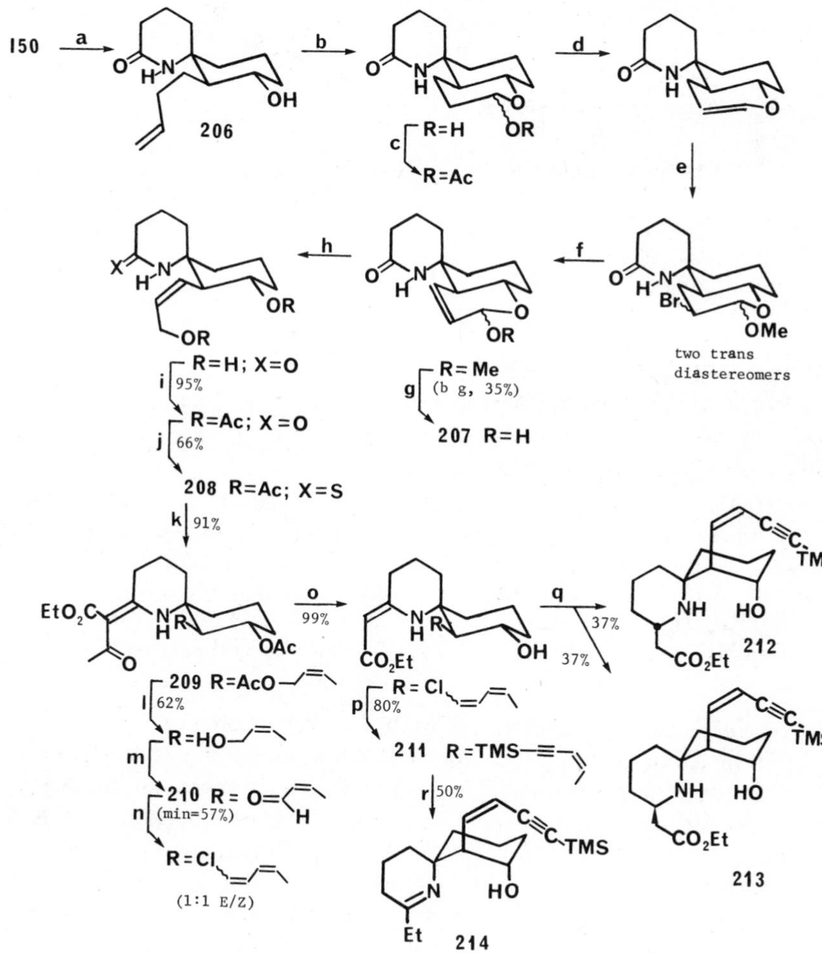

Scheme 54. (*Continued on next page*)

obtained. ($\pm$)-*epi*-Histrionicotoxin was also synthesized by the same route from **213**. Extensive studies on the Wittig reaction carried out on analogous systems indicated, quite surprisingly, greater Z/E-ratios at room temperature than at $-78°$ and the insensitivity of these ratios to dissolved LiBr.

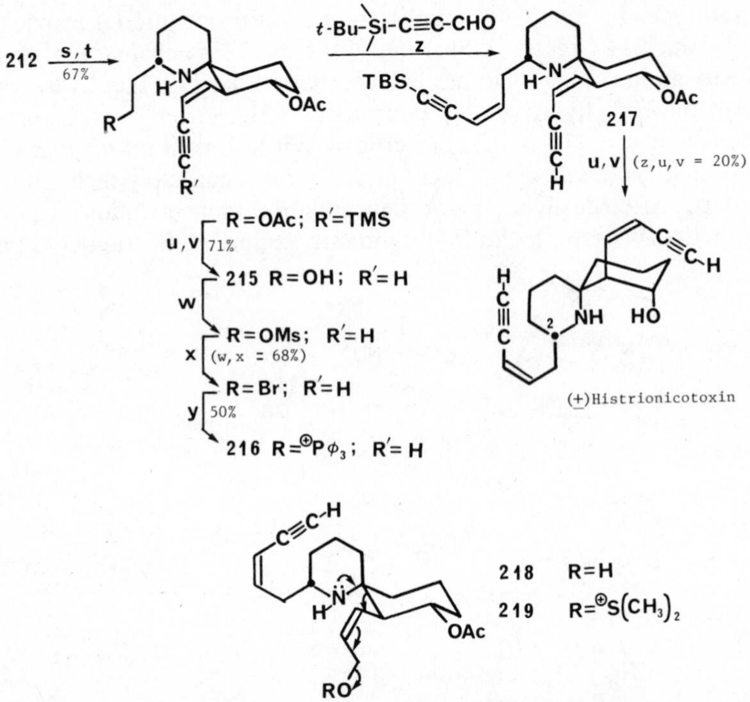

Scheme 54. The synthesis of ($\pm$)-histrionicotoxin [265]. (*a*) Li/NH$_3$, $-78°$; (*b*) O$_3$/ MeOH, $-78°$ then Me$_2$S; (*c*) Ac$_2$O/Py; (*d*) 180°, 11 mm Hg; (*e*) Br$_2$/CH$_2$Cl$_2$-MeOH; (*f*) DBU-DMSO, 140°, 24 hr; (*g*) aq. HOAc, 60°; (*h*) NaBH$_4$, aq. THF; (*i*) Ac$_2$O/ Py; (*j*) P$_2$S$_5$/Py, 100°; (*k*) AcCHBrCO$_2$Et/CH$_2$Cl$_2$-NaHCO$_3$; (*l*) NaOH/MeOH, $-20°$; (*m*) PCC/CH$_2$Cl$_2$; (*n*) ϕ_3P$^{\oplus}$CH$_2$Cl Cl$^{\ominus}$, *n*-BuLi/THF, $-78°$; (*o*) NaOEt/EtOH, 50°; (*p*) MeLi then TMS—Cl; (*q*) NaCNBH$_3$/C$_6$H$_{12}$, r.t. several d; (*r*) Red-Al, ϕCH$_3$, 0°; (*s*) LiAlH$_4$/THF; (*t*) Ac$_2$O/Py; (*u*) Bu$_4$N$^{\oplus}$F$^{\ominus}$/THF; (*v*) NaOH/MeOH; (*w*) MsCl then HCl; (*x*) LiBr/DMF; (*y*) ϕ_3P/CH$_3$CN, 160°, 5 d; (*z*) LiN(*i*-Pr)$_2$/THF.

3.2.9. Chemical Modifications of Natural Histrionicotoxins.

"Natural" *l*-perhydrohistrionicotoxin has not been detected as yet in extracts from some 40 species of dendrobatid frogs, but instead has been prepared for biological studies by reduction of histrionicotoxin or dihydroisohistrionicotoxin. Reduction of a variety of histrionicotoxins with hydrogen and either palladium on charcoal or platinum oxide had yielded this common perhydro derivative [247,248]. However, in one case with palladium on charcoal, histrionicotoxin yielded a partially reduced mixture of tetrahydro and octahydro derivatives, which were isolated as the co-crystallized hydrochloride salt [247]. Reduction of histrionicotoxin with hydrogen and Lindlar's palladium catalyst yielded tetrahydrohistrionicotoxin, hexahydrohistrionicotoxins, and a dihydrohistrionicotoxin in which the addition of hydrogen had occurred at the acetylenic terminus of the five-carbon side chain to yield dihydrohistrionicotoxin, a compound later isolated from a dendrobatid frog [247] (Fig. 2).

N-Methylation of histrionicotoxins has been investigated. Perhydrohistrionicotoxin was readily converted to the mono-N-methyl derivative with methyl iodide in acetonitrile or methanol [247]. Histrionicotoxin has been converted to its N-methyl derivative by reductive formylation using formaldehyde followed by NaCNBH$_3$ [304]. The formation of the methiodide of perhydrohistrionicotoxin also was reported [247], but this conversion has proven difficult to reproduce under a variety of conditions with methyl iodide [unpublished results]. Conversion to the methiodide appears to proceed with great difficulty because of hydrogen bonding from the alcohol function to the nitrogen [see 262,293]. The O-acetyl derivative of histrionicotoxin and of N-methylhistrionicotoxin was prepared [304]. O-Acetyl-N-methylhistrionicotoxin, but not N-methylhistrionicotoxin, was readily converted to the methiodide with methyl iodide in acetonitrile, presumably because of the lack of the hydrogen bond in the O-acetyl derivative. Hydrolysis of O-acetylhistrionicotoxin methiodide to histrionicotoxin methiodide with K$_2$CO$_3$ was unsuccessful under a variety of conditions.

Perhydrohistrionicotoxin has been converted to the deoxy derivative in a sequence, involving reaction with thionyl chloride in pyridine, to yield a 7,8-dehydro derivative followed by reduction with hydrogen and platinum oxide [247,249]. The mass spectrum of deoxyperhydrohistrionicotoxin was as follows: m/z 279 (28), 250 (12), 237 (16), 236 (71), 222 (10), 209 (15), 208 (39), 194 (12), 181 (40), 180 (83), 167 (100), 152 (16), 138 (8), 124 (9), 123 (10), 110 (23), 96 (80). Thus the fragment ion at m/z 96 may be typical of both histrionicotoxins and deoxyhistrionicotoxins. As yet it remains uncertain whether deoxyhistrionicotoxins occur naturally in dendrobatid frogs (see [24] and Section 7).

3.3. Biological Activity

3.3.1. Toxicity and Pharmacological Activities.
Histrionicotoxins have relatively low toxicity in mammals and the toxin designation is, indeed, a misnomer. At 2 mg/kg isodihydrohistrionicotoxin has virtually no effect in mice, while a 5 mg/kg subcutaneous dose of either histrionicotoxin or isodihydrohistrionicotoxin causes only slight locomotor difficulties and prostration [23,245].

Pharmacologically, histrionicotoxins affect the function of at least three channels in electrogenic membranes: The first is the nicotinic acetylcholine receptor channel complex of vertebrate neuromuscular endplate, where histrionicotoxins in a time- and stimulus-dependent manner block the conductance of the channel and in a more rapid process shorten the time that channels remain open. The binding of histrionicotoxins to this receptor-channel complex has been studied extensively in ray (*Torpedo*) electroplax preparations. The second is the voltage-dependent sodium channel of nerve and muscle, where histrionicotoxins reduce conductances in a manner reminiscent of local anesthetics. The third is the voltage-dependent potassium channel of nerve and muscle, where histrionicotoxins reduce conductances in a time- and stimulus-dependent manner. Structure-activity profiles at the three channels differ [305].

Analogs of histrionicotoxin, namely the 2-depentylperhydrohistrionicotoxin and its *N*-methyl derivative, had virtually no analgesic activity [293] and were stated to have "little effect on blood pressure" and to have "some diuretic effect with a concomitant inhibition of carbonic anhydrase" [38].

3.3.2. Acetylcholine Receptor Channel Complex.

The effects of histrionicotoxins on the acetylcholine receptor channel have been investigated in detail in neuromuscular preparations from various vertebrates. As with the batrachotoxins, the fundamental aspects of the pharmacological action of histrionicotoxins were defined by Albuquerque and colleagues in the early 1970s [306–309]. The data strongly indicated that histrionicotoxins affect the function of nicotinic receptor–modulated channels through interactions with channel sites rather than with the receptor recognition site itself. Thus histrionicotoxins have no effect, except at high concentrations, on the binding of nicotinic antagonists, such as *d*-tubocurarine and α-bungarotoxin, to the acetylcholine receptor. Instead, the histrionicotoxins appear to interact with other sites on the receptor channel complex, thereby either blocking the open form of the channel, preventing its activation, or accelerating inactivation or desensitization. The voltage, time, and stimulus dependence of the blockade by histrionicotoxin [305,306,310] provides evidence suggestive of an interaction with a site on the open channel. The most rapid effect of histrionicotoxins on the acetylcholine receptor channel complex is to shorten the decay time, thus appearing to increase the rate of inactivation of the channel. This effect may reflect binding of histrionicotoxin to a site on the closed or desensitized form of the channel. The dual effects of histrionicotoxins on the acetylcholine receptor channel complex are reversible on washing. Furthermore, the stimulus-dependent blockade of receptor-elicited increases of conductances is reversed in the continued presence of the alkaloid. However, the blockade quickly reinstates itself when stimuli are repeated. Perhydrohistrionicotoxin at very low concentrations (<0.1 μM) blocks endplate currents elicited by iontophoretic acetylcholine in rat neuromuscular preparations, while having no effect on spontaneous miniature endplate currents or endplate currents evoked by nerve stimulation [311,312]. Further studies will be required to clarify the reason for the remarkable potency of perhydrohistrionicotoxin versus responses to ionto-phoretically applied acetylcholine. "Histrionicotoxin" blocks carbamylcho-line-elicited increases in conductance even with a purified acetylcholine receptor complex reconstituted into vesicular preparations [313]. Thus such a purified and reconstituted preparation still retains a functional histrio-nicotoxin site. Although the compound used was stated to be histrionicotoxin, it appears more likely that it was dl-perhydrohistrionicotoxin. Quotation marks are used in the present review when the nature of the histrionicotoxin used in a biological study has not been precisely defined.

Histrionicotoxins have been proposed to convert the acetylcholine re-ceptor channel complex in chick muscle cells to a state similar to, if not identical with, the so-called desensitized state [314]. Certainly, the presence of low concentrations of histrionicotoxins appear to increase the nicotinic agonist–elicited phenomenon known as desensitization. In addition, the presence of histrionicotoxin, like classical desensitization, increases the ap-parent affinity of nicotinic agonists for the receptors [314–316]. However, in one study, with electroplax membranes, dl-perhydrohistrionicotoxin was reported to have no significant effect on desensitization processes [317]. In muscle cells, octahydrohistrionicotoxin (referred to as histrionicotoxin in the paper but stated to be synthetic octahydrohistrionicotoxin in *Materials*) increases the affinity of the acetylcholine receptor for the nicotinic agonist carbamylcholine and increases the rate of desensitization of agonist-induced sodium influx [318]. The data suggest an allosteric cooperative mechanism in which at least two sites for octahydrohistrionicotoxin exist per receptor channel complex.

A variety of agents that include the histrionicotoxins stabilize the high affinity "desensitized" state of the acetylcholine receptor channel complex. The term "noncompetitive blockers" has been used to describe such agents [319,320] and is used in this review. Perhydrohistrionicotoxin is one of the more potent of the noncompetitive blockers. It initially appeared to interact with saturable sites in the receptor channel complex with a stoichiometry of about one per two α-bungarotoxin sites [307]. However, the ratio of α-bungarotoxin/acetylcholine–binding sites to perhydrohistrionicotoxin bind-ing sites in electroplax membranes has ranged widely in different studies [321 and references therein]. Although very potent as a noncompetitive blocker, perhydrohistrionicotoxin increases the proportion of desensitized receptors only slightly, while other less potent noncompetitive blockers are much more efficacious [316, 319]. Noncompetitive blockers, such as chlorpromazine and trimethisoquin, appear to interact with at least two classes of sites: Per-hydrohistrionicotoxin competes with these agents for the higher affinity, low density sites, but not for the lower affinity, high density sites. The ace-tylcholine receptor channel complex consists of subunits in the following stoichiometry: $\alpha_2\beta\gamma\delta$. Radioactive chlorpromazine, when used as a photo-affinity label, reacts with all four protein subunits of the acetylcholine receptor channel complex and perhydrohistrionicotoxin reduces irreversible labeling of all subunits [319,322]. Other noncompetitive blockers, including radioactive perhydrohistrionicotoxin, appear to photoaffinity label primarily the 65,000 (δ) dalton subunit [323]. Photoaffinity labeling of the 65,000 dalton subunit by 5-azidotrimethisoquin is blocked by histrionicotoxin [324]. Other data, however, suggest that the functional sites of action of noncompetitive blockers are at the α- and β-subunits [325]. Selective labeling of the α- and β-subunit by quinacrine azide is blocked by histrionicotoxin [326]. The α-subunit contains the recognition site for acetylcholine. Perhydrohistrionicotoxin has no effect on labeling of various subunits by a radioactive azido derivative of ethidium bromide except in the presence of carbamylcholine, where it reduced labeling of the 40,000 (α) dalton subunit and completely blocked labeling of the 65,000 dalton subunit [327]. Fluorescence studies with ethidium bromide provide evidence for conformational changes in the receptor-channel elicited by perhydrohistrionicotoxin [328]. The effects of various noncompetitive blockers, such as local anesthetics and detergents, on binding of fluorescent cholinergic agonists can be competitively reduced by perhydrohistrionicotoxin, which, although potent, is less efficacious [320]. "Histrionicotoxin" was stated to have no effect on photoaffinity labeling of a 43,000 dalton protein with a radioactive azidoprocaine amide and the azido derivative was stated to have no effect except at very high concentrations on binding of radioactive perhydrohistrionicotoxin to electroplax membranes [329, see Section 3.3.3]. This 43,000 dalton protein is apparently not an essential component of the receptor-channel complex. "Histrionicotoxin" had no effect on the potency or efficacy of carbamylcholine-elicited increases in fluorescence of salicylate or nitrobenz-2-oxa-1,3-diazole probes covalently linked to electroplax acetylcholine receptor channel complex [330,331].

Histrionicotoxin has similar effects in neuromuscular preparations from *Rana pipiens* and from the frog *Dendrobates histrionicus*, which produces the alkaloids [309]. Muscle contractions are potentiated and muscle action potentials prolonged in both species, but blockade of neuromuscular transmission requires higher concentrations in the *Dendrobates* than in the *Rana*. Thus the site of action of histrionicotoxin is retained in the frog that produces the alkaloid. Since histrionicotoxins have relatively low toxicity, dendrobatid frogs probably have little reason to protect themselves from the actions of this class of alkaloid.

Histrionicotoxin at low concentrations blocked carbamylcholine-evoked depolarization in eel electroplax [209,210]. It was more potent at lower pH values. Histrionicotoxin noncompetitively inhibits nicotine-stimulated secretion of catecholamines from adrenal medulla cells [152,154], suggesting similar sites of action for the alkaloid on the nicotinic acetylcholine receptor channel complex in adrenal medulla, in muscle, and in electroplax.

Histrionicotoxins antagonize both acetylcholine and glutamate-elicited excitation of central neurons [315], and have been cited as antagonizing glutamate-responses in invertebrate muscles [305]. Histrionicotoxin also has a depressant effect on spontaneous activity of cortical and spinal neurons [315,332]. Histrionicotoxin blocks acetylcholine-induced currents in cockroach motoneurons [333].

Histrionicotoxin inhibits competitively the binding of ^{125}iodo-α-bungarotoxin to chick brain membranes, but has no effect on potency of carbamylcholine as an inhibitor of binding [334]. The nature of the central receptors that bind α-bungarotoxin is not clearly defined.

Isodihydrohistrionicotoxin at 4 μM has little or no effect on muscarinic responses in ileum or atrium [128]. Histrionicotoxins are, however, noncompetitive muscarinic antagonists in neural cell lines [335], suggestive of interaction with a nonreceptor site, perhaps at a channel component. Perhydrohistrionicotoxin is a very weak antagonist of binding of quinuclidinyl benzilate to muscarinic receptors in rat brain membranes [336]. Decahydro(pentenyl)histrionicotoxin inhibited binding of a muscarinic antagonist to receptors in atrial membranes and solubilized preparations [337].

Recently, preliminary structure activity relationships have been reported for various histrionicotoxins [305]. The onset of the time- and stimulus-dependent blockade of evoked conductances by the alkaloids increases threefold in a series including histrionicotoxin, and dihydro, tetrahydro, octahydro and perhydro derivatives. Thus potency in terms of onset of action apparently decreases with increased saturation in the side chains. A related synthetic compound, namely azaspiro[5.5]undecan-8-ol (**87**, Scheme 33), which has no side chains, has very low activity. The 2-depentyl analog of perhydrohistrionicotoxin is less active than perhydrohistrionicotoxin itself [262,293]. In the *Torpedo* electroplax, histrionicotoxin is slightly more potent than isodihydrohistrionicotoxin in blocking carbamylcholine-elicited depolarization [315]. Quantitative assessment of potencies of histrionicotoxins at acetylcholine receptor channel complexes, at sodium channels (Section 3.3.4), and at potassium channels (Section 3.3.5) is difficult, since the effects are dependent not only on concentration but also on time of exposure and rate of stimulation [305]. The potencies of four histrionicotoxins with respect to inhibition of endplate currents are as follows: isotetrahydrohistrionicotoxin > histrionicotoxin ≃ perhydrohistrionicotoxin > octahydrohistrionicotoxin. Histrionicotoxins increase the rate of closing of channels, but there are insufficient data to provide a reliable rank order for potency. Isotetrahydrohistrionicotoxin appears, however, more potent than octahydrohistrionicotoxin, histrionicotoxin, and perhydrohistrionicotoxin [338]. Azaspiro[5.5]undecan-8-ol (**87**, Scheme 33) is very weak in reducing endplate currents and appears to *decrease* the rate of closing of channels [305]. The latter effect might, however, be due to high concentrations of ethanol, the solvent in which the azaspiro compound was added.

It has been stated that the *dl*-perhydrohistrionicotoxin is *not* significantly less potent in reducing endplate currents and increasing rate of closure of channels than optically active *l*-perhydrohistrionicotoxin obtained by reduction of histrionicotoxin or isodihydrohistrionicotoxin [305]. The chemical literature [264,267] also contains statements suggesting that *dl*-perhydro-histrionicotoxin has biological activity "identical" with that of *l*-perhydro-histrionicotoxin. Studies have now been carried out with *d*- and *l*-enantiomers of perhydrohistrionicotoxin: The natural *l*-enantiomer has a potency identical with or even slightly less than the unnatural *d*-enantiomer [262,293,339]. Furthermore, the two enantiomeric 2-depentylperhydrohistrionicotoxins are equipotent in blocking indirect evoked muscle twitch.

The depentyl analog of perhydrohistrionicotoxin, a *N*-benzyl azaspiro-[5.5]undec-7-ene and related *N*-benzylazaspiroalkenes, appear to interact primarily with the open form of the acetylcholine receptor channel complex [340,341], in contrast to perhydrohistrionicotoxin, which appears to interact with both closed and open forms.

No biological data are available for the 2,7-epimer of perhydrohistrio-nicotoxin, but a corresponding "dioxo"-2,7-epimer of perhydrohistrionico-toxin (side chains: $2\text{-}CH_2CH_2OCH_2CH_3$ and $7\text{-}CH_2OCH_2CH_3$, see Scheme 42) was stated [294] to have "*ca* one fourth the biological activity of the naturally derived perhydrohistrionicotoxin." The activity was ascertained "using murine nerve/diaphragm preparation by Dr. E. X. Albuquerque and associates." A synthetic 2'-pentenyl-7'-butyl analog of histrionicotoxin causes a time-dependent inhibition of responses to iotophoretic acetylcholine in neuromuscular preparations [342].

All of the studies of the various histrionicotoxins indicate that the nature of the side chains and even their configuration on the spiro-ring system are not critical to biological activity. Furthermore, the lack of stereoselectivity for *d*- and *l*-perhydrohistrionicotoxin suggests that only two points of binding of such alkaloids to a histrionicotoxin site on the acetylcholine receptor channel complex are critical.

3.3.3. Binding of Radioactive Perhydrohistrionicotoxin.

A radioactive per-hydro derivative of histrionicotoxin ([^{3}H]H_{12}-HTX) [307,343,344] has proven a very useful tool for investigation of the binding sites for noncompetitive blockers in ray (*Torpedo*) electroplax preparations [304,317,323,336,345–381]. It is likely that [^{3}H]$_{12}$-HTX will not be useful as a ligand in preparations, such as muscle, where the density of nicotinic receptor channel complexes are many fold lower than in *Torpedo* electroplax. Indeed, specific binding was not detected in neuromuscular preparations [307] nor in the electroplax of the Egyptian electric catfish (*Malapterurus electricus*) [212], where the density of nicotinic receptors is relatively low compared to *Torpedo* electroplax. In membranes from *Torpedo* electroplax [^{3}H]H_{12}-HTX binds to a site with an affinity constant (K_D) of about 0.4 μM [343]. In this initial study the density of the binding sites for [^{3}H]H_{12}-HTX was twofold higher than the density of the acetylcholine binding sites in electroplax membranes, but this ratio has varied considerably in different studies. The rate of binding of [^{3}H]H_{12}-HTX is greatly accelerated by the presence of nicotinic agonists although the apparent binding constant and density of sites appear unaffected

[352,365]. Nicotinic antagonists at pharmacologically relevant concentrations do not block binding of [^{3}H]H$_{12}$-HTX, indicating that the histrionicotoxin binding site is not at the antagonist recognition site of the receptor. A potent nicotinic agonist, anatoxin-a, as expected, stimulates binding of [^{3}H]H$_{12}$-HTX to electroplax membranes [354]. A variety of cholinergic agonists including carbamylcholine, succinylcholine, and nicotine increase binding of [^{3}H]H$_{12}$-HTX [363]. Decamethonium and nicotine at higher concentrations markedly inhibit binding. Although carbamylcholine enhances and accelerates binding of [^{3}H]H$_{12}$-HTX, prolonged incubation under desensitizing conditions with carbamylcholine decreases binding [358]. Pyridostigmine enhances [^{3}H]H$_{12}$-HTX binding at low concentrations and inhibits binding at high concentrations [372]. Nereistoxin (4-dimethylamino-1,2-dithiolane), a partial agonist-antagonist of the cholinergic receptor, has no effect on binding of [^{3}H]H$_{12}$-HTX [362]. Tubocurarine at low concentrations increases binding of [^{3}H]H$_{12}$-HTX, but inhibits binding at higher concentrations [378]. This suggests that this classical cholinergic antagonist also exhibits both agonist activity and channel effects. Aliphatic alcohols increase binding of [^{3}H]H$_{12}$-HTX [359].

The binding sites for [^{3}H]H$_{12}$-HTX were first thought to be separable from acetylcholine receptor sites [344] and to be mainly associated with a 43,000 dalton subunit of the acetylcholine receptor channel complex [382,383]. This subunit was subsequently found to be nonessential to either binding of [^{3}H]H$_{12}$-HTX [371] or function of the receptor channel complex [384]. Some studies indicate that [^{3}H]H$_{12}$-HTX binds mainly to the 65,000 dalton (δ) subunit [323,324], but more recent results suggest that functional sites of action of "noncompetitive blockers" may be at the α- or β-subunits [325,326]. Irreversible binding of [^{3}H]H$_{12}$-HTX after intense photolysis has been reported [323] to occur mainly to the 65,000 (δ) dalton subunit of the acetylcholine receptor complex. Antibodies to acetylcholine receptor had no effect on binding of [^{3}H]H$_{21}$-HTX, but did inhibit the effects of carbamylcholine on binding [373].

Binding of [^{3}H]H$_{12}$-HTX to electroplax membranes is antagonized by a variety of compounds that have activity as noncompetitive blockers of the acetylcholine receptor channel complex. These include local anesthetics, amantidine, phencyclidine, quinacrine, antidepressants, phenothiazines, and tetraethylammonium ions [317,336,343–348,350–352,357,367,368,370,371, 374–377,379]. The potency of many of these compounds as antagonists of [^{3}H]H$_{12}$-HTX binding is increased by carbamylcholine. This is not true for a few compounds, such as tetracaine, piperocaine, and amantidine, whose potency is reduced by carbamylcholine (see [183] for a summary of data). The potencies of pyrethroids as noncompetitive antagonists of [^{3}H]H$_{12}$-HTX binding did not correlate with their potency as insecticides [355,356]. Electrophysiological studies in neuromuscular preparations suggest that many of the noncompetitive blockers have actions similar to those of the histrionicotoxins. As yet, the histrionicotoxins are the most potent ligands for such channel sites. The potency of four histrionicotoxins versus binding of [^{3}H]H$_{12}$-HTX appears as follows: perhydrohistrionicotoxin > isotetrahydrohistrionicotoxin > octahydrohistrionicotoxin > histrionicotoxin [360]. This rank ordering does *not* correspond to the rank order with respect to reduction of indirect elicited twitch or of endplate currents in frog neuromuscular preparations (see Section 3.3.2). Azaspiro[5.5]undecan-8-ol (**87** Scheme 33) is a very weak antagonist to binding of [^{3}H]H$_{12}$-HTX. Perhydrohistrionicotoxin antagonizes the binding of a radioactive local anesthetic, meproadifen [385], and of radioactive phencyclidine [368] to electroplax membranes.

Recently, the effects of histrionicotoxin and 22 analogs on binding of [^{3}H]H$_{12}$-HTX and of [^{3}H]phencyclidine to electroplax membranes have been reported [304]. Histrionicotoxin is threefold more potent versus binding of [^{3}H]phencyclidine than versus binding of [^{3}H]H$_{12}$-HTX. All of the other histrionicotoxin analogs are also more potent versus [^{3}H]phencyclidine than versus [^{3}H]H$_{12}$-HTX. Carbamylcholine increases potency of histrionicotoxin as an antagonist versus [^{3}H]H$_{12}$-HTX binding by about threefold, and increases the potency of certain analogs by as much as eight- to tenfold. *O*-Acetylation and *N*-methylation of histrionicotoxin, although not affecting potency in the absence of carbamylcholine, do enhance selectivity for the carbamylcholine-activated state. Variations in the side chains had marked effects on potency and can either reduce or enhance selectivity for the carbamylcholine-activated state. The binding data do not correlate well with physiological data, perhaps because binding data reflect composite interactions of these alkaloids with sites on open, closed, inactive, and desensitized states of the complex. The most potent alkaloid in the binding assay is isotetrahydrohistrionicotoxin. Carbamylcholine has little effect on its potency. The most selective alkaloids for the carbamylcholine-activated state are *O*-acetylhistrionicotoxin, in the case of [^{3}H]H$_{12}$-HTX binding, and *N*-methyl-*O*-acetylhistrionicotoxin, in the case of [^{3}H]phencyclidine binding. The methiodide of *O*-acetylhistrionicotoxin has very low activity, while the methiodide of depentylperhydrohistrionicotoxin retains moderate activity. Other dendrobatid alkaloids, namely gephyrotoxin [380] and pumiliotoxin C [381], antagonize binding of [^{3}H]H$_{12}$-HTX.

[^{3}H]H$_{12}$-HTX binds to low affinity, high density, relatively heat-stable sites in chick brain synaptic membranes [334]. Binding is inhibited by high concentrations of local anesthetics and phencyclidine. The nature of these sites are unknown.

Histrionicotoxin (Ki 6 μM) inhibits binding of [^{3}H]phencyclidine to sites in brain membranes [386]. These sites may represent those responsible for the central effects of phencyclidine. The central sites of action of phencyclidine may be associated with one class of potassium channels [387].

Histrionicotoxin at 10 μM enhances binding of radioactive nitrendipine to brain membranes [388]. It shares this property with phencyclidines and certain other drugs.

3.3.4. Sodium Channels.

The histrionicotoxins show only weak interactions with voltage-dependent sodium channels, thereby reducing the rate of rise of action potentials in muscle [305]. The profile of activity for seven histrionicotoxins in this regard is as follows: isotetrahydrohistrionicotoxin > tetrahydrohistrionicotoxin > perhydrohistrionicotoxin > isodihydrohistrionicotoxin > neodihydrohistrionicotoxin > octahydrohistrionicotoxin > histrionicotoxin. Thus histrionicotoxin appears to be the least potent alkaloid of this class with respect to classical local anesthetic activity. In frog sciatic nerve, histrionicotoxin at 500 μM causes only a slight blockade of action potentials [332]. Histrionicotoxin at high concentrations blocks, in a stimulus-dependent manner, directly evoked action potentials in the eel electroplax [210,211]. In this action, in contrast to the more effective blockade of carbamylcholine responses that occurs at lower pH values, histrionicotoxin was more effective at higher pH values. Histrionicotoxin at high concentrations antagonizes batrachotoxin-elicited depolarization of electroplax. Histrionicotoxin inhibits binding of [^{3}H]BTX-B to sodium channels in brain preparations with an IC$_{50}$ value of 17 μM [183]. Perhydrohistrionicotoxin is much more potent with an IC$_{50}$ value of about 0.3 μM.

3.3.5. Potassium Channels.

Histrionicotoxins at low concentrations in a time- and stimulus-dependent manner block conductances through the potassium channels responsible for termination of action potentials in nerve and muscle [305,306]. Delayed rectification, a reliable indicator of increased potassium conductances, is blocked by histrionicotoxin in frog muscles. The antagonism of potassium channels by histrionicotoxins results in a marked prolongation of action potentials. The profile of activity for eight histrionicotoxins with respect to prolongation of muscle action potentials is as follows: tetrahydrohistrionicotoxin > histrionicotoxin > neodihydrohistrionicotoxin > isodihydrohistrionicotoxin $\simeq$ perhydrohistrionicotoxin > N-methylperhydrohistrionicotoxin [306]. Presynaptic prolongation of action potentials by histrionicotoxins will result in greater release of neurotransmitter. Postsynaptically, the resultant enhanced amount of released acetylcholine and the histrionicotoxin-induced prolongation of muscle action potentials probably accounts for the initial potentiation of muscle contractures seen during stimulation in the presence of histrionicotoxins. Soon, however, the stimulus-dependent blockade of acetylcholine-receptor complexes ensues and nerve stimulation no longer elicits a response. However, direct stimulation of muscle still causes an enhanced contraction due probably to the prolonged action potentials in muscle. Potentiative effects of histrionicotoxin on directly elicited muscle contractions are reversible after washing, but the reversal appears to occur on a slower time course than does the reversal of the histrionicotoxin-elicited blockade of the acetylcholine receptor channel complex. The relative potency of histrionicotoxins with respect to potentiation of muscle contracture has not been quantitated. Perhydrohistrionicotoxin is very weak with respect to prolongation of action potentials and causes only a marginal potentiation of muscle contracture [305]. Thus the two effects appear to be correlated for this analog.

3.3.6. Summary. The various studies on pharmacological activity of histrionicotoxins are summarized in Table 5. *This class of dendrobatid alkaloids are not really "toxins," since even at relatively high dosages in mammals, death is not elicited* [23]. There are several reviews related entirely or in part to the pharmacology and mechanism of action of histrionicotoxins [193,309,321,338,349,360,392–395].

3.4. Addendum

The 7-debutyl analog of **110** (Scheme 36) and its 8-epimer resulted in a 4:1 mixture (42%) from a modified Speckamp procedure using glutarimide and

Table 5. Effects of Histrionicotoxins on Biological Systems

System	Species	Comments	References
Whole animal	Mouse	Relatively nontoxic	[23,245]
Neuromuscular preparation (striated)	Frog, rat, mouse	Blockade of neuromuscular transmissioin and either stimulus or iontophoretically evoked endplate currents; enhancement of rate of closing of endplate channels; prolongation of action potentials; potentiation of directly evoked muscle contraction	[128,262,293, 305–312,315, 332,338–343, 365,389,390]
Muscle cells	Chick	Enhancement of nicotinic agonist–induced desensitization	[314,318]
Neuronal cells	Mouse neuroblastoma, mouse neuroblastoma–rat glioma hybrid	Noncompetitive inhibition of binding of muscarinic antagonist	[335]

Table 5. (*continued*)

System	Species	Comments	References
Nerve	Insect	Blockade of acetylcholine-induced currents	[333]
Cardiac preparations	Guinea pig atria	No effect on spontaneous activity or on carbamylcholine responses	[128]
	Pig atrial membranes	Competitive inhibition of binding of muscarinic antagonist	[337]
Smooth muscle	Guinea pig ileum	No effect on tension or on acetylcholine-elicited contracture	[128]
Adrenal medulla cells	Bovine	Noncompetitive inhibition of nicotine-elicited release of catecholamines	[152,154]
Brain	Cat	Depression of spontaneous activity cortical and spinal neurons; blockade of excitatory responses to acetylcholine and glutamate	[315,332]
	Rat synaptosomes	No effect on sodium conductances	[94]
	Rat membranes	Virtually no effect on binding of muscarinic antagonist	[336]
		Inhibition of binding of phencyclidine	[386]
		Enhancement of binding of nitrendipine	[388]
	Chick membranes	Inhibition of binding of α_1-bungarotoxin; binding of $[^3H]H_{12}$-HTX	[234]

Table 5. (*continued*)

System	Species	Comments	References
Electroplax	Ray membranes, microsacs	Inhibition of binding of certain local anesthetics, and fluorescent probes; blockade of carbamylcholine-elicited ion flux and depolarization; increased affinity of agonist for receptor	[313,315,316, 319,320,322, 324,326–328, 330,331,382, 383,391]
		$[^3H]H_{12}$-HTX: inhibition of binding by local anesthetics, phencyclidine, benzomorphans, tricyclic antidepressants, amantadine, phenothiazines, pyrethroids, pumiliotoxin C, gephyrotoxin and other agents; enhancement and acceleration of binding by cholinergic agonists and alcohols	[304,317,323, 336,343–380]
	Fish	Binding $[^3H]H_{12}$-HTX not detectable	[212]
	Eel	Blockade of carbamylcholine-evoked depolarization and of directly evoked action potentials	[209,210]

4-pentenyl magnesium bromide. The major formate after hydrolysis and reduction afforded **87** (Scheme 33); the minor formate gave the previously
unreported *trans* amino alcohol diastereomer. The 7-debutyl analog of **5**
(Scheme 14) arising by formate hydrolysis was converted via thiolactam,
iminothioether, and 2-alkylimine intermediates to a 3:2 mixture of ($\pm$)7-
debutylperhydrohistrionicotoxin and its 2-pentyl epimer [W. Gessner, K.
Takahashi, B. Witkop, A. Brossi, and E. X. Albuquerque, *Helv. Chem.
Acta*, **68**, 49 (1985)].

4. GEPHYROTOXINS

4.1. Structures

One of the major alkaloids from *Dendrobates histrionicus*, first described
in 1974 as "HTX-D" [247], ultimately was found not to be a histrionicotoxin,
but instead a unique tricyclic alkaloid [248]. After elucidation of its structure,
it was named gephyrotoxin. The name is derived from the Greek *gephyra*,
meaning bridge, and literally refers to the "bridge" that has been presumed
to be formed biosynthetically by addition of the nitrogen to one of two side
chains of a proposed parent 2,6-disubstituted piperidine (see [24] and Section
11). This bridge forms a bicyclic indolizidine ring system. A further biosynthetic cyclization—analogous to the cyclization that has been presumed to
form the decahydroquinoline pumiliotoxin C from a 2,6-disubstituted piperidine—would then form the tricyclic gephyrotoxin. The term *gephyra*
could also, in the case of the parent compound, be considered to refer to
the fact that this alkaloid bridges various classes of dendrobatid alkaloids
by having the vinylacetylene side chain of the histrionicotoxins, a decahydroquinoline moiety like pumiliotoxin C and in retrospect, an indolizidine
moiety like the pumiliotoxin-A class. The term gephyrotoxin has been applied both to simple bicyclic indolizidines and to the parent tricyclic perhydrobenzoindolizidines [23,24]. In view of the apparent large number and
variety of simple indolizidines in dendrobatid frogs (see Section 5) it now
appears preferable to refer to only the tricyclic compounds as gephyrotoxins
and to refer to the bicyclic compounds as indolizidines.

Gephyrotoxin, isolated in early studies from extracts of one population
of *Dendrobates histrionicus*, had an empirical formula of $C_{19}H_{29}NO$, which
was identical with that of certain congeneric histrionicotoxins [247]. Analysis
of the proton magnetic resonance spectra revealed the presence of a *cis*-
$CH_2CH{=}CHC{\equiv}CH$ side chain, like that in histrionicotoxin. However, the
mass spectrum of gephyrotoxin, referred to at that time as HTX-D, was
distinctively different from those of the histrionicotoxins. The base peak
resulted from the loss of what proved to be a $-CH_2CH_2OH$ moiety. This
moiety allowed formation of *O*-acetyl and *O-p*-bromobenzoyl derivatives.
Hydrogenation of the compound led to a hexahydro derivative containing a

saturated five-carbon side chain and three rings. Finally, X-ray analysis of
a crystal of the hydrobromide salt revealed its structure as that of a unique
tricyclic alkaloid [248] and the name gephyrotoxin was coined (see above).
The absolute configuration was assigned from the X-ray analysis
as [1S,3aS,5aS,6S(Z),9aR,10R]dodecahydro-6-(2-penten-4-yl)pyrrolo[1,2-a]-
quinoline-1-ethanol (Fig. 4A). This absolute configuration has recently been
questioned, based on a synthesis of this enantiomer from L-glutamate [396,
see Section 4.2]. The synthetic compound was dextrorotatory, while the
natural compound was levorotatory. Thus the structure of natural gephyrotoxin, based on synthesis, would be as in Fig. 4B. No satisfactory explanation for the contradictory X-ray and chemical data is apparent and there
remains, therefore, a question as to the absolute configuration of natural
gephyrotoxin. Consideration of the known absolute stereochemistry of other
dendrobatid alkaloids, such as histrionicotoxin (Fig. 2), indolizidine **223AB**
(Figure 5), and pumiliotoxin C (Figure 11) does not allow an unambiguous
choice between the two possible configurations for gephyrotoxin (Fig. 4A
and B). Three explanations can be considered:

1. The X-ray crystallographic analysis might have led to an incorrect
   absolute stereochemistry. However, no errors could be found in the
   analysis [I. Karle, personal communication]. The agreement factor
   R is 8.1% for the absolute configuration assigned and 10.5% for the
   mirror image [248]. Gephyrotoxin is a very labile compound and successful crystallization has not been repeated and the original crystals
   are no longer satisfactory.

2. There might have been an inversion of configuration at some early
   stage of the synthesis. This does not, however, appear likely, and the
   synthesis was repeated to ensure that a minor isomeric by-product
   was not used at any stage [Y. Kishi, personal communication].

3. Frogs from the same population might have produced some d-enantiomer in the early 1970s. If so, a crystal of the d-enantiomer might

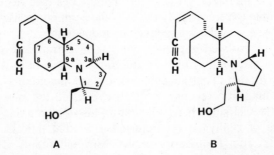

A   B

Figure 4. Structure of natural gephyrotoxin (**287C**). (A) Based on X-ray analysis
[248]. (B) Based on synthesis of d-gephyrotoxin [396]. Natural gephyrotoxin was
found to be levorotatory (see text).

Table 6. **Physical and Spectral Properties of Gephyrotoxin [248,396].**

Gephyrotoxin	
m.p. 231–232°	
Ultraviolet (C₂H₅OH)	λ_{max} 225 nm, ϵ 8400
Infrared (HCCl₃)	2120 cm^{-1} (acetylene)
Synthetic d-enantiomer	
Optical rotation (c .0, C₂H₅OH)	$[\alpha]_D^{25}$ +50.0°
Natural l-enantiomer	
Optical rotation (c 1.0, C₂H₅OH)	$[\alpha]_D^{25}$ −51.5°

have been used for X-ray analysis. The crystals of gephyrotoxin for X-ray analysis were from alkaloid isolated from 1100 skins of *Dendrobates histrionicotoxin* obtained in 1971, and no optical rotation was measured. Optical rotation values were measured on gephyrotoxin isolated from extracts obtained from 3200 skins of the same population of *D. histrionicus* obtained in 1974, and were equal but opposite in sign to the rotation of synthetic *d*-gephyrotoxin (see Table 6).

Finally, it should be noted that gephyrotoxin is somewhat anomalous among dendrobatid alkaloids of defined structures, regardless of its absolute configuration. The others, which include the histrionicotoxins, pumiliotoxin C, and the indolizidine **223AB**, can be considered to be derived from a 2,6-disubstituted piperidine with **2R,6S** or **2S,6R** stereochemistry. Gephyrotoxin must be considered to be derived from a 2,6-disubstituted piperidine with **2R,6R** or **2S,6S** stereochemistry depending on whether X-ray or chemistry, respectively, is correct.

A minor congener, dihydrogephyrotoxin, was also isolated from *Dendobates histrionicus* and differed only in the presence of a *cis*-CH₂CH=CHCH=CH₂ side chain. Hydrogenation of these gephyrotoxins yielded perhydrogephyrotoxin. Properties of these two alkaloids are reported below in the standard format for dendrobatid alkaloids (see Section 3.1).

Gephyrotoxin Class.   (Perhydrobenzoindolizidines = dodecahydropyrrolo[1,2-a]quinolines.)

287C.   Gephyrotoxin, C₁₉H₂₉NO, 0.20, 218°, *m/z* 287 (5), 286 (3), 242 (100), 222 (45), 122 (14). H₆ derivative, *m/z* 293 (5), 292 (30), 250 (16), 248 (100), 222 (32). 1D. *O*-Acetyl derivative, *m/z* 329 (3), 264 (45), 242 (100). Structure: Fig. 4.

289B.   Dihydrogephyrotoxin, C₁₉H₃₁NO, 0.25, 217°, *m/z* 289 (4), 288 (3), 244 (100), 222 (49). H₄ derivative. 1D.

Further physical and spectral properties of gephyrotoxin are in Table 6. The mass spectra of gephyrotoxin, dihydrogephyrotoxin, and perhydrogephyrotoxin have been presented in detail [24,247,248]. The proton magnetic resonance spectrum of natural gephyrotoxin and of synthetic *d*-gephyrotoxin are presented in [247] and in the microfilm supplement of [396], respectively. The spectrum of *dl*-perhydrogephyrotoxin is presented in the microfilm supplement of [397]. Proton magnetic resonance spectral assignments are reviewed in [24]. The carbon-13 magnetic resonance peaks of perhydrogephyrotoxin are reported in [398].

The tricyclic gephyrotoxins have a very limited distribution in dendrobatid frogs (see Section 11).

4.2. Syntheses

The syntheses of the tricyclic gephyrotoxins have been furthered by synthetic approaches developed earlier for histrionicotoxin (Section 3.2) and pumiliotoxin C (Section 7.2). Among these are the 2-azonia[3,3]sigmatropic rearrangement, Diels-Alder reactions with 1-carbobenzyloxyaminobutadiene and 1,3-bistrimethylsilyloxybutadiene and the acyliminium ion cyclization. Because the C(3a) configuration of gephyrotoxin, relative to the decahydroquinoline ring-junction hydrogens, is (**E**) and not (**Z**) as with the equivalent C(2) of pumiliotoxin C, it was much more difficult to create the natural chirality at C(3a), subsequent to forming the *cis*-fused ring junction. Reductions, for example, would have to proceed from the more congested concave face.

4.2.1. Stereoselective Hydrogenation. In the first synthesis of gephyrotoxin, Kishi and colleagues [399] circumvented the problem of the configuration at C(3a) by forming the A-B ring junction subsequent to building the B-C ring (indolizidine) system. The syntheses of perhydrogephyrotoxin and gephyrotoxin [399] also illustrate the imaginative use of a remote side chain hydroxyl group to direct two stereoselective catalytic hydrogenations, whereby three contiguous chiral centers [C(9a), C(5a), C(6)] are created (Scheme 55, steps *n,u*). In step *n*, by the adsorption of the hydroxyethyl group of **1** to the catalyst surface (a "haptophilic" effect), hydrogen is added preferentially to the enone system from that same molecular face producing **2** (see Section 4.2.2 for another approach to this intermediate), while in step *u*, after conversion of the hydroxyl group to a nonpolar bulky silyl ether (*t*), the addition of hydrogen to the exo olefinic bond proceeds from the opposite face with 9:1 stereoselectivity. The two remaining chiral centers of gephyrotoxin were also created by catalytic hydrogenation (*d*) [399]. After oxidation of the alcohol **3** to the aldehyde **4** with pyridinium chlorochromate (*w*), steps of Wittig olefination, reduction, and desilylation with fluoride, provided perhydrogephyrotoxin in 9% overall yield. The aldehyde **4** was

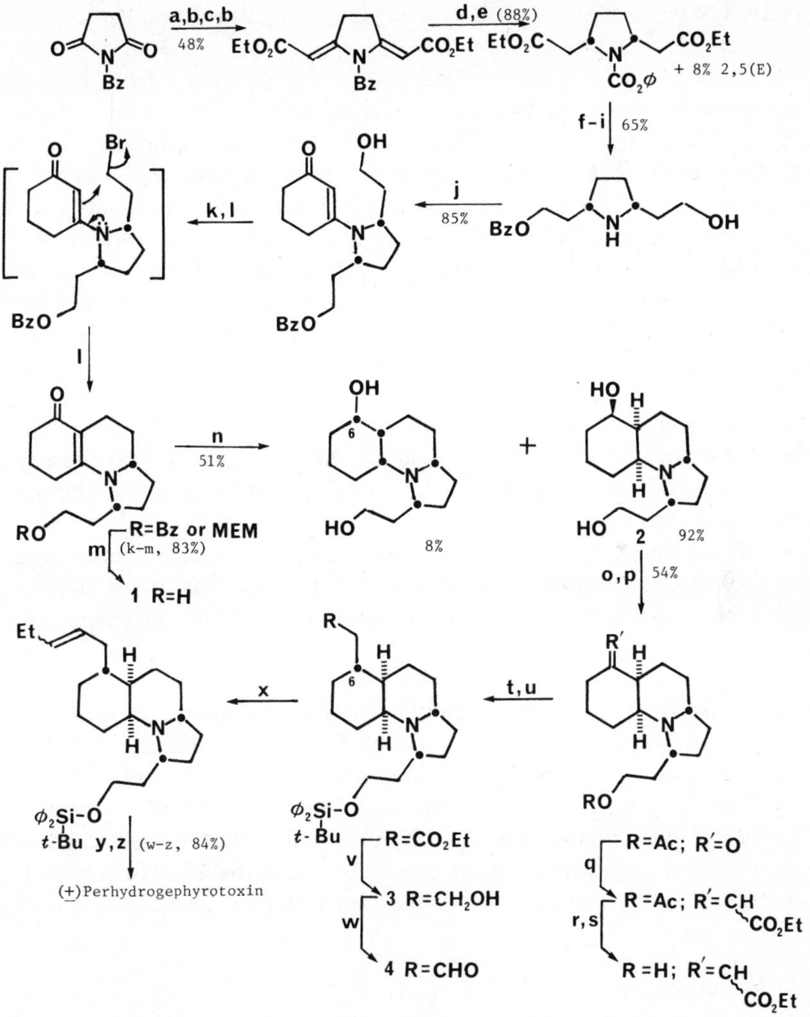

Scheme 55. The synthesis of perhydrogephyrotoxin [399]. (*a*) ClMgC≡COEt/THF; (*b*) 5% HCl, 0°; (*c*) BrMgC≡COEt/THF; (*d*) H₂ (4 atm.)/Pd-C/MeOH-HClO₄; (*e*) φCOCl/Py/CH₂Cl₂, chromatographic separation; (*f*) LiBH₄/THF; (*g*) KH/THF; (*h*) φCH₂Br/DMF; (*i*) Ba(OH)₂/H₂O, Δ; (*j*) cyclohexan-1,3-dione, TsOH·Py/φH, Δ; (*k*) MsCl/Et₃N/CH₂Cl₂; (*l*) LiBr/DMF; (*m*) H₂/Pd-C/MeOH-HClO₄; (*n*) H₂ (4 atm.)/Pt-Al₂O₃/EtOAc (12-deoxy, 12-acetyl, or 12-trimethylsilyl derivatives of **1** were *not* reduced under these conditions; 19% of a 6-deoxy hydrogenolysis product also accompanies the indicated products); (*o*) Ac₂O; (*p*) DMSO-(COCl)₂/Et₃N, −65°; (*q*) ClMgC≡COEt/THF; (*r*) MeMgBr; (*s*) aq. HCl; (*t*) *t*-Buφ₂SiCl/imidazole/DMF; (*u*) H₂/Rh-Al₂O₃/hexane, −20°; (*v*) LiAlH₄; (*w*) PCC/CH₂Cl₂; (*x*) EtCH=Pφ₃/THF, 0°; (*y*) H₂ (4 atm.)/Rh-Al₂O₃/EtOH; (*z*) *n*-Bu₄N⊕F⊖/DMF.

converted to gephyrotoxin itself in 45% yield using the Corey-Ruden method (Scheme 56) (5% overall).

Using the 2,5-*cis* substituted pyrrolidine **6** synthesized from *L*-pyroglutamic acid **5** as shown in Scheme 57 and the sequences shown in Schemes 55 and 56, Fujimoto and Kishi [396] carried through an asymmetric synthesis of gephyrotoxin. The absolute configuration of natural gephyrotoxin had been as determined by X-ray diffraction [248] and is that indicated as **7**. The portion derived from the pyrrolidine chiral synthon with its two chiral centers, C(1) and C(3a), is emphasized and its relationship to natural L-(S)pyroglutamic acid is indicated (Scheme 57). Natural levorotatory gephyrotoxin accordingly should have the same S-configuration at C(1) as the α-carbon of natural pyroglutamic acid. After conversions of the optically active pyrrolidine **6** to the analogous 1,3-cyclohexanedione adduct (Scheme 55, step *j*), ring formation (*k,l*), and deblocking to **1**, this intermediate was carried through the steps of Schemes 55 and 56. Surprisingly, dextrorotatory gephyrotoxin resulted, implying that the absolute configuration of the natural gephyrotoxin must be opposite to that determined by X-ray analysis; that is, the mirror image of **7**. This conclusion assumes, as seems likely, that no chiral interchange occurred during one of the 17 steps involved in the synthesis of **6** (e.g., steps *o* or *p*) (see Section 4.1 for a discussion of this paradox).

4.2.2. Diels-Alder Additions.

The Diels-Alder approach, so useful in synthesis of pumiliotoxin C, was applied by Overman and Freerks [397] to a practical 13-step synthesis of perhydrogephyrotoxin (~15% overall yield) as outlined in Scheme 58. In step *c*, the initial presence of trifluoroacetic acid prevents the formation of the imine by protonation of the deblocked amine function. Subsequent basification generates the imine **15**, which is reduced preferentially from the less accessible concave face by lithium aluminum

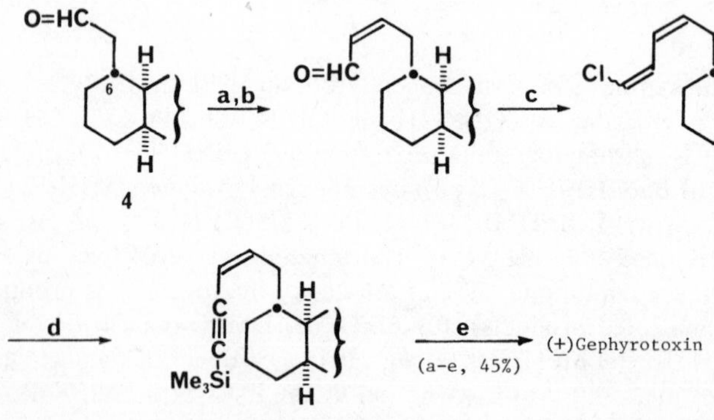

Scheme 56. The synthesis of (±)-gephyrotoxin [399]. (*a*) EtOCH=CHPφ$_3^{\oplus}$Br$^{\oplus}$/ NaOEt; (*b*) TsOH/aq. Me$_2$CO, 0°; (*c*) ClCH$_2$Pφ$_3^{\oplus}$Cl$^{\ominus}$/*n*-Buli/THF; (*d*) MeLi/THF then TMS—Cl; (*e*) *n*-Bu$_4$N$^{\oplus}$F$^{\ominus}$/DMF.

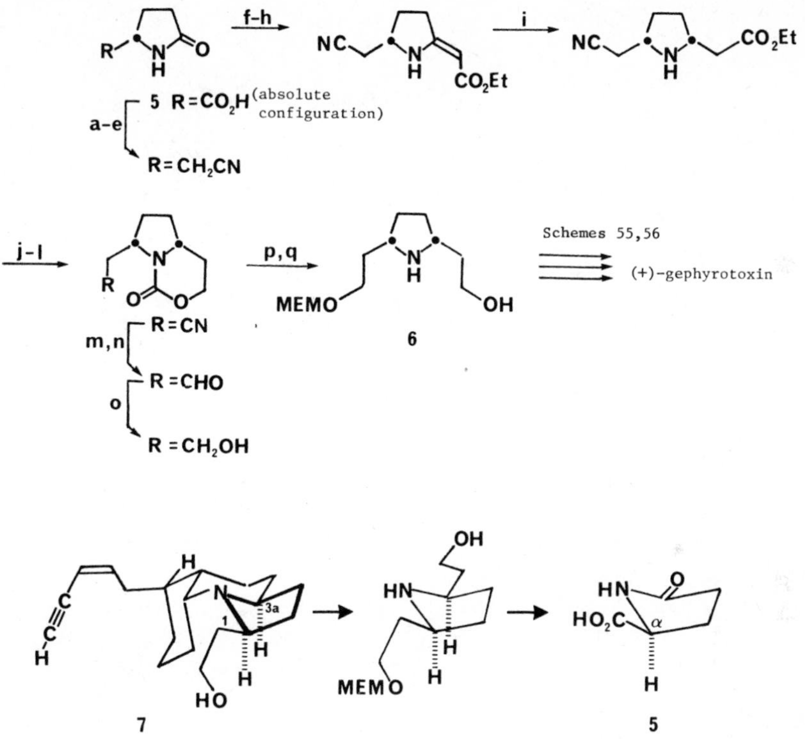

Scheme 57. Enantiospecific synthesis of (+)-gephyrotoxin [396]. (*a*) SOCl$_2$/EtOH; (*b*) LiBH$_4$; (*c*) aq. HOAc; (*d*) ϕ_3P, CBr$_4$/CH$_3$CN; (*e*) KCN-Al$_2$O$_3$/ϕCH$_3$; (*f*) P$_2$S$_5$/ Py, 80°; (*g*) AcCH(Br)CO$_2$Et, NaHCO$_3$, Δ; (*h*) KOH/EtOH, 60°; (*i*) H$_2$/Pd-C/MeOH-HClO$_4$; (*j*) ϕCOCl/Py/CH$_2$Cl$_2$; (*k*) LiBH$_4$/THF; (*l*) KH/THF; (*m*) Al(*i*-Bu)$_2$H/THF-ϕCH$_3$, -105°; (*n*) 3*N* HCl; (*o*) NaBH$_4$/DME; (*p*) MeOCH$_2$Br(MEM—Br)/(*i*-Pr)$_2$NEt/ CH$_2$Cl$_2$; (*q*) Ba(OH)$_2$/H$_2$O, Δ.

hydride at low temperature (*d*). The high stereoselectivity (9:1) in this key step is considered a result of the stereoelectronic requirement that the hydride nucleophile and the developing lone pair on nitrogen be in a *trans*-diaxial orientation. Conformation **16**, in which steric interactions between C(9) and the nitrogen complexing group are minimized, probably best represents the reactive intermediate. Other key steps are the oxidation (*h*) of the enol silyl ether to the *trans*-enal by Saegusa and Ito's procedure and the highly stereoselective intramolecular Michael cyclization (*l*) producing **17**.

Similar methodology was used by Overman and colleagues [400] in a synthesis of gephyrotoxin itself (Scheme 59). The ready epimerization of the formyl group in **18** necessitated a low temperature olefination (*b*) with the lithium salt of the phosphonate. An improved synthesis (*g*) of the enal function was developed using (formylmethylene)triphenyl phosphorane. The same sequence (*j,k,l*) as used for perhydrogephyrotoxin effected ring closure

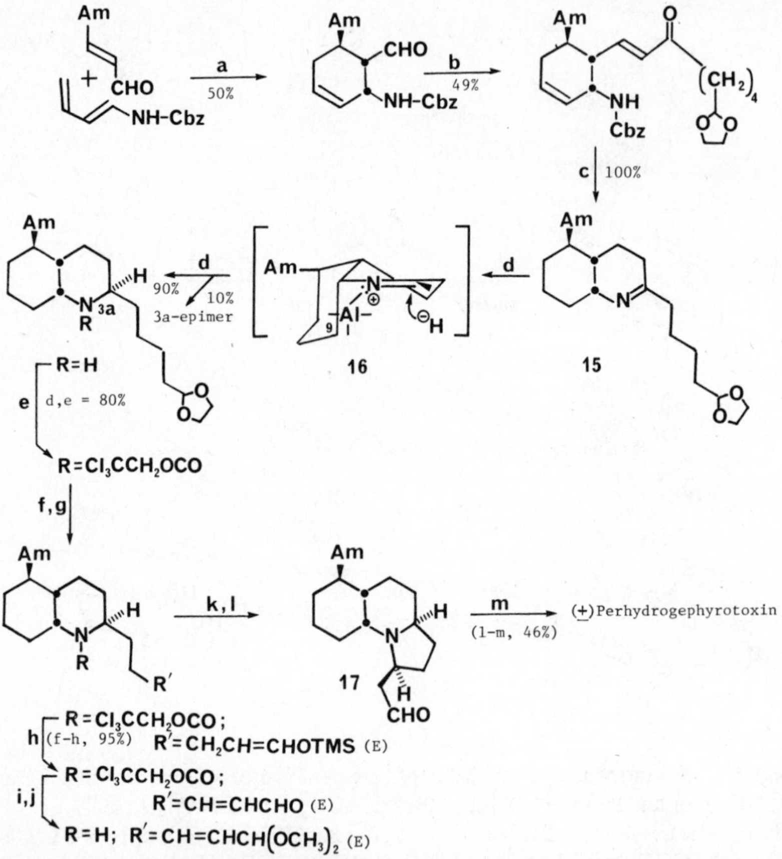

Scheme 58. A Diels-Alder approach to perhydrogephyrotoxin [397]. (*a*) 100°, 3 hr;
(*b*) (MeO)$_2$P(=O)CH$_2$CO(CH$_2$)$_4$CH(OCH$_2$)$_2$/THF, −50° → 25°; (*c*) H$_2$/Pd-C/EtOAc,
TFA then OH$^\ominus$; (*d*) LiAlH$_4$/Et$_2$O, −15°; (*e*) Cl$_3$CCH$_2$OCOCl/1,2,2,6,6-pentame-
thylpiperidine/CCl$_4$; (*f*) 1*N* HCl/THF-HOAc; (*g*) TMS—OSO$_2$F/(*i*-Pr)$_2$NEt/φCH$_3$;
(*h*) Pd(OAc)$_2$ (1.5 eq.)/CH$_3$CN-DMF; (*i*) TsOH·Py/MeOH; (*j*) Zn-Pb couple/
NH$_4$OAc/aq. THF; (*k*) 1*N* HCl (2 eq.); (*l*) NaOMe (20 eq.)/MeOH; (*m*) NaBH$_4$.

and reduction. After blocking the resulting hydroxyl group with the acid-
stable diphenyl tertiary-butylsilyl group (*m*), the methoxymethylene (MEM)
group was removed with acid (*n*), the liberated hydroxyl group was oxidized
to the aldehyde using Swern's reagent (*o*), and the Corey-Rücker procedure
(*p*) (see Section 3.2.7) was used to introduce the *cis*-enyne side chain. Fluoride-ion deblocking of both silyl groups provided gephyrotoxin.

Ito and colleagues [401] recently have utilized an intramolecular Diels-
Alder reaction for a short and efficient synthesis of the Kishi intermediate
1. Their key intermediate, a reactive quinone methide *N*-alkylimine **10**, is

generated *in situ* by a fluoride-ion induced 1,4-elimination (Scheme 60). The bistrimethylsilyl intermediate **9** is prepared in two steps from **8**, available in turn (69%, three steps) from commercial materials. The cycloaddition proceeds with good stereoselectivity to a 5:1 mixture of diastereomers **11** and **12**, whose separation is deferred until after *e*. Birch reduction of this mixture failed, but a high pressure hydrogenation (*d*) succeeded in delivering the enone mixture **13** and **14** and two minor by-products (<10%) of over-reduction. Another catalytic reduction step (*e*) provided the diol mixture from which the desired stereoisomer, **1**, could be separated [401]. The Kishi intermediate **1** has been converted to gephyrotoxin in 14 steps [399].

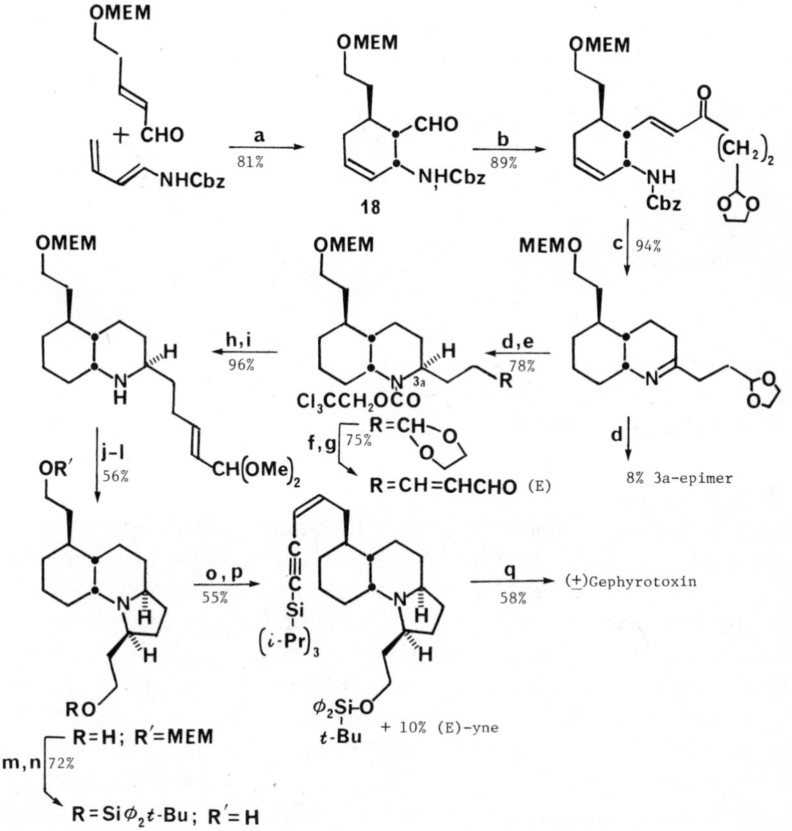

Scheme 59. The synthesis of gephyrotoxin by a Diels-Alder route [400]. (*a*) 110°, 2.5 hr; (*b*) (MeO)$_2$P(=O)CH$_2$CO(CH$_2$)$_2$CH(OCH$_2$)$_2$, LiN(TMS)$_2$, −70°; (*c*) H$_2$/Pd-C/EtOAc, TFA then OH$^\ominus$; (*d*) LiAlH$_4$/Et$_2$O, −19°; (*e*) Cl$_3$CCH$_2$OCOCl/1,2,2,6,6-pentamethylpiperidine; (*f*) HClO$_4$-THF; (*g*) ϕ_3P=CHCHO/CHCl$_3$, Δ; (*h*) TsOH·Py/MeOH; (*i*) KOH/aq. *i*-PrOH; (*j*) 1N HCl/THF; (*k*) NaOMe/MeOH; (*l*) NaBH$_4$/MeOH; (*m*) *t*-Buϕ_2SiCl/Et$_3$N/DMAP/CH$_2$Cl$_2$; (*n*) HBr/DME, 50°; (*o*) DMSO-(COCl)$_2$; (*p*) (*i*-Pr)$_3$SiCH(Li)C≡CSi(iPr)$_3$, −78°; (*q*) (*n*-Bu)$_4$N$^\oplus$F$^\ominus$/DMF.

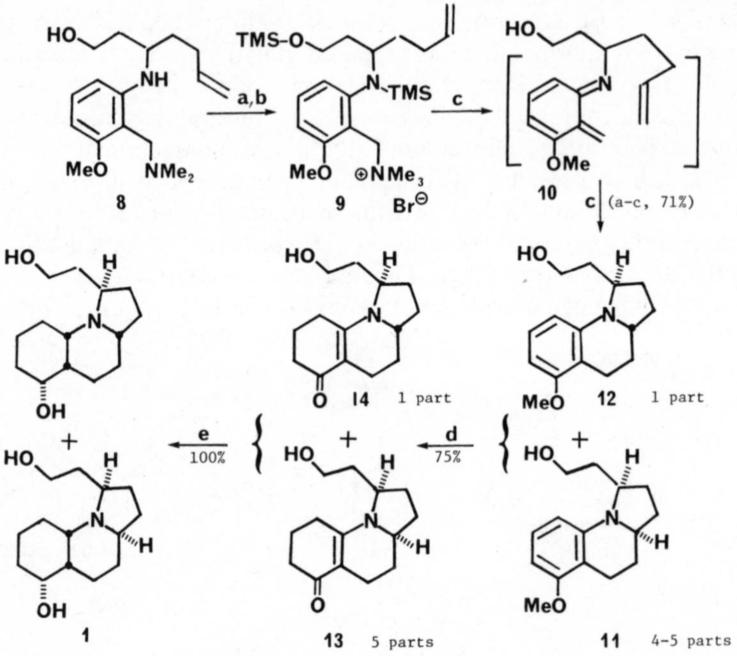

Scheme 60. Synthesis of the Kishi intermediate **1** by an intramolecular Diels-Alder reaction [401]. (*a*) *n*-BuLi/THF-hexane then TMS—Cl/THF (DMAP catalyst); (*b*) MeBr/CH₃CN, 0° → r.t.; (*c*) CsF/CH₃CN, 65° then Δ; (*d*) H₂ (70 atm.)/Rh-Al₂O₃/ EtOH; (*e*) H₂/Pt-Al₂O₃/EtOAc.

4.2.3. Sigmatropic Rearrangement.

Overman's initial approach to per-hydrogephyrotoxin [398], although less satisfactory than his later synthesis [397], demonstrated the elegant and practical use of the 2-azonia[3,3]-sigmatropic rearrangement coupled with the Diels-Alder reaction to generate expeditiously four of the five chiral centers of the alkaloid (Scheme 61). 1-Carbobenzyloxy aminobutadiene and 4-benzyloxycrotonal provided the Diels-Alder adduct **19** with only 10% *exo*-isomer. This was converted to the key intermediate **20** by reductive alkylation (*e,f*) with 2-methyl-2-methoxy but-3-enal (three steps from pyruvaldehyde dimethyl acetal). After hydrolysis of the acetal, the reactive iminium salt **21** forms *in situ* and undergoes a [3,3]sigmatropic rearrangement stereospecifically on the less sterically hindered convex face, followed by an intramolecular Mannich reaction to give the tricyclic intermediate **22** in an impressive overall yield of 32%. The ensuing transformation to perhydrohistrionicotoxin (3% overall) is a lengthy *tour de force* of 18 steps. A Hoffman elimination (*i*) opens ring C and an orthoester Claisen rearrangement (*k*) on the allylic alcohol produced by re-duction (*j*) lengthens the side chain and shifts the oxygen functionality to the terminal position as required. A thiolytic debenzylation generates the

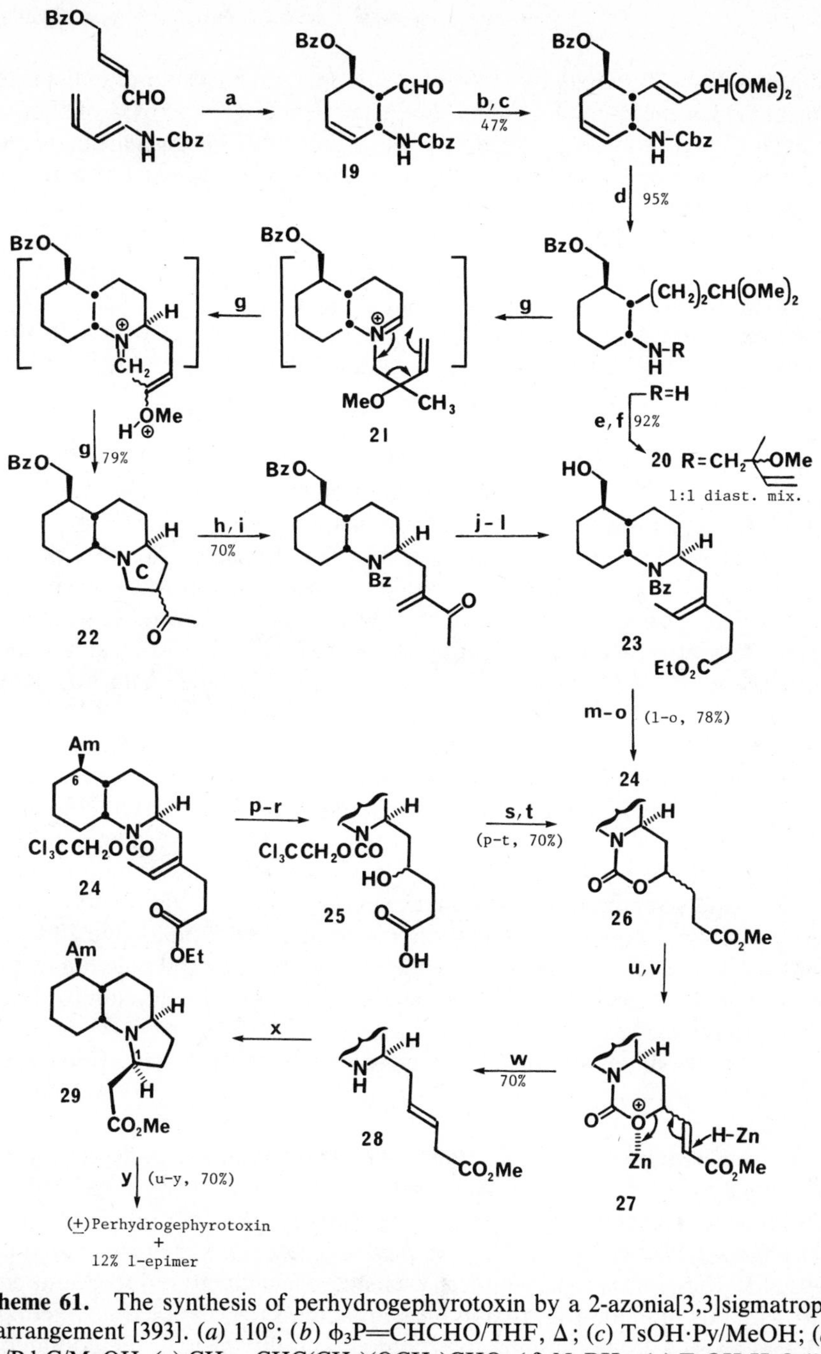

Scheme 61. The synthesis of perhydrogephyrotoxin by a 2-azonia[3,3]sigmatropic rearrangement [393]. (a) 110°; (b) ϕ_3P=CHCHO/THF, Δ; (c) TsOH·Py/MeOH; (d) H_2/Pd-C/MeOH; (e) CH_2=CHC(CH_3)(OCH_3)CHO; (f) $NaBH_4$; (g) TsOH·H_2O (0.9 eq.)/ϕH, 80°, 6 hr; (h) ϕCH_2Br/CHCl$_3$, Δ; (i) 2% NaOH; (j) $NaBH_4$; (k) $CH_3CH(OEt)_3$, Δ; (l) EtSH/BF$_3$·Et$_2$O; (m) TsCl/Py; (n) LiBu$_2$Cu/Et$_2$O, $-20°$; (o) Cl$_3$CCH$_2$OCOCl; (p) O$_3$; (q) NaBH$_4$; (r) LiOH; (s) NaH; (t) CH$_2$N$_2$/Et$_2$O; (u) LiN(i-Pr)$_2$ then ϕSeBr; (v) H$_2$O$_2$; (w) Zn/HOAc, 95°; (x) NaOMe/MeOH; (y) LiAlH$_4$.

105

primary alcohol **23**, which by steps of tosylation and butyl cuprate displacement provides the 6-amyl product **24**. After ozonolytic removal of the extraneous two carbons, introduced by the orthoester Claisen rearrangement (*p*), reduction and saponification yield **25**. An intramolecular urethane exchange and diazomethane esterification provides **26**. Phenylselenation (*u*) and selenoxide elimination (*v*) then create the (E)α,β-unsaturated ester **27**, which undergoes an unusual vinylogous reductive elimination (*w*) with decarboxylation, to afford the (E)γ,δ-unsaturated ester **28**. Reisomerization to the α,β-unsaturated ester and a Michael-addition yielded **29** with high stereospecificity (8:1). A final reduction produced the first synthetic sample of perhydrogephyrotoxin [398].

4.2.4. Diels-Alder/Michael Addition.

Ibuka and colleagues [402] also employed a Diels-Alder approach to perhydrogephyrotoxin using 1,3-bistrimethylsilyloxybutadiene, the same diene used in their syntheses of perhydrohistrionicotoxin and pumiliotoxin C. The initial adduct **30** of 1,3-bistrimethylsilyloxybutadiene with ethyl (E)-oct-2-enoate was transformed to the ketolactam **31** by steps *b–h* (Scheme 62), in which the correct stereochemistry at its three chiral centers was created by the Michael addition (*h*). After removal of the superfluous keto group, the side chain was added by a variant (*l*) of Eschenmoser's method. Reduction (*m*) afforded **32**, which could be equilibrated with base (retro-Michael ⇌ Michael addition) to the desired epimer (**33**) as the minor equilibrium component (1:3). Subsequent steps of reduction, lactonization (*o*), N-phenylcarbamoylation (*p*), and phenylselenation (*q*), produced **34**. This was converted to Overman's α,β-unsaturated ester **27** by steps of hydrolysis, cyclic urethane formation, esterification, and selenoxide elimination (*r,s*). The use of lithium dibutylcuprate effected a novel reductive elimination and decarboxylation to give **28** which could be cyclized with good stereoselectivity to **29** (6:1) and reduced to give perhydrogephyrotoxin (overall 3%) [402]. Interestingly, reduction (*o*) is evidently stereospecific as a single lactone **27**, with 2-H and 2′-H in a (Z)-relationship as shown by nuclear-Overhauser effects, resulted after steps *p* through *s*.

4.2.5. Acyliminium Ion Cyclization.

An acyliminum ion cyclization route has been applied elegantly by Hart and colleagues [403,404] to syntheses of gephyrotoxin and the naturally occurring dihydro congener in overall yields of 1.5 and 2.5%, respectively. The *trans*-fused cyclohexenone–butadiene adduct **35** (Scheme 63) of Wenkert was stereoselectively reduced and coupled with succinimide by Mitsunobu's procedure (*b*), then ozonolyzed and reduced to give **36**, possessing the required chirality at its three contiguous centers. After conversion of the hydroxyethyl side chains to vinyl groups by Grieco's method, diisobutylaluminium hydride reduction afforded the carbinol amide **37** whose proton magnetic resonance spectrum indicated a *trans*-diaxial orientation of vinyl groups. Model studies [405] on a simpler

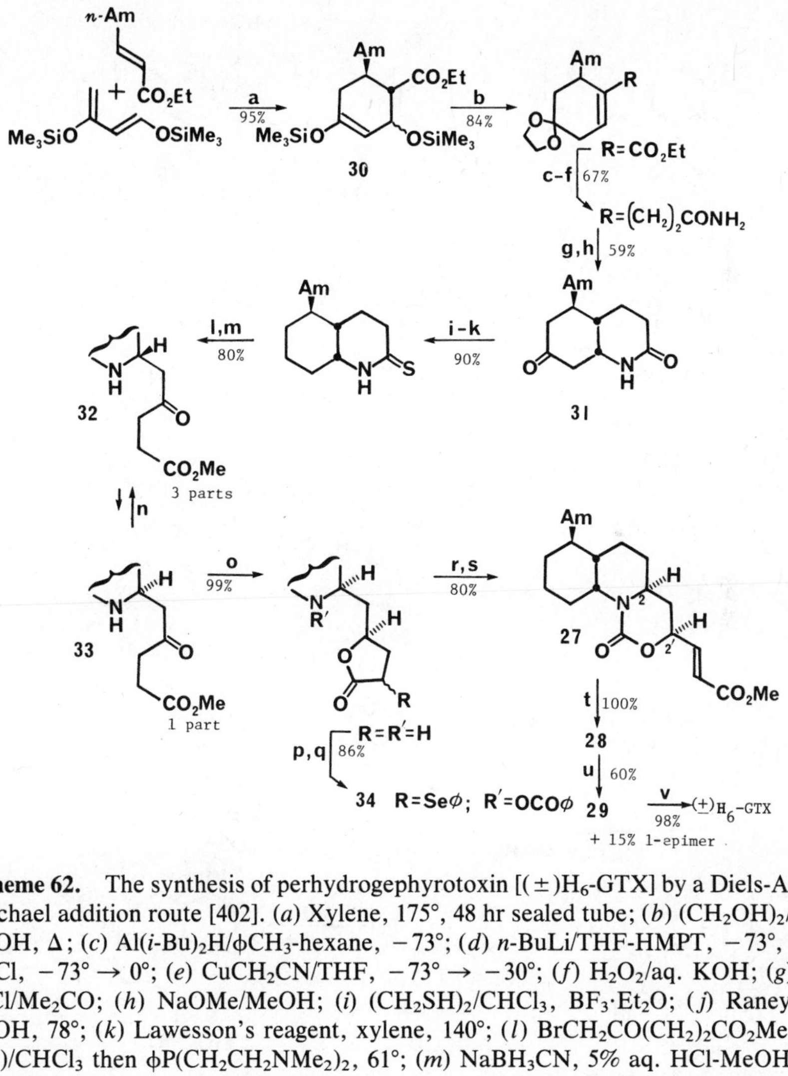

Scheme 62. The synthesis of perhydrogephyrotoxin [(±)H₆-GTX] by a Diels-Alder/ Michael addition route [402]. (*a*) Xylene, 175°, 48 hr sealed tube; (*b*) (CH₂OH)₂/φH, TsOH, Δ; (*c*) Al(*i*-Bu)₂H/φCH₃-hexane, −73°; (*d*) *n*-BuLi/THF-HMPT, −73°, then TsCl, −73° → 0°; (*e*) CuCH₂CN/THF, −73° → −30°; (*f*) H₂O₂/aq. KOH; (*g*) aq. HCl/Me₂CO; (*h*) NaOMe/MeOH; (*i*) (CH₂SH)₂/CHCl₃, BF₃·Et₂O; (*j*) Raney Ni/ EtOH, 78°; (*k*) Lawesson's reagent, xylene, 140°; (*l*) BrCH₂CO(CH₂)₂CO₂Me (1.4 eq.)/CHCl₃ then φP(CH₂CH₂NMe₂)₂, 61°; (*m*) NaBH₃CN, 5% aq. HCl-MeOH; (*n*) Et₃N/MeOH, 65°; (*o*) NaBH₄/MeOH, −20° then TsOH/φH, Δ; (*p*) φOCOCl/Py/ DMAP, 2 d; (*q*) Li(*i*-Pr)NC₆H₁₁/THF-HMPT, −73° → 40° then φSeCl, −73°; (*r*) LiOH/aq. MeOH, Δ then HCl and CH₂N₂; (*s*) H₂O₂/Py/CH₂Cl₂; (*t*) Li(*n*-Bu)₂Cu/ THF, −73°; (*u*) NaOMe/MeOH, Δ; (*v*) Al(*i*-Bu)₂H/φCH₃-hexane, −60° → −30°.

system lacking the 6-vinyl group indicated that cyclization proceeded stereospecifically via the chair conformation analogous to **38a** rather than to **38b**, which is destabilized by a C(9):carbonyl interaction [(A¹,³)-strain]. This assumption was corroborated by the production of a single formate ester **39** in good yield on cyclization (*h*) in formic acid. The unwanted C(5)-formyloxy group is removed by hydrolysis and Barton's procedure (*j–m*). After con-

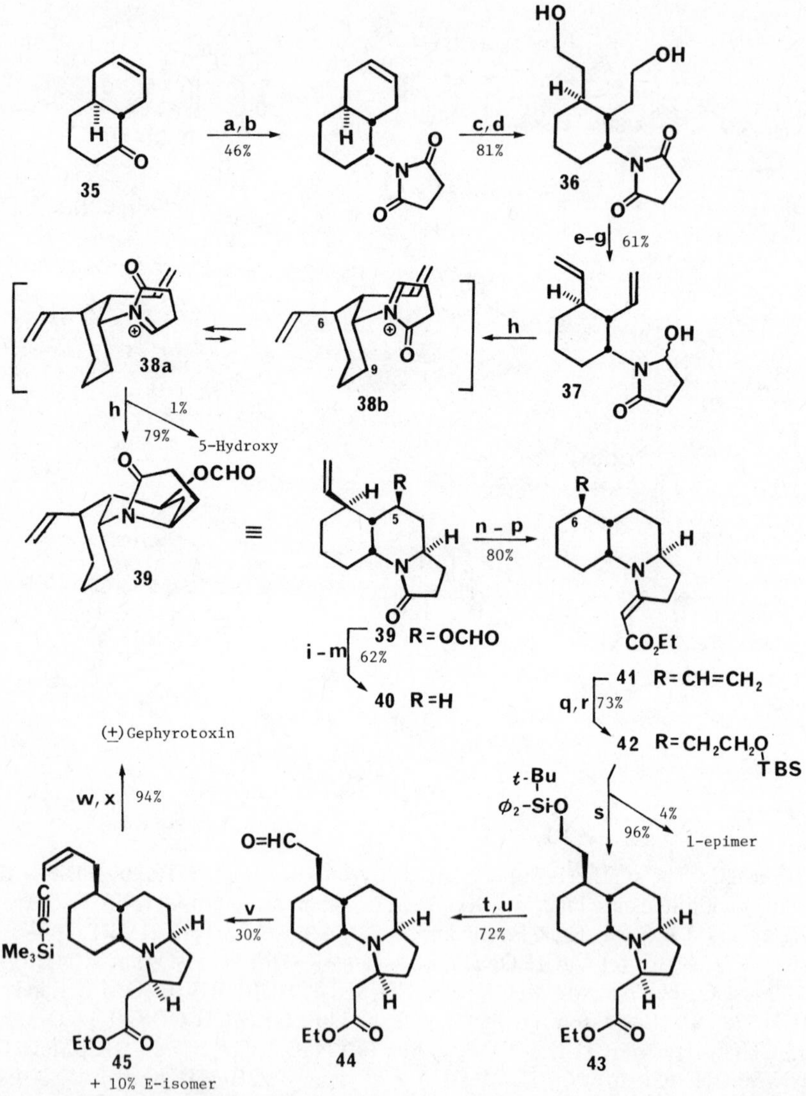

Scheme 63. The synthesis of gephyrotoxin by the acyliminium ion cyclization route [403,404]. (*a*) LiAlH$_4$/THF; (*b*) ϕ_3P/EtO$_2$CN=NCO$_2$Et, succinimide; (*c*) O$_3$/MeOH, $-70°$; (*d*) NaBH$_4$/MeOH, $-70° \rightarrow 25°$; (*e*) *o*-NO$_2\phi$SeCN/(*n*-Bu)$_3$P/THF; (*f*) H$_2$O$_2$; (*g*) Al(*i*-Bu)$_2$H/ϕCH$_3$, $-65°$; (*h*) HCO$_2$H/CH$_2$Cl$_2$, 0°; (*i*) NaOH/aq. MeOH; (*j*) NaH/ imidazole/THF, 60°; (*k*) CS$_2$; (*l*) MeI; (*m*) (*n*-Bu)$_3$SnH/ϕCH$_3$, Δ; (*n*) Lawesson's reagent; (*o*) BrCH$_2$CO$_2$Et; (*o'*) MeI; (*p*) ϕ_3P/Et$_3$N; (*p'*) EtO$_2$CCH=C(OMgX)$_2$; (*q*) disiamylborane then H$_2$O$_2$, OH$^\ominus$; (*r*) *t*-Buϕ_2SiCl; (*s*) H$_2$/Pt-Al$_2$O$_3$/EtOAc-hexane; (*t*) (*n*-Bu)$_4$N$^\oplus$F$^\ominus$/THF; (*u*) DMSO-(COCl)$_2$/Et$_3$N; (*v*) *t*-BuMe$_2$SiC$\equiv$C—TMS, *t*-BuLi; (*w*) Al(*i*-Bu)$_2$H; (*x*) (*n*-Bu)$_4$N$^\oplus$F$^\ominus$/DMF.

108

version of **40** to the thiolactam with Lawesson's reagent (*n*), the Eschenmoser episulfide contraction procedure (*o,p*) or an alternative (*o',p'*) developed by Hart, provided the urethane vinylog **41**. This could be converted
to gephyrotoxin by two routes as shown in Schemes 63 and 64. Cyanoborohydride reduction (*a*) gave a 2:1 mixture of diastereomers **46** and **47**
(Scheme 64). The major isomer was converted (*b–d*) to Kishi's intermediate
3 [403], which has been converted to gephyrotoxin in 45% yield. A superior
stereoselective sequence (*q–x*) (Scheme 63) improves the modest stereoselectivity of reduction step *a* by adopting a tactic used by Kishi and colleagues [399] for an analogous reduction (step *u*, Scheme 55). A hydroxyethyl
side chain at C(6) is produced by hydroboration hydrolysis and then converted to the silyl ether **42**. When this ether is catalytically reduced under
critical conditions, hydrogen adds stereoselectively (23:1) to the molecular
face opposite the bulky silyl ether group to give **43** [404]. Fluoride-ion desilylation and Swern oxidation provided the aldehyde **44**, which was converted to **45** using Yamamoto's *cis*-enyne procedure (see Section 3.2.7).
Reduction and deblocking provided gephyrotoxin [404].

The sole synthesis to date of dihydrogephyrotoxin was accomplished
(Scheme 65) by conversion of the aldehyde ester **44** to the β-hydroxysilane
48 using the Matteson-Tsai procedure. After ester reduction (*b*), the product
(**49**) was treated under the usual conditions for *syn*-elimination in the Peterson olefination (*c*) and produced dihydrogephyrotoxin in excellent yield
[404].

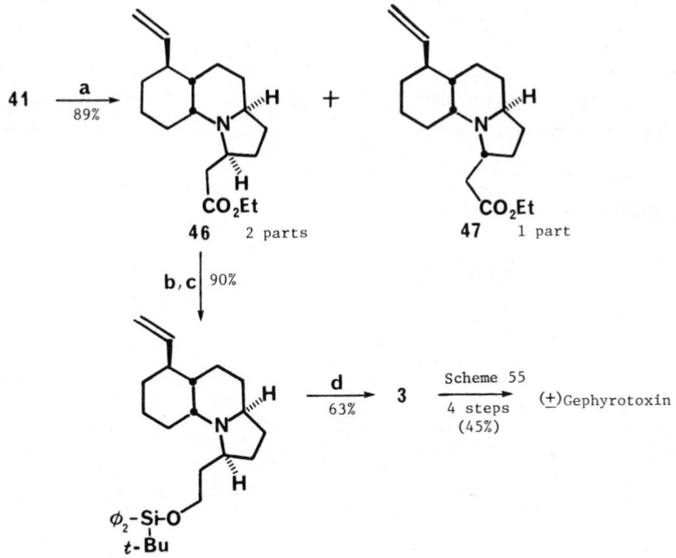

Scheme 64.   A formal synthesis of gephyrotoxin [403]. (*a*) NaCNBH$_3$; (*b*) LiAlH$_4$;
(*c*) *t*-Buφ$_2$SiCl; (*d*) disiamylborane then H$_2$O$_2$, OH$^\ominus$.

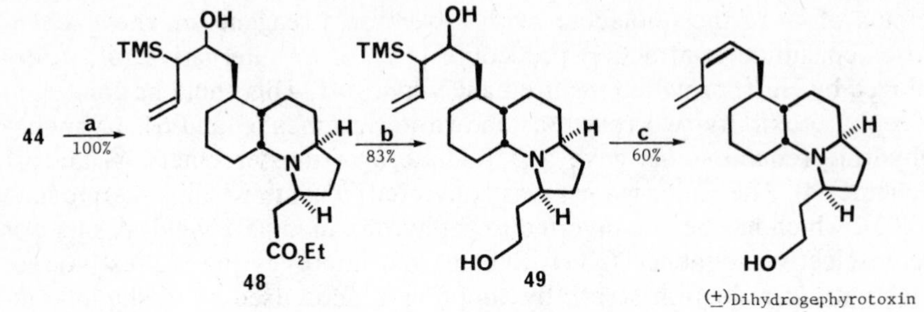

Scheme 65. The synthesis of dihydrogephyrotoxin [404]. (*a*) (E)-TMS—CH=CHCH$_2$B(OC(CH$_3$)$_2$)$_2$/CH$_2$Cl$_2$; (*b*) Al(*i*-Bu)$_2$H/ϕCH$_3$, $-78°$ → $-20°$; (*c*) KH/THF.

Hart and colleagues [404,406] also studied radical cyclizations as a route to gephyrotoxin (Scheme 66). When the thio ether lactam diastereomeric mixture **50** was exposed to radical initiators, cyclization occurred to give the gephyran system **40** in fair yield with a preponderance of the unwanted regioisomer **51**. The stereochemistry of the gephran product may indicate that the radical cyclization is under the same conformational and steric constraints as the cationic pathway.

4.2.6. The Synthesis of Gephran. Habermehl and Thurau [407], using an enamine exchange–intramolecular alkylation sequence analogous to that used in their synthesis of pumiliotoxin C (see Section 7.2.4), prepared the parent tricyclic ring (gephran) system (**53**) of gephyrotoxin in 35–40% overall yield (Scheme 67). The minor *trans*-fused by-product, **54**, was the major product when the iminium salt **55**, produced from the intermediate enamine mixture **52** by stereospecific protonation, was reduced by borohydride.

4.3. Biological Activity

4.3.1. Toxicity. Gephyrotoxin itself is relatively nontoxic and hence does not truly deserve the toxin designation. Gephyrotoxin at subcutaneous doses

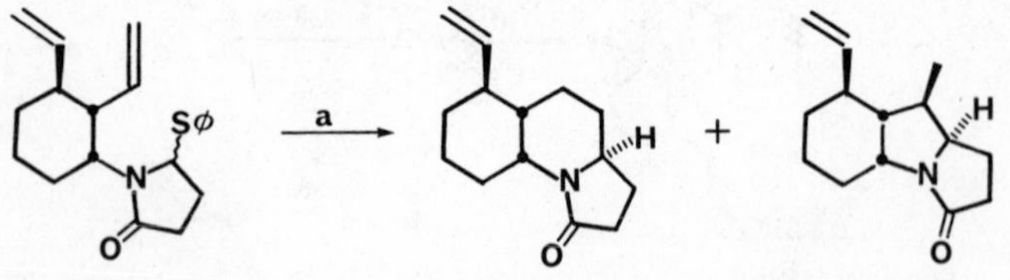

Scheme 66. A radical cyclization route to gephyrotoxin [404,406]. (*a*) (*n*-Bu)$_3$SnH/AIBN/ϕH, Δ.

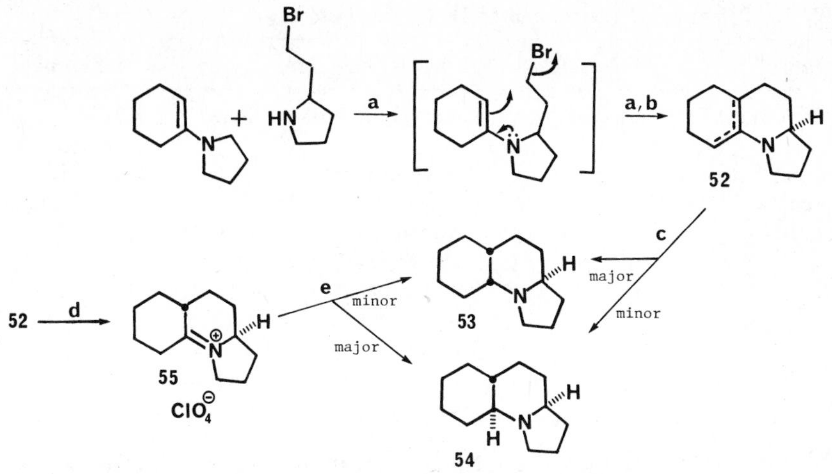

Scheme 67. An intramolecular enamine alkylation approach to the gephyran system [407]. (*a*) DMF, 95°; (*b*) OH$^{\ominus}$; (*c*) H$_2$/Pd-C; (*d*) HClO$_4$; (*e*) NaBH$_4$.

of 10 mg/kg in mice has little effect beyond a slight reduction in spontaneous activity [23 and unpublished results].

4.3.2. Pharmacological Activity. Gephyrotoxin at 10 μ*M* was found in preliminary studies to have virtually no effect on direct or indirect elicited twitch in rat phrenic nerve diaphragm [128]. Subsequent studies indicated that gephyrotoxin at 20 μ*M* potentiates directly elicited twitch in frog neuromuscular preparations, while causing a slow blockade of indirectly elicited twitch [380,408]. The potentiation appears to be due to inhibition of potassium conductances, while the blockade appears to result from noncompetitive blockade of the acetylcholine receptor channel complex. The effects of gephyrotoxin on the receptor complex are stimulus-dependent, suggesting interaction with open conformations of the acetylcholine receptor channel complex. The lack of voltage and time dependence further suggests that, unlike histrionicotoxins, gephyrotoxin has little effect on closed channels. Desensitization of responsiveness to acetylcholine occurs rapidly with gephyrotoxin, while channel blockade and reversal occurs relatively slowly.

Gephyrotoxin inhibits binding of [^{3}H]H$_{12}$-HTX and [^{3}H]phencyclidine in electroplax membranes [380]. The natural levorotatory enantiomer is equivalent in potency to the synthetic dextrorotatory enantiomer in these binding studies. Gephyrotoxin is three- to fivefold more potent as a binding antagonist in the presence of carbamylcholine than in the absence of a nicotinic receptor agonist. Conversely, gephyrotoxin enhances potency of carbamylcholine for the receptor.

Gephyrotoxin at 5 μ*M* reduces acetylcholine-elicited, but not histamine-elicited, contractures of guinea pig ileum, and at 10 μ*M* causes a reduction

Table 7.   Effects of Gephyrotoxin on Biological Systems

System	Species	Comments	References
Whole animal	Mouse	Relatively nontoxic (20 mg/kg)	[23]
Neuromuscular preparation (striated)	Frog	Antagonism of neuromuscular transmission through blockade of open channel and enhancement of desensitization; prolongation of action potentials and potentiation of directly elicited muscle contraction	[380,408]
	Rat	No effect at 10 μM; stimulus-dependent blockade of acetylcholine responses at 20 μM	[128,380]
Heart	Guinea pig atria	Weak muscarinic antagonist	[128]
Smooth muscle	Guinea pig ileum	Weak muscarinic antagonist	[128]
Electroplax	Ray	Blockade [^{3}H]H$_{12}$-HTX binding, increases affinity of agonists for receptors	[380]

in the response to acetylcholine in guinea pig atrium [128]. Thus gephyrotoxin appears to act as a relatively weak muscarinic antagonist. Gephyrotoxin O-p-bromobenzoate is also a relatively weak antagonist of acetylcholine in ileum. The effects of gephyrotoxin on various biological systems are summarized in Table 7.

5.   INDOLIZIDINES (BICYCLIC GEPHYROTOXINS)

5.1.   Structures

Three simpler bicyclic indolizidine alkaloids from dendrobatid frogs were originally included in the gephyrotoxin class [23]. These indolizidines, like the tricyclic gephyrotoxins, contain in a literal sense a bridge formed by the addition of the nitrogen to one side chain of a 2,6-disubstituted piperidine. It is now apparent that dendrobatid alkaloids include a great variety of simple indolizidines, and therefore, it appears preferable to no longer use the term gephyrotoxin for these alkaloids, but to refer to them simply as indolizidines. Thus gephyrotoxin **223AB** will be termed indolizidine **223AB** in this review.

Initially, the mass spectrum of this alkaloid was interpreted as due to the presence of two isomeric compounds (223A and 223B) not separable by gas chromatography, with one (223A) losing a propyl side chain during mass spectral fragmentation, the other (223B) losing a butyl side chain. Later when it was realized that the gas chromatographic peak was due to a single compound losing almost equally either a propyl or a butyl side chain, this alkaloid was given the AB designation. The same history led to the 239AB and 239CD designations (see below). Alkaloids 223A, 223B, 239A, 239B, 239C, and 239D do occur as trace constitutents in certain dendrobatid extracts (see Sections 10.1 and 11).

The original proposed structures for the three indolizidines 223AB, 239AB, and 239CD were tentative in nature, although supported by a number of observations: The compounds were saturated bicyclic alkaloids; they did not form N-acetyl derivatives; for each alkaloid, two fragments dominated the mass spectrum [23]. These fragments corresponded to loss of either a four-carbon or a three-carbon side chain. In the simplest member of this group, indolizidine 223AB ($C_{15}H_{29}N$), the side chains were C_3H_7 and C_4H_9, while indolizidine 239AB ($C_{15}H_{29}NO$) contained a hydroxyl group in the three-carbon side chain and indolizidine 239CD ($C_{15}H_{29}NO$) contained a hydroxy group in the four-carbon side chain. This data and biosynthetic considerations led to the proposal that the compounds were 3,5-disubstituted indolizidines with the four-carbon substituent at the 3-position and the three-carbon substituent at the 5-position [23]. At that time, the structures of all the simpler dendrobatid alkaloids, that is, the histrionicotoxins, the tricyclic gephyrotoxins, and pumiliotoxin C, were consonant with a biosynthetic origin from a 2,6-disubstituted piperidine with side chains having lengths of three, five, seven, or nine carbons (see Section 11). Structures of the indolizidines 223AB, 239AB, and 239CD were, therefore, in part based on a postulated biogenetic pathway leading via cyclization from a 2,6-disubstituted piperidine with a three-carbon side chain and a seven-carbon side chain to the indolizidine. The nitrogen was to have formed the "bridge" by cyclization to carbon-3 of the seven-carbon side chain. All four possible stereoisomers of the proposed indolizidine 223AB were subsequently synthesized (see Section 5.3). Indolizidine 223AB proved to be identical by gas chromatographic and mass spectral analysis to (5E,9E)-3-butyl-5-propylindolizidine [409] (Fig. 5). The designations E or Z for such indolizidines indicate whether the hydrogen at the 5- and 9-positions is *trans* (E) or *cis* (Z) to the hydrogen at position 3 [410]. The other three diastereomers of 3-butyl-5-propylindolizidine differed in gas chromatographic and mass spectral properties from indolizidine 223AB. Natural indolizidine 223AB has now been isolated in sufficient quantities for proton and ^{13}C magnetic resonance spectroscopy [251]. The spectra proved identical to synthetic (5E,9E)-3-butyl-5-propylindolizidine (see Section 5.3). Recently, one enantiomer of indolizidine 223AB, namely 3R,5R,9R-3-butyl-5-propylindolizidine, has been synthesized (see Section 5.3) and shown to be levorotatory [411]. Natural in-

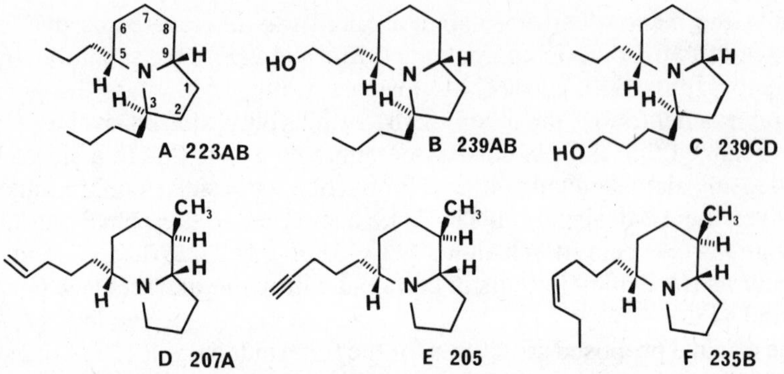

Figure 5. Structures of indolizidine alkaloids from dendrobatid frogs. (*A*) Indolizidine **223AB**. (*B*) Indolizidine **239AB**. (*C*) Indolizidine **239CD**. (*D*) Indolizidine **207A**. (*E*) Indolizidine **205**. (*F*) Indolizidine **235B**. Absolute stereochemistry of *D*, *E*, and *F* is undefined.

dolizidine **223AB** is also levorotatory [251], indicating that the absolute stereochemistry of indolizidine **223AB** is 3R,5R,9R (Fig. 5).

Indolizidines **239AB** and **239CD** have now been isolated in sufficient quantities for nuclear magnetic resonance spectroscopy [251]. The structures of **239AB** and **239CD** were, thereby, defined as 3-butyl-5-(3′-hydroxypropyl)indolizidine and 3-(4′-hydroxybutyl)-5-propylindolizidine, respectively (Fig. 5). The ^{13}C resonances were most compatible with the same 5E,9E stereochemistry found in the parent indolizidine **223AB**. Both **239AB** and **239CD** are levorotatory, suggesting that their absolute stereochemistry is the same as that of indolizidine **223AB**.

Dendrobatid frogs produce an array of simple bicyclic alkaloids with tertiary nitrogen, many of which seem likely to be indolizidines. Only three of these simple indolizidines, namely **223AB**, **239AB**, and **239CD**, have been structurally defined in detail. One further alkaloid was included in this group in a previous review [24], namely indolizidine **167B**, which, based on mass spectral properties and biosynthetic considerations, was considered likely to be a 5-*n*-propylindolizidine. Another bicyclic tertiary amine, **209D**, also affords a base peak at *m/z* 124, and has been tentatively added to the indolizidine class (see below). Another indolizidine alkaloid, namely **195B**, has been isolated in a sufficient amount for magnetic resonance spectroscopy. The spectrum indicates it to be a 3-butyl-5-methylindolizidine [412] (Fig. 6).

Recently, one of the many formerly unclassified dendrobatid alkaloids [24], namely **207A**, was isolated in sufficient quantity from *Dendrobates speciosis* (Panama) for nuclear magnetic resonance spectroscopy and its structure proposed as 5-(pent-5-enyl)-8-methylindolizidine [258] (Fig. 5). Another alkaloid **205** has been isolated from *Dendrobates pumilio* and shown to be the ynyl congener of **207A**; the relative stereochemistry, based on magnetic resonance spectroscopy, appeared to be as in Fig. 5, with both the

8-methyl and the 5-(pent-5-ynyl) substituents equatorial [412]. Thus extracts from dendrobatid frogs appear to have a series of congeneric indolizidines (**203A, 205, 207A, 209B**) differing only in the degree of unsaturation in the five-carbon side chain. All of these reduce to a perhydro derivative with a molecular weight of 209 and a mass spectral base peak at m/z 138. The mass spectrum and properties of **207A** and **205** are similar to many of the unclassified dendrobatid alkaloids, and it appears likely that many but not all of these alkaloids, namely bicyclic tertiary amines yielding a base peak at m/z 138, will prove to be similar monosubstituted-8-methylindolizidines. However, alkaloid **195B** is of this type and from magnetic resonance spectra it appears to be a 3,5-disubstituted indolizidine [412] (Fig. 6). One other alkaloid of this type, namely **235B**, has been isolated and characterized by magnetic resonance spectroscopy as 8-methyl-5-(hept-4,5-*cis*-enyl)indolizidine [412] with the probable stereochemistry shown in Fig. 5. All the other bicyclic tertiary amine alkaloids with a base peak at m/z 138 are included, tentatively, in this review in the indolizidine class of dendrobatid alkaloids. Two alkaloids, namely **225D** and **239G**, included in this class, appear to contain, on the basis of deuteroammonia chemical ionization mass spectrometry, a hydroxyl group in the side chain. It remains possible that some of these alkaloids will prove to be monosubstituted quinolizidines. Conversely, many of the presently unclassified alkaloids from dendrobatid frogs (Section 10.1) will probably prove to be indolizidines.

Properties of the dendrobatid alkaloids of the indolizidine class or tentatively assigned to the indolizidine class are reported below in the standard format (see Section 3.1). Empirical formulas that are based only on analogy and on chemical and chromatographic properties and that have not been confirmed by high resolution mass spectrometry are in quotations.

Indolizidine Class (Bicyclic Gephyrotoxins)

167A. $C_{11}H_{21}N$,—, 151°, m/z 167 (1), 166 (1), 138 (100). H_0 derivative. 0D. Tentative structure: a 5-ethyl-8-methylindolizidine.

167B. "$C_{11}H_{21}N$,"—, 151°, m/z 167 (12), 166 (5), 124 (100). H_0 derivative. 0D. Tentative structure: a 5-propylindolizidine.

181B. "$C_{12}H_{23}N$,"—, 153°, m/z 181 (2), 180 (2), 138 (100). H_0 derivative. 0D. Tentative structure: a 5-propyl-8-methylindolizidine.

195B. $C_{13}H_{25}N$, 0.28, 156°, m/z 195 (2), 194 (1), 138 (100). H_0 derivative. 0D. Tentative structure: Fig. 6.

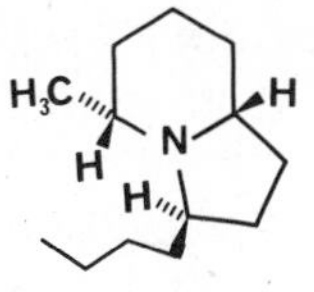

195B

Figure 6. Tentative structure for indolizidine **195B**.

203A. $C_{14}H_{21}N$, 0.33, 158°, m/z 203 (1), 202 (2), 138 (100). H_6 derivative, m/z 209 (1), 138 (100). 0D. Tentative structure: a 5-pentenynyl-8-methylindolizidine. An isomer **203A′** has been detected at 157°.

205. $C_{14}H_{23}N$, —, 158°, m/z 205 (1), 204 (2), 138 (100). H_4 derivative, m/z 209, 138. 0D. Structure: Fig. 5. An isomer **205′** has been detected at 156°.

207A. $C_{14}H_{25}N$, 0.35, 158°, m/z 207 (1), 206 (1), 138 (100). H_2 derivative, m/z 209, 138. 0D. Structure: Fig. 5. An isomer **207A′** has been detected at 156°.

209B. "$C_{14}H_{27}N$," —, 162°, m/z 209 (5), 138 (100). H_0 derivative. 0D. Structure: a dihydro-**207A**.

209D. "$C_{14}H_{27},N$," —, 159°, m/z 209 (2), 124 (100). H_0 derivative. 0D. Tentative structure: a 3- or 5-hexylindolizidine.

221A. "$C_{15}H_{27}N$," —, 162°, m/z 221 (<1), 220 (2), 138 (100). H_2 derivative, m/z 223,138. 0D. Tentative structure: a 5-hexenyl-8-methylindolizidine.

223AB. $C_{15}H_{29}N$, 0.30, 160°, m/z 223 (1), 222 (2), 180 (85), 166 (100). H_0 derivative. 0D. Structure: Fig. 5.

223D. "$C_{15}H_{29}N$," 0.3, 159°, m/z 223 (1), 222 (2), 138 (100). H_0 derivative. 0D. Tentative structure: a 5-hexyl-8-methylindolizidine.

225D. $C_{14}H_{27}NO$, —, 164°, m/z 225 (<1), 138 (100). H_0 derivative. 1D (not on nitrogen). Tentative structure: a 5-hydroxypentyl-8-methylindolizidine.

231C. $C_{16}H_{25}N$, —, 171°, m/z 231 (3), 138 (100). H_6 derivative, m/z 237, 138. 0D. Tentative structure: a 5-(hepten-6-ynyl)-8-methylindolizidine.

235B. $C_{16}H_{29}N$, —, 166°, m/z 235 (1), 234 (1), 138 (100). H_2 derivative, m/z 237, 138. 0D. Structure: Fig. 5.

237D. "$C_{16}H_{31}N$," —, 163°, m/z 237 (1), 236 (2), 138 (100). H_0 derivative. 0D. Tentative structure: a 5-heptyl-8-methylindolizidine.

239AB. $C_{15}H_{29}NO$, 0.22, 178°, m/z 239 (2), 238 (3), 182 ($C_{11}H_{20}NO$, 100), 180 ($C_{12}H_{22}N$, 90). H_0 derivative. 1D. O-acetyl derivative. Structure: Fig. 5.

239CD. $C_{15}H_{29}NO$, 0.16, 179°, m/z 239 (4), 238 (3), 196 ($C_{12}H_{22}NO$, 100), 166 ($C_{11}H_{10}N$, 60). H_0 derivative. 1D, O-acetyl derivative. Structure: Fig. 5.

239G. $C_{15}H_{29}NO$, —, 178°, m/z 239 (1), 238 (3), 138 (100). H_0 derivative. 1D (not on nitrogen). Tentative structure: a 5-hydroxyhexyl-8-methylindolizidine.

251B. "$C_{16}H_{29}NO$," —, 184°, m/z 251 (2), 234 (4), 138 (100). H_2 derivative m/z 253, 138. 1D. Tentative structure: a 5-hydroxyheptenyl-8-methylindolizidine.

253B. "$C_{16}H_{31}NO$," —, 192°, m/z 253 (1), 138 (100). H_0 derivative. 1D. Tentative structure: a 5-hydroxyheptyl-8-methylindolizidine.

257C. "$C_{18}H_{27}N$," —, 190°, m/z 257 (<1), 138 (100). H_8 derivative m/z 265, 138. 0D. Tentative structure: a 5-nonadienynyl-8-methylindolizidine.

Table 8. Physical and Spectral Properties of Indolizidine 223AB.

Indolizidine 223AB		
Infrared (racemic synthetic)	1190 cm^{-1}	[413]
Optical rotation		
HCl (synthetic) (c 0.62, CH_3OH)	$[\alpha]_D^{20} -91°$	[411]
HCl (natural)	$[\alpha]_D^{16} -35°{}^a$	[251]
Indolizidine 239AB		
Optical rotation		
HCl (c 1.0, CH_3OH)	$[\alpha]_D -38°$	[251]
Indolizidine 239CD		
Optical rotation		
Free base (c 0.2, CH_3OH)	$[\alpha]_D -52°$	[251]

^{a} The apparent difference in magnitude of rotation of synthetic and natural **223AB** requires further investigation.

Other physical and spectral properties of indolizidine **223AB** are reported in Table 8. The mass spectra of indolizidines **223AB**, **239AB**, and **239CD** were shown in a review on the initial structural studies of these alkaloids [24, see also 251]. The mass spectra of the various diastereomers of synthetic indolizidine **223AB** are shown in [409]. The proton and ^{13}C magnetic resonance spectral properties of indolizidine **223AB**, both synthetic [411,413–415] and natural [251], have been reported. The ^{13}C magnetic resonance spectra of natural and synthetic **223AB** appear identical. However, the ^{13}C magnetic resonance spectrum obtained on the first preparation of synthetic **223AB** [413] evidently shows a calibration error. The proton and ^{13}C magnetic resonance spectral properties of indolizidines **239AB** and **239CD** [251] and of indolizidines **207A** [258], **205** [412], and **235B** [412] have been reported.

The 3,5-disubstituted indolizidines **223AB**, **239AB**, and **239CD** occur in only a limited number of dendrobatid species (see Section 11). These three alkaloids have the (5E,9E) stereochemistry. Ants produce a range of (5Z,9Z) 3-alkyl-5-methylindolizidines [see 416 and references therein]. The 5-substituted-8-methylindolizidines, such as **205**, **207A**, and **235B**, occur in many dendrobatid species (see Section 11). Indolizidine **207A** has recently been reported from the Madagascan mantellid frog, *Mantella madagascariensis* [261].

5.2. Syntheses

In order to confirm the gross structure of indolizidine **223AB** (formerly referred to as gephyrotoxin **223AB**, see Section 5.1) and to establish the stereochemistry of the 3- and 5-substitutents, a number of laboratories developed syntheses of varying diastereoselectivity of the four possible stereoisomers.

A cooperative effort by three groups led to the assignment of the (5E,9E) structure 4 to the natural material and confirmed the provisional gross structure [409]. Three stereoselective syntheses of 4 have been completed [411,413–415]. X-ray analysis [417] of the hydrobromide of 4, prepared by Edwards and colleagues [415], has confirmed the earlier assignment of structure [409], based upon comparison of isomers by gas chromatography and mass spectrometry. An enantioselective synthesis of (−)-indolizidine 223AB has recently been completed [411]. The natural alkaloid is also levorotatory [251]. The rotations are not in good agreement, but such indolizidines are frustratingly labile to autoxidation during isolation, thereby perhaps complicating quantitative comparisons.

5.2.1. Synthesis of (5E,9E)-3-n-Butyl-5-n-propylindolizidine (Indolizidine 223AB).

Macdonald [413], using a stereoselective (>95%) (E)-bisalkylation of the Δ^3-pyrroline carbamate 1 (Scheme 68), followed by deblocking and ring closure, obtained a 1:1 mixture of dehydroindolizidines 2 and 3. Hydrogenation unexpectedly provided a single diastereomer of the (5E,9E) configuration 4 as proved by an analogous synthesis [413] of (5E,9E)-3-n-butyl-5-methylindolizidine 5, previously prepared by Sonnett and colleagues [410]. Evidently the (5Z,9E) diastereomers, which should accompany the (5E,9E) compounds, decompose or undergo hydrogenolysis in some as yet unexplained fashion. Gas chromatographic–mass spectral comparison of 4 with the natural indolizidine 223AB showed them to have identical retention times and fragmentation patterns [409].

Natsume and co-workers [414] prepared indolizidine 223AB from the 2,6-cis-Δ^3-dehydropiperidine intermediate 7 (Scheme 69), originally prepared for their synthesis of 18-deoxypalustrine [418]. Homologization of the ketal side chain (i–k), deblocking (l), hydrolysis (m), and iminium salt formation produced 8, which was converted (n) to the cyanide adduct 9. The ensuing

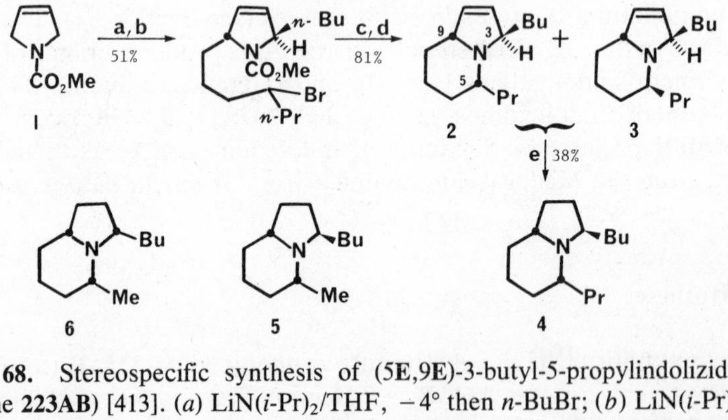

Scheme 68. Stereospecific synthesis of (5E,9E)-3-butyl-5-propylindolizidine (indolizidine 223AB) [413]. (a) LiN(i-Pr)$_2$/THF, − 4° then n-BuBr; (b) LiN(i-Pr)$_2$/THF, − 4° then 1,4-dibromo-n-heptane; (c) TMS—I/CHCl$_3$, 70°; (d) Na$_2$CO$_3$/MeOH, Δ; (e) H$_2$/Pt/HOAC.

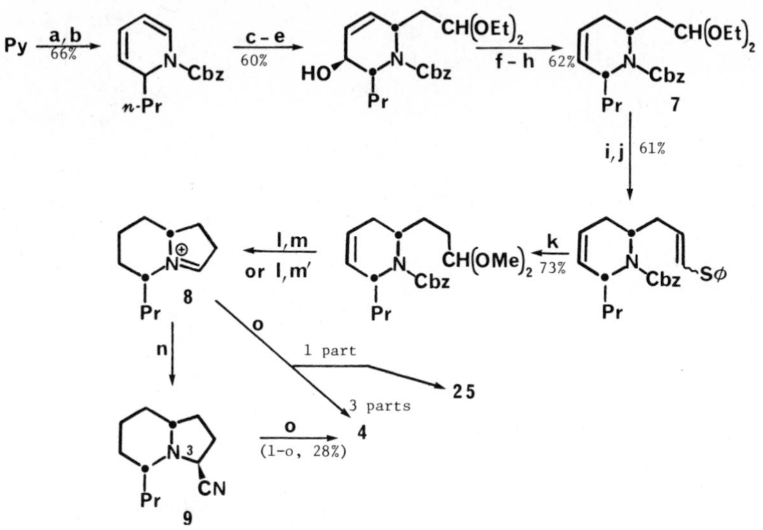

Py $\xrightarrow[66\%]{a,b}$ [structure, *n*-Pr, N-Cbz] $\xrightarrow[60\%]{c-e}$ [structure, HO, Pr, N-Cbz, CH(OEt)$_2$] $\xrightarrow[f-h\ 62\%]{}$ [structure, Pr, N-Cbz, CH(OEt)$_2$, **7**]

$\biggm\downarrow i,j\ 61\%$

[structure **8**, iminium salt, N$^{\oplus}$, Pr] $\xleftarrow[\text{or } l,m']{l,m}$ [structure, Pr, N-Cbz, CH(OMe)$_2$] $\xleftarrow[73\%]{k}$ [structure, Pr, N-Cbz, Sϕ]

[structure **9**, N, CN, Pr]

n ↓

o

1 part → 25

3 parts → 4

$\xrightarrow[(1-o,\ 28\%)]{o}$ 4

Scheme 69. Stereospecific synthesis of (5E,9E)-3-butyl-5-propylindolizidine (indolizidine **223AB**) [414]. (*a*) CbzCl; (*b*) *n*-PrMgBr; (*c*) O$_2$, hv; (*d*) CH$_2$=CHOEt, SnCl$_2$; (*e*) EtOH; (*f*) H$_2$/Pd-C/DME; (*g*) TsCl/Py; (*h*) DBU/ϕCH$_3$, 100°; (*i*) H$^+$/aq. DME; (*j*) ϕ_2P(=O)CH$^{\ominus}$Sϕ/THF, $-78°$ → r.t.; (*k*) TsOH/MeOH-HgCl$_2$, Δ; (*l*) H$_2$/Pd-C/MeOH; (*m*) HCl; (*m'*) TFA; (*n*) KCN; (*o*) *n*-BuMgBr.

Grignard reaction (*o*) proceeded stereospecifically with retention of configuration at C(3), as Stevens and Lee [420] had previously observed with Grignard reactions on the related six-membered ring nitriles. Surprisingly, the Grignard addition directly upon the iminium salt **8** produced by *m'* gave a 3:1 mixture (46%) of (5E,9E) (**4**) and (5Z,9Z) (**25**) diastereomers.

Recently, Edwards and colleagues [415] have utilized the Diels-Alder reaction of the acyliminium salt **10** with (E)-1,3-*n*-heptadiene (Scheme 70), producing the desired 5,9-(Z) stereochemistry in the cycloadduct **11** as a result of the normal endoselection of this reaction. After reduction (*b*) and conversion to the thiolactam **13a** with Lawesson's reagent (*c*), the butyl side chain is added using the Eschenmoser procedure (*d*), followed by steps of reduction (*e*), separation of the 1:1 mixture of isomers **14a** and **15a**, and removal of the side chain keto group from **14a** by Raney nickel desulfurization (*g*) of the derived dithiolone **16**. The ^{13}C magnetic resonance spectrum of **4** prepared by the groups of Natsume [414], Edwards [415], and Husson [411] were virtually identical with the spectrum obtained by Tokuyama [251] on natural material.

5.2.2. Synthesis of Other Stereoisomers of 3-Butyl-5-propylindolizidine.

Hart and Tsai [419] prepared the (5E,9Z) and (5Z,9E) diastereomers, **26** and **27**, by a related route also shown in Scheme 70. Commencing with a stereospecific acyliminium ion cyclization (*a'*) (via **17A**) of the easily prepared

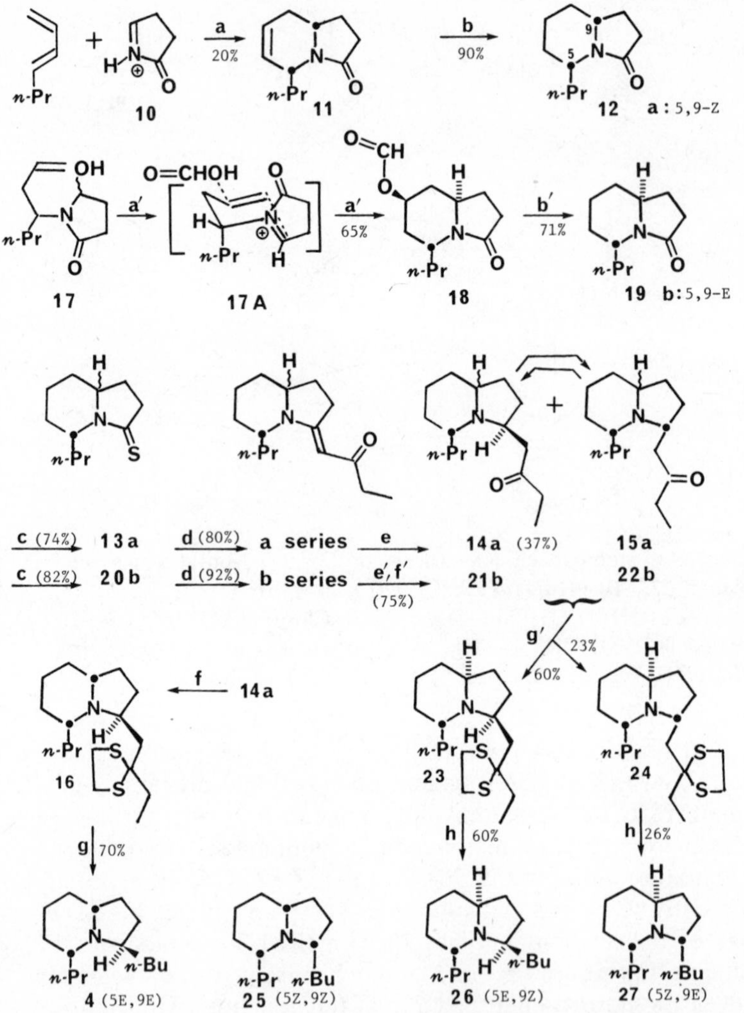

Scheme 70. The synthesis of 3-butyl-5-propylindolizidine diastereomers [415,418,419]. (*a*) MeOH/CH$_2$Cl$_2$, 0°, 5 d; (*a'*) HCO$_2$H; (*b*) H$_2$/Pd-C/EtOH; (*b'*) NaOH/aq. MeOH then steps *j–m*, Scheme 63; (*c*) Lawesson's reagent/DME; (*d*) BrCH$_2$COEt/ϕ$_3$P-Et$_3$N/Et$_2$O; (*e*) NaCNBH$_3$/H$^+$; chromatographic separation; (*e'*) NaCNBH$_3$ or H$_2$/Pt/HOAc; (*f*) (CH$_2$SH)$_2$/HCl/BF$_3$·Et$_2$:O; (*f'*) CrO$_3$/Me$_2$CO; (*g*) Raney Ni/EtOH; (*g'*) (CH$_2$SH)$_2$/HCl/CHCl$_3$; chromatographic separation; (*h*) Li/EtNH$_2$, −78°.

carbinolamido olefin **17**, a single lactam formate **18** was obtained, with the 5,9-(**E**) configuration as shown by proton magnetic resonance spectroscopy. After removal of the extraneous 7-formyloxy group by hydrolysis and Barton's procedure (*b'*, see steps *j–m*, Scheme 63), **19b** resulted. The butyl side chain was added to the derived thiolactam **20b** using Eschenmoser's method (*d*) and cyanoborohydride reduction (*e'*). The mixture of ketoamines, **21b** and **22b**, arising after reoxidation (⇆ *f'*), was configurationally unstable (retro Michael ⇆ Michael addition), but could be converted (*g'*) to the dithiolone mixture **23** and **24** for separation. A lithium ethylamine reduction (*h*) provided the two stereoisomers **26** and **27**, *trans*-substituted in the piperidine ring and epimeric at C(3). These compounds and structural assignments were important to the original assignment of configuration to the natural indolizidine **223AB** [409].

Stevens and Lee [420] prepared the (5E,9Z) isomer **26** (Scheme 71), using a stereospecific catalytic hydrogenation (*d*) of the pyrroline **29** generated *in situ* from the nitroketone **28**, followed by cyclic iminium salt formation (*e*) and reaction of propylmagnesium bromide (*g*) with the derived cyanide adduct **31** (Scheme 71). The stereospecificity of the Grignard addition is thought to arise from the stereoelectronic requirement that the alkyl nucleophile and developing lone pair on nitrogen be in a *trans*-diaxial orientation as the Grignard reagent reacts with the iminum salt **30** regenerated *in situ* from the nitrile **31**.

Since it initially seemed likely that indolizidine **223AB** might have the same *cis*-substituted pyrrolidine moiety as found in gephyrotoxin, a synthesis was devised by Spande [421] to provide the (5Z,9Z) and (5E,9Z) stereoisomers, **25** and **26**. Applying a new 1,2-dialkylpyrrole synthesis developed

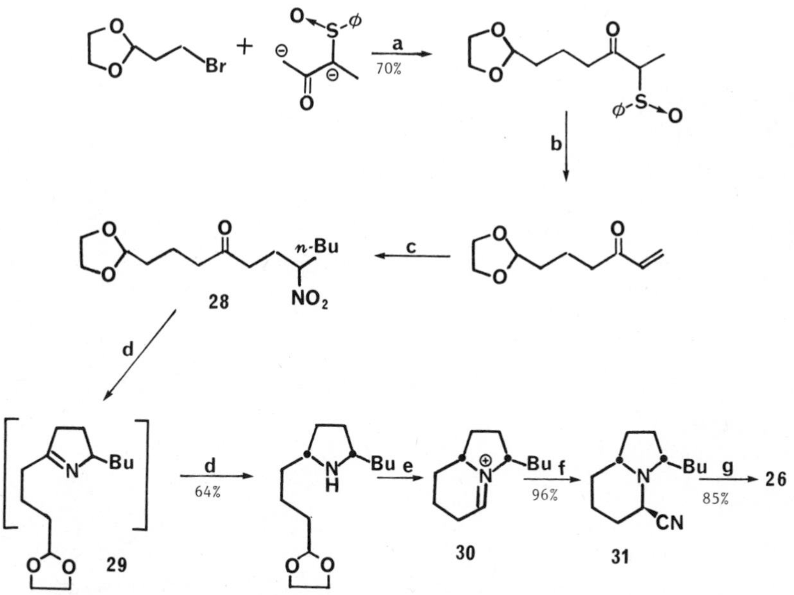

Scheme 71. The stereoselective synthesis of (5E,9Z)-3-butyl-5-propylindolizidine [420]. (*a*) THF/HMPT; (*b*) CCl$_4$, Δ; (*c*) *n*-AmNO$_2$, (Me$_2$N)$_2$C=NH; (*d*) Raney Ni/H$_2$; (*e*) H$_3$O$^\oplus$; (*f*) CN$^\ominus$; (*g*) *n*-PrMgBr/Et$_2$O, 0°.

using the γ-ketoaldehyde equivalent **32** and the amino acetal **33** followed by an intramolecular condensation, the dihydroindolizine **34** was prepared (Scheme 72). This provided all four stereoisomers **4**, **25–27** on catalytic reduction; the two 9(Z)-isomers predominating over the two 9(E)-isomers by about four to one. This mixture was used in conjunction with the diastereomers prepared by Hart and Tsai [419] and the isomer prepared by Macdonald [413] to assign the unexpected (5E,9E) configuration to the natural material [409]. Indolizidine **223AB** does, however, share the same (**Z**)-2,6-substituted piperidine configuration as found in pumiliotoxin C (Section 7.1). However, it now appears that the absolute stereochemistry at the 5-propyl substituent (5**R**) is opposite (see above) to that at the 2-propyl substituent (2**S**) of pumiliotoxin.

Using butyryl 2-methoxycyclopropane **36** and the aminoacetal **35**, derived from valeryl 2-methoxycyclopropane (**32**), a synthesis of the four "iso" stereoisomers **37**, with propyl and butyl side chains interchanged, was carried

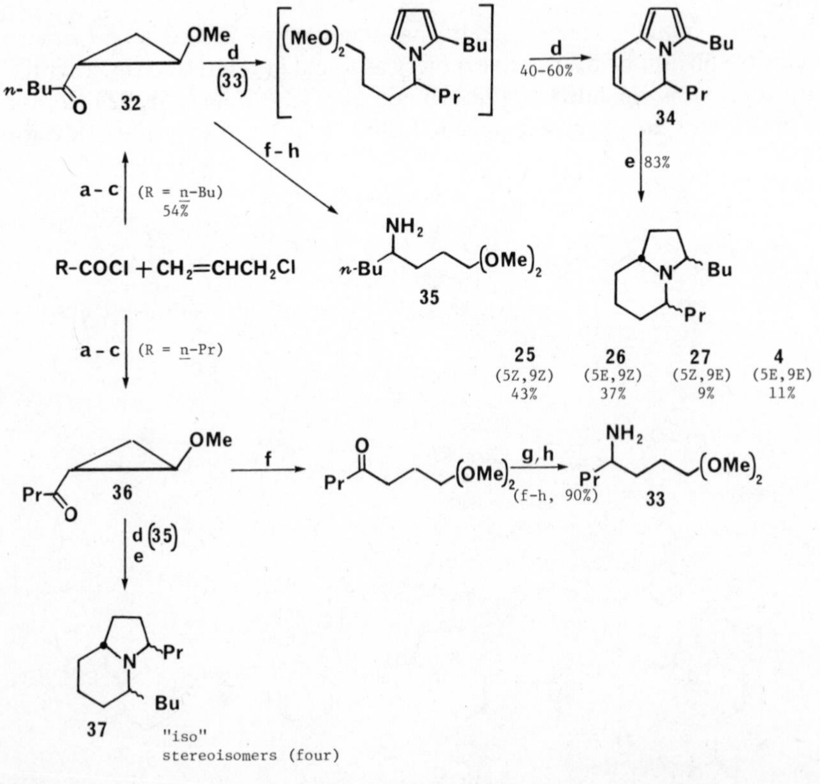

Scheme 72. Synthesis of 3-butyl-5-propylindolizidine diastereomers [421]. (a) AlCl₃/CH₂Cl₂, −15°; (b) MeOH, Δ; (c) NaOMe/MeOH, Δ; (d) **33** or **35**, 2,6-lutidine/TFA/xylene, Δ; (e) H₂/Rh-Al₂O₃/EtOH-HOAc; (f) MeOH/TFA; (g) NH₂OH·HCl/NaOAc/EtOH; (h) Na/n-BuOH, Δ.

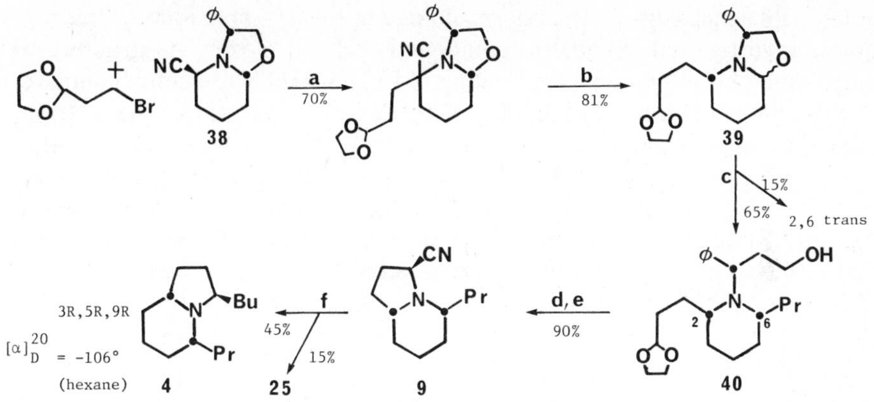

Scheme 73. Enantioselective synthesis of (−)-3-butyl-5-propylindolizidine (indolizidine **223AB**) [411]. (a) LiN(i-Pr)₂/THF, −78°; (b) AgBF₄/tetramethylguanidine then Zn(BH₄)₂, −50°; (c) n-PrMgBr/Et₂O, −50°; (d) H₂/Pd-C/MeOH; (e) HCl/CH₂Cl₂ then KCN; (f) n-BuMgBr/Et₂O, 0°.

out [42]. The resulting mixture co-chromatographed, peak for peak, with the normal mixture. However, the mass spectrum of the (5E,9E) isomer with the same retention time as **4** differed significantly from the natural alkaloid, allowing the (5E,9E) 3-n-propyl-5-n-butylindolizidine structure to be excluded for indolizidine **223AB** [409].

5.2.3. Enantioselective Synthesis of (−)-Indolizidine 223AB. Employing their versatile 2-cyanopiperidine alkylation procedure and the chiral synthon **38**, Husson's group [411] has recently completed an enantioselective synthesis of the 3R,5R,9R enantiomer of 3-butyl-5-propylindolizidine (Scheme 73), apparently identical in absolute configuration with the naturally occurring alkaloid [251]. The Grignard addition to the carbinolamine ether **39** gives chiefly the desired 2,6-*cis*-substituted piperidine **40** (*cis/trans* = 4.3/1). The chiral N-substituent was removed by hydrogenolysis, the aldehyde deblocked, and the resulting iminium salt treated with cyanide (e) to produce **9**. The second Grignard reaction (f) proceeded with less stereoselectivity, providing the desired (5E,9E) enantiomer **4** in a three-to-one excess over the (5Z,9Z) by-product **25**. Curiously, Natsume [414] reports no (5Z,9Z) by-product in the same reaction (see above) but does obtain a similar 3:1 mixture when the Grignard reaction is performed directly on the iminium salt intermediate.

5.3. Biological Activity

5.3.1. Toxicology. Toxicity data for the indolizidines are limited. Indolizidine **239CD** causes long-lasting locomotor difficulties and prostration after subcutaneous administration to mice at 4 mg/kg [23].

5.3.2. Pharmacology. The pharmacology of these simple indolizidines requires investigation. Synthetic compounds related closely in structure to indolizidine **223AB** antagonize binding of [^{3}H]H$_{12}$-HTX to acetylcholine receptor channel complexes [422]. Thus these indolizidines, like the histrionicotoxins, gephyrotoxin, and pumiliotoxin C, probably represent another structural class of noncompetitive blockers of neuromuscular transmission. A (5Z,9Z) 3-methyl-5-butylindolizidine (monomorine) related in structure to the (5E,9E) dendrobatid indolizidines has been isolated from Pharoah ants and serves as a component of the trail pheromone for this invertebrate [423, see also 410,416].

5.4. Addendum

A stereospecific synthesis of indolizidine 223AB (**4,** Scheme 68) proceeded using a [4 + 2] intramolecular cycloaddition reaction of an acylnitroso intermediate that was obtained on periodate oxidation of N-(5E,7E)-dodeca-5,7-dienoyl hydroxylamine. Subsequent key reactions of the cycloaddition product (82%) were stereospecific and proceeded with generally good yields [H. Iida, Y. Watanabe, and C. Kibayashi, *J. Am. Chem. Soc.,* **107,** 5534 (1985)].

6. PUMILIOTOXIN-A CLASS

6.1. Structures

In 1965, the investigation of active principles from skins of dendrobatid frogs was extended to *Dendrobates pumilio*, a small, brightly colored frog of Panama. C. W. Myers, a herpetologist at that time studying the reptile and amphibian fauna of Panama, became aware through a news release [441] of the research of Daly and Witkop on batrachotoxin isolated from a Colombian dendrobatid frog. Prompted by his interest in the extreme variability in color and habit of the Panamanian poison frog, *D. pumilio*, Myers contacted Daly. A collaboration on the nature of the poison in *D. pumilio* and on possible correlations of toxicity of different populations with their brightness of hue developed between Daly and Myers. In these initial studies on *D. pumilio*, batrachotoxin was not detected, but instead simpler alkaloids were found [424]. The two scientists have collaborated in both field work and investigation of skin extracts of dendrobatid frogs from 1965 until the present time. As a result of these collaborations, over 200 new alkaloids have been detected and some nine new species of dendrobatid frogs discovered [2].

Pumiliotoxin A and B were the first alkaloids isolated from the Panamanian poison frog *Dendrobates pumilio*; approximately 1.5 mg of each compound was isolated from extracts of 20 frog skins [424]. The molecular formulas were C$_{19}$H$_{33}$NO$_2$ and C$_{19}$H$_{33}$NO$_3$, respectively, and the nuclear magnetic resonance spectra were thought to be reminiscent of those expected of steroidal alkaloids. This interpretation was influenced by the fact

that the only known alkaloids from amphibians at that time were the batrachotoxins and the samandarines, both of which were steroidal in nature.
Pumiliotoxin A and B formed O-methyl ethers with methanolic hydrogen
chloride. Initial studies suggested the formation of an N-acetyl derivative
with, in addition, either one or two O-acetyl groups for pumiliotoxin A and
B, respectively. Further quantities of pumiliotoxin A and B were isolated
from later collections of $D.$ $pumilio$: From 250 skins, 17 mg of pumiliotoxin
A, 20 mg of pumiliotoxin B, and, in addition, 16 mg of a simpler alkaloid,
pumiliotoxin C, were isolated [425]. A more detailed investigation now revealed that pumiliotoxin A and B were bicyclic alkaloids with two double
bonds, only one of which underwent facile hydrogenation [23]. Pumiliotoxin
C proved unrelated to the other two alkaloids in structure and was soon
found to be a simple 2,5-dialkyl-cis-decahydroquinoline [426] (see Section
7.1). The formation of neutral N-acetyl derivatives from pumiliotoxin A and
B could not be confirmed. Analysis of nuclear magnetic resonance spectra
and mass spectra indicated that pumiliotoxin A and B differed only in the
terminal portion of a side chain; $—CH\!=\!CCH_3CHOHCH_2CH_3$ in pumiliotoxin A and $—CH\!=\!CCH_3CHOHCHOHCH_3$ in pumiliotoxin B. The allylic
hydroxyl group of the side chain was undoubtedly the source of the frus-
trating instability of pumiliotoxin A and B under acidic conditions and the
site of formation of O-methyl ethers. This instability thwarted repeated efforts to prepare a crystalline salt or derivative. Because of this, ten years
after their initial isolation, the structures of pumiliotoxin A and B were still
undeciphered.

During this period, structures for other dendrobatid alkaloids such as
pumiliotoxin C, the various histrionicotoxins and gephyrotoxin were determined. In addition, alkaloid profiles were delineated in many species of
dendrobatid frogs. Thin-layer and gas chromatographic characteristics, mass
spectra, and chemical properties with respect to hydrogenation and acetylation had been determined for over 100 alkaloids. Some two dozen of these
alkaloids [23] appeared to be related in structure to pumiliotoxin A and B,
based on mass spectra similarities, namely a prominent ion of C_4H_8N (m/z
70) accompanied usually by either a prominent ion of $C_{10}H_{16}NO$ (m/z 166)
or $C_{10}H_{16}NO_2$ (m/z 182). Pumiliotoxin A and B and 8 other alkaloids exhibited the $C_{10}H_{16}NO$ ion, while 11 alkaloids exhibited the $C_{10}H_{16}NO_2$ ion. The
11 alkaloids that exhibited the $C_{10}H_{16}NO_2$ ion (m/z 182) were deemed an
allo-series. Another four alkaloids showed a prominent C_4H_8N ion, but the
other most intense fragment ions were at m/z 84, 110, 138, 150, or 168, not
at 166 or 182. All of these compounds were first assigned tentatively to the
pumiliotoxin-A class [23], but in a later, more conservative review [24] only
the alkaloids exhibiting the major peaks at 70, 166, or 182 were retained as
members of the pumiliotoxin-A class or allopumiliotoxin subclass. The alkaloids that did not exhibit a major peak at 166 or 182 were documented
under "other alkaloids" [24].

During screening of various extracts we hoped that a source of a simpler alkaloid related to pumiliotoxin A and B would be discovered and that ample supplies of such a simpler alkaloid would provide the key to this class of dendrobatid alkaloids. One such compound was alkaloid **251D**. It was certainly less complex than pumiliotoxin A and B, having an empirical formula of $C_{16}H_{29}NO$ and only one double bond [23]. The alkaloid **251D** occurred in a number of species, including an Ecuadorian poison frog, *Dendrobates tricolor*. It did not, however, appear to be a major alkaloid in extracts from some eight specimens of this frog collected by Daly and Myers in 1974. At this time the remarkable volatility of alkaloid **251D** was not appreciated and **251D** was, indeed, a major component of the alkaloids in extracts from *D. tricolor*. Further extracts of this frog were obtained, not as a source of **251D**, but instead as a source of another alkaloid, present in trace amounts, but exhibiting remarkable pharmacological properties. This alkaloid, whose molecular formula is still uncertain (see Section 10.5), was a potent analgesic. Extracts from skins of some 750 *D. tricolor* provided much less than 1 mg of the analgesic alkaloid, but by serendipity did provide some 21 mg of pumiliotoxin **251D** [425], the long-sought simpler analog of pumiliotoxin A and B.

X-ray analysis of a crystal of the hydrochloride of **251D** provided the structure and absolute configuration: Pumiliotoxin **251D** is 8-hydroxy-8-methyl-6-(2'-methylhexylidene)-1-azabicyclo[4.3.0]nonane [425] (Fig. 7C). Unlike most of the other dendrobatid alkaloids, this compound could not be simply derived by postulated biosynthetic ring closures of a precursor 2,6-disubstituted piperidine (Section 11). It was an indolizidine, but contained an anomalous substituent at what would have corresponded to the 5-position of a precursor 2-substituted piperidine. The hydroxyl group in pumiliotoxin **251D** is very unreactive and provides acetyl and trimethylsilyl derivatives

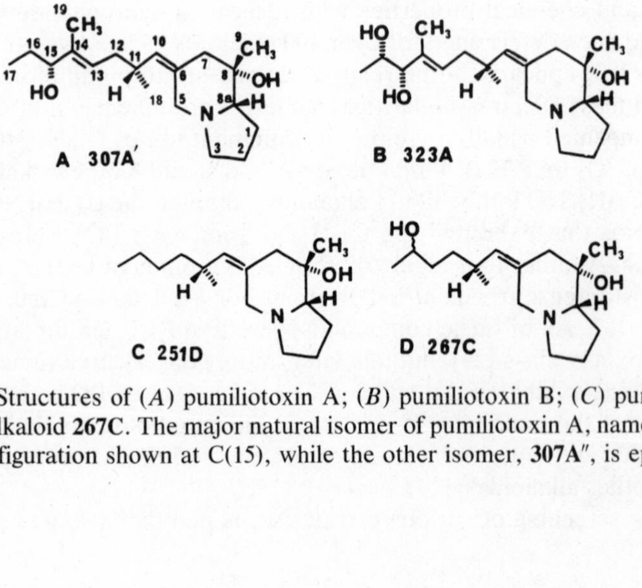

Figure 7. Structures of (*A*) pumiliotoxin A; (*B*) pumiliotoxin B; (*C*) pumiliotoxin **251D**; (*D*) alkaloid **267C**. The major natural isomer of pumiliotoxin A, namely **307A'**, has the configuration shown at C(15), while the other isomer, **307A''**, is epimeric at C(15).

only with great difficulty [425]. The cryptic nature of the hydroxyl group probably contributed to the volatility of pumiliotoxin **251D** and to relatively high R_f values (0.52) on silica gel thin-layer chromatography. Hydrogenation of **251D** yields a dihydro derivative with a more reactive hydroxyl group and a low R_f value of 0.26 ($CHCl_3:CH_3OH$, 9:1). Acetyl and trimethylsilyl derivatives of dihydropumiliotoxin **251D** are readily formed. Reduction of the double bond from the less hindered aspect should yield a dihydro derivative with the 8-methyl group axial and the 6-hexyl and the 8-hydroxyl groups now equatorial.

Analysis of nuclear magnetic resonance spectra now provided the long sought key to the structures of pumiliotoxin A and B. Comparison of their ^{13}C spectra to that of pumiliotoxin **251D** allowed identification of all the carbons of the indolizidine portion and of initial carbons of the side chain, namely, $=CHCHCH_3—$. The remaining seven-carbon moieties of pumiliotoxin A and B were readily formulated from proton and ^{13}C magnetic resonance spectra, although some question remained as to the configuration of the $\Delta^{13,14}$- double bond. It was initially tentatively assigned the less stable (**Z**)-configuration [424]. Nuclear-Overhauser effects, namely a marked intensity enhancement in the C(13) olefin proton resonance peak by irradiation of the C(15) proton resonance peak of pumiliotoxin B, indicated that the 13,14-double bond was actually in the more stable (**E**)-configuration [427]. Pumiliotoxin B was oxidized at the glycol function to an aldehyde ($C_{17}H_{27}NO_2$), a portion of which was oxidized with Corey's procedure to the carboxylic acid methyl ester. The chemical shifts of the C(13) olefin resonance peak of the aldehyde (δ 6.40) and the carboxylic acid methyl ester (δ 6.67) were compatible with the (**E**)-configuration [427]. The relative configuration of the two side chain hydroxyl groups in pumiliotoxin B has been defined [427,428]. Proton magnetic resonance spectra of pumiliotoxin B and its phenylboronides were compared to those of model compounds and the 15,16-diol in pumiliotoxin B was assigned the threo configuration. The absolute configuration of the hydroxyl groups was later defined as **15R,16R** by identification of the ozonolysis product from pumiliotoxin B di-*O*-acetate with authentic di-*O*-acetyl derivative of 3,4-dihydroxy-2-pentanone synthesized from (−)tartaric acid [429]. Thus the structures of pumiliotoxin A and B appeared to be completely defined (Fig. 7). However, pumiliotoxin A was recently shown to consist of a mixture of 15-hydroxy epimers [430, see below].

Interpretation of the ^{13}C and proton magnetic resonance spectra of pumiliotoxin A, B and **251D**, prior to X-ray analysis of a crystal of **251D**, had not been successful in large part due to the misinterpretation of resonances due to the carbon at C(8a): The ^{13}C resonance peak for C(8a) is at 71.8 ppm. It was assumed incorrectly that this carbon had an oxygen rather than a nitrogen substituent. Once this misassignment was recognized, interpretation of spectra of pumiliotoxin A, B, and **251D** was relatively straightforward.

Boronate and dimethylsilanate derivatives form very readily at the 15,16-diol function of pumiliotoxin B. Indeed, a portion of pumiliotoxin B often was converted to a dimethylsilanate during gas chromatography on OV-1 columns. This led to an incorrect postulate of the presence of three dendrobatid alkaloids 379, 381, and 395 [23]. All of these compounds were subsequently shown to represent artifacts formed by variable formation during gas chromatography of dimethylsilanates from pumiliotoxin B, a dihydropumiliotoxin B, and an alkaloid now designated pumiliotoxin 339A.

One further example of the pumiliotoxin-A class of alkaloids was recently isolated and its structure defined on the basis of nuclear magnetic resonance spectroscopy [261]. This is pumiliotoxin 267C (Fig. 7), which was isolated from a Brazilian bufonid toad, *Melanophryniscus moreirae*. This alkaloid has not been detected in any of the dendrobatid species.

The structures of the allopumiliotoxins could now be postulated based on the mass spectral fragmentation patterns. The typical mass spectra of the pumiliotoxin-A class of compounds with major ions of C_4H_8N (m/z 70) and either $C_{10}H_{16}NO$ (m/z 166) for the pumiliotoxins or of $C_{10}H_{16}NO_2$ (m/z 182) for the allopumiliotoxins would appear to derive from the indolizidine portion of the molecule. Thus the allopumiliotoxins must contain an additional hydroxyl group in the indolizidine system. The hydroxy is not at the 9 (methyl)-position, since this situation would result in a major loss of CH_2OH, a result not observed for allo compounds. The additional hydroxyl group undergoes facile acylation, and hence is a secondary hydroxyl. The tertiary hydroxyl at the 8-position of pumiliotoxin 251D, as mentioned previously, is highly hindered and forms O-acetyl and O-trimethylsilyl derivatives only with difficulty. Analysis of the mass spectra of perhydro derivatives of the allopumiliotoxin-A series suggests that the additional hydroxyl of the indolizidine ring in this series is at the 7-position adjacent to the other ring hydroxyl.

The initial magnetic resonance spectra of small amounts of two isomeric allopumiliotoxins (323B′ and 323B″) isolated as minor alkaloids from *Dendrobates auratus* confirmed the position of the additional hydroxyl at position 7 and suggested that both 7-hydroxy epimers occur in this alkaloid series [431]. The interpretation provéd to be incorrect when larger quantities of allopumiliotoxin 323B′ and 323B″ were isolated from skin extracts of *D. auratus* [430]. The designation allopumiliotoxin 323B had undoubtedly been applied previously to both isomers [23], since they do not separate on gas chromatography or thin-layer chromatography, but only on high pressure liquid chromatography. The nuclear magnetic resonance spectra and the failure of allopumiliotoxin 323B′ and 323B″ to form phenyl boronides or to react with periodic acid led to the conclusion that both compounds contained a *trans*-diaxial 7,8-diol in the indolizidine ring (Fig. 8). It thus appeared most likely that both differed only in the configuration of the 15-hydroxy group [430]. Pumiliotoxin A (307A) was at the same time found to consist of a pair of isomers that also could not be separated except by high-pressure liquid chromatography [430]. These were designated 307A′ and 307A″ and also

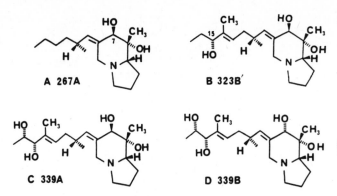

Figure 8. Structures of (*A*) allopumiliotoxin **267A**; (*B*) allopumiliotoxin **323B'**; (*C*) allopumiliotoxin **339A**; (*D*) allopumiliotoxin **339B**. The configurations at the 15-hydroxy group in the two epimers of allopumiliotoxin **323B**, namely **323B'** (major) and **323B''** (minor), are unknown. The configuration of the major isomer **323B'** is presumably the same as that of pumiliotoxin B (**323A**) and the major 15-hydroxy epimer of pumiliotoxin A (**307A'**).

appeared likely to be epimeric at the 15-hydroxy group. Pumiliotoxin A from *D. pumilio* consisted of about a 3:2 mixture of **307A'** and **307A''**. Comparison of **307A'** and **307A''** to synthetic 15-S-pumiliotoxin A [432] by Overman and Lin revealed that the major epimer **307A'** had the same configuration at the 15-hydroxyl group as pumiliotoxin B (Fig. 8).

Other allopumiliotoxins were isolated from skin extracts of 1050 *Dendrobates pumilio* [430]. Allopumiliotoxin **267A** and **339A** (Fig. 8) had the same *trans*-diaxial 7,8-diol as had **323B'** and **323B''**. However, an isomer **339B** (Fig. 8) was also isolated; this isomer readily formed a diboronide with phenylboronic acid, indicating a *cis*-7,8-diol. The nuclear magnetic resonance spectrum confirmed this conclusion: The juxtaposition of the hydroxyl group and H(10) resulted in a downfield shift of H(10) compared to allopumiliotoxins that contain the *trans*-diaxial 7,8-diol moiety.

Magnetic resonance spectral analysis has allowed formulation of two more pumiliotoxins: The first, an alkaloid with properties, similar to those reported for **307B**, has been isolated from *Dendrobates pumilio* [412]: The magnetic resonance spectra indicate that it is 13,14-dihydro-15-ketopumiliotoxin A (Fig. 9). It has a small fragment ion corresponding to loss of an OH group. It has now proven different from **307B** and will be named **307F** [412]. Another trace alkaloid, namely **321**, has been isolated from *D. pumilio* [412] and appears, based on magnetic resonance spectroscopy, to be 15-*O*-methylpumiliotoxin A (Fig. 9). It is possible that it actually represents an artifact formed by reaction at the allylic hydroxyl of pumiliotoxin A during isolation. *O*-Methyl ethers can be prepared from the pumiliotoxin A and B with methanolic HCl [424].

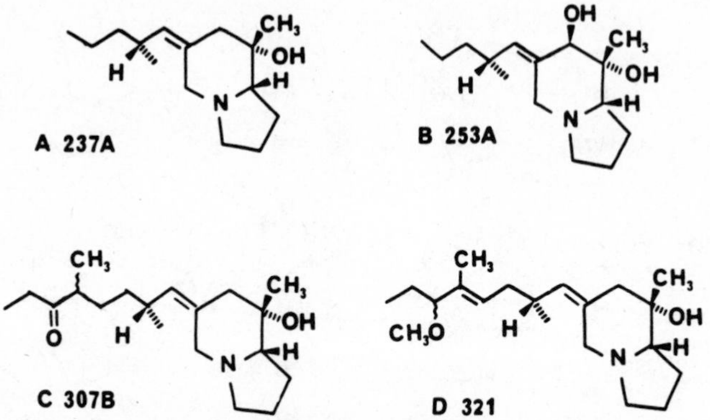

Figure 9. Tentative structures for other pumiliotoxins. (*A*) Pumiliotoxin **237A**; (*B*) allopumiliotoxin **253A**. (*C*) pumiliotoxin **307B** (Structure shown is that of a new alkaloid to be named **307F**, see text). (*D*) pumiliotoxin **321**. The last alkaloid may be an artifact formed from methanolysis of pumiliotoxin A during extraction and isolation.

Complete structural elucidation of the remaining compounds of the pumiliotoxin-A class and allopumiliotoxin-A class will require isolation of sufficient quantities for nuclear magnetic resonance spectroscopy. However, based on mass spectra of the alkaloids and their perhydro derivatives and on biosynthetic considerations, structures can be proposed for two of these compounds, namely pumiliotoxin **237A** and allopumiliotoxin **253A** (Fig. 9).

Properties of dendrobatid alkaloids of the pumiliotoxin-A and allopumiliotoxin classes are reported in the following section in the standard format (see Section 3.1). Empirical formulas that are based only on analogy and on chemical and chromatographic properties and that have not been confirmed by high resolution mass spectrometry are in quotations. Certain of the trivial names [23,24] are no longer used now that the structures of these alkaloids are better understood. Two alkaloids, namely **267C** and **267D**, are included even though they have not been detected in dendrobatid frogs [261]: Both are pumiliotoxins based on structure elucidation (**267C**) and mass spectra (**267D**).

Pumiliotoxin Class

207B. $C_{13}H_{21}NO$, 0.47, 161°, *m/z* 207 (10), 190 (15), 166 (100), 70 (80). H_2 derivative, *m/z* 209, 70. 1D.

237A. $C_{15}H_{27}NO$, 0.58, 167°, *m/z* 237 (4), 236 (3), 220 (3), 194 (12), 166 (40), 70 (100). H_2 derivative, *m/z* 239 (5), 196 (5), 168 (10), 110 (30), 84 (100), 70 (45). 1D. Tentative structure: Fig. 9.

251D. $C_{16}H_{29}NO$, 0.52, 172°, *m/z* 251 (5), 250 (3), 236 (2), 234 (1), 194 (5), 176 (3), 166 (78), 70 (100). H_2 derivative, *m/z* 253 (18), 110 (70), 84 (100), 70 (55). 1D. Structure: Fig. 7.

265D. "$C_{16}H_{27}NO_2$," —, 191°, m/z 265 (5), 222 (14), 166 (100), 70 (35). H_4 derivative. 2D.

267C. $C_{16}H_{29}NO_2$, 0.28, 190°, m/z 267 (16), 194 (12), 166 (100), 84 (18), 70 (75). H_2 derivative. 2D. Structure: Fig. 7. From a nondendrobatid anuran [261].

267D. $C_{16}H_{29}NO_2$, —, 178°, m/z 267 (13), 250 (10), 194 (16), 166 (100), 70 (80). H_2 derivative. 2D. From nondendrobatid anurans [261].

281A. $C_{17}H_{31}NO_2$, 0.28, 205°, m/z 281 (4), 280 (2), 264 (2), 194 (12), 166 (72), 70 (100). H_2 derivative, m/z 283 (1), 282 (2), 266 (4), 208 (40), 138 (10), 110 (10), 84 (100), 70 (85).

297B. $C_{18}H_{35}NO_2$, 0.35, 222°, m/z 297 (10), 166 (92), 70 (100). H_4 derivative.

307A. Pumiliotoxin A, $C_{19}H_{33}NO_2$, 0.36, 216°, m/z 307 (5), 290 (3), 278 (4), 206 (10), 194 (16), 176 (10), 166 (85), 70 (100). H_2 derivative, m/z 309 (3), 308 (2), 210 (8), 166 (30), 110 (45), 84 (100), 70 (40). H_4 derivative, m/z 311 (3), 110 (32), 84 (100), 70 (55). 2D. O-acetyl derivative. Structure: Fig. 7. Pumiliotoxin A from *Dendrobates pumilio* consisted of a 3:2 mixture of isomers **307A'** and **307A"**, which do not separate on thin-layer or gas chromatography; hence their relative proportions in pumiliotoxin A from other species are unknown.

307B. $C_{19}H_{33}NO_2$, 0.36, 211°, m/z 307 (12), 306 (4), 290 (2), 194 (24), 193 (45), 166 (100), 70 (56). H_2 derivative, m/z 309 (1), 280 (3), 208 (10), 128 (25), 110 (60), 84 (100), 70 (45). H_4 derivative, m/z 311 (1), 110 (25), 84 (100), 70 (30). 1D. Structure shown in Fig. 9 is that of a new alkaloid **307F** (see text).

307D. "$C_{18}H_{31}NO_3$," —, 234°, m/z 307 (8), 306 (5), 292 (3), 290 (5), 264 (4), 262 (11), 206 (11), 194 (18), 166 (100), 70 (85). H_4 derivative, m/z 311, 110, 84, 70.

309A. "$C_{19}H_{35}NO_2$," 0.38. 218°, m/z 309 (9), 308 (3), 292 (2), 280 (4), 206 (4), 194 (15), 176 (5), 166 (100), 110 (10), 84 (20), 70 (51). H_2 derivative. 2D.

309C. "$C_{19}H_{35}NO_2$," —, 210°, m/z 309 (3), 308 (2), 292 (1), 280 (4), 266 (4), 194 (15), 166 (100), 70 (90). H_2 derivative, m/z 311 (3), 110 (25), 84 (100), 70 (30).

321. $C_{20}H_{35}NO_2$, —, 223°, m/z 321 (3), 304 (8), 166 (65), 70 (100). H_2 derivative. 1D. Structure: Fig. 9.

323A. Pumiliotoxin B, $C_{19}H_{33}NO_3$, 0.17, 230°, m/z 323 (10), 306 (5), 290 (2), 278 (12), 260 (2), 206 (15), 194 (26), 193 (22), 176 (15), 166 (75), 70 (100). H_2 derivative, m/z 325 (5), 166 (25), 110 (24), 84 (100), 70 (38). H_4 derivative, m/z 327 (3), 326 (20), 312 (1), 310 (2), 282 (14), 264 (12), 110 (30), 84 (100), 70 (43). 3D. Di-O-acetyl derivatives. Structure: Fig. 7.

325B. $C_{19}H_{21}NO_3$, —, 228°, m/z 325 (6), 309 (8), 166 (85), 70 (100). H_2 derivative. 3D.

353. "$C_{19}H_{31}NO_5$,"—, 240°, m/z 353 (4), 338 (10), 336 (5), 194 (20), 166 (80), 70 (100). 3D.

Allopumiliotoxin Class

237B. $C_{14}H_{23}NO_2$, 0.58, 168°, m/z 237 (11), 236 (2), 182 (60), 114 (30), 112 (25), 70 (100). 2D.

251I. "$C_{15}H_{25}NO_2$,"—, 180°, m/z 251 (7), 236 (4), 210 (15), 209 (20), 182 (11), 70 (100). H_4 derivative. 2D.

253A. $C_{15}H_{27}NO_2$, 0.30, 179°, m/z 253 (4), 252 (1), 236 (22), 210 ($C_{12}H_{20}NO_2$, 3), 208 (2), 192 (5), 182 ($C_{10}H_{16}NO_2$, 16), 114 ($C_6H_{12}NO$, 27), 112 ($C_6H_{10}NO$, 26), 70 (C_4H_6N, 100). H_2 derivative, m/z 255 (3), 238 (8), 110 (50), 84 (100), 70 (65). 2D. Tentative structure: Fig. 9.

267A. $C_{16}H_{29}NO_2$, 0.31, 186°, m/z 267 (15), 250 (33), 182 (18), 114 (52), 112 (44), 70 (100). H_2 derivative, m/z 269 (6), 252 (18), 110 (50), 84 (100), 70 (60). 2D. O-acetyl derivative. Structure: Fig. 8.

297A. $C_{17}H_{31}NO_3$, 0.13, 225°, m/z 297 (3), 296 (4), 280 (9), 236 (2), 210 (3), 194 (4), 193 (3), 182 (21), 114 (27), 112 (16), 70 (100). H_2 derivative, m/z 299 (4), 282 (12), 256 (6), 224 (4), 110 (50), 84 (100), 70 (75). 3D.

305. "$C_{19}H_{32}NO_2$," —, 214°, m/z 305 (23), 288 (10), 182 (43), 70 (100). H_6 derivative. 2D.

307C. $C_{19}H_{33}NO_2$, 0.39, 214°, m/z 307 (9), 290 (11), 182 (62), 70 (100). H_4 derivative. m/z 311 (3), 294 (3), 268 (5), 236 (2), 110 (30), 84 (100), 70 (35). 2D.

309D. "$C_{19}H_{35}NO_2$,"—, 208°, m/z 309 (4), 292 (10), 182 (16), 70 (100). H_2 derivative. 2D.

323B. $C_{19}H_{33}NO_3$, 0.20, 228° m/z 323 (5), 306 (10), 210 (4), 209 (3), 182 (50), 114 (20), 70 (100). H_4 derivative. 3D. Structure: Fig. 8. Allopumiliotoxin **323B** from *Dendrobates auratus* consisted of a mixture of two isomers, **323B'** > **323B''**, which do not separate on thin-layer or gas chromatography and hence their relative proportions in **323B** from other species is unknown. Only **323B'** was isolated from *D. pumilio*.

325A. $C_{19}H_{35}NO_3$, 0.20, 232°, m/z 325 (3), 308 (12), 280 (3), 210 (23), 182 (56), 114 (16), 112 (9), 70 (100). H_2 derivative. 3D.

339A. $C_{19}H_{33}NO_4$, 0.07, 243°, m/z 339 (3), 322 (3), 210 (6), 209 (3), 192 (14), 182 (75), 114 (25), 70 (100). H_4 derivative, m/z 343 (3), 342 (2), 324 (3), 138 (10), 110 (25), 84 (100), 70 (65). 4D. Structure: Fig. 8.

339B. $C_{19}H_{33}NO_4$, 0.12, 243°, m/z 339 (3), 322 (3), 294 (3), 210 (5), 209 (3), 192 (10), 182 (70), 114 (25), 70 (100). H_4 derivative. 4D. Structure: Fig. 8.

341A. $C_{19}H_{35}NO_4$, 0.48, 222 °, m/z 341 (4), 324 (3), 323 (1), 306 (1), 298 (3), 266 (4), 254 (7), 210 (5), 182 (10), 114 (10), 112 (40), 84 (22), 70 (100). H_2 derivative, m/z 343 (1), 342 (2), 328 (2), 266 (10), 138 (100), 110 (5), 84 (15), 70 (15). 3D.

341B. "$C_{19}H_{35}NO_4$,"—, 223°, m/z 341 (1), 324 (4), 182 (60), 114 (20), 112 (20), 70 (100).

357. $C_{19}H_{35}NO_5$, 0.30, 240°, m/z 357 (3), 340 (8), 272 (4), 182 (20), 128 (10), 110 (20), 84 (20), 70 (100). 4D.

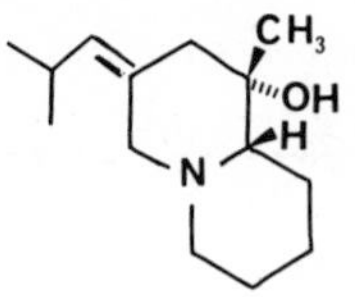

223G

Figure 10. Tentative structure for a homopumiliotoxin alkaloid **223G** [412].

One of the atypical dendrobatid alkaloids, namely **223G**, has recently been isolated from *Dendrobates pumilio* and a structure was proposed by Tokuyama based on mass spectrometry and magnetic resonance spectroscopy. Instead of an indolizidine ring, typical of the pumiliotoxin-A class, alkaloid **223G** has a quinolizidine ring (Fig. 10). Its mass spectrum has a base peak at m/z 84 ($C_5H_{10}N$) instead of at m/z 70. There are other piperidine-based dendrobatid alkaloids (see Section 10.1) that have a major fragment ion at m/z 84, but at present they are not included in a possible "homopumiliotoxin" class of dendrobatid alkaloids.

Homopumiliotoxins

 223G. $C_{14}H_{27}NO$,—, 163°, m/z 223 (18), 190 (22), 180 (39), 98 (27), 84 (100). H_2 derivative. 1D. Structure: Fig. 10.

Other physical and spectral properties of pumiliotoxins and allopumiliotoxins are reported in Table 9. The mass spectra of various pumiliotoxins and allopumiliotoxins have been reported in detail in [23,24,424,425,430]. The mass spectrum of pumiliotoxin A is depicted in [424]. Fragmentation pathways for pumiliotoxins, allopumiliotoxins, and hydrogenated derivatives are discussed in [23,425]. However, the suggested origin of the fragment of pumiliotoxins at m/z 166 ($C_{10}H_{16}NO$) must be questioned, since a synthetic analog of **237A** [386] lacking the side chain methyl [433] still exhibits a major fragment at m/z 166. The proton magnetic resonance spectra of pumiliotoxin A, B, and **251D** and of allopumiliotoxins **323B′**, **323B″**, **339A**, and **339B** have been depicted in [425,430]. The carbon-13 magnetic resonance peaks of pumiliotoxin A (**307A′** and **307A″**), pumiliotoxin B, pumiliotoxin **251D**, and allopumiliotoxins **267A**, **323B′**, **323B″**, **339A**, and **339B** have been reported [425,430]. Magnetic resonance spectral assignments have been discussed [24,425,430]; for the most recent assignments see [430]. The mass spectral data and magnetic resonance spectral assignments for the pumiliotoxin **267C**, detected as yet only in a bufonid toad, have been presented [261].

 Three alkaloids, namely **247**, **251F**, and **265B**, from the frog *Dendrobates bombetes* [254], were previously included in the pumiliotoxin-A class based primarily on the presence of a significant fragment ion of C_4H_8N (m/z 70) [23]. However, all three lack a major ion at m/z 166 or 182 and instead show

Table 9. Physical and Spectral Properties of Pumiliotoxins and Allopumiliotoxins [261,430,433][a]

	$[\alpha]_D^{25°}$	
Pumiliotoxin **251D**		
m.p. 205–206° (synthetic)		
HCl (synthetic)	+31.4°	(c 0.62, CH_3OH)
HCl (natural)	+17°	(c 0.15, CH_3OH)
Free base (synthetic)	−3.1°	(c 1.6, $CHCl_3$)
Pumiliotoxin A		
Natural mixture (**307A′** and **A″**)	+22.7°	(c 1.0, CH_3OH)
307A′	+14.3°	(c 0.74, $CHCl_3$)
307A″	+14.4°	(c 0.52, $CHCl_3$)
Pumiliotoxin B	+20.5°	(c 1.0, CH_3OH)
	+1.5°	(c 1.0, $CHCl_3$)
Allopumiliotoxin **267A**	+24.7°	(c 0.17, CH_3OH)
Allopumiliotoxin **323B′**	+22.3°	(c 1.0, CH_3OH)
323B″	+55.0°	(c 0.1, CH_3OH)
Allopumiliotoxin **339A**	+29.4°	(c 1.0, CH_3OH)
Allopumiliotoxin **339B**	+4.5°	(c 0.5, CH_3OH)
Alkaloid **267C**	+7.2°	(c 0.8, CH_3OH)

[a] For additional physical and spectral data on synthetic pumiliotoxins and allopumiliotoxins see [432–435].

anomalous base peaks with odd integer masses in their mass spectra. In addition, some four other alkaloids had been tentatively placed in the pumiliotoxin-A class based primarily on a major ion of C_4H_8N (m/z 70) [23]. None exhibit a major ion at m/z 166 or 182, but instead show other major ions ranging from m/z 84 to m/z 168. These alkaloids, **251E**, **281B**, **341A**, and **351**, and the above-mentioned alkaloids **247**, **251F**, **265B**, and several other dendrobatid alkaloids with major ions at m/z 70 are for the present included under piperidine-based alkaloids (Section 10.1) until further characterization is possible. It should be noted that upon saturation of the exo-cyclic 9,10 double bond, pumiliotoxins exhibit a major fragment at m/z 84, as do several of the piperidine-based alkaloids (see Section 10.1). However, the quinolizidine homopumiliotoxin **223G** (see above) also affords a base peak at m/z 84.

The pumiliotoxin-A class of alkaloids is widely distributed in frogs of the genus *Dendrobates* (see Section 11). Pumiliotoxin B was recently detected in a mantellid frog, *Mantella madagascariensis*, while a unique pumiliotoxin **267C** was a major alkaloid in a Brazilian bufonid toad, *Melanophryniscus moreirae* [261]. Another pumiliotoxin isomer (**267D**) was detected in *Mantella aurantiaca* and in a myobatrachid frog *Pseudophryne semimarmorata* from Australia [261]. The allopumiliotoxin subclass is also widely distributed in frogs of the genus *Dendrobates* (see Section 11). Allopumiliotoxin **323B** has now been detected in the Brazilian bufonid toad, in two species of mantellid frogs from Madagascar, and in the Australian myobatrachid frog [261]. Allpumiliotoxin **339A** was detected in *Mantella aurantiaca*.

6.2. Syntheses

Overman's group at the University of California, Irvine, has developed a novel, highly efficient asymmetric synthesis of the pumiliotoxin-A class of alkaloids, which applies two unusual reactions: (1) the nucleophilic ring opening of highly hindered epoxides by α-trimethylsilyl vinyl alanates, and (2) the intramolecular electrophilic substitution of a vinyl silane. A simpler analog of a naturally occurring pumiliotoxin **237A**, namely 11-desmethyl-**237A**, was the initial synthetic goal (Scheme 74) [433]. When the epoxide

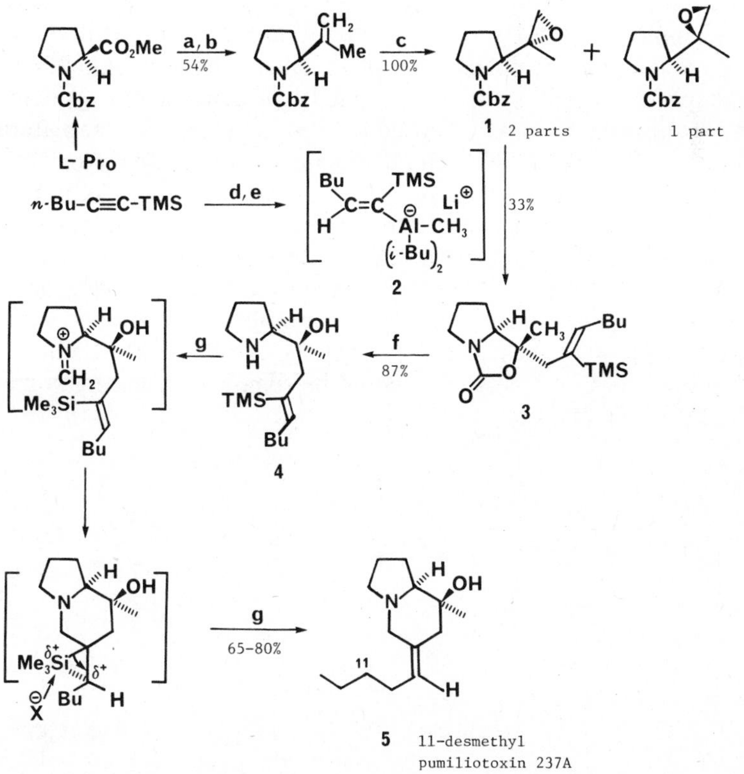

Scheme 74. Enantiospecific synthesis of a desmethyl analog of pumiliotoxin **237A** using an iminium ion–vinylsilane cyclization [433]. (*a*) MeMgI (2.2 eq.)/Et$_2$O; (*b*) SOCl$_2$/Et$_3$N/THF, −45°; (*c*) *m*-ClφCO$_3$H/hexane (CH$_2$Cl$_2$ is used in scaled-up preparations but gives a 1:1 mixture of epoxide diastereomers); (*d*) Al(*i*-Bu)$_2$H/Et$_2$O, Δ; (*e*) MeLi/Et$_2$O, r.t. then Δ, 48 hr; (*f*) KOH/MeOH, Δ; (*g*) (CH$_2$O)$_n$/EtOH, *d*-camphorsulfonic acid.

diastereomer 1 derived from L-proline was treated with the vinyl alanate 2, derived in two steps (d,e) from 1-trimethylsilylhexyne, and refluxed in ether for 48 hr, the adduct 3 was obtained in 33% yield. The cyclic urethane was hydrolyzed and the resulting amino alcohol 4 treated with paraformaldehyde and acid to provide solely the alkylidene indolizidine 5 in 65–80% yield. No (E)-isomer could be detected; evidently the stereochemical integrity of the trimethylsilyl olefin is completely maintained during the electrophilic substitution by partial bonding between the silicon atom and the β-cationic center [433].

Further simple pumiliotoxin analogs 8 and 9 with unbranched side chains were later prepared for investigation of cardiac activity (Scheme 75) [436]. The carbamate intermediate 6 could be carried through step e (70%) and f (78%) to provide 8 or first hydroborated and oxidized to 7, then converted by steps e (57%) and f (60%) to 9.

6.2.1. Pumiliotoxin 251D.

The synthetic approach used for 5 now was directed to the more complex, naturally occurring pumiliotoxin 251D, where a ten-step asymmetric synthesis supplied this dendrobatid alkaloid in 6% overall yield from 1 [433]. The product was, as expected, the natural (+)-enantiomer (Scheme 76). The chiral alanate 12 is generated from 11, which was obtained by enantioselective reduction (82%) of n-butylethynyl ketone (10) with the chiral borane derived from 9-borabicyclo[3.3.1]nonane and (−)-α-pinene, followed by silylation, carbonate formation, and a cuprate displacement procedure developed by Macdonald. Epoxide ring opening as before with ensuing steps h–j, produced natural (+)-pumiliotoxin 251D. Cyclization was effected best using preformed oxazolidine 13. A small amount of the unwanted C(11) (S)-epimer could be removed by chromatography of the final product [433].

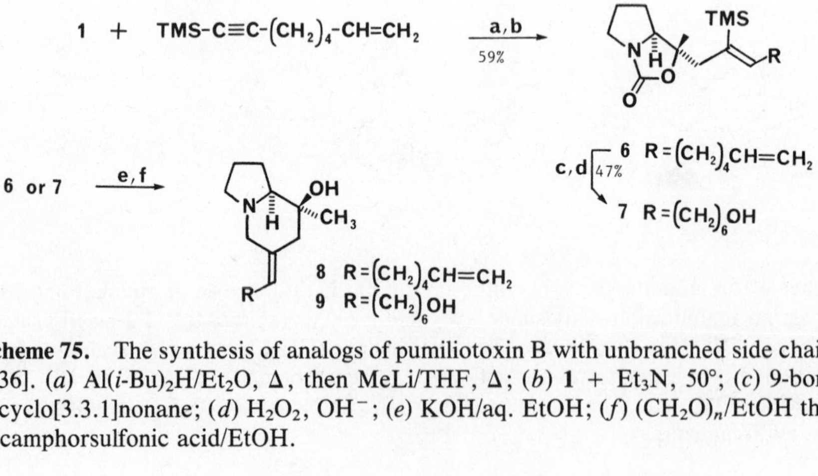

Scheme 75. The synthesis of analogs of pumiliotoxin B with unbranched side chains [436]. (a) Al(i-Bu)$_2$H/Et$_2$O, Δ, then MeLi/THF, Δ; (b) 1 + Et$_3$N, 50°; (c) 9-borobicyclo[3.3.1]nonane; (d) H$_2$O$_2$, OH$^-$; (e) KOH/aq. EtOH; (f) (CH$_2$O)$_n$/EtOH then d-camphorsulfonic acid/EtOH.

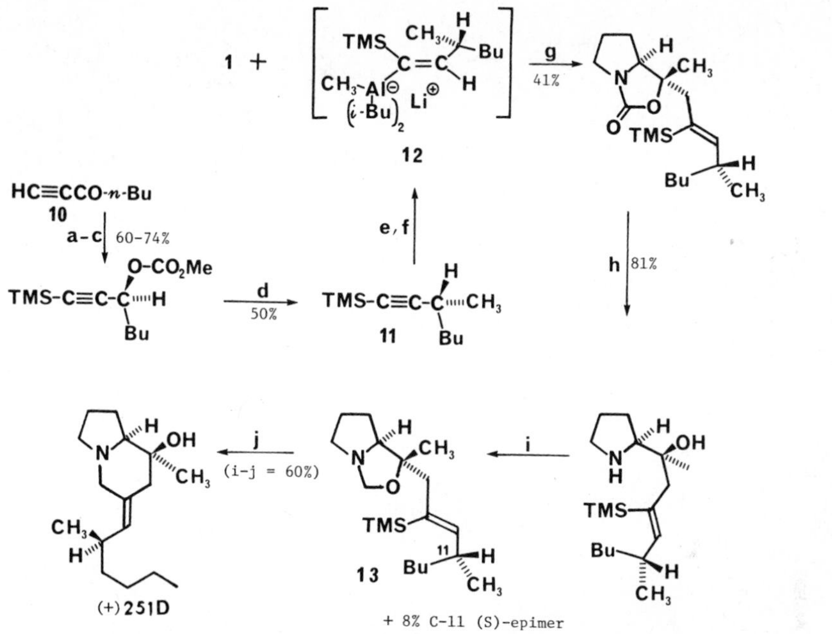

Scheme 76. Enantioselective synthesis of the (+)-pumiliotoxin **251D** [433]. (*a*) B-(3-pinanyl)-9-borabicyclo[3.3.1]nonane/THF; (*b*) MeLi; TMS—Cl; dil. HCl; (*c*) MeOCOCl; (*d*) MeMgBr-CuI complex/Et$_2$O-THF, 0°; (*e*) Al(*i*-Bu)$_2$H; (*f*) MeLi; (*g*) Et$_2$O, Δ; (*h*) KOH/MeOH, Δ; (*i*) (CH$_2$O)$_n$/EtOH, Δ; (*j*) *d*-camphorsulfonic acid/EtOH, Δ.

6.2.2. Pumiliotoxin A. Overman and Lin have recently reported [432] the enantioselective synthesis of a natural dextrorotatory pumiliotoxin A (Scheme 77). The synthesis both confirmed the tentative structure and established the absolute stereochemistry of the alkaloid.

The 1-trimethylsilyloctyne intermediate **18** possessing the required (**R**) chirality at C(3) and (**S**)-chirality at C(7) was synthesized efficiently and elegantly by steps *a*–*h* (Scheme 77). The allylic alcohol, 2-methyl-3-hydroxypent-1-ene, derived from methacrolein and ethyl magnesium bromide, is kinetically resolved by the method of Sharpless whereby the 3(**R**)-alcohol is preferentially epoxidized by a chiral reagent. The remaining optically pure 3(**S**)-alcohol **14** is converted to the benzyl ether and ozonized (*b*) to the methyl ketone. This undergoes a chelation-controlled diastereoselective addition of vinyl magnesium bromide and *in situ* propionylation to yield the threo-3(**R**)-allylic ester **15**, which is then subjected to the conditions of Ireland's enolate-Claisen rearrangement. A stereospecific sigmatropic rearrangement of the (**Z**)-enolate silyl enol ether, **16**, followed by esterification (*d*), provides the 4(**E**)-olefinic-2(**R**)-methyl ester **17** as the major product. This intermediate is transformed in a straightforward, high-yield sequence

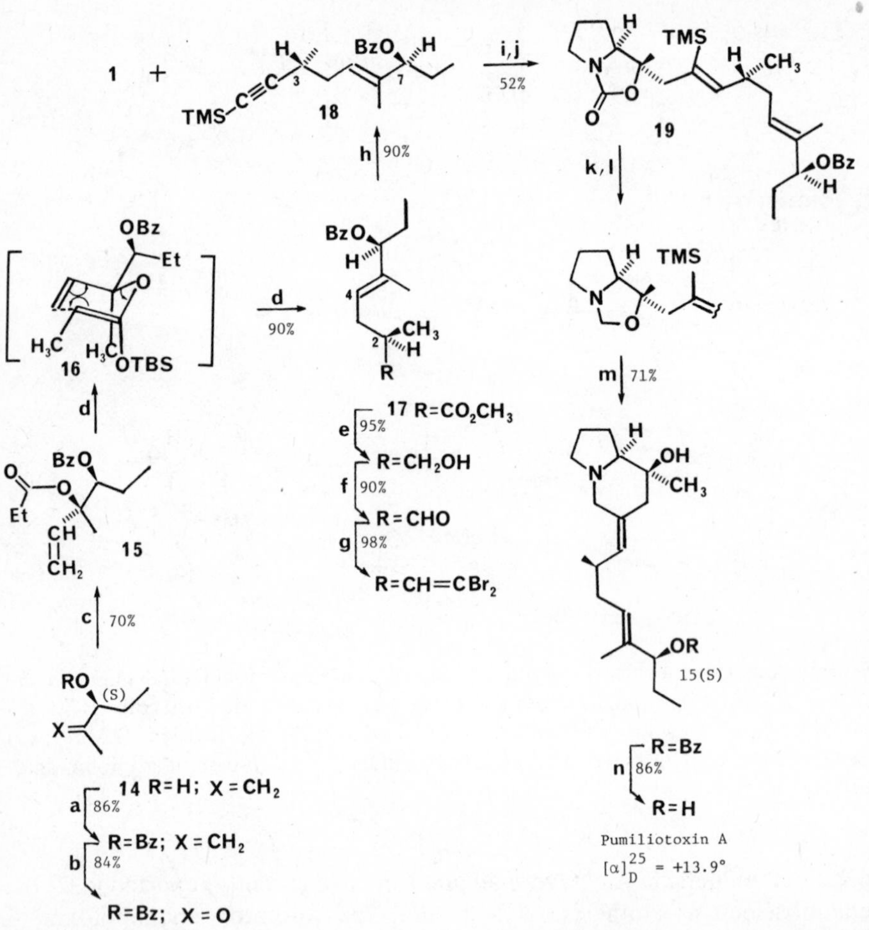

Scheme 77. Enantioselective synthesis of (+)-15S-pumiliotoxin A [432]. The configuration at carbon-15 is the same as in pumiliotoxin B. (a) NaH/THF then BzBr; (b) O_3/MeOH, $-78°$ then Me_2S; (c) vinylmagnesium bromide/THF then EtCOCl/HMPT; (d) LiN(i-Pr)$_2$/THF-HMPT, $-78°$ then t-BuMe$_2$SiCl(TBS—Cl)/THF, r.t.; (e) LiAlH$_4$/Et$_2$O, $0°$; (f) DMSO/Py·SO$_3$/Et$_3$N; (g) ϕ_3P/CBr$_4$/CH$_2$Cl$_2$/K$_2$CO$_3$; (h) n-BuLi/hexane-THF, $-78°$ then TMS—Cl; (i) Al(i-Bu)$_2$H/hexane then MeLi/Et$_2$O; (j) **1** in THF, $60°$, 17 hr; (k) KOH/aq. EtOH, $80°$, 70 hr; (l) formalin/aq. MeOH; (m) (CH$_2$O)$_n$/TsOH·Py/MeOH-Py, $80°$, 91 hr; (n) Li/NH$_3$-THF, $-78°$.

to **18**. After alanate formation, reaction with the epoxide **1** provides the oxazolone **19**, which is converted in subsequent steps as indicated to pumiliotoxin A. In order to prevent destruction of the C(15) allylic ether, the iminium ion–vinyl silane cyclization is conducted at pH 4.5.

6.2.3. Pumiliotoxin B. The absolute configuration was unknown, but it was reasonably assumed that the absolute chirality of centers shared with pumiliotoxin **251D** would be the same. The *threo*-stereochemistry of the 15,16-diol was known [427,428] and the absolute configuration was 15R,16R [293]. Accordingly, a chiral benzyloxy trimethylsilyl acetylene **20** was synthesized by steps *a–f* as shown (Scheme 78) using an alcohol resolution step (*c*) with (+)-α-methylbenzylamine, a carbamate cleavage procedure developed by Pirkle, and a cuprate displacement of acetate (*f*) devised by Macdonald. This sequence provided **20** with the required **R** configuration at C(3), equivalent to C(11) in pumiliotoxin B. An improved procedure for epoxide

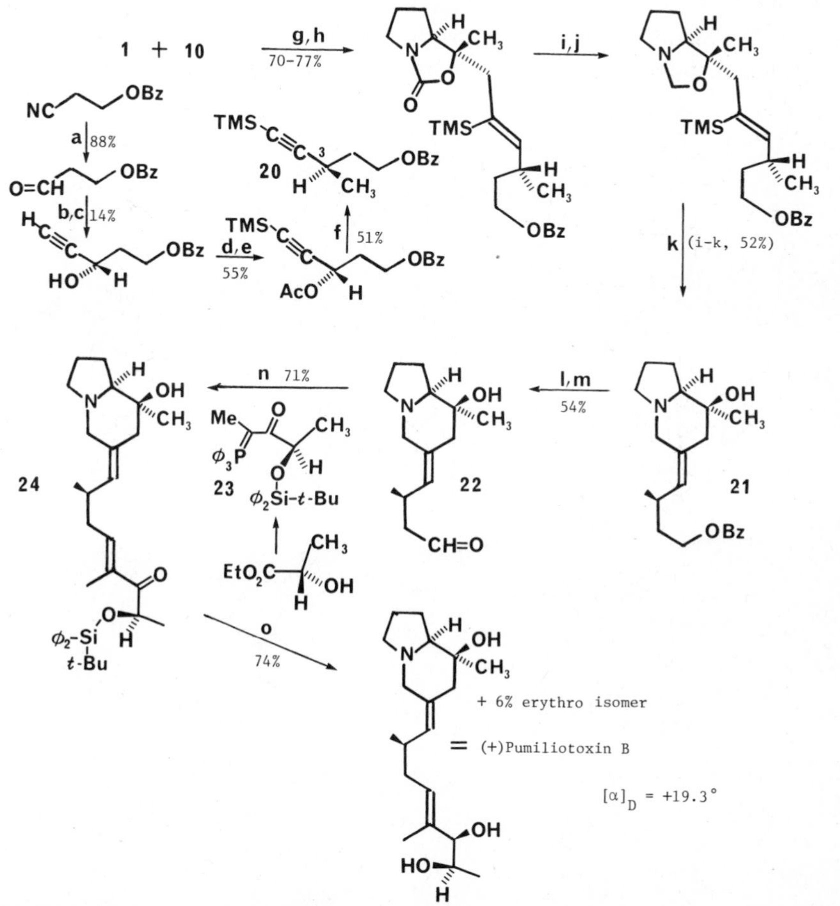

Scheme 78. Enantioselective synthesis of (+)-15R,16R-pumiliotoxin B [434]. (*a*) Al(*i*-Bu)$_2$H/hexane-THF; (*b*) LiC≡CH, THF, −75° → r.t.; (*c*) COCl$_2$, then (+)-α-methylbenzylamine, resolve and cleave with Cl$_3$SiH; (*d*) MeLi; TMS—Cl; dil. HCl; (*e*) Ac$_2$O/Py/DMAP; (*f*) MeMgBr/CuI/THF; (*g*) Al(*i*-Bu)$_2$H/hexane; (*h*) MeLi/THF, 60°, 25 hr; (*i*) KOH/aq. EtOH; (*j*) aq. formalin/MeOH; (*k*) (CH$_2$O)$_n$/*d*-camphorsulfonic acid/CH$_3$CN, 80°, 13 hr; (*l*) Li/NH$_3$-THF, −78°; (*m*) DMSO-(COCl)$_2$; (*n*) CH$_2$Cl$_2$, Δ; (*o*) LiAlH$_4$/THF, −20°.

ring opening (*h*) was developed with steps *i–k* proceeding as before to give **21**. Debenzylation with lithium-in-ammonia, followed by Swern oxidation, provided the aldehyde **22**. This compound was treated with the phosphorane **23** derived in five steps (18%) from ethyl-L-lactate, and the resulting enone derivative **24** was reduced under conditions favoring the *threo*-product. Surprisingly, the silyl ether was also deblocked (evidently by the AlH₃ generated *in situ*) and pumiliotoxin B, with optical rotation identical with that of the natural alkaloid, resulted in 1.8% overall yield from *N*-carbobenzyloxy L-proline [434].

6.2.4. Allopumiliotoxins. For syntheses of the allopumiliotoxins, where ring vicinal diols are required, Overman and Goldstein [435] developed a

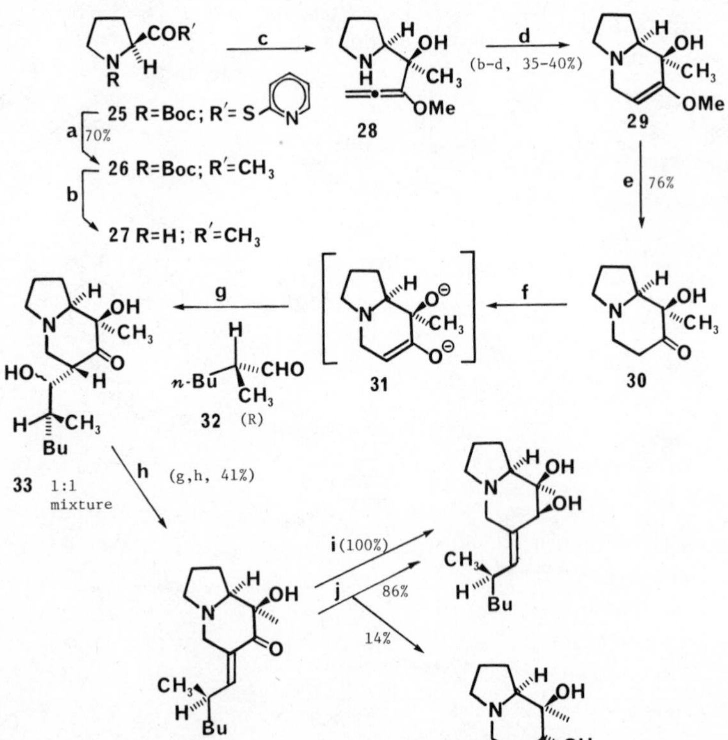

Scheme 79. Enantiospecific synthesis of (+)-allopumiliotoxin **267A** [435]. (*a*) LiMe₂Cu; (*b*) TFA/anisole/CH₂Cl₂; (*c*) CH₂=C=C(OCH₃)Li (5 eq.)/THF, −78°; (*d*) TsOH/CH₃CN; (*e*) 5% HCl; (*f*) TrLi/Et₂O, 0°; (*g*) **32**/Et₂O, 0°; (*h*) (CF₃CO)₂O/DBU/DMAP, 0°; (*i*) NaBH₄-CeCl₃; (*j*) LiAlH₄/THF, 0°.

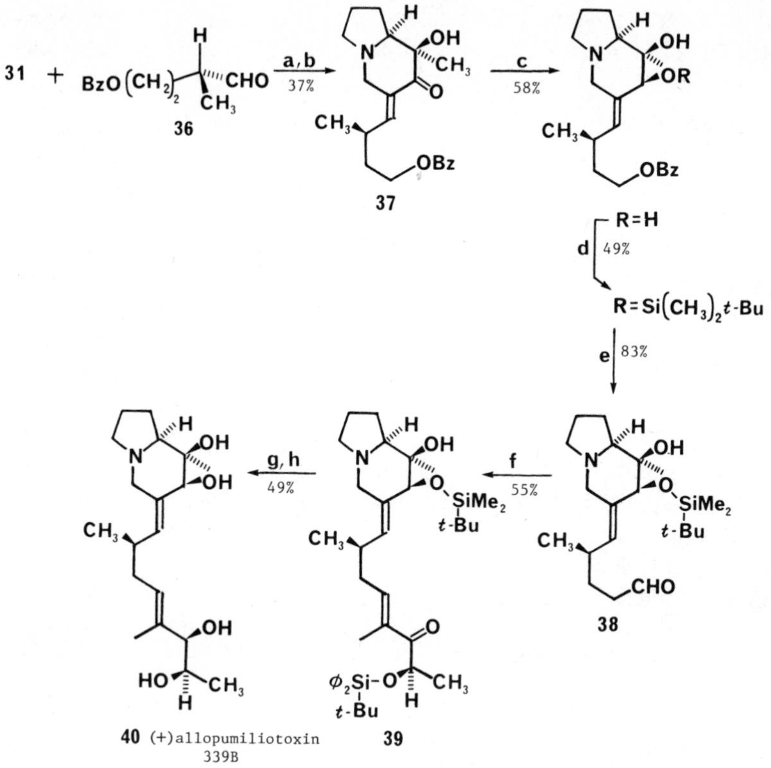

Scheme 80. Enantiospecific synthesis of the allopumiliotoxin 339B [435]. (a) Et$_2$O, 0°; (b) (CF$_3$CO)$_2$O/DBU/DMAP, 0°; (c) CeCl$_3$-NaBH$_4$; (d) n-BuLi/HMPT, −78°, then t-BuMe$_2$SiCl; (e) debenzylation and Swern oxidation; (f) **23** (Scheme 78); (g) LiAlH$_4$/THF, −20°; (h) (n-Bu)$_4$N$^\oplus$F$^\ominus$/THF, chromatographic purification.

new indolizidine synthesis using the addition of 1-lithio-1-methoxyallene to the methyl ketone **27** derived from the L-proline derivative **25** via **26** (Scheme 79). The cyclic Cram (or chelation-controlled) product is formed exclusively and the resulting amino allene **28** is cyclized with acid to the enol ether **29**, which on hydrolysis provides the key ketoindolizidine intermediate **30**. The lithium dianion **31**, formed with trityllithium (f), is treated with (**R**)-2-methylhexanal **32** and the intermediate diastereomeric aldol mixture **33** is dehydrated stereoselectively to the more stable (**E**)-enone **34**. The minor *threo*-diastereomer **35**, obtained by reduction with lithium aluminum hydride (j), proved identical with naturally occurring (+)-allopumiliotoxin **267A**. Using (**R**)-4-benzyloxy-2-methylbutanal **36** with similar aldol reaction and dehydration steps (a,b), the enone **37** resulted (Scheme 80). Erythro-selective reduction (c), blocking of the secondary alcohol (d), and debenzylation and oxidation (e) provided the aldehyde **38**, which was converted to the dienone **39** using the Wittig reaction with the chiral phosphorane **23**, used in the pumiliotoxin B synthesis. *Threo*-selective reduction (*g*) and desilylation (*h*) gave the tetrol **40** identical with natural (+)-allopumiliotoxin **339B** [435].

6.3. Biological Activity

6.3.1. Toxicity. Pumiliotoxins A and B were the first alkaloids after the batrachotoxins to be isolated from dendrobatid frogs. Both are relatively toxic compounds with minimum lethal doses for mice estimated at 2.5 and 1.5 mg/kg, respectively [424]. For this reason further alkaloids from dendrobatid frogs were termed toxins, which is unfortunate since pumiliotoxin C, the histrionicotoxins, and gephyrotoxin later proved to be relatively nontoxic. Pumiliotoxins A and B at lethal doses in mice elicit locomotor difficulties, clonic convulsions, and death in less than 20 min. Even at 1 mg/kg, pumiliotoxin B causes death in less than 20 min [23], indicating a toxicity greater than was first estimated. Pumiliotoxin **267A**, a member of the allo-pumiliotoxin-A subclass of dendrobatid alkaloids containing an axial 7-hydroxyl group, is less toxic, and at 2 mg/kg in mice causes locomotor difficulties but no deaths [23]. No toxicity data have been obtained for pumiliotoxin **251D**, the 7-deoxy congener of allopumiliotoxin **267A**.

6.3.2. Pharmacological Activity. Pumiliotoxin B, a representative member of the pumiliotoxin-A class of dendrobatid alkaloids, potentiates both direct and indirect evoked contractions of striated muscle [128,437–439]. In addition, pumiliotoxin B prolongs muscle contraction. The potentiation of muscle contracture by this alkaloid appears linked to mechanisms involving facilitation of calcium influx into the muscle fiber and/or a facilitation of release of calcium from the sarcoplasmic reticulum. The effect of pumiliotoxin B on the magnitude of muscle twitch is time- and stimulus-dependent. No blockade of neuromuscular transmission occurred with a low concentration of pumiliotoxin B in rat neuromuscular preparations [128], while in a frog neuromuscular preparation a concentration- and stimulus-dependent blockade of both direct and indirect elicited twitch follows the transient period of potentiated twitch [437]. A sixfold potentiation of twitch occurs in frog preparations at about 3 μM pumiliotoxin B. Pumiliotoxin B is even more potent in the preparations of individual muscle fibers, where maximal effects occur at less than 3 μM.

In the presence of high concentrations of calcium ion, the onset of potentiation of muscle contraction by pumiliotoxin B is delayed and the maximal effect is reduced [437]. Furthermore, no blockade of twitch occurs in the presence of high external calcium. In the absence of external calcium, pumiliotoxin B does not potentiate directly elicited muscle twitch. The results suggest that the potentiative effects of pumiliotoxin B are in some way linked to a facilitation of influx of calcium. Pumiliotoxin B restores directly evoked contractures in a glycerol-shocked frog neuromuscular preparation in which excitation-contraction coupling has been virtually disrupted. This result suggests that pumiliotoxin B may facilitate the process leading to evoked release of calcium from the sarcoplasmic reticulum. Pumiliotoxin B also enhances muscle contractures evoked by caffeine, an agent that directly causes release of calcium from the sarcoplasmic reticulum.

Although the site or sites of action at which pumiliotoxin B acts to potentiate striated muscle contracture require further delineation, the results strongly suggest that the mechanism involves transport systems and/or channels involved in calcium fluxes across the plasma and sarcoplasmic reticulum membranes. Effects of pumiliotoxin B appear to be readily reversible, although in whole muscle the lipid solubility of the pumilitoxin B may impede the removal of excess alkaloid [437]. Blockade of twitch by pumiliotoxin B during repetitive stimulation of muscle has been proposed to be linked to depletion of critical pools of calcium in the sarcoplasmic reticulum, due perhaps to the facilitation of evoked release by pumiliotoxin B. That pumiliotoxin B may displace critical internal pools of calcium was further suggested in experiments with the calcium-chelator EGTA. After treatment of muscle with EGTA and pumiliotoxin B, reintroduction of calcium can restore the potentiative actions of the alkaloid only if pumiliotoxin B is removed and then subsequently reapplied. Thus no marked potentiation of contractions occurs in media containing EGTA and pumiliotoxin B or during subsequent incubation with media containing calcium and pumiliotoxin B. Marked potentiation of contractions occurs only after washing and incubation with media containing calcium, but not pumiliotoxin B, followed by reinstatement of alkaloid.

Pumiliotoxin B has no apparent effect on sodium, potassium, or chloride conductances or on resting membrane potential [437]. The potentiative effects of pumiliotoxin B are fully manifest in preparations in which acetylcholine receptors are blocked by α-bungarotoxin. Dantrolene, an agent purported to interfere with release of calcium from the sarcoplasmic reticulum, and methoxyverapamil, a purported antagonist of calcium channels, only partially prevent the effects of pumiliotoxin B on muscle.

Pumiliotoxin B potentiates and prolongs muscle contractions in chick neuromuscular preparations [438]. Remarkably, in muscles from a dystrophic strain of chicks, pumiliotoxin B has no effect on contractions. Thus in confirmation of other data, calcium mechanisms appear aberrant in the dystrophic muscles of this strain of chicks.

Structure-activity correlations for pumiliotoxin-A class alkaloids have not been delineated fully in neuromuscular preparations because of the limited supplies of these alkaloids available from nature. Pumiliotoxin B is more active in potentiating frog muscle contraction than is pumiliotoxin A, which in turn is much more active than pumiliotoxin 251D [cited in 437]. Thus the two hydroxyl groups in the side chain appear important to the biological activity of this class of alkaloids. No data on allopumiliotoxin-A subclass of alkaloids, such as allopumiliotoxin 267A, are available in neuromuscular preparations.

Further evidence for facilitation by pumiliotoxin B of calcium translocation across membranes derives from studies on crayfish skeletal muscle [437]. Pumiliotoxin B enhances calcium-dependent action potentials in crayfish muscle with a half maximal effect at about 25 μM. Pumiliotoxin B also markedly potentiates direct evoked contractions of crayfish muscle.

Pumiliotoxin B at 10 μM enhances binding of radioactive nitrendipine to brain membranes [388]. It shares this property with histrionicotoxin, phencyclicine, and certain other drugs.

In guinea pig atrial strips, pumiliotoxin B (1.5–7.5 μM) causes a dose-dependent potentiation of spontaneous contractions and increases the rate of contractions [128]. The effects are readily reversible and are not blocked by a β-adrenergic antagonist.

Structure-activity correlations for pumiliotoxins and allopumiliotoxins have now been investigated in spontaneously beating guinea pig atrial strips [436]. Pumiliotoxin B increases force of contractures by three- to fivefold with half maximal effects at about 3 μM and increases rates by two- to threefold with half maximal effects at about 6 μM. The presence of an axial 7-hydroxyl substituent in allopumiliotoxin **339A** decreases the efficacy, but not the potency as a positive inotropic agent, while having only slight effects on activity as a positive chronotropic agent. The presence of an equatorial 7-hydroxyl substituent in allopumiliotoxin **339B** greatly decreases efficacy and potency as a positive chronotropic and inotropic agent. Pumiliotoxin A, which lacks one of the side chain hydroxyl groups of pumiliotoxin B, causes only a twofold increase in force of contracture at 54 μM, while having minimal effects on rate. The presence of an axial 7-hydroxyl substituent in allopumiliotoxin **323B'** and **323B"**, which are epimeric at the side chain 15-hydroxyl group, markedly enhances positive inotropic and chronotropic effects compared to pumiliotoxin A. Another congener, pumiliotoxin **251D**, which has a 6-(2'-methylhexylidene) side chain, and a synthetic analog with a 6-(6'-heptylenylidene) side chain, are cardiac depressants. Both lack hydroxyl groups in the side chain. The presence of an ω-1 hydroxyl group in the side chain of **251D** yields an alkaloid **267C**, which has weak positive inotropic effects and minimal chronotropic effects. The presence of an axial 7-hydroxyl group in the indolizidine ring of **251D** results in allopumiliotoxin **267A**, which has very weak positive inotropic effects, while retaining the negative chronotropic effects of **251D**. A synthetic analog with a 6-(7'-hydroxyheptylidene) side chain is a cardiac depressant even though it contains a side chain hydroxyl corresponding in position to the 16-hydroxyl group of the side chain of pumiliotoxin B (see Scheme 75). The positive chronotropic and inotropic effects of pumiliotoxin B in atrial strips are reversed only by relatively high concentrations of the calcium channel blockers nifedipine and verapamil, suggesting that pumiliotoxin B may owe its cardiotonic activities to effects on internal mobilization of calcium.

Pumiliotoxin B causes dose-dependent contractions and peristaltic movements in guinea pig ileum segments [128]. The effects are blocked by tetrodotoxin, suggesting that nerve activity and perhaps facilitation of transmitter release by pumiliotoxin B are involved. Neither muscarinic nor histaminergic antagonists prevent the effects of pumiliotoxin B in ileum.

Pumiliotoxin B both potentiates and prolongs contractions in frog neuromuscular preparations [437,438]. The prolongation of muscle contractions occurs with pumiliotoxin B even in calcium-free medium and is still present, even after the potentiative effects on twitch have reversed during repetitive stimulation and have been replaced by a marked reduction in twitch. The prolongation of twitch elicited by pumiliotoxin B results in a fusion of muscle twitches during tetanic stimulation. The mechanism involved in prolongation of contractions was first thought to involve inhibition of the calcium-dependent ATPase of sarcoplasmic reticulum [440]. This enzyme is primarily responsible for re-uptake of calcium released into the cytoplasm and, thus termination of muscle contractions. Subsequent studies revealed that the inhibition of the calcium-dependent ATPase by pumiliotoxins was due to an impurity, perhaps present in the solvent used for silica gel chromatography or formed during passage of solvent over silica gel [unpublished results]. Removal of this impurity, which inhibits the ATPase, did not affect the physiological effects of pumiliotoxin B (unpublished results). The reported inhibition of calcium-dependent ATPase of chick sarcoplasmic reticulum [438] is due undoubtedly to the impurity in the pumiliotoxin B.

Pumiliotoxin B has calcium-dependent effects on evoked release of acetylcholine from nerve terminals in frog neuromuscular preparations [438]. The alkaloid in a concentration- and stimulus-dependent manner causes repetitive endplate potentials in response to a single stimulation of nerve. This effect on neurotransmitter release is strongly calcium-dependent and may involve facilitation by pumiliotoxin B of evoked calcium transport through plasma membranes and/or membranes of the endoplasmic reticulum of the nerve terminal.

6.3.3. Summary.

Pumiliotoxin B and its congeners appear to have selective effects on nerve and muscle, linked in the case of pumiliotoxin B to apparent facilitation of evoked calcium translocation across both plasma membranes and internal membranes of calcium storage organelles. Certain congeners appear to have opposite inhibitory effects. Further studies now indicate that pumiliotoxin B causes repetitive firing after stimulation of nerve or muscle. This may be responsible for the apparent facilitation of calcium translocation.

6.4. Addendum

Analogs of pumiliotoxin A have been produced in high yield by a Lewis-acid catalyzed ene cyclization reaction using 2-acetylpyrrolidines, N-substituted with allyl derivatives (e.g., N—$CH_2C(=CH_2)CH_2$-n-Bu) and easily prepared from proline. A mixture solely of E- and Z-alkylidene cycloproducts is produced. A model synthesis afforded three parts E- to one part Z-stereoisomers (91% total) with two equivalents $AlCl_3$ in CH_2Cl_2 at 25°. The minor isomer proved identical with **5** (Scheme 74), synthesized earlier by the iminium ion–vinyl silane cyclization. Even greater E-selectivity resulted when bulkier N-allyl groups were employed, for example, E/Z = 6 for N—CH₂C(=CH₂)CH₂-*i*-Pr [L. E. Overman and D. Lesuisse, *Tetrahedron Lett.*, **26**, 4167 (1985)].

7. DECAHYDROQUINOLINES (PUMILIOTOXIN-C CLASS)

7.1. Structures

In the initial studies in the late 1960s on *Dendrobates pumilio*, three alkaloids were isolated [424,426] and designated as pumiliotoxins A, B, and C. Pumiliotoxins A and B were quite toxic and closely related in structure (see Section 6.1). Pumiliotoxin C was relatively nontoxic and quite different in structure. A total of 16 mg of pumiliotoxin C was isolated from skin extracts of 250 frogs [426]. The compound proved to be a 2,5-dialkyl-*cis*-decahydroquinoline. The absolute stereochemistry of pumiliotoxin C ([2S,4aS,5R,8aR]5-methyl-2-*n*-propyl-*cis*-decahydroquinoline; Fig. 11) was determined by X-ray crystallography of the hydrochloride salt [426]. It should be noted that the absolute stereochemistry at C(2) differs from that at C(2) of the histrionicotoxins. The mass spectrum of pumiliotoxin C was extremely simple, consisting, as would be expected, of a base peak due to loss of the *n*-propyl substituent at m/z 152 (C₁₀H₁₈N). The parent ion peak and other fragment ion peaks were very weak. A large number of other dendrobatid alkaloids were originally tentatively classified, based primarily on their mass spectra, as of a pumiliotoxin-C class or hydroxypumiliotoxin-C class [23]. Classification as a pumiliotoxin-C alkaloid was based primarily

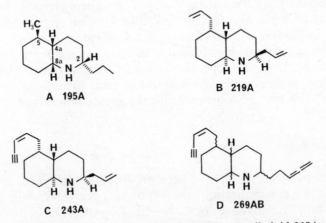

Figure 11.   Structures of (*A*) pumiliotoxin C (**195A**), (*B*) alkaloid **219A** and (*C*) alkaloid **243A**, and tentative gross structure of (*D*) alkaloid **269AB**.

on a mass spectrum exhibiting a dominant fragment ion at m/z 152, as in pumiliotoxin C, or at m/z 138, 166, or 180 consonant with the tentative formulations of decahydroquinoline alkaloids having, respectively, a hydrogen, ethyl, or propyl substituent in place of the 5-methyl group of pumiliotoxin C. Classification as a hydroxypumiliotoxin-C alkaloid was based primarily on a mass spectrum exhibiting a dominant fragment ion at m/z 154, 168, 182, or 196. Recognition of the presence of a variety of indolizidines and hydroxyindolizidines in dendrobatid extracts led to a more conservative approach, namely the tabulation of all of these compounds as "other alkaloids" [24] until further characterization was forthcoming. The use of deuteroammonia to assess exchangeable hydrogens in gas chromatographic–chemical ionization mass spectrometry [183,251] and the isolation of certain alkaloids in sufficient quantities to define close structural relationships to pumiliotoxin C [442] now allow a reasonable reassignment of several alkaloids to the decahydroquinoline class. The older terminology "pumiliotoxin-C class" is no longer used because of the likelihood of confusion with the structurally dissimilar pumiliotoxin-A class. Instead "decahydroquinoline" is used to describe pumiliotoxin C and structurally related dendrobatid alkaloids. All of these are bicyclic secondary amines, as shown by the gas chromatographic–mass spectral analytical protocols described in Section 3.1.

Some of these alkaloids, like pumiliotoxin C, are bicyclic secondary amines that exhibit a mass spectrum with a very small parent ion peak and a single major fragment ion at m/z 152 ($C_{10}H_{18}N$). Others, such as **219A** and **243A** (formerly **243**), are also bicyclic secondary amines, but unlike pumiliotoxin C, afford major fragment ions at m/z 178 and 202, respectively, rather than at m/z 152 (see tabulation below). Structures for three of the decahydroquinolines to be discussed below are presented in Fig. 11.

Certain alkaloids were previously isolated from *Dendrobates histrionicus* in sufficient quantities for partial characterization by nuclear magnetic resonance spectroscopy and were tentatively assigned decahydroquinoline structures [248]. These were three trace compounds designated **I**, **II**, and **III** that had been isolated in small amounts from extracts of some 3200 skins of *D. histrionicus*. All three compounds were bicyclic, and based on their mass spectra and proton magnetic resonance spectra, they were proposed as members of the decahydroquinoline class of dendrobatid alkaloids. These assignments remained rather tentative. At that time, the deuteroammonia method for determination of exchangeable hydrogens had not been developed, and subsequent studies have suggested that compounds **I** and **II** are actually tertiary amines, probably indolizidines.

Compound **I** (mol. wt. 195, $C_{13}H_{25}N$) is a saturated bicyclic compound that loses a C_4H_9 fragment to yield a base peak at m/z 138. The data, including a proton magnetic resonance spectrum, have been stated to be compatible with its structural formulation as a 2-*n*-butyl-*cis*-decahydroquinoline [248]. No evidence for the presence of an alkaloid with a secondary amine function and corresponding in other properties to compound **I** has been obtained with dendrobatid skin extracts, including several populations of *Dendrobates his-trionicus*, and therefore compound **I** probably corresponds to the tertiary amine alkaloid **195B**, which is tabulated with the indolizidines in Section 5.1.

Compound **II** (mol. wt. 223, $C_{15}H_{29}N$) is a saturated bicyclic compound that loses a C_3H_7 fragment to yield a base peak at m/z 180 [248]. No evidence for the presence of an alkaloid with a secondary amine moiety and corresponding in other properties to compound **II** has been obtained with dendrobatid skin extracts, and therefore compound **II** probably corresponds to the tertiary amine alkaloid **223A**, which is tabulated with piperidine alkaloids in Section 10.1.

Compound **III** (mol. wt. 269, $C_{19}H_{27}N$) contains two unsaturated five-carbon side chains based on its proton magnetic resonance spectrum and yields on catalytic reduction a decahydro derivative [248]. It is felt to be a decahydroquinoline even though its mass spectrum differs from other compounds proposed as belonging to the decahydroquinoline class, since it exhibits *two* major fragment ions corresponding to loss of C_5H_5 and C_5H_7. The decahydro derivative of compound **III** also differs from an isomeric semisynthetic deoxydodecahydrohistrionicotoxin (see Section 3.2.9), both in mass spectrum and proton magnetic resonance spectrum [248]. The structure has been tentatively proposed as that of a 2-(3,4-pentadienyl)-5-(2-penten-4-ynyl)-*cis*-decahydroquinoline (Fig. 11*D*). The possibility that compound **III** is a mixture of two isomeric decahydroquinolines, one that has 2-(3',4'-pentadienyl)-5-(2'-penten-4'-ynyl) substituents and that loses a C_5H_5 fragment on mass spectrometry, and the other that has 2-(2'-penten-4'-ynyl)-5-(3',4'-pentadienyl) substituents and that loses a C_7H_7 substituent on mass spectrometry, has been proposed [23]. Reduction of these proposed isomers would yield a common 2,5-dipentyl-*cis*-decahydroquinoline. The question of the structure and homogeneity of compound **III** will require isolation of sufficient quantities for detailed magnetic resonance spectral and chromatographic characterization. Even capillary columns do not resolve compound **III** [unpublished results], and the relative intensity of the m/z 204 and m/z 202 fragments are constant at about 2:1 in almost all extracts in which it has been detected. Compound **III** is, for the present, tabulated below as **269AB** in the decahydroquinoline class of dendrobatid alkaloids. In a few extracts, alkaloids with molecular weights of 269 afford mainly either m/z 204 or m/z 202 fragment ions. These alkaloids, namely **269A** and **269B**, are included under piperidine-based alkaloids (Section 10.1).

Another dendrobatid alkaloid, designated **219A**, had been, tentatively, proposed to be 2,5-diallyl-*cis*-decahydroquinoline [23]. However, the perhydro derivative of **219A** separated on gas chromatography from a synthetic 2,5-dipropyl-*cis*-decahydroquinoline (prepared by Overman and Jessup [443]) having the same stereochemistry at the 2- and 5-positions as pumiliotoxin C (see Section 7.2.2). The emergent temperature of perhydro-**219A** was 168°C on a 1.5% OV-1 column, while the synthetic dipropyl compound emerged at 165°C [unpublished results]. Alkaloid **219A** has now been isolated in milligram quantities from *Dendrobates histrionicus*, and the proton and ^{13}C magnetic resonance spectra are consonant with those expected of a 2,5-diallyl-*trans*-decahydroquinoline [442]. X-ray analysis of the hydrochloride revealed its structure as [2S, 4aS, 5S, 8aS] 2,5-diallyldecahydroquinoline (Fig. 9) [442]. A minor isomer of **219A**, termed **219A'**, occurs in some dendrobatid species and emerges prior to **219A** on gas chromatography. Both **219A** and **219A'** are tabulated below as decahydroquinolines.

The final major alkaloid, presently included in the decahydroquinoline class, is **243A**. This alkaloid yields a single major fragment by loss of an allyl side chain and seems likely to be a 2-allyl-5-(2'-penten-4'-ynyl)decahydroquinoline, as tentatively formulated in 1978 [23]. The proton and ^{13}C magnetic spectra are consonant with a *trans*-isomer [442]. There is a minor isomer, termed **243A'**, which occurs in certain species, and emerges prior to **243A** on gas chromatography. The stereochemistry of **243A** does, as in the case of **219A**, differ from that of pumiliotoxin C.

Alkaloids **219A**, **243A**, and **269AB** have also been discussed as perhaps representing members of a deoxyhistrionicotoxin class [26]. It is true that they usually occur together with the histrionicotoxins in the dendrobatid species. However, **219A** and **243A** have now proven to be *trans*-decahydroquinolines and alkaloid **269AB** is also now tentatively included in this group. The possible occurrence of deoxyhistrionicotoxins in dendrobatid extracts remains an attractive, but as yet unsupported hypothesis.

The properties of pumiliotoxin C and the other postulated decahydroquinolines are presented below in the standard format (Section 3.1). It is probable that certain of the so-called piperidine alkaloids (see Section 10.1) also will prove to be decahydroquinolines or hydroxydecahydroquinolines, but at present there is insufficient data or precedent for their inclusion. Both *cis*- and *trans*-decahydroquinolines occur naturally: Magnetic resonance spectra will be required to distinguish them. The mass spectra of *cis*- and *trans*-isomers would be expected to be nearly identical. *Trans*-fused isomers of pumiliotoxin C have been prepared (see Section 7.2).

Decahydroquinolines

153A. "$C_{10}H_{19}N$," —, 154°, *m/z* 153 (100), 152 (60). H_0 derivative. 1D. Tentative structure: a 5-methyldecahydroquinoline.

167D. "$C_{11}H_{21}N$," —, 154°, *m/z* 167 (100), 166 (53). H_0 derivative. 1D. Tentative structure: a 5-ethyldecahydroquinoline.

181D. "$C_{12}H_{23}N$," —, 156°, *m/z* 181 (3), 152 (100). Tentative structure: a 2-ethyl-5-methyldecahydroquinoline.

181E. "$C_{12}H_{23}N$," —, 156°, *m/z* 181 (100), 180 (46). H_0 derivative. 1D. Tentative structure: a 5-propyldecahydroquinoline.

195A. Pumiliotoxin C, $C_{13}H_{25}N$, 0.20, 157°, *m/z* 195 (3), 194 (5), 180 (1), 152 (100), 109 (8). H_0 derivative. 1D. *N*-acetyl derivative. Structure: Fig. 11.

219A. $C_{15}H_{25}N$, 0.32, 165°, m/z 219 (1), 218 (2), 178 ($C_{12}H_{20}N$, 100). H_4 derivative, m/z 223, 180. 1D. N-acetyl derivative. Structure: Fig. 11. A minor isomer **219A'** emerges at 164°.

219C. $C_{15}H_{25}N$, —, 170°, m/z 219 (<1), 152 (100). H_4 derivative, m/z 223, 152. 1D. Tentative structure: a 2-pentadienyl-5-methyldecahydro-quinoline.

219D. $C_{15}H_{25}N$, —, 175°, m/z 219 (3), 180 (100). H_4 derivative, m/z 223, 180. 1D. Tentative structure: a 2-propynyl-5-propyldecahydroquinoline.

221C. $C_{15}H_{27}N$, —, 166°, m/z 221 (2), 152 (100). H_2 derivative 223, 152. 1D. Tentative structure: a 2-pentenyl-5-methyldecahydroquinoline.

221D. $C_{15}H_{27}N$, —, 168°, m/z 221 (3), 180 ($C_{12}H_{22}N$, 100). H_2 derivative. 1D. Tentative structure: a 2-propenyl-5-propyldecahydroquinoline.

223F. "$C_{15}H_{29}N$," —, 163°, m/z 223 (3), 180 (100). H_0 derivative. 1D. Tentative structure: a 2,5-dipropyldecahydroquinoline.

231E. "$C_{16}H_{25}N$," —, 160°, m/z 231 (<1), 232 (3), 152 (100). H_6 derivative, m/z 237, 152. 1D. Tentative structure: a 2-hexenynyl-5-methyldecahydroquinoline.

243A. $C_{17}H_{25}N$, 0.36, 182°, m/z 243 (2), 242 (1), 202 ($C_{14}H_{20}N$, 100). H_8 derivative, m/z 251, 208. 1D. N-acetyl derivative. Structure: Fig. 11. A minor isomer **243A'** emerges at 179°.

251A. "$C_{17}H_{33}N$," —, 170°, m/z 251 (2), 208 (6), 152 (100). H_0 derivative. 1D. Tentative structure: a 2-heptyl-5-methyldecahydroquinoline.

269AB. $C_{19}H_{27}N$, 0.35, 207°, m/z 269 (4), 268 (12), 204 ($C_{14}H_{22}N$, 100), 202 ($C_{14}H_{20}N$, 50). H_{10} derivative, m/z 279, 208. 1D, N-acetyl derivative. Tentative structure: Fig. 11. However, the occurrence of two major fragment ions appears remarkable for such a structure.

293. "$C_{20}H_{35}N$," —, 192°, m/z 293 (2), 152 (100). H_0 derivative. 1D. Tentative structure: a 2-decanyl-5-methyldecahydroquinoline.

Further properties of pumiliotoxin C are documented in Table 10. A proton magnetic resonance spectrum is presented for pumiliotoxin C in [426]: Assignments for proton and ^{13}C resonance peaks have been summarized [24]. The carbon-13 resonance peaks of a *trans*-2-epi-isomer of pumiliotoxin C have been reported [444].

Pumiliotoxin C itself is not widely distributed in dendrobatids, occurring in only four species, where often it is a major alkaloid (see Section 11). Alkaloids **219A** and **243** occur in a wider range of species. Alkaloid **269AB**

Table 10.   Physical and Spectral Properties of Pumiliotoxin C [444,445]

(−)Pumiliotoxin C HCl	
m.p. 230–240°	$[\alpha]_D^{20}$ −13.1°
Optical rotation (c 1.0, CH_3OH)	$[\alpha]_{436}^{20}$ −27.6°

always occurs with the histrionicotoxins. Recently, pumiliotoxin C was reported present in a mantellid frog, *Mantella madagascarensis* [261].

7.2. Syntheses

A number of syntheses of pumiliotoxin C have been reported. Of the nine routes, six feature the Diels-Alder reaction; these are considered first. The required *cis* ring fusion of the decahydroquinoline and the (Z)-configuration of the C(5) methyl relative to the ring fusion hydrogens, H_{4a} and H_{8a}, is a natural consequence of appropriately chosen substrates and this concerted endoselective process. The *cis*-fused intermediates exhibit great stereoselectivity in nucleophilic additions or catalytic hydrogenations, the approach of reagent being restricted exclusively to the convex face of the molecule. Most syntheses of pumiliotoxin exploit this fact. Inubushi and Ibuka [446] reviewed syntheses of pumiliotoxin C in 1977. Two other reviews have appeared [24,38]. The early literature relevant to syntheses of decahydroquinolines and octahydroquinolines is cited in Oppolzer and Frostl [447].

7.2.1. Intramolecular Diels-Alder Reactions.
Oppolzer's group has developed three such approaches to pumiliotoxin C (Schemes 81, 82, and 83). In the first [448], the triene carbamate, **1**, synthesized as shown in Scheme 81, is heated in a sealed tube to produce the olefin carbamate **2**, possessing

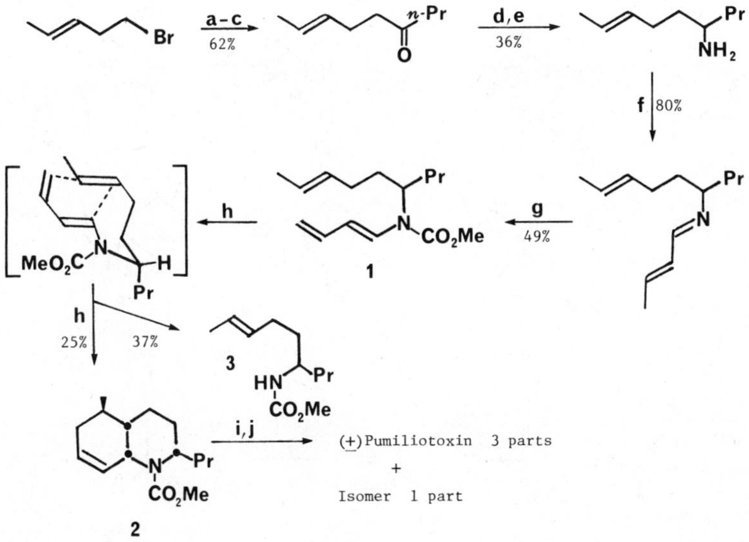

Scheme 81. Synthesis of ($\pm$)-pumiliotoxin C by an intramolecular Diels-Alder reaction [448]. (*a*) Mg/Et$_2$O; (*b*) *n*-PrCN/Et$_2$O; (*c*) H$_3$O$^{\oplus}$; (*d*) NH$_2$OH; (*e*) LiAlH$_4$/Et$_2$O; (*f*) crotonal; (*g*) ClCO$_2$CH$_3$/(Me$_3$Si)$_2$/ϕCH$_3$, $-40°$; (*h*) ϕCH$_3$, 215°, 20 hr; (*i*) H$_2$/Pd-C/MeOH; (*j*) HCl/aq. HOAc, 30 hr.

the desired stereoconfiguration at all four centers, in 25% yield accompanied by the cleavage product 3 in 37% yield. Subsequent reduction and hydrolysis steps provided (±)-pumiliotoxin C in 1.6% overall yield and minor amounts of an unspecified isomer.

In route two, Oppolzer and Flaskamp [445] synthesized (+)- and (−)-pumiliotoxin C using the enantiomeric trienes (e.g. 5) prepared from commercial R- or S-norvaline (Scheme 82). S-norvaline, 4, yielded natural (−)-pumiliotoxin C 6, confirming the S-configuration at C(2) originally shown by X-ray analysis of natural (−)-pumiliotoxin C·HCl [248,426].

The cycloaddition step yield (i) was improved by suppressing the base-catalyzed elimination of vinyl acetylene with added silylating reagent. The isobutyryl blocking group is smoothly removed by reduction with Al(i-Bu)$_2$H and afforded 97% enantiomerically pure (−)-pumiliotoxin C, identical in rotation with the natural material [445]. Habermehl's erroneous assignment [444] of the 2R-configuration to natural pumiliotoxin C (see below) was shown to arise from the inadvertent mislabeling of his R- and S-bromoamines

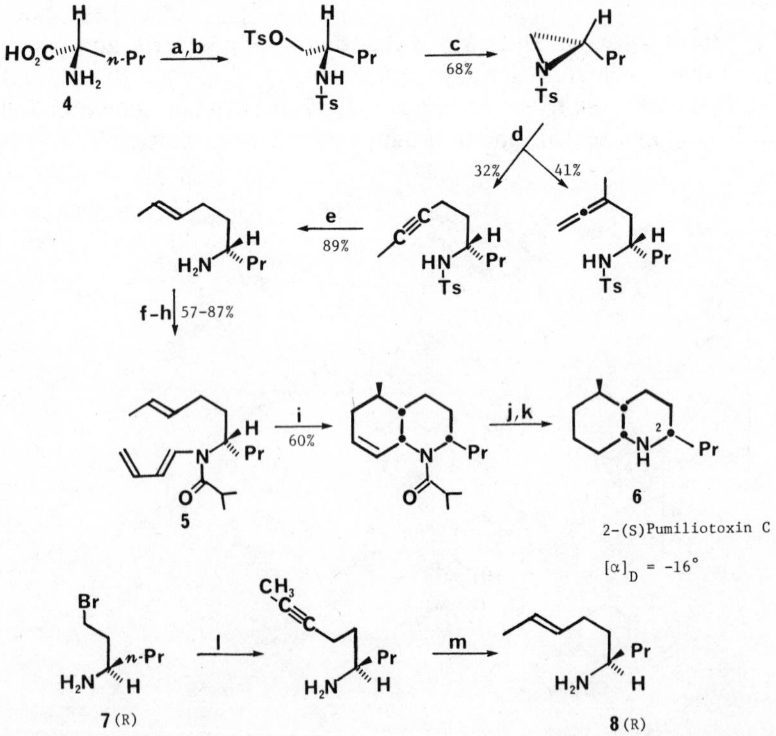

Scheme 82. Enantiospecific synthesis of (−)-pumiliotoxin C [445]. (a) LiAlH$_4$; (b) TsCl/Py; (c) KOH/MeOH; (d) MeC≡CCH$_2$MgBr; (e) Na/NH$_3$, −78°; (f) crotonal; (g) NaH/DME, −30°; (h) i-PrCOCl; (i) φCH$_3$, 2% AcN(TMS)$_2$, 230°, 16 hr; (j) H$_2$/Pd-C/MeOH; (k) Al(i-Bu)$_2$H; (l) MeC≡CNa/NH$_3$, −78°; (m) Na/NH$_3$, −78°.

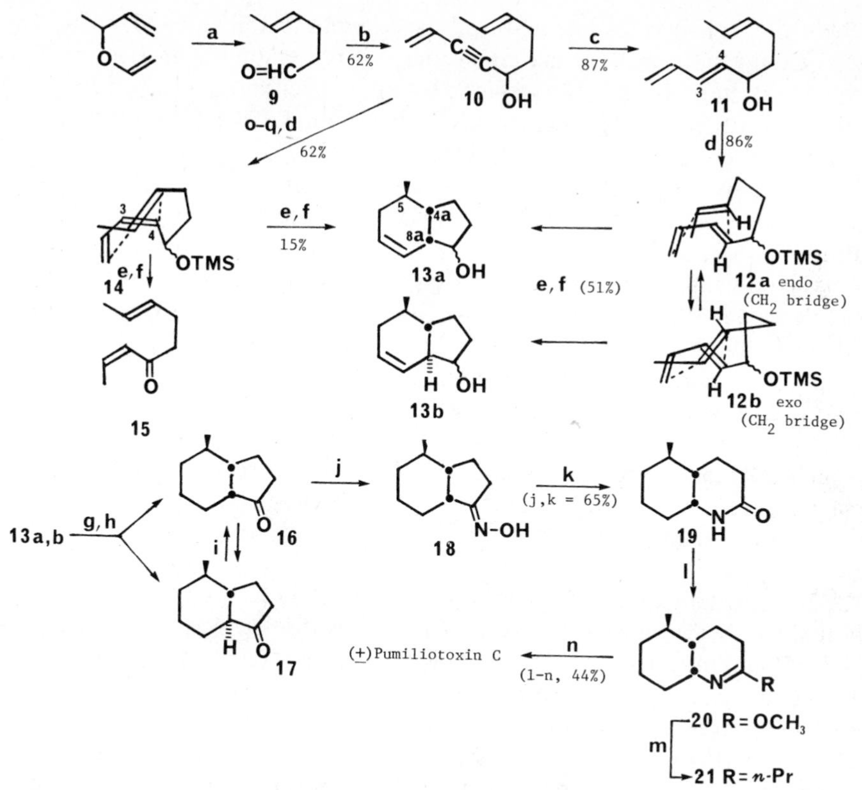

Scheme 83. Synthesis of (±)-pumiliotoxin C via an indanone [449]. (a) Δ; (b) $CH_2=CHC\equiv CMgBr/Et_2O$; (c) $LiAlH_4/NaOMe/THF$; (d) $AcN(TMS)_2/C_6H_{14}$, Δ; (e) φ-CH$_3$, 245°–250°, 16 hr; (f) KF/MeOH; (g) $H_2/Pt/MeOH$; (h) $CrO_3/aq.$ H_2SO_4; (i) t-BuOK/t-BuOH, 50°; (j) $NH_2OH \cdot HCl/NaOAc/MeOH$; (k) TsCl/aq. NaOH/dioxane; (l) $Et_3O^{\oplus}BF_4^{\ominus}/(i-Pr)_2NEt/CH_2Cl_2$, 10°; (m) n-PrMgBr/Et$_2$O-φH; (n) $H_2/Pt-C/MeOH$; (o) t-BuMe$_2$SiCl/imidazole/DMF, 35°; (p) Zn/KCN/aq. i-PrOH, 0°; (q) $(n\text{-}Bu)_4N^{\oplus}F^{\ominus}/$ THF.

(Scheme 89) and not from a misassignment of configuration, since his R-bromoamine 7 could be converted by steps *l* and *m* to Oppolzer's R-amino olefin 8 [cited in 248].

In their third approach [449], Oppolzer and co-workers reversed the order of control of chirality utilized in the routes shown in Schemes 81 and 82, where the amine chirality controls that produced at C(5), C(4a), and C(8a). Instead these latter three centers are now used to control the configuration at C(2). Their key intermediate, the trienol TMS ether 12, is synthesized (Scheme 83) by Grignard addition of vinylacetylene magnesium bromide to the aldehyde 9 (resulting from an oxy-Cope rearrangement), followed by an (E)-specific reduction of the resulting acetylenic alcohol 10. Whereas the trienol 11 itself cyclized poorly, the TMS ether 12 did so satisfactorally, albeit to a 2:1 mixture of ring-fused diastereomers **13a** and **13b**, indicating little energy difference in the endo- and exo-transition states **12a** and **12b**. The Δ^3 (**Z**)-stereoisomer **14**, prepared by a (**Z**)-specific reduction of **10**, gives solely, after deblocking, the *cis*-fused hydrindenol **13a** (15%), but unfortunately is accompanied by a preponderance (32%) of the ketone **15** arising from a 1,5-hydrogen shift. Models indicate that only the exo-bridged transition-state conformation shown is possible. Since the diastereomeric hydrindenols **13a** and **13b**, after reduction (*g*), could be oxidized (*h*), separated, and the unwanted *trans*-ketone **17**, isomerized (*i*) with base to provide more **16** (*cis*:*trans*, 3:2 at equilibrium), the Diels-Alder route from the *trans*-diene **12**, even though nonstereoselective, was judged more practical than that from the *cis*-diene **14**. The *cis*-oxime, **18**, which formed much more slowly than that from the *trans*-ketone (another way of separating ketones **16** and **17**), was subjected to the Beckmann rearrangement. After converting the resulting lactam **19** to the imino ether **20**, Grignard reaction with *n*-propyl magnesium bromide and the steroeospecific hydrogenation of the resulting imine **21** afforded (±)-pumiliotoxin C [449].

7.2.2. Intermolecular Diels-Alder Reactions. Three such routes to pumiliotoxin C have been developed. The two by Ibuka, Mori, and Inubushi [450,451] employed the acyclic and cyclic bistrimethylsilyloxydienes **22** and **27** with reactive dienophiles to produce good yields of the Diels-Alder adducts with excellent regio- and stereoselectivity. In their first route [450,451] (Scheme 84), the butadiene derivative **22** and ethyl crotonate, heated in xylene at 170°C gave an 80% yield of pure endo-product **23**. This was converted to the ester ketal **24** and a propionamido side chain appended (*c–f*). An acid-catalyzed Michael addition provides the key *cis*-fused lactam **26**. The required chirality at C(5), C(4a), and C(8a) results evidently because of the preference for the *trans*-diequatorial orientation of ring substituents in intermediate **25**, and the more stable *cis*-ring fusion. The keto group is removed by Mosingo's procedure. The 2-propylimine **21** produced by Mundy's method is then reduced stereospecifically to give (±)-pumiliotoxin C.

The second route of Ibuka, Mori, and Inubushi [450,451] produced the mixture of bicyclic bistrimethylsilyloxy diastereomers **28** by Diels-Alder reaction of the cyclohexadiene **27** and acrylonitrile (Scheme 85). Hydrolysis followed by bromination, epoxide formation (*d*), and deoxygenation (*e*) gave the olefin **29**. On heating in acid, a retro aldol reaction, nitrile hydrolysis, and Michael reaction (presumably via **30** and **25**) ensue, yielding **26**. This could be converted to (±)-pumiliotoxin-C as indicated in Scheme 84 [450,451].

A third intermolecular Diels-Alder approach was developed by Overman and Jessup [443,452] and is one of the most efficient routes to (±)-pumiliotoxin C to date (Scheme 86). 1-Carbobenzyloxy (Cbz)-aminobutadiene **31** and crotonal cyclize readily to the endo-adduct **32**. Phosphonate olefination produces the (**E**)-unsaturated ketone **33**, which on hydrogenation in one step

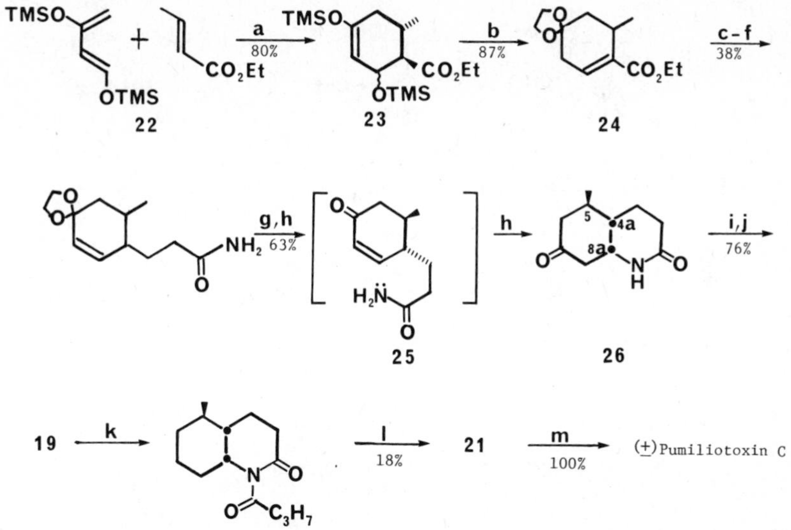

Scheme 84. A Diels-Alder Michael addition route to (±)-pumiliotoxin C [450,451]. (a) Xylene, 170°; (b) (CH₂OH)₂/φH/TsOH, Δ; (c) LiAlH₄/THF; (d) TsCl; (e) CuCH₂CN; (f) H₂O₂, aq. OH⊖; (g) 1% HCl; (h) NaOMe/MeOH; (i) (CH₂SH)₂, H⊕; (j) Raney Ni; (k) n-PrCOCl/NaH/THF-HMPT; (l) CaO, Δ; (m) H₂/Pt/2N HCl.

undergoes: (1) deblocking, (2) imine formation, and (3) the reduction of three double bonds to produce (±)-pumiliotoxin C in an impressive overall yield of 51%. Other pumiliotoxin C analogs **34** and **35** were also prepared in this manner [443].

Masamune et al. [453] used the chiral dienophile **36** with Overman's diene **31** in a remarkably facile enantioselective low-temperature Lewis-acid cat-

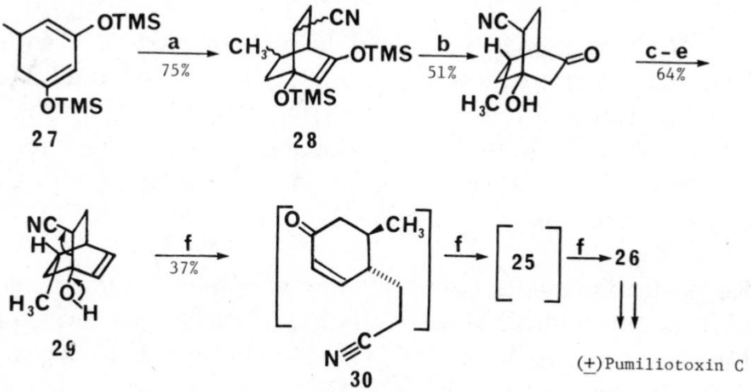

Scheme 85. An intermolecular Diels-Alder route to (±)-pumiliotoxin C [450,451]. (a) CH₂=CHCN; (b) 10% HCl, 0°; (c) Py·HBr·Br₂; (d) NaBH₄; (e) Zn/HOAc; (f) 15% HClO₄/HOAc, 100°.

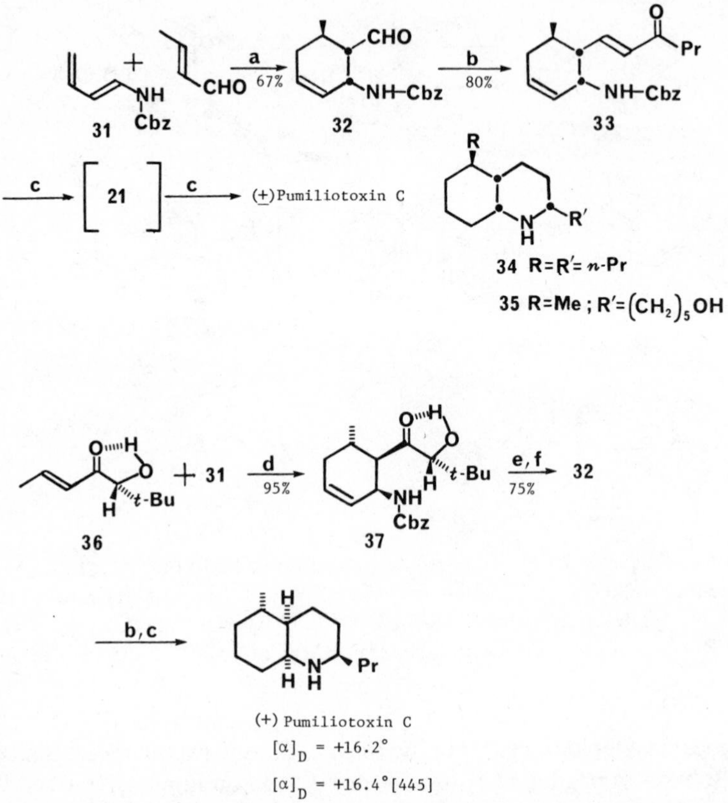

Scheme 86. The synthesis of racemic and unnatural (+)-pumiliotoxin C using an intermolecular Diels-Alder reaction [443,452,453]. (*a*) 110°, 2.5 hr; (*b*) (MeO)$_2$P(=O)CHNaCOPr/THF, −10°→65°; (*c*) H$_2$Pd-C/EtOH-HCl; (*d*) ϕCH$_3$/ BF$_3$·Et$_2$O (1 eq.), −78°, 0.5 hr; (*e*) LiBH$_4$/THF; (*f*) NaIO$_4$/aq. MeOH.

alyzed Diels-Alder reaction and obtained the (+)-endoadduct **37** with greater than 98% diastereomeric excess (Scheme 86). The endoadduct **37** was reduced (*e*) to the diol and cleaved with periodate (*f*) to afford the (+)-enantiomer of Overman's compound **32**. Using Overman's sequence (*b,c*), it was converted to the unnatural (+)-enantiomer of pumiliotoxin C, earlier prepared by Oppolzer and Flaskamp [445].

7.2.3. Tetrahydroindanones. Two syntheses of pumiliotoxin C have been developed from tetrahydroindanones: Ibuka, Inubushi, and their colleagues [454,455] in one of the first routes to (±)-pumiliotoxin C (Scheme 87) employed the Beckmann rearrangement on the mixture of tetrahydroindanones **38** to obtain the lactam mixture **39**. After separation of isomers, N-benzylation, and stereospecific epoxidation (convex face) afforded the lactam epoxide **40**. Steps *e–g* created the enone lactam **41**, which underwent ster-

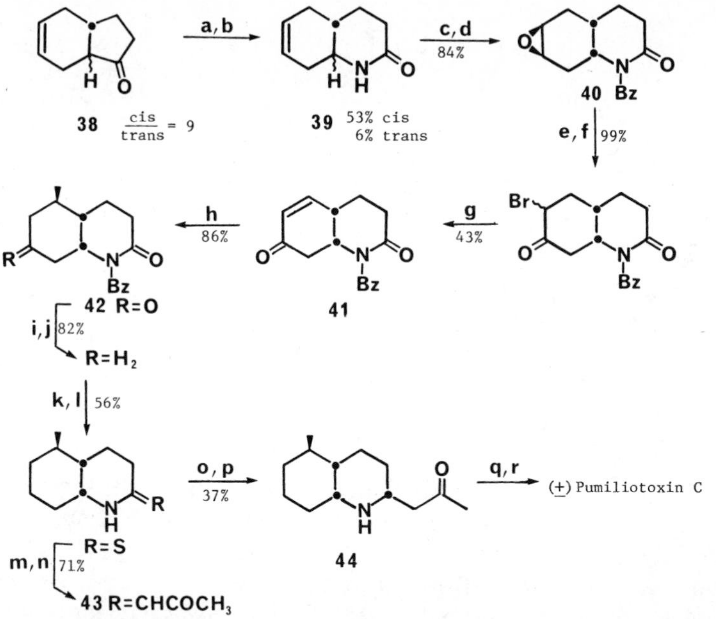

Scheme 87. The synthesis of (±)-pumiliotoxin C from a tetrahydroindanone [454,455]. (*a*) NH$_2$OH·HCl/NaOAc/MeOH; (*b*) TsCl/Py, separation of lactam diastereomers; (*c*) BzCl/ϕH, NaH; (*d*) *m*-ClϕCO$_3$H; (*e*) HBr/CHCl$_3$; (*f*) CrO$_3$/Me$_2$CO; (*g*) LiBr/Li$_2$CO$_3$/DMF; (*h*) LiMe$_2$Cu/Et$_2$O-THF; (*i*) (CH$_2$SH)$_2$/CHCl$_3$/BF$_3$·Et$_2$O; (*j*) Raney Ni/MeOH, Δ; (*k*) Na/NH$_3$-Et$_2$O, −78°; (*l*) P$_2$S$_5$/ϕH; (*m*) BrCH$_2$COCH$_3$/CH$_2$Cl$_2$; (*n*) *t*-BuOK/*t*-BuOH-ϕH/ϕ_3P; (*o*) H$_2$/Pt/HOAc; (*p*) CrO$_3$/Me$_2$CO; (*q*) (CH$_2$SH)$_2$/CHCl$_3$/BF$_3$·Et$_2$O; (*r*) Raney Ni/MeOH, Δ.

eospecific axial cuprate addition, again from the convex face, to give **42**, whereupon the ketone functionality was removed (*i,j*). After debenzylation (*k*) and thiolactam formation (*l*), the method of Eschenmoser (*m,n*) was used to attach a 3-carbon side chain, yielding **43**. This was stereospecifically transformed (*o,p*) to **44** and its carbonyl group removed (*q,r*) to produce (±)-pumiliotoxin C, whose stereochemistry was confirmed by X-ray analysis of the hydrochloride.

Yamamoto and colleagues [456] have recently introduced a very efficient route to (±)-pumiliotoxin C commencing with the reduction of the tetrahydro indanone **45** (Scheme 88) under critically chosen conditions. The oxime tosylate (**46**) of the *cis*-ring fused product **16** is treated with tripropylaluminum (*d*) to effect Beckmann rearrangement and alkylation in one step. Reduction of the resulting imine **21** with Al(i-Bu)$_2$H without isolation, gave (±)-pumiliotoxin C in 48% overall yield.

7.2.4. Intramolecular Enamine Cyclizations.

Habermehl and colleagues [444,457] employed enamine alkylation methodology in one of the first

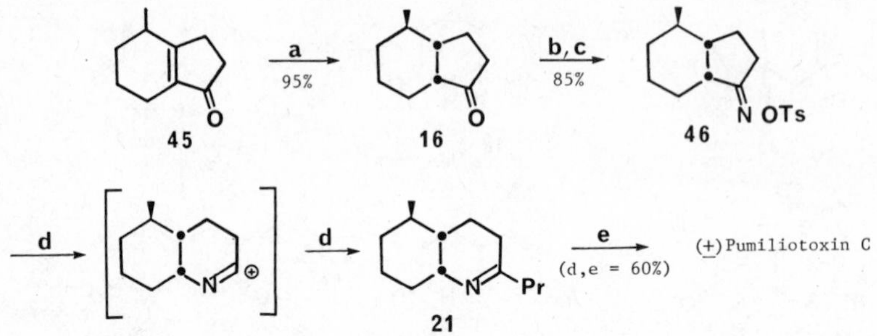

Scheme 88. The synthesis of (±)-pumiliotoxin C from a tetrahydroindanone using a novel Beckmann rearrangement–alkylation procedure [456]. (a) H_2 (1 atm.)/Pd/dioxane-12% $EtCO_2H$, 12 hr; (b) $NH_2OH\cdot HCl/NaOAc/MeOH$; (c) TsCl/Py, $-20° \rightarrow 0°$; (d) $(n\text{-Pr})_3Al$ (3 eq.)/CH_2Cl_2; (e) $Al(i\text{-Bu})_2H$ (4 eq.)/CH_2Cl_2.

syntheses of (±)-pumiliotoxin C (Scheme 89). Enamine **48**, formed by exchange with **47**, cyclizes on heating in dimethylformamide by a stereospecific *trans*-alkylation process via **48a** to the imine ester **49**. Hydrolysis, decarboxylation, and reduction afford (±)-pumiliotoxin C in 25% overall yield with the *trans*-fused epi-2-propyl-5-methyl isomer **50** as a major by-product. Presumably, since an X-ray analysis was stated to have been obtained [cited in 444], this is not the 2,7-disubstituted *trans*-decahydroquinoline obtained

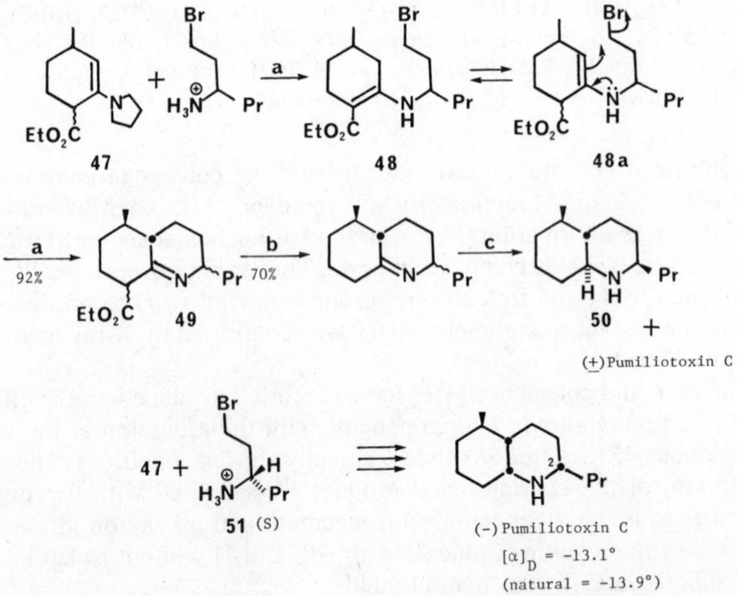

Scheme 89. Synthesis of racemic pumiliotoxin C and its natural enantiomer using an intramolecular enamine alkylation reaction [444,457]. (a) DMF, 100°, 6 hr; (b) 20% HCl, Δ; (c) H_2/Pd-C/EtOH.

in Habermehl's laboratory [458] "as a by-product of the synthesis of pum-iliotoxin C" [cited in 459] and shown by X-ray analysis to be 2-*n*-propyl-7-methyl-*trans*-decahydroquinoline [459] with the 2- and 7-substituents E with respect to the hydrogen at the ring junction 8a. The *trans*-fused epi-2-propyl-5-methyl isomer **50** has also been synthesized by Husson and co-workers [460] (See the discussion of Scheme 90, following) and the ^{13}C magnetic resonance signals were the same as reported for **50** by Habermehl and col-leagues [444]. Using optically active S-bromoamine **51**, produced by syn-thesis and resolution in the above sequence, Habermehl accomplished the first asymmetric synthesis of natural (−)-pumiliotoxin C [444]. The *trans*-isomer **50** was obtained as a by-product and had an optical rotation of $[\alpha]_D^{19}$ + 27.5° (c 0.6, CH$_3$OH). The inadvertent mislabeling of the S-bro-moamine as the **R**-enantiomer, however, led to the initial erroneous assign-ment of the **R**-configuration to C(2) of natural (−)-pumiliotoxin C. This as-

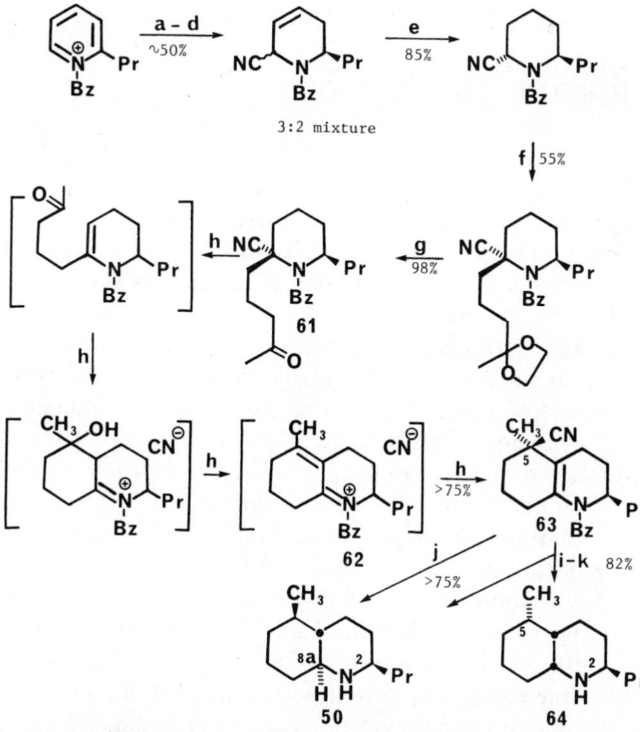

Scheme 90. The synthesis of 2,5- and 2,8a-epimers of a pumiliotoxin C: a possible biomimetic enamine condensation [460]. (*a*) NaBH$_4$/MeOH; (*b*) H$_2$O$_2$ (30%)/EtOH, 60°; (*c*) (CF$_3$CO)$_2$O/CH$_2$Cl$_2$, 0°; (*d*) KCN/CH$_2$Cl$_2$/H$_2$O, pH 4; (*e*) H$_2$/Pd-C/EtOH; (*f*) LiN(*i*-Pr)$_2$/THF, −30° then Cl(CH$_2$)$_3$C(CH$_3$)(OCH$_2$)$_2$, r.t.; (*g*) HCl/MeOH; (*h*) Al$_2$O$_3$ (act. II–III)/CH$_2$Cl$_2$; (*i*) NaBH$_4$/MeOH, 20 hr; (*j*) Na/NH$_3$-THF, −78°; (*k*) H$_2$/Pd-C/MeOH, H$^+$.

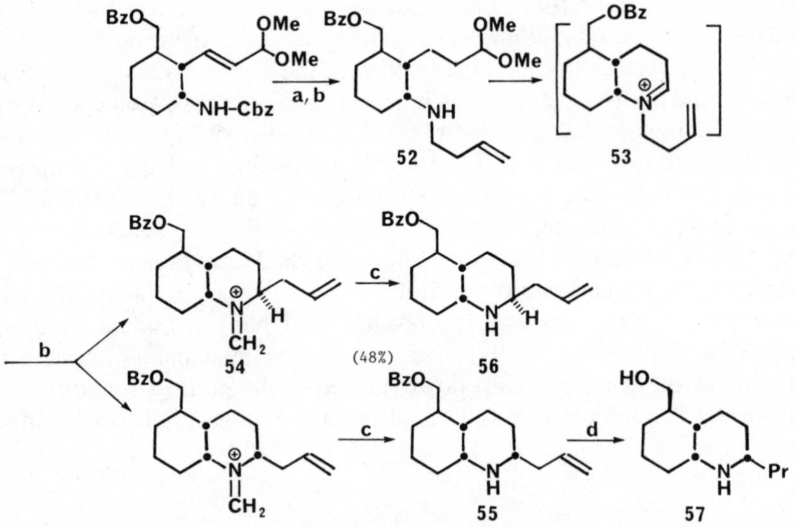

Scheme 91　Synthesis of a hydroxy analog of pumiliotoxin C [462]. (*a*) H_2/Pd-C; (*b*) $BrCH_2CH_2CH{=}CH_2$, then $H_3O^{\oplus}$; (*c*) $H_3O^{\oplus}$, Δ; (*d*) H_2/Pd-C.

signment was corrected by Oppolzer's asymmetric synthesis [445] and the error in assignment acknowledged by Habermehl [461].

Husson's group [460] has recently developed an interesting and possibly biomimetic transformation of the alkylated 2-cyanopiperidine **61** to a 2,5-epi isomer of pumiliotoxin C **64** (Scheme 90). The enamine equivalent **61** cyclizes spontaneously and stereoselectively on contact with alumina to the 5-cyano enamine **63** via the presumptive intermediates shown. The 5-cyano enamine (90% of product, 10% epimeric material also results) is thought to arise by axial attack of cyanide on intermediate **62**, where the 2-propyl group is constrained in the axial conformation to avoid A_{1-2} strain with the *N*-benzyl group. Reduction of this material with sodium borohydride (*i*), followed by reductive decyanation (*j*) and catalytic debenzylation (*k*), provided an 85:15 mixture of the decahydroquinolines **64** and **50**, respectively. The major isomer **64**, a 2,5-epi isomer of pumiliotoxin C, has both substituents in the equatorial configuration and has the same relative stereochemistry as that of the C(2), C(9), and C(9a) carbons of gephyrotoxin (Section 4.1). When the cyano enamine **63** is treated first with sodium in liquid ammonia, a 1,4 elimination of cyanide and benzyl anion occurs, providing a conjugated imine that is stereospecifically reduced to the *trans*-ring-fused product **50**, also having both C(2) and C(5) substituents in equatorial configurations [C(5) is inverted during step *j*].

7.2.5. [3,3]Sigmatropic Rearrangement.　Overman and Yokomatsu [462] synthesized (Scheme 91) a hydroxy analog **57** of pumiliotoxin C by hydro-

genation of the minor by-product **55** (10% obtained), from the 2-azonia[3,3]sigmatropic rearrangement of intermediate **53**, produced from precursor **52**. The major product **56** (90%) results by preferential rearrangement of the allyl group across the convex face to C(2) (via **54**) (Scheme 91). The hydroxy analog **57** is related to structures previously but tentatively proposed for trace alkaloids of dendrobatid frogs [23], some of which, like **57**, afforded a mass spectral base peak at m/z 154. These dendrobatid alkaloids are currently tabulated under piperidine-based alkaloids (Section 10.1).

7.2.6. Unsuccessful Routes. Bennett and Minor [463] have reported the surprising resistance of the dihydroquinolone **58** and certain of its 5-keto derivatives to catalytic hydrogenation. Reduction of the 5-substituent and partial reduction only of the pyridine ring to the dihydro stage occurs even at high pressure (40–70 atm) and elevated temperatures (Scheme 92). Only with the 5-diethylketal was complete reduction of the pyridine ring possible.

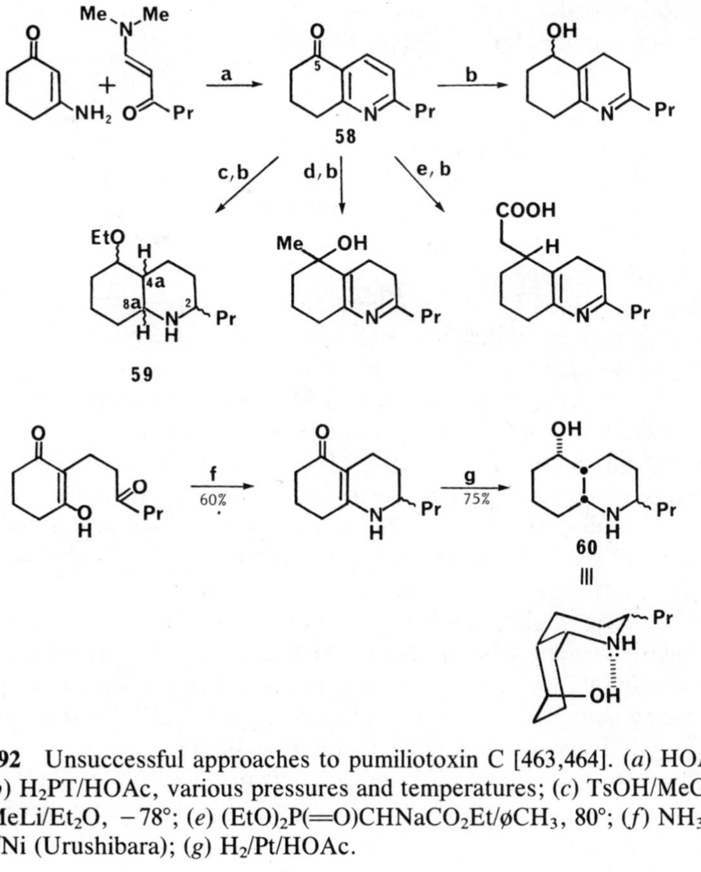

Scheme 92 Unsuccessful approaches to pumiliotoxin C [463,464]. (*a*) HOAc, 85°, 10 hr; (*b*) H_2PT/HOAc, various pressures and temperatures; (*c*) TsOH/MeC(OEt)$_3$, Δ; (*d*) MeLi/Et$_2$O, −78°; (*e*) (EtO)$_2$P(=O)CHNaCO$_2$Et/øCH$_3$, 80°; (*f*) NH$_3$/MeOH then H$_2$/Ni (Urushibara); (*g*) H_2/Pt/HOAc.

However, the ethyl ether product **59** was a complex mixture of diastereomers. Consequently this promising route to pumiliotoxin C and analogs, where the desired (**Z**)-configuration at C(2), C(4a), and C(8a) would be created as a result of catalytic hydrogenation, had to be abandoned.

Habermehl and Kissing [464] prepared the *cis*-fused decahydroquinolin-5-ol (**60**) as an intermediate in an unsuccessful approach to pumiliotoxin C (Scheme 92). Evidence was presented for an intramolecular hydrogen bond.

7.3. Biological Activity

7.3.1. Toxicity. Pumiliotoxin C is a relatively nontoxic compound and the toxin designation is, therefore, a misnomer. A subcutaneous dose of 5 mg/kg in white mice elicits only locomotor difficulties and 20 mg/kg was estimated as the minimum lethal dose [23,250]. An enormous dose of 125 mg/kg elicits convulsions and death in 10 min (454). Alkaloid **219A** causes paralysis of hind limbs, salivation, and piloerection in mice when administered subcutaneously at 4 mg/kg [23]. A synthetic lower homolog of pumiliotoxin C, 2-*n*-propyl-*cis*-decahydroquinoline, was stated to cause headache and stupefaction in humans, but no details were given [465].

7.3.2. Pharmacological Activity. Effects of pumiliotoxin C on biological systems appear to require relatively high concentrations. Pumiliotoxin C at concentrations up to 10 μM has no effect on basal tension nor on acetylcholine-induced contracture of guinea pig ileum [128]. At these concentrations pumiliotoxin C also has no effect on spontaneous activity of guinea pig atria, nor on direct or indirect elicited twitch in rat phrenic nerve diaphragm preparations [128]. Pumiliotoxin C and three synthetic analogs at somewhat higher concentrations block indirectly elicited twitches in rat neuromuscular preparations [381]. The most potent analog was the 2,5-di-*n*-propyl-*cis*-decahydroquinoline. At 40 μM it blocks responses to iontophoretic acetylcholine in denervated preparations. Pumiliotoxin C and the three analogs inhibit binding of [^{3}H]H$_{12}$-HTX to acetylcholine receptor channel complexes in electroplax membranes [381]. Carbamylcholine enhances their potency as inhibitors. The results indicate that decahydroquinolines represent another class of noncompetitive blockers of the acetylcholine receptor channel complex (see Section 3.3.2).

The synthetic analog 2,5-di-*n*-propyl-*cis*-decahydroquinoline inhibits the falling phase of action potentials in rat neuromuscular preparations [381], suggesting an inhibition of potassium channels. Pumiliotoxin C inhibits specific binding of [^{3}H]BTX-B to brain membranes with IC$_{50}$ at 45 μM [183], suggesting activity as a local anesthetic at sodium channels.

Another synthetic analog, 2-*n*-propyl-*cis*-decahydroquinoline, was suggested to have pharmacological properties in heart preparations similar to the plant alkaloid coniine [465]. The compound was stated to have no effect on striated muscle preparations, but no details were reported.

7.4. Addendum

The reductive amination of an easily prepared diketone provided pumiliotoxin C and equivalent amounts of a stereoisomer of undetermined configuration [K. Abe, T. Tsugoshi, and N. Nakamura, *Bull. Chem. Soc. (Japan)*, **57**, 3351 (1984)].

8. SAMANDARINES

8.1. Structures

Since ancient times, salamanders of Europe have been considered to be poisonous. In 1866 toxic principles from the German fire salamander were shown to be alkaloidal and an isolated material was named samandarine [466]. These alkaloids were further investigated at the turn of the century by Faust [467,468] and by Netolitzky [469], and their presence was reported in two species of the salamander, the fire salamander (*Salamandra salamandra*) of Western Europe and the alpine salamander (*S. atra*). The two subspecies of fire salamander used in the chemical investigations were formerly referred to as *S. maculosa taeniata* (western Europe) and *S. maculosa maculosa* (Balkan states). Both are now classified as *S. salamandra*. It is not always clear exactly which subspecies were investigated in some studies. A German pharmacologist, O. Gessner, and his colleagues became interested in the salamander alkaloids in the late 1920s. Although earlier workers had used extracts of whole animals, Gessner used the gummy poison expressed from the parotid glands of fire salamanders. After isolation and pharmacological investigation of samandarines [470], Gessner interested a chemist, C. Schöpf, in purification and characterization of the constituent alkaloids. Animal alkaloids were not known at that time and their structures were, therefore, of some interest. Schöpf and colleagues over the next few years isolated and characterized the major alkaloid samandarine, and, in addition, samandarone, samandaridine and cycloneosamandione [471–473]. These studies were done prior to the emergence of mass spectrometry and nuclear magnetic resonance spectroscopy and relied on classical methods of chemical conversions, analysis, and infrared and ultraviolet spectroscopy [471–476]. Isolations were by crystallization and later by counter-current partitions and thin-layer chromatography. Some 17 and 24 mg of samandarine were isolated from each fire salamander from Belgium or Spain, respectively, and 4–5 mg from each alpine salamander from Austria. In all extracts samandarine represented 60–70% of the alkaloids and samandarone about 10%. By 1961, chemical and finally X-ray crystallographic studies [477] led to establishment of structures for the major salamander alkaloids and were reviewed in depth at this time by Schöpf [478].

Samandarine ($C_{19}H_{31}NO_2$) was a saturated secondary amine with a secondary alcohol group, which could be oxidized with chromic acid to a ketone alkaloid that proved identical with samandarone. The carbonyl peak at 1740

cm^{-1} in the infrared spectrum of samandarone indicated a cyclopentanone. The second oxygen in both alkaloids appeared to be present as an ether, based on stability to strong base and lack of carbonyl or hydroxyl reactions. A diacetyl derivative was formed, consonant with titrametric evidence for the presence of two active hydrogens. Chemical analysis of C-methyl groups indicated the presence of two such groups. Two key conversions, one the Hoffman degradation of the methiodide, the other the formation of phenyl and methyl samandiols with Grignard reagents and later the reduction with $LiAlH_4$ to samandiol, followed by incisive investigation of the properties of the products, led to formulation of a five-membered oxazolidine ring. The reactions and the structure of the oxazolidine are depicted in Fig. 12. *N*-Methylation yielded a methiodide, which on Hoffman degradation yielded a tertiary amine containing all of the carbons present in samandarine methiodide. Hence the nitrogen must be present in a heterocyclic ring. The Hoffman degradation product appeared to contain an enol ether based on catalytic reduction to an ether and conversion with acid to a tautomeric mixture of hydroxyaldehyde and hemiacetal. This mixture could be oxidized with chromium trioxide to a lactone or converted with acidic methanol to a methyl acetal or reconverted via an *O*-acetylhemiacetal to the methiodide on treatment with acetic anhydride. Formulation as an oxazoline was consonant with the infrared spectrum, with formation of methyl and phenyl samandiols with Grignard reagents and reduction to samandiol with $LiAlH_4$. The methyl and phenylsamandiols were each oxidized to a pair of isomeric methyl or phenyl samandiones, while samandiol was oxidized to a samandione, indicating that the new hydroxyl group in these diols was secondary in nature. The conversion of samandiol with one equivalent of lead tetraacetate to a keto alkaloid and formaldehyde was limited by other data to the existance of a —$NHCH_2CHOH$ moiety. Thus samandarine contained the partial structure of the oxazolidine (Fig. 12*A*) and an additional three carbocyclic rings. Aromatization with selenium dioxide led to isolation of a $C_{15}H_{16}$ hydrocarbon whose ultraviolet spectrum was consonant with a 1,2,5,6-tetrasubstituted naphthalene. Based on the probability that samandarine had a steroidlike carbocyclic skeleton, a structure was postulated for the hydrocarbon (Fig. 13*A*). Synthesis and comparison confirmed this postulate. Based on analogies to steroids, three structures (Fig. 13*B,C,D*) now appeared likely for samandarine with the hydroxyl group in the cyclopentyl ring [476]. Fortunately, a crystal of samandarine HBr was obtained from methanol and the oxazolidine portion of samandarine was shown to correspond to structure *B* [477]. Samandarine and samandarone are interconvertible by chromic acid oxidation and stereoselective reduction with sodium-in-alcohol. The absolute configuration of samandarone, and hence samandarine, was confirmed from the optical rotatory dispersion curve that, like other natural 16-keto steroids, showed a strong negative Cotton effect. The structures of samandarine and other salamander alkaloids are depicted in Fig. 14.

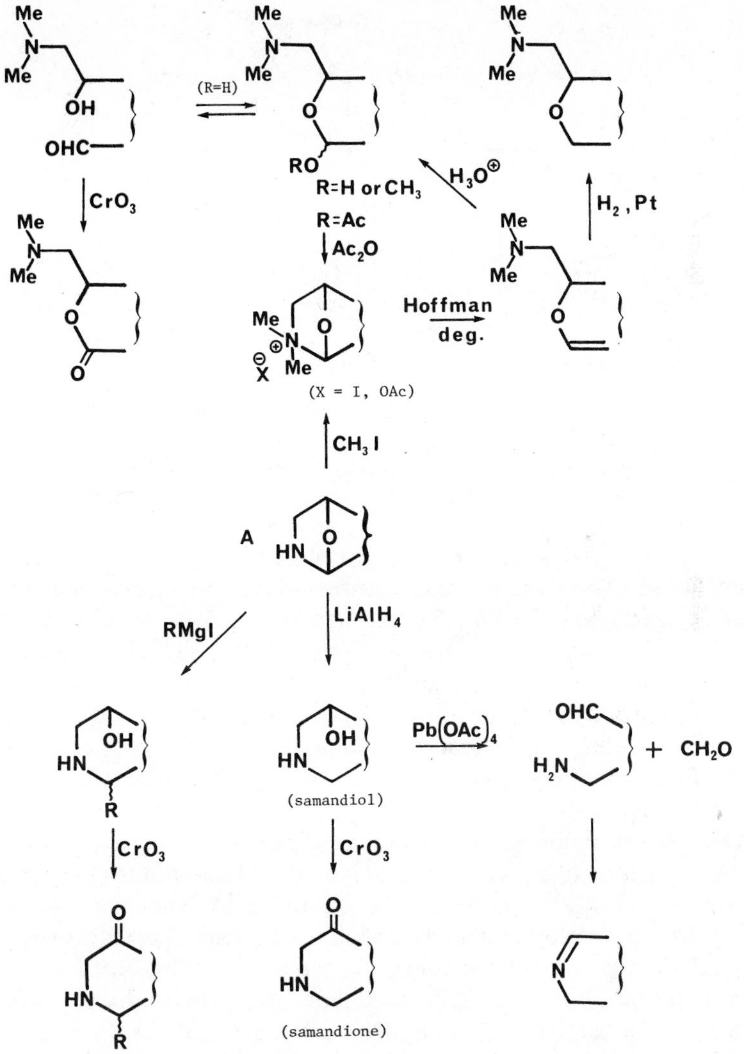

Figure 12. Chemical reactions and a proposed partial structure (*A*) of the oxazolidine moiety in samandarine [478].

The subsequent structure elucidation of other samandarine alkaloids was in large part due to G. Habermehl, a student of Schöpf, and relied heavily on X-ray crystallographic analyses.

Samandaridine, a minor alkaloid from the fire salamander, was isolated by Schöpf and shown to have the same oxazolidine ring as samandarine/samandarone [478]. Samandarine also contained, based on chemical properties and an infrared spectrum, a five-membered lactone ring. The structure

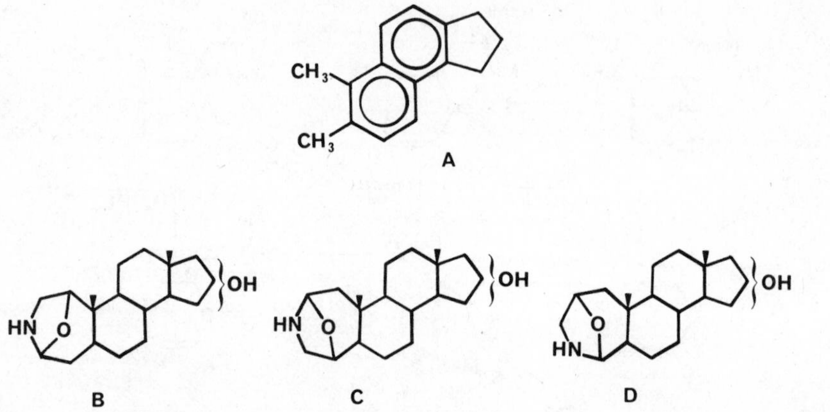

Figure 13. The selenium dioxide dehydrogenation product (*A*) of samandarine and proposed structures (*B,C,D*) for samandarine [476].

(Fig. 14) was finally elucidated by X-ray crystallographic analysis of the hydrobromide [479]. The absolute configuration was determined by partial synthesis from samandarine [480, see Section 8.2.5].

Cycloneosamandione, another minor alkaloid (~3% of total alkaloids) from the fire salamander, was isolated by Schöpf and shown to be isomeric with samandarone and to have the 16-keto moiety [476]. It did not contain an oxazolidine ring, but instead a tautomeric aminocarbonyl/carbinolamine moiety. Two structures were proposed: one a carbinolamine formed with a β-C(19)-aldehyde, the other a carbinolamine formed with a 6-keto moiety. The former was favored. However, a Wolff-Kishner reduction of the cycloneosamandione had given a neosamane ($C_{19}H_{33}N$) differing in properties from 3-aza-*A*-homo-5β-androstane as reported by Shoppee and Krueger [481]. It was speculated that an inversion at C(5) might have occurred during the Wolff-Kishner reduction. Later the same gross structure for cycloneosamandione, but with an α-C(19)-aldehyde, was proposed, based on X-ray crystallographic analysis of the hydroiodide [482,483]. The α-configuration at C(10) of this steroidal alkaloid was considered very unusual and the corresponding 10-β-isomer was synthesized [484] (see Section 8.2.3). This 10-β-isomer proved, unexpectedly, to be identical with natural cycloneosamandione. The X-ray analysis was reevaluated and found nearly equivalent for either 10α or 10β configurations [484, note added in proof]. The earlier choice of the 10α-configuration had been influenced by evidence that neosamane obtained by Schöpf and Müller [476] from cycloneosamandione was not identical in properties with the 3-aza-*A*-homo-5β-androstane, reported by Shoppee and Krueger [481]. It was stated by Habermehl and Haaf [484, note added in proof] that their own preparation of 3-aza-*A*-homo-5β-androstane was identical with neosamane. The apparent identity of 3-aza-*A*-homo-5β,10α-androstane synthesized by Eggart and Wehrli with the neo-

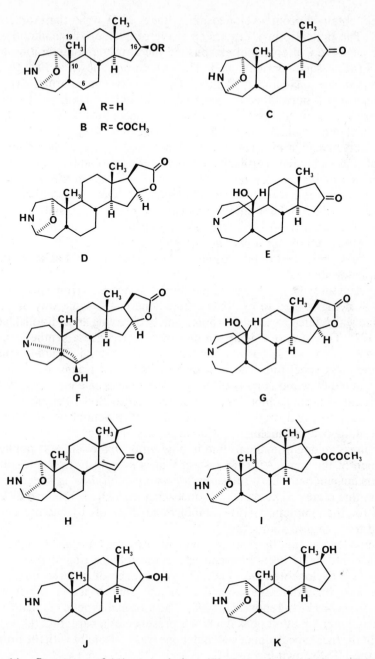

Figure 14. Structures of (*A*) samandarine; (*B*) *O*-acetylsamandarine; (*C*) samandarone; (*D*) samandaridine; (*E*) cycloneosamandione; (*F*) isocycloneosamandaridine; (*G*) synthetic cycloneosamandaridine (not detected in nature); (*H*) samandenone; (*I*) samandinine; (*J*) samanine; and (*K*) Hara-Oka alkaloid (not detected in nature, see text).

167

samane obtained from cycloneosamandione [485] was, therefore, questioned. The optical rotary dispersion curve of cycloneosamandione with a strong negative Cotton effect established the absolute configuration as the same as other samandarines [483]. The structure is shown in Fig. 14.

Other minor salamander alkaloids are O-acetylsamandarine, cycloneosamandaridine, samandenone, samandinine, and samanine. Cycloneosamandaridine, samandenone, and samanine appear to be the "Nebenalkaloids" isolated originally by Schöpf and colleagues [478].

O-Acetylsamandarine had not been detected in the earlier studies by Schöpf. Undoubtedly hydrolysis occurred during prolonged treatment with pepsin/HCl of the gummy poison, after expression from the salamander's parotid gland, prior to extraction of alkaloids. Habermehl [486] eliminated the acid treatment in later studies and isolated O-acetylsamandarine, whose identity was confirmed by hydrolysis to samandarine. In the subspecies, referred to as *Salamandra maculosa taeniata*, it was stated that most of the samandarine appeared to be present as the O-acetate and that samandarone was a major alkaloid.

The structure of cycloneosamandaridine was deduced from spectral properties as being related to cycloneosamandione in the same way as samandaridine was related to samandarone, namely in having an additional lactone ring [487]. This structure is apparently incorrect, since a synthetic sample [488, see Section 8.2.3] was not identical with the natural alkaloid in melting point or mass spectrum. The lack of a M-29 (CHO) fragment in the mass spectrum led Oka and Hara [488] to propose an alternate structure (Fig. 14) in which the carbinolamine function was formed with a 6-keto group. Such a carbinolamine had originally been proposed by Schöpf and Müller [476] as one of the two carbinolamine structures possible for cycloneosamandione. The name cycloneosamandaridine is now unfortunate, since it implies the presence of the same AB ring system that is present in its namesake, cycloneosamandione. The term *isocycloneosamandaridine* is suggested and used in this review. The name cycloneosamandaridine is retained as appropriate for the synthetic C(19)-carbinolamine (Fig. 14). It has not been reported from natural sources.

Samandenone, another minor alkaloid from the fire salamander, contained, based on its infrared spectrum, an oxazolidine ring (830, 875 cm^{-1}) and an α,β-cyclopentenone [489]. The occurrence of infrared peaks, diagnostic for oxazolidines in samandarines and their N-acetyl and N-methyl derivatives, has been reported in [490]. The proton magnetic resonance spectra (two doublet methyl peaks) and the presence of a parent-43 (C_3H_7) fragment in the mass spectrum of samandenone was consonant with the proposed structure [489] shown in Fig. 14.

Samandinine, another minor alkaloid (0.4%) from fire salamander appeared related in structure to samandenone, since it also contained an apparent isopropyl moiety: It was an oxazolidine with an ester function and the structure shown in Fig. 14 was proposed [491].

Samanine, a final minor alkaloid from the fire salamander, did not have the infrared band that is typical of oxazolidines, nor did it show the peaks at m/z 85 and 86 typical of oxazolidine-containing salamander alkaloids [492]. A structure lacking an oxygen function in the A-ring was proposed and was confirmed by synthesis [see Section 8.2].

An alkaloid isomeric with samandarine has been stated to be a constituent of another salamander, referred to as "*Cryptobranchus maximus*" [493]. However, S. Hara has personally informed us that their laboratory did *not* isolate this alkaloid from a natural source and that the Chemical Abstract citation [493] and perhaps even the patent itself are in error in this regard. Because of the Chemical Abstract citation, others have referred to this alkaloid as being isolated from "*Cryptobranchus maximus*" [38] and have termed it the "Cryptobranchus alkaloid" [505,506]. It should be noted that the North American genus of salamanders *Cryptobranchus* does not have a *Cryptobranchus maximus*. The alkaloid, which contains the 17β-hydroxy group rather than the 16β-hydroxy group of the isomeric samandarine, is referred to in the present review as the Hara-Oka alkaloid.

The structural elucidation of the various samandarines was the subject of several reviews in the late 1960s and early 1970s [495–498]. Certain physical properties of samandarine alkaloids are presented in Table 11. Infrared spectra have been presented for various salamander alkaloids [476,477,485 and references therein]. Proton magnetic resonance spectra for samandarone and cycloneosamandione have been presented [482,489]. Mass spectra of various samandarine alkaloids and derivatives have also been reported

Table 11. Physical and Spectral Properties of Samandarine Alkaloids [471–473,478,480,486,488,492]

Alkaloid	m.p.	Optical Rotation
Samandarine ($C_{19}H_{31}NO_2$)	HCl 312–313°	
	Free base 188°	$[\alpha]_D^{21}$ +29.5°
O-Acetylsamandarine ($C_{21}H_{31}NO_3$)	Free base 158–159°	—
Samandarone ($C_{19}H_{29}NO_2$)	Free base 190°	$[\alpha]_D^{21}$ −115.7°
Samandaridine ($C_{21}H_{31}NO_3$)	HCl 338–340°	
	Free base 287–288°	$[\alpha]_{578}^{24}$ +14.1°
Cycloneosamandione ($C_{19}H_{29}NO_2$)	Free base 118–119°	—
Isocycloneosamandaridine ($C_{21}H_{31}NO_3$)	Free base 282°	—
Samandenone ($C_{22}H_{33}NO_2$)	HCl 310–311°	
	Free base 189–191°	—
Synthetic cycloneosamandaridine	Free base 270–272°	—
Samandinine ($C_{24}H_{39}NO_3$)	—	—
Samanine ($C_{19}H_{33}NO$)	Free base 196–197°	—
Hara-Oka alkaloid (synthetic) ($C_{19}H_{31}NO_2$)	Free base 191–193°	—

[482,499]. The spectra of the oxazolidine-containing salamander alkaloids are dominated by base peaks at m/z 85 and m/z 86 (C_4H_7NO and C_4H_8NO). Samandarine, samandarone, and O-acetylsamandarine have elution temperatures of 220°, 223°, and 243°, respectively, on gas chromatography under conditions (see Fig. 3) used for dendrobatid alkaloids [unpublished results].

Samandarine alkaloids occur in the two species that make up the genus *Salamandra*, namely *S. atra* and *S. salamandra* [478,496]. The Chemical Abstract citation [493] to a samandarine alkaloid as a "toxic constituent in *Cryptobranchus maximus*" is apparently incorrect [S. Hara, personal communication]. Other salamandrid genera, namely *Taricha, Notophthalmus, Triturus*, and *Cynops*, contain tetrodotoxin. Samandarines were not detected in European newts of the genus *Triturus* [500]. Alkaloids were detected from skin of the Australian myobatrachid frog, *Pseudophryne corroboree*, and it was tentatively suggested, based primarily on chromatographic properties, that such alkaloids were of the samandarine class [501]. Recently, a pumiliotoxin **267D** ($C_{16}H_{29}NO_2$) and allopumiliotoxin **323B** ($C_{19}H_{33}NO_5$) were identified in extracts from a related species, *Pseudophryne semimarmorata* [261].

8.2. Syntheses

The salamander alkaloids can be classified into three main types on the basis of their A-ring structures. All feature a common 3-aza-A-homo-5β-androstane structure, but can be segregated into: (1) those containing an additional 1α-4α oxido bridge (oxazolidines, or 6-aza-8-oxabicyclo[3.2.1]octanes) as represented by samandarine itself; (2) those in which a carbinolamine has been formed by intramolecular addition of the 3-aza atom to a C(19)-aldehyde or a 6-ketone as represented, respectively, by cycloneosamandione and isocycloneosamandaridine; and (3) the unmodified aza steroid, as exemplified by samanine. Syntheses are discussed in this order with an additional section devoted to miscellaneous interconversions involving the D-ring. All syntheses discussed are partial in that natural chiral starting materials are employed. In all cases where optical rotary dispersion spectra have been examined, the salamander alkaloids have been shown to have the same absolute configuration as the synthetically derived substance. Most of the synthetic operations will be seen to involve the A-ring with some attention also given to the D-ring where the chief synthetic requirement is the transposition of the usual 17-oxy steroid substituent to the C(16)-position shared by most salamander alkaloids.

8.2.1. Samandarone and the Hara-Oka Alkaloid.
Samandarone, the first of the salamander alkaloids to be synthesized, was prepared in 1967 by Hara and Oka [493,502] according to Schemes 93 and 94. Commencing with 1-formyl-A-nor-5β-androst-1-en-17β-ol (**1**) prepared from testosterone by a

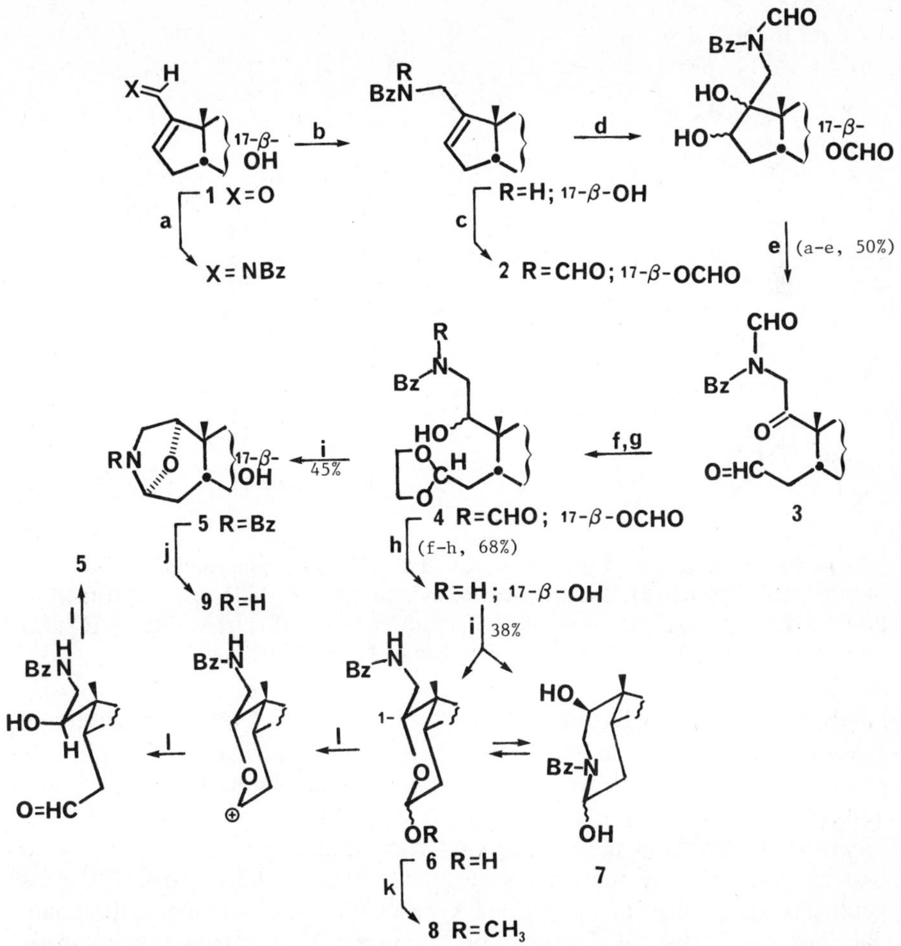

Scheme 93. The synthesis of the Hara-Oka alkaloid [494,502]. (*a*) BzNH$_2$; (*b*) NaBH$_4$; (*c*) formylation; (*d*) OsO$_4$/Et$_2$O-Py; (*e*) Pb(OAc)$_4$; (*f*) (CH$_2$OH)$_2$, H$^+$; (*g*) NaBH$_4$; (*h*) NaOH/aq. EtOH; (*i*) 75% HOAc, 100°; (*j*) H$_2$/Pd-C; (*k*) MeOH, HCl;(*l*) 20% HCl/aq. Me$_2$CO.

multistep sequence, they reduced the derived benzylamine Schiff's base and formylated it to obtain **2**. The A-Ring was then cleaved (*d,e*) to the keto aldehyde **3**, the aldehyde group selectively ketalized, and the keto group reduced to provide **4** as a 1:1 mixture of epimers. After deformylation and ketal hydrolysis, the desired oxazolidine **5** resulted in 45% yield accompanied by a 1:1 mixture (38%) of the isomeric hemiacetal **6** and carbinolamine **7**, having C(1)-stereochemistries, precluding the second cyclization step essential for oxazolidine formation. When the equilibrium mixture of **6** and **7** was treated with methanolic HCl, a single product, the acetal **8**, resulted

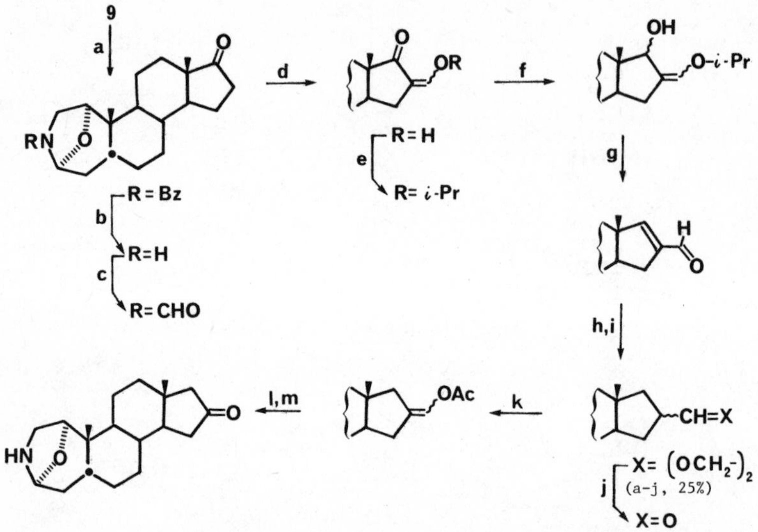

Scheme 94. Conversion of the Hara-Oka alkaloid to samandarone [494]. (*a*) CrO₃/
Me₂CO, aq. H₂SO₄; (*b*) H₂/Pd-C; (*c*) formylation; (*d*) EtOCH=O, NaOEt; (*e*) *i*-
PrOH/TsOH; (*f*) NaBH₄; (*g*) dilute HCl; (*h*) (CH₂OH)₂, H⁺; (*i*) H₂/Pt; (*j*) H⁺; (*k*)
MeC(OAc)=CH₂, H⁺; (*l*) O₃; (*m*) HCl (cleaves N—CHO).

[494]. Quite surprisingly, when the acetal is hydrolyzed (*l*), the oxazolidine
5 results, evidently by a Walden inversion at C(1), producing, thereby, an
aminoalcohol of the proper configuration for oxazolidine formation [502].
After hydrogenolytic *N*-debenzylation (*j*), the Hara-Oka alkaloid, 9, was
obtained. Conversion of 9 to samandarone required 13 steps outlined in
Scheme 94, whereby the C(17)-oxygen function is shifted to C(16). This
synthesis also constitutes a formal synthesis of samandarine and samandaridine, since those alkaloids can be chemically derived from samandarone
(Section 8.2.5).

   Shimizu [503] in 1972, observing that Hara and Oka's synthesis relies on
a laborious multistep preparation of the starting material, applied a procedure
from the corynantheine synthesis of Autrey and Scullard to form the oxazolidine ring of samandarine from a more accessible starting material
(Scheme 95). The 17β-hydroxy-5β-androstan-3-one hydroxymethylene derivative 10 is converted to the 2-methylthioketone 11 with methyl *p*-toluene
thiosulfate (*a*) and acetylation, then to the oxime mixture 12, which is subjected to Beckmann cleavage to yield the seco thioenol ether nitrile 13 as a
mixture of diastereomers [503,504]. After desulfurization (*e*) and stereospecific epoxidation (*f*), the oxide is regiospecifically opened with azide to yield
15, which on reduction with borohydride affords, in one step, the oxazolidine
9 presumably via either the amidine 16 or the imino ether 17. During the
reduction, the 17β-acetate is cleaved, yielding the Hara-Oka alkaloid 9 directly. Since this had been earlier converted to samandarone by Hara and

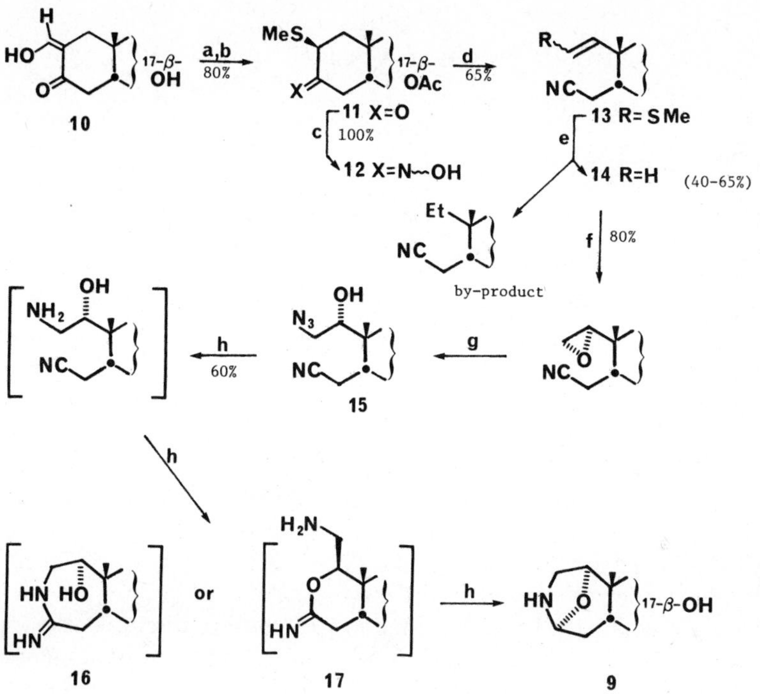

Scheme 95. The synthesis of the Hara-Oka alkaloid: a formal synthesis of samandarone [503,504]. (*a*) Ts-SCH₃/EtOH-KOAc, Δ; (*b*) Ac₂O/Py; (*c*) TsCl/Py, Δ; (*e*) Raney Ni/MeOH, Δ; (*f*) *m*-ClØCO₃H/CHCl₃; (*g*) NaN₃-NH₄Cl/aq. MeOCH₂CH₂OH, Δ; (*h*) NaBH₄, *i*-PrOH, Δ, 15 hr.

Oka [494] (Scheme 94), this synthesis is equivalent to a synthesis of samandarone.

Shimizu [504] has also reported a route to the isomeric α- and β-oxido bridged aza-A-homo steroids **21** commencing from the testosterone oxide mixture **18** (Scheme 96). Tosylhydrazine and acid treatment lead to the seco

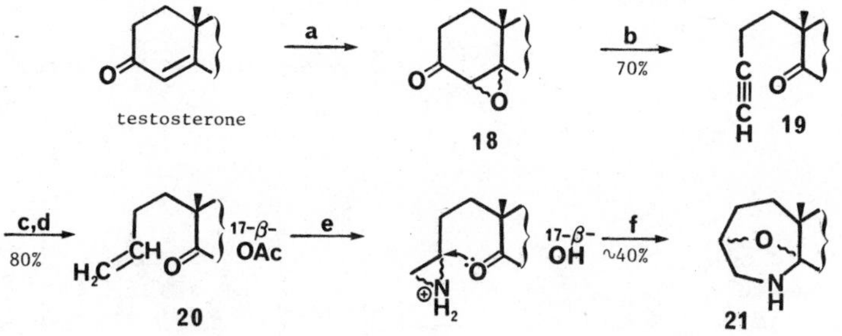

Scheme 96. Isomeric A-ring oxazolidine analogs of samandarine [504]. (*a*) H₂O₂-NaOH; (*b*) TsNHNH₂/CH₂Cl₂-HOAc; (*c*) H₂/Lindlar catalyst/ØH; (*d*) Ac₂O/Py; (*e*)Cl₂NCO₂Et/ØH, Δ then (i) aq. NaHSO₃; (ii) KOH/EtOH, Δ; (*f*) 2*N* H₂SO₄, Δ.

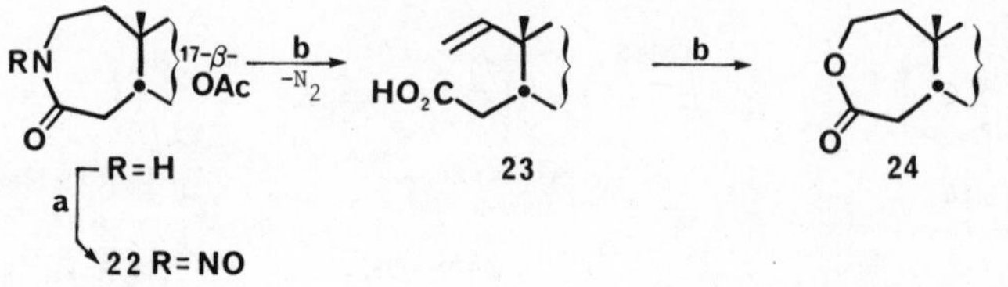

Scheme 97. A proposed route to A-seco intermediates for conversion to A-homo aza steroids [504]. (a) $NaNO_2/Ac_2O-HOAc$, 0°; (b) DMSO, 80°, 2 hr.

acetylenic ketone **19**, which on Lindlar reduction provides the keto olefin **20**. The mixture of aziridines, resulting after step *e*, is treated with acid to produce the mixture of two oxazolidines **21**. Mass spectra of these and **9** were reported [504].

A promising route to intermediates of the type **14** was abandoned on discovery that the thermal decomposition of the *N*-nitrosolactam **22**, failed to produce useful amounts of the seco olefinic acid **23** but gave predominantly or exclusively the lactone **24** (Scheme 97) [504]. Only 5% of **23** could be isolated under the conditions of step *b*.

Benn and Shaw [505,506] reported in 1974 a fairly direct, stereoselective approach to the azaoxobicyclo ring-A system using the α-oxido dienone **25** as a point of departure (Scheme 98). The hydrogenation procedure of Pelk and Hodkova provided the 1α-hydroxy-3-keto-5β-androstane derivative **26**, which on Schmidt rearrangement produced the isomeric lactam alcohol mixture **27** and **28**. The major lactam, after separation by preparative thin-layer chromatography, could be converted to the Hara-Oka alkaloid **9** using a modified Birch reduction in 50% yield. A less satisfactory route employed reduction of **27** with $LiAlH_4$ to the amino alcohol **29**, then the use of oxidants such as silver picolinate. A mixture of seven products resulted, among which **9** could be identified.

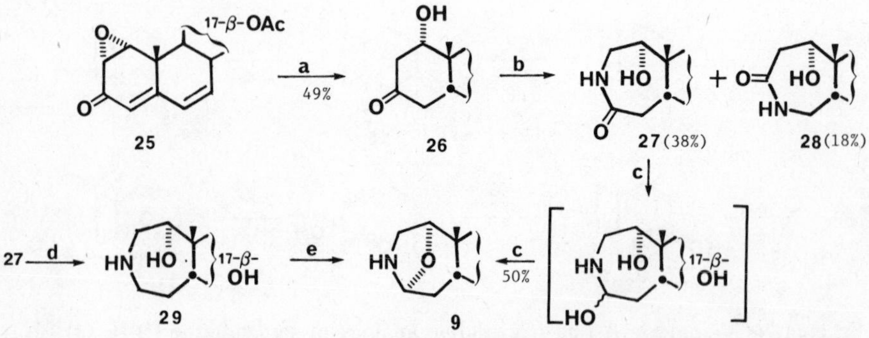

Scheme 98. The synthesis of the Hara-Oka alkaloid **9** [505,506]. (a) $H_2/Pd-C/i-PrOH$; (b) HN_3, H^+; (c) $Li/EtNH_2-i-BuOH$; (d) $LiAlH_4/THF$; (e) silver picolinate/DMSO.

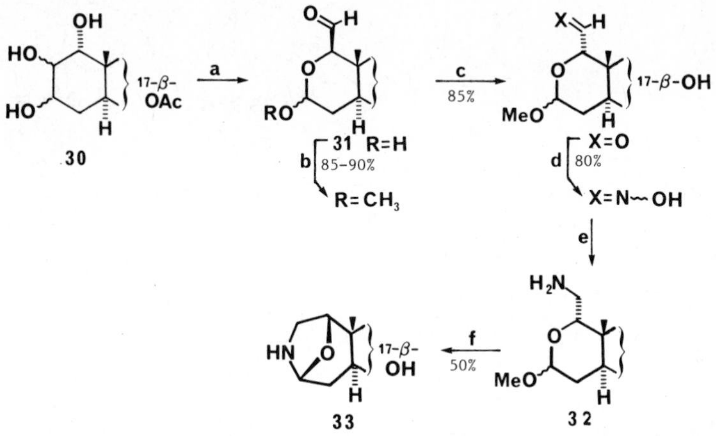

Scheme 99. An epioxido bridged analog of samandarine [507]. (*a*) HIO$_4$; (*b*) MeOH/
HCl; (*c*) NaOMe/MeOH, Δ; (*d*) NH$_2$OH; (*e*) LiAlH$_4$; (*f*) 75% aq. HOAc, Δ.

Wehrli and co-workers [507] prepared the 1β,4β-oxido bridged-5α-dias-
tereomer **33** of the Hara-Oka alkaloid by the route shown in Scheme 99.
Periodate cleavage of the triol **30** generated the aldehyde hemiacetal **31**,
proceeding by selective cleavage of the *cis*-2,3-glycol to the seco dialdehyde
and ring closure. Conversion to the acetal (*b*), epimerization of the C(1)-
formyl group (*c*), formation of the oxime (*d*), and reduction (*e*) afforded the
aminoacetal **32**. On treatment with acid, the oxazolidine **33** resulted, having
a 1,4-oxide bridge epimeric with that found in the salamander alkaloids and
also having the unnatural *trans*-fused A-B ring junction.

8.2.2. The Structure of Neosamane. As a preface to the chemistry of cycloneosamandione, it is instructive to focus on one aspect of salamander
alkaloid chemistry, namely the elusive structure of "neosamane," the
Wolff-Kishner reduction product obtained from cycloneosamandione in 1960
by Schöpf and Müller [476].
    In an attempt to prove the structure of this aza steroid, Shoppee and
Krueger [481] in 1961 prepared the four most likely possibilities, the 3- and
4-aza-A-homo-5β- and 5-α-androstanes **34–37** (Scheme 100). Although isomeric with neosamane, none of the four appeared to be identical with it, nor
were their derivatives identical with corresponding derivatives of neosamane.
    In 1963, Habermehl and Göttlicher [482, 483], examining a Fourier pro-
jection of the X-ray crystallographic data, which had been collected on cycloneosamandione, noted that it failed to clearly distinguish whether the
C(19)-aldehydic group was in an α- or β-orientation relative to C(10). This
ambiguity, an optical rotatory dispersion model study, and the nonidentity
of neosamane with any of Shoppee and Krueger's four aza steroids led Habermehl to conclude that the C(19)-masked aldehyde of cycloneosamandine
must possess the unprecedented 10α-configuration [482,483].

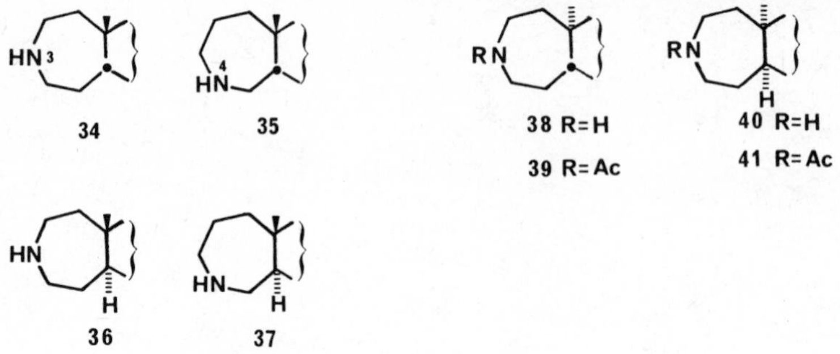

34　　　　　35　　　　　　　　38 R=H　　　40 R=H

　　　　　　　　　　　　　　　　　　39 R=Ac　　　41 R=Ac

36　　　　　37

Scheme 100. Neosamane **34** and isomers.

On the basis of this conclusion, Eggart and Wehrli [485] in 1966 reported the synthesis of the most likely candidate, **38**, for neosamane, as well as its 5α-H isomer **40**, employing a regiospecific Beckmann rearrangement of a 2-keto oxime. N-Acetyl derivatives **39** or **40** did not agree in melting point with N-acetylneosamane. However, the melting point of the hydroiodide of **38** did agree fairly well with that reported for neosamane hydroiodide. The lack of any other reliable data for comparison prevented them from either excluding or accepting structure **38** for neosamane.

Two years later, Habermehl and Haaf [484] repeated the preparation of 3-aza-A-homo-5β-androstane **34** and found its properties, in contrast to those reported by Shoppee and Krueger, to be identical with those reported for neosamane. This fact and their synthesis of cycloneosamandione with a 10β-configuration (see below) forced them to revise the structure of that alkaloid and the supposedly related alkaloid, cycloneosamandaridine. The structure of the latter was later revised further [488] and still requires confirmation. An iso designation is presently suggested for cycloneosamandaridine (see Section 8.1).

8.2.3. Cycloneosamandione and Cycloneosamandaridine. Habermehl and Haaf [484,487], as alluded to above, embarked on a synthesis of the supposed 10β-isomer of cycloneosamandione, planning to show its nonidentity with **43**, but instead, to their surprise, found it to be identical with that carbinolamine alkaloid. Their route is shown in Scheme 101 and features the usual Beckmann rearrangement, chromatographic separation of lactams, and reduction of the desired 3-aza isomer with LiAlH₄. A final Jones' oxidation of **42**, neosamandiol, provided product **43**, which is identical with natural cycloneosamandione, conclusively proving the structure for that alkaloid and obligating a structural revision.

Oka and Hara [488] in 1978 also reported a synthesis of cycloneosamandione (Scheme 102). Preferring the easier separation of syn and anti 3-oximes, they subjected only the pure syn isomer **44** to the Beckmann rearrangement, thereby obtaining solely the 3-aza lactam **45**. In this way the

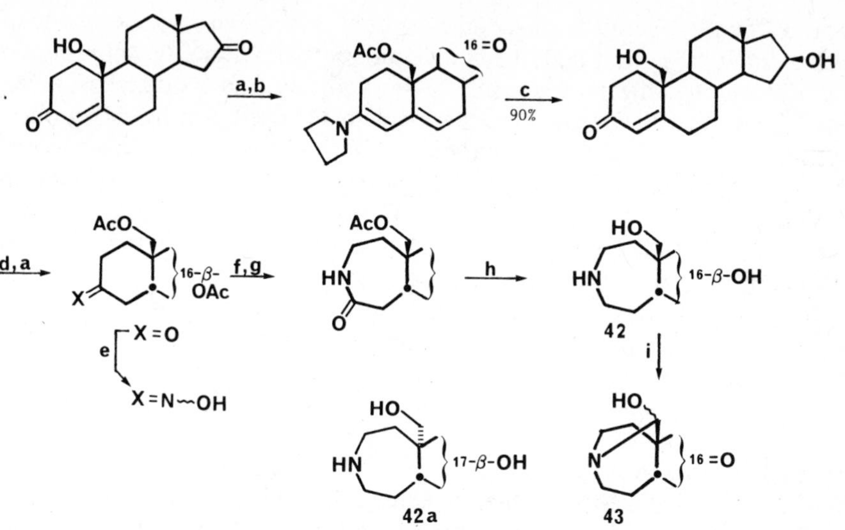

Scheme 101. The synthesis of cycloneosamandione [487]. (*a*) Ac$_2$O/Py; (*b*) pyrrolidine/MeOH; (*c*) LiAlH$_4$/THF then H$_2$O; (*d*) H$_2$/Pd-C/EtOH-KOH; (*e*) NH$_2$OH; (*f*) SOCl$_2$, −20°; (*g*) Al$_2$O$_3$ chromatography; (*h*) LiAlH$_4$/THF; (*i*) CrO$_3$/dil. H$_2$SO$_4$.

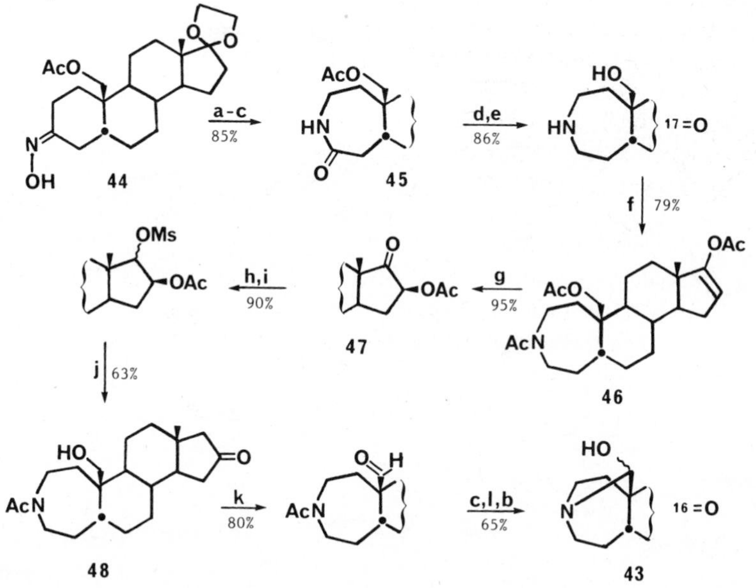

Scheme 102. The synthesis of cycloneosamandione [488]. (*a*) TsCl/Py; (*b*) dil. HCl; (*c*) (CH$_2$OH)$_2$/øH, TsOH, Δ; (*d*) LiAlH$_4$/Et$_2$O-THF, Δ; (*e*) dil. HCl; (*f*) MeC(OAc)=CH$_2$, H$_2$SO$_4$, Δ; (*g*) Pb(OAc)$_4$/HOAc-Ac$_2$O; (*h*) NaBH$_4$/MeOH; (*i*) MsCl/Py; (*j*) KOH/EtOH, Δ; (*k*) CrO$_3$/H$_2$SO$_4$, 0°; (*l*) KOH/aq. *n*-BuOH, Δ.

177

more difficult separation of lactam mixures is avoided and the unwanted *anti* oxime can be recycled to *syn* by equilibration. After steps of reduction and deblocking, the enol acetate **46** is produced and oxidized with lead tetraacetate to provide the key intermediate **47**, used also in the synthesis of the putative cycloneosamandaridine structure (see below). Reduction of the 17-keto group, mesylation, elimination, and hydrolysis produces the 16-ketone **48**. Jones' oxidation of the C(19)-hydroxymethyl group and its protection as a ketal, then deacetylation and ketal hydrolysis provide cycloneosamandione **43**. At the time Habermehl's revised structure for cycloneosamandione was published, Oka, Ike, and Hara had nearly completed a synthesis of the originally proposed 19-retro structure. Work was abandoned at the stage of intermediate **42a** [488].

As required for their synthesis of the supposedly related cycloneosamandaridine, Oka and Hara [508] devised a regiospecific synthesis of steroid D-ring lactones from the ketol acetates **49** and **50** as sketched briefly in Scheme 103. Two series of intermediates result with either 16α-acetoxy or 16β-acetoxy substituents, *but* each series converges to the same end product: **54**. The intermediate **51** with the 16β-substituent is converted under much milder conditions than the 16α, **52**, which requires an inversion at C(16) in the hydroxy acid intermediate **53** to form the *cis*-fused lactam ring. Oka and Hara applied this procedure to a synthesis of the presumed structure of cycloneosamandaridine **55** as outlined in Scheme 104. The properties of the synthetic material differed significantly from the natural alkaloid. In particular the synthetic compound, but not the natural alkaloid, exhibited a large

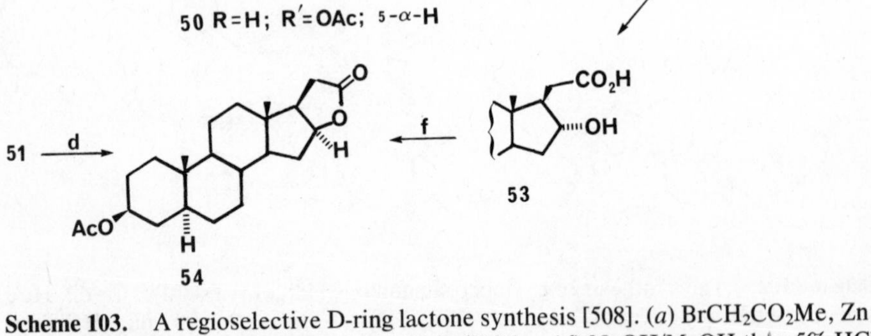

Scheme 103. A regioselective D-ring lactone synthesis [508]. (*a*) BrCH$_2$CO$_2$Me, Zn; (*b*) Ac$_2$O/Py then TsOH/øCH$_3$, Δ; (*c*) H$_2$/Pt/HOAc; (*d*) NaOH/MeOH then 5% HCl then Ac$_2$O/Py; (*e*) KOH/MeOH, Δ; (*f*) HCl/HOAc, Δ then Ac$_2$O.

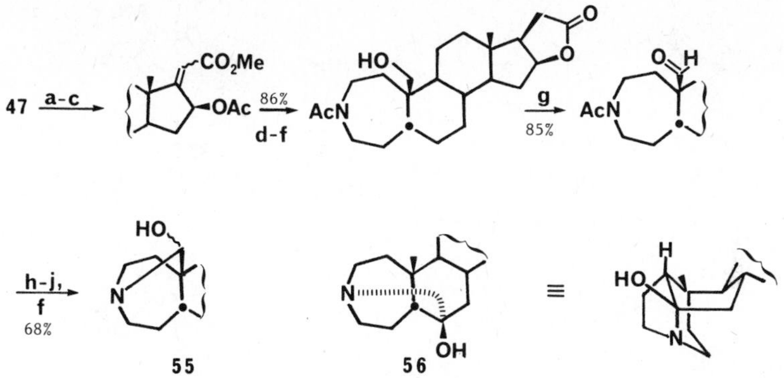

Scheme 104. The synthesis of the proposed structure for cycloneosamandaridine and a revised structure **56** for this alkaloid [488], now termed isocycloneosamandaridine. (a) $BrCH_2CO_2Me/Zn/I_2$, Et_2O-ϕH, Δ; (b) Ac_2O/Py; (c) $TsOH/\phi CH_3$; (d) $H_2/Pt/HOAc$; (e) $NaOH/MeOH$; (f) $TsOH/\phi H$, Δ; (g) CrO_3/H_2SO_4, $0°$; (h) $(CH_2OH)_2/\phi H$, $TsOH$, Δ; (i) $KOH/aq.$ n-$BuOH$; (j) dil. HCl.

M-29 peak (—CHO) in its mass spectrum, as expected for an aldehyde-amine adduct. Its absence in the natural alkaloid led Oka and Hara [488] to propose the alternative nonaldehydic structure **56**, a 3,6-cyclic carbinolamine, for the natural alkaloid, which is referred to in this review as isocycloneosamandaridine.

8.2.4. Samanine. The two earliest syntheses of samanine, reported in 1969, feature Beckmann rearrangement of oximes of 3-keto-5β-steroidal substrates. In the synthesis used by Oka and Hara [502] (Scheme 105), the 3:2 mixture of *syn*- and *anti*-oximes, prepared from **59** (in turn available in eight steps from **58** [502]), are separated and the *syn*-isomer **60** alone rearranged to the pure 3-aza lactam **62**. The unwanted *anti*-isomer can be equilibrated in acetone to form more *syn*-isomer. A final reduction with hydride affords samanine (**63**).

Habermehl and Haaf [492] employed the same starting material **59** as used by Oka and Hara, deriving it from **57** by the steps of Scheme 105 [488]. They chose, however, not to separate the oxime mixture, but rather to separate the 1:1 lactam mixture, **61** and **62**, by repeated chromatography. Reduction of lactam **62** with $LiAlH_4$ provided samanine, used in their assignment of structure to the natural product.

Rao and Weiler [509], utilizing a 2,3-seco-5β-steroidal starting material **64**, achieved a regiospecific synthesis of the 3-aza-A-homo ring of samanine as shown in Scheme 106. The 17-oxy function is shifted to C(16) by the efficient sequence shown with good yields reported for most steps, to afford the intermediate **68**, useful for conversion to samanine and derivatives. Interestingly, the conversion of the amine tosylate **65** to the *N*-benzyl hexahydroazepine **67**, requires one equivalent of benzoic anhydride in a reaction

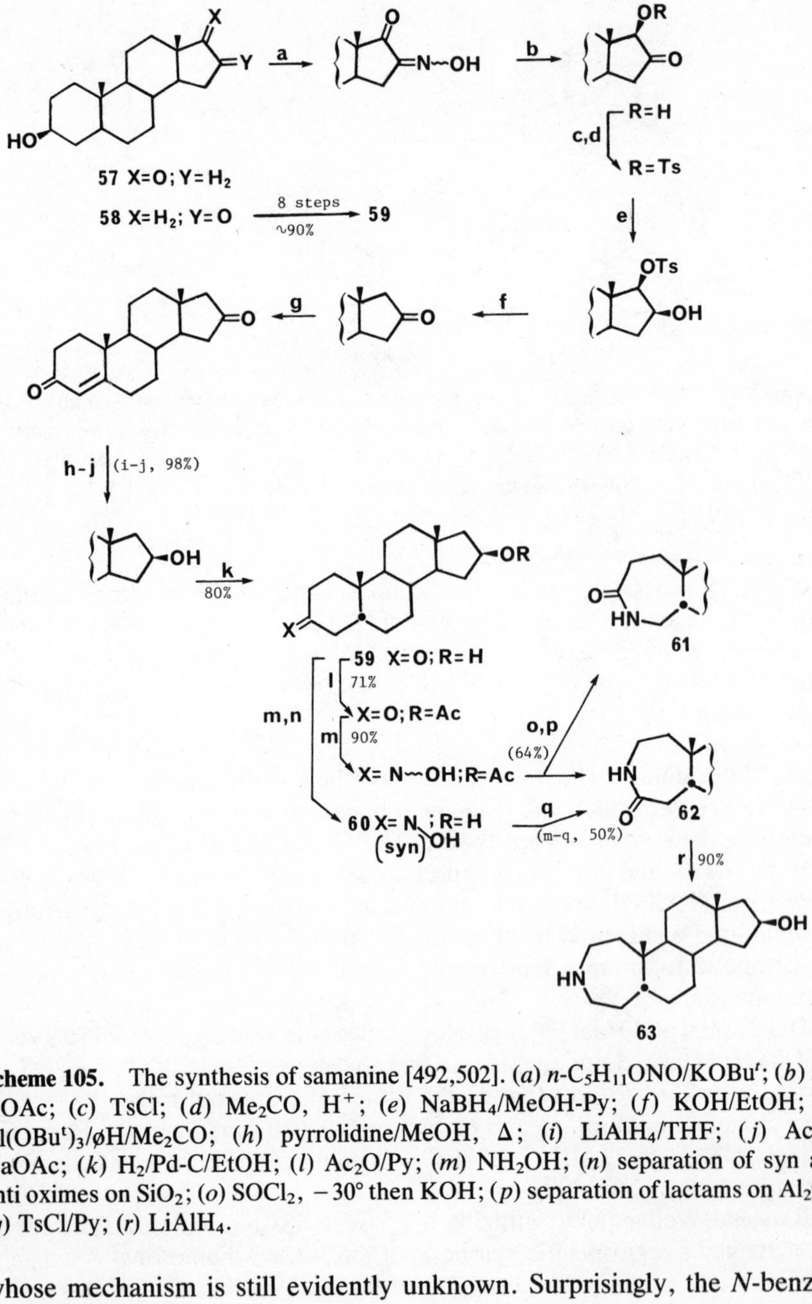

Scheme 105. The synthesis of samanine [492,502]. (*a*) *n*-C$_5$H$_{11}$ONO/KOBu^t; (*b*) Zn/HOAc; (*c*) TsCl; (*d*) Me$_2$CO, H$^+$; (*e*) NaBH$_4$/MeOH-Py; (*f*) KOH/EtOH; (*g*) Al(OBu^t)$_3$/øH/Me$_2$CO; (*h*) pyrrolidine/MeOH, Δ; (*i*) LiAlH$_4$/THF; (*j*) Ac$_2$O/NaOAc; (*k*) H$_2$/Pd-C/EtOH; (*l*) Ac$_2$O/Py; (*m*) NH$_2$OH; (*n*) separation of syn and anti oximes on SiO$_2$; (*o*) SOCl$_2$, $-30°$ then KOH; (*p*) separation of lactams on Al$_2$O$_3$; (*q*) TsCl/Py; (*r*) LiAlH$_4$.

whose mechanism is still evidently unknown. Surprisingly, the *N*-benzoyl tosylate **66** is not an intermediate.

8.2.5. Interconversions of Samandarine Alkaloids.

Samandarone has been converted to samandaridine **71** by Habermehl [480] using a glyoxalic acid

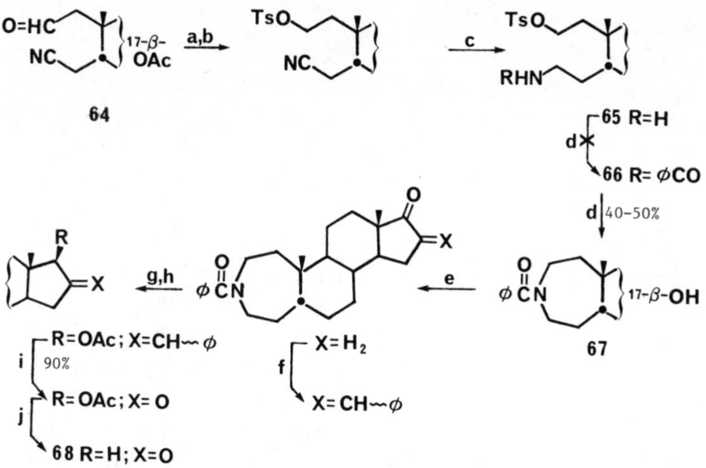

Scheme 106. Synthesis of an *N*-benzoyl 16-keto samanine analog [509]. (*a*) NaBH$_4$; (*b*) TsCl; (*c*) B$_2$H$_6$; (*d*) (øCO)$_2$O/Py; (*e*) CrO$_3$/Me$_2$CO, aq. H$_2$SO$_4$; (*f*) øCHO/MeOH-KOH; (*g*) NaBH$_4$; (*h*) Ac$_2$O; (*i*) O$_3$ then Zn reduction; (*j*) Zn/HBr/CH$_2$Cl$_2$.

condensation with *N*-acetyl samandarone followed by two steps of reduction and a final lactonization step (Scheme 107). Both condensation regioisomers **69** and **70** were carried through the sequence without separation to provide the isomeric lactones, one of which, **71**, could be obtained by crystallization in modest yield. It proved identical with samandaridine, whose structure rests upon X-ray analysis.

Samandarone hydrochloride has been stereospecifically reduced to samandarine by reduction with borohydride [480] or sodium-in-alcohol [478]. Samandarine has been oxidized to samandarone by chromic acid oxidation [478] and acetylated to the naturally occurring *O*-acetylsamandarine [486].

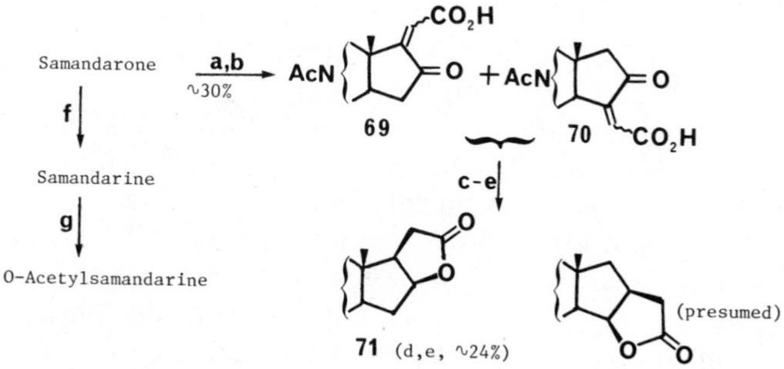

Scheme 107. Interconversions of some salamander alkaloids [480]. (*a*) Ac$_2$O; (*b*) O=CHCO$_2$H/NaOH/aq. MeOH; (*c*) NaBH$_4$/aq. MeOH, 0° (reduction and *N*-deacetylation); (*d*) H$_2$/Pt/HOAc-aq. HCl; (*e*) 2*N* HCl, Δ; (*f*) NaBH$_4$/MeOH, 0°; (*g*) Ac$_2$O.

8.3. Biological Activity

Samandarine and samandarone are potent toxins for a variety of vertebrates [466]. The salamander itself is apparently not resistant to the toxin. Minimum lethal doses of samandarine and samandarone are about 3.4 µg/kg and 6.2 µg/kg, respectively, for subcutaneous administration to mice [510]. Various derivatives were much less toxic. Convulsions, weakening of respiration and reflexes, cardiac arrhythmias, and partial paralysis precede death [466,511]. After respiratory failure, the heart continues to function for several minutes. Many of these effects are stated to be centrally mediated.

The exact mechanisms of such biological actions of samandarines need to be defined. Samandarine has potent local anesthetic activity [512]. Indeed, binding of a radioactive batrachotoxin analog to a site on the sodium channel is blocked by low concentrations of samandarine [183], providing biochemical confirmation of the potent local anesthetic activity of such alkaloids observed by Gessner in the 1930s [512]]. In isolated heart and smooth muscle preparations, samandarine can have depressant effects [410,512].

9.   TETRODOTOXINS

9.1.   Structures

9.1.1.   Tetrodotoxin.   The original source of tetrodotoxin was puffer fishes of the suborder Tetraodontidae, which had been known from antiquity to contain toxic principles, particularly in ovary, liver, and other internal organs. Recently, it has been realized that this complex, highly polar compound occurs elsewhere in the animal kingdom, in particular in a number of amphibian species. The first detection of such a toxin in amphibia was during salamander transplant experiments by Twitty and colleagues [513,514] in 1934. The toxin was discovered because of paralysis of *Amblystoma* host embryos after transplantation of eye grafts from *Taricha* embryos. The paralysis persisted in the *Amblystoma* embryo until the egg sac was absorbed and feeding and swimming normally occurred. The toxin was present in all parts of the embryonic *Taricha torosa* [515]. At that time, this Californian newt was classified as *Triturus torosa*. The toxin was also in the blood of some adults, females but not males. The toxin was present in eggs and persisted during embryonic development. Twitty [515] found it present in embryos of all three species of *Taricha* from the western United States and at lower levels in embryos of the Japanese newt *Cynops pyrrhogaster* and the newt *Notophthalmus viridescens* from the eastern United States. The toxin was water-soluble and appeared somewhat selectively toxic to motor neurons. In the early 1960s, the active principle was purified by Mosher and colleagues from eggs of *Taricha torosa* and named tarichatoxin [516]. Earlier purifications by others had yielded preparations only one-hundredth as active [517,518]. The purification used by Mosher was briefly as follows. Ninety-five kilograms of eggs were homogenized in salt solution, filtered, concentrated, and dialyzed. The dialysates were evaporated and the active principles extracted into methanol. Silica gel chromatography yielded a hundredfold purification, followed by preparative electrophoresis and crystallization from aqueous acetic acid by addition first of ethanol, then ether to yield 13 mg of tarichotoxin. The spectral and chemical properties and biological activity of the tarichotoxin were realized as quite similar to those of tetrodotoxin. A sample of tetrodotoxin provided by K. Tsuda proved identical to tarichotoxin in spectral properties, optical rotation, thin-layer chromatographic properties, conversion to identical hepta-, penta-, and diacetates and biological activity [519].

The presence of a guanidinium group in tetrodotoxin, and its facile conversion to a quinazoline (Fig. 15A) and oxalic acid on base treatment had provided key elements of its structure [see 520–523 for reviews]. The empirical formula of the highly polar tetrodotoxin molecule had remained in doubt for years, but was finally deduced from mass spectral analysis and

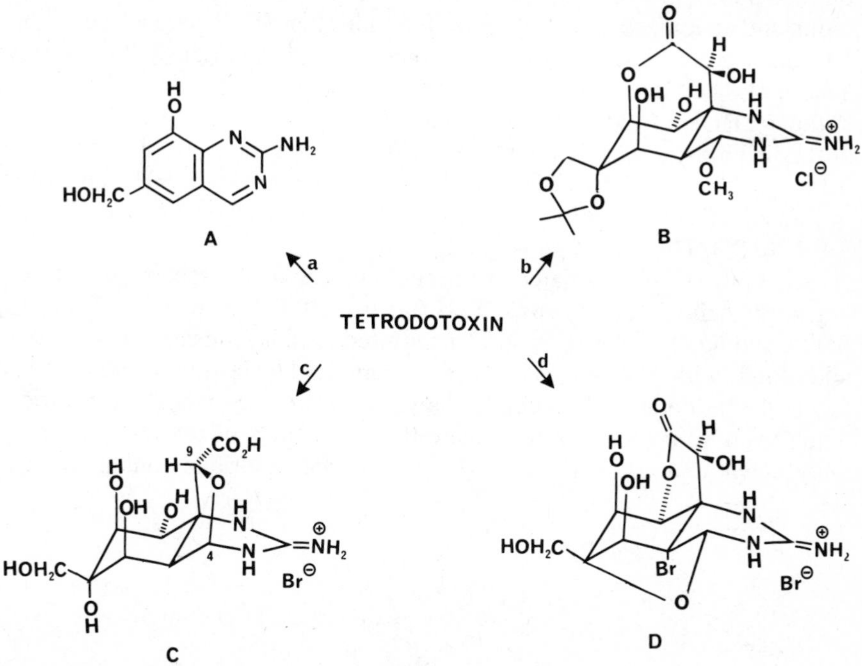

Figure 15. Conversions of tetrodotoxin to a quinazoline (*A*) and three crystalline derivatives; (*B*) *O*-methyl-*O'*, *O''*-isopropylidene tetrodotoxin HCl; (*C*) tetrodoic acid HBr; and (*D*) bromoanhydrotetrodoic lactone·HBr. (*a*) 0.7*N* NaOH, 100°, 45 min; (*b*) HCl, (CH₃)₂C=O, MeOH, 25° 2 d; (*c*) H₂O, Δ, then dil. HBr; (*d*) 5% BaOH, r.t., then aq. Br₂.

combustion analysis of a nonacetyl derivative to be $C_{11}H_{17}O_8N_3 \cdot 4H_2O$. Proton magnetic resonance spectra of a crystalline *O*-methyl-*O'*,*O"*-isopropylidene derivative provided sufficient information, coupled with chemical properties, to deduce its structure (Fig. 15*B*), which was confirmed by X-ray analysis of the hydrochloride salt [see 521]. At the same time the structures of a hydrolysis product, tetrodoic acid·HBr (Figure 15*C*) [see 522] and another derivative, a bromoanhydrotetrodoic lactone·HBr (Figure 15*D*) [see 524] were confirmed by X-ray analysis. The conformations at C(9) and C(4) were not the same in all these derivatives and inversion was shown to have occurred at C(9) and at C(4) of the carbinolamine moiety in formation of the tetrodoic acid. Tetrodotoxin·HCl appeared likely to be a lactone corresponding to the methylisopropylidene derivative B. But the stable free base could not be a lactone, since it lacked the requisite infrared bands and its pK_a 8.5 was too low for a guanidinium compound.

In 1964, the structure of tetrodotoxin was deduced from this data, independently by three laboratories, namely those of Goto [520], Woodward [521], and Tsuda [522]. Mosher's laboratory [523] also deduced the structure of tetrodotoxin (tarichotoxin) at that time. All four laboratories concluded that tetrodotoxin itself must exist as a zwitterion containing a guanidinium group and an ionized hemiorthoester function (Fig. 16). The reactions of this alkaloid were compatible with a tautomeric equilibrium between hydroxylactone and hemiorthoester, and with facile reactions, including dehydration, at the 4-CHOH of the carbinolguanidine function. Methosulphate formation at the ortho ester was demonstrated. Dimeric forms suggested earlier were incorrect. The X-ray analysis of the bromolactone (Figure 15*D*) had provided the absolute configuration of tetrodotoxin, which is levorotatory: $[\alpha_D]^{25}$ $-8.5°$ (c 3.3, D_2O).

Tetrodotoxin was isolated from egg clusters of all species of western newts, namely *Taricha torosa, T. rivularis*, and *T. granulosa* [523]. Tetrodotoxin in adult *T. torosa* was concentrated in skin, muscle, ovaries, ova, and blood, with only traces in liver, viscera, and testes [525]. Females had higher levels than males. Crude dialysates from homogenates of various amphibians were used to investigate the occurrence of tetrodotoxin [525]. Only newts and salamanders of the family Salamandridae contained tetro-

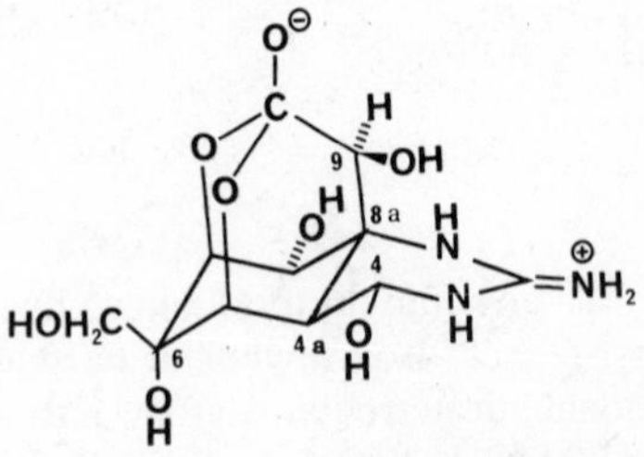

Figure 16.   Structure of (−)-tetrodotoxin.

dotoxin. High levels (13–47 μg/g tissue) were found in extracts of three species (*T. torosa, T. rivularis,* and *T. granulosa*) from western United States; modest levels (2–4 μg/g tissue) were found in a newt (*Notophthalmus viridescens*) from the eastern United States and from two species (*Cynops pyrrhogastor, C. ensicauda*) from the northwestern Pacific region, and low levels were found in four European species of newts (*Triturus vulgaris, T. cristatus, T. alpestris, T. marmoratus*). One species (*Salamandra salamandra*) of this family did not contain tetrodotoxin, but does contain samandarine alkaloids (Section 8). Brodie and colleagues [526–528] have examined *Notophthalmus viridescens, Cynops pyrrogaster,* and various species of the genus *Taricha,* for toxicity of skin and ova. Both juveniles (efts) and adults of *Notophthalmus* were toxic, but the brightly colored efts were much more toxic. The *Cynops,* in contrast to earlier studies [525], were much more toxic than adult *Notophthalmus.* The Asian salamander, *Paramesotriton hongkongensis,* also of the family Salamandridae, was quite toxic [528]. The toxin present in *Notophthalamus* is identical with tetrodotoxin in thin-layer chromatographic properties, chemical characteristics, and biological activity [529]. All of the newts or salamanders that contained tetrodotoxin were at one time classified as belonging to a single genus *Triturus.* Six other families of the order Caudata were examined for tetrodotoxin and none (<0.15 μg/ whole animal) was detectable [528]. These were Cryptobranchidae (*Cryptobranchus alleganiensis*), Proteidae (*Necturus maculosus*), Sirenidae (*Siren lacertina, S. intermedia*), Ambystomidae (*Ambystoma tigrinum*), Amphiumidae (*Amphiuma means*), and Plethodontidae (*Batrachoseps attenuatus, Ensatina eschscholtzi, Aneides lugubris*). Eggs of another species from the family Plethodontidae (*Desmognathus quadramaculatus*) were not toxic [528]. Tetrodotoxin was also not detectable in species from four families of the amphibian order Anura, namely Bufonidae (*Bufo boreus*), Hylidae (*Hyla cinerea*), Ranidae (*Rana pipiens*), and Pipidae (*Xenopus laevis*).

Tetrodotoxin has now been found in skin of South American frogs of the bufonid genus, *Atelopus* [530]: Tetrodotoxin was detected in two subspecies of *Atelopus varius,* namely *A. varius varius* (Costa Rica) and *A. varius ambulatorius,* and in *A. chiriquiensis* (Costa Rica). It was not detectable in extracts from another subspecies, *A. varius zeteki* (Panama), although this frog contains large amounts of another water soluble toxin, namely atelopidtoxin (see below). This frog may actually be a distinct species and is referred to in the present review as *A. zeteki.* The other recognized species of central American atelopid frog, namely *A. senex,* was apparently not studied. In an earlier publication, skins of *A. varius varius, A. varius ambulatorius,* and the South American species, *A. cruciger,* had been stated to contain "30–150 mouse units per skin of a toxin that produces symptoms identical with those produced by atelopid toxin" [531]. Apparently, the toxic effects of atelopidtoxin and tetrodotoxin could not be distinguished in these studies. Skins of another South American species, *A. planispina,* were stated to contain about 10 mouse units per skin of a toxin that produced marked depression of activity and cessation of respiration before cardiac arrest. Interpretation of toxicity studies of this type are complicated by the fact that atelopid frogs, like toads of the genus *Bufo*, can contain large amounts of bufodienolides [532]. Tetrodotoxin was isolated and identified from *A. varius* by its chemical and optical properties including a proton magnetic resonance spectrum [530]. It was present in skin, but not in muscle, bone, or viscera. It was the only water-soluble toxin from *A. varius varius* and *A. varius ambulatorius*, while representing only 30% of the water-soluble toxins in *A. chiriquiensis*. The major toxin in skins of the latter frog was separated from tetrodotoxin by high-pressure liquid chromatography and named chiriquitoxin, after the specific name of the frog. The levels of toxins in atelopid frogs are presented in Table 12. Recently, tetrodotoxin or a closely related compound was reported from the brachycephalid toad *Brachycephalus ephippium* [533].

9.1.2. Chiriquitoxin. This toxin, isolated from both skin and eggs of *Atelopus chiriquiensis*, had similar properties to its congener, tetrodotoxin [530,534]. Extraction of chiriquitoxin from eggs occurred with dilute acetic acid, but not with water, suggesting acid hydrolysis of a bound form. Aqueous acetic acid also appeared to extract more from skins than had water. A proton magnetic resonance spectrum of chiriquitoxin differed from that of tetrodotoxin primarily in the absence of the 2-proton peak at 4.44 ppm for the —CH_2OH moiety and apparent replacement by a 2-proton peak at 4.80 ppm. A peak at 4.68 ppm in the tetrodotoxin spectrum appears to be present in chiriquitoxin, but is much sharper. The coupled doublets ($J = 9.5$) at 5.91 ppm for proton 4 and at 2.76 ppm for proton 4a of tetrodotoxin (Fig. 16) are present at 5.91 and 2.71 in the spectrum of chiriquitoxin. In a later paper [535] it was stated that, based on mass spectrometry and proton and ^{13}C magnetic resonance spectroscopy, chiriquitoxin differs from tetro-

Table 12.   Tetrodotoxins[a] in Atelopid Frogs [530]

Species	Locale	Toxicity per Frog (mouse units)	% of Toxin		
			TTX	CTX	ZTX
Atelopus varius varius	San Antonio de Patarra, San Jose Province, Costa Rica	100	~100	0	0
A. varius ambulatorius	Valle de Parrazy and San Carlos, Costa Rica	120	~100	0	0
A. zeteki	El Valle, Panama	1200	<2.5	<2.5	~100
A. chiriguiensis	Cerro de la Muerte, Costa Rica	350	30	70	0

[a] 100 mouse units ≃ 14 μg tetrodotoxin (TTX) or chiriquitoxin (CTX) and ~30 μg of zetekitoxin (atelopid toxin, ZTX).

dotoxin only in having the —CH$_2$OH of tetrodotoxin replaced by an unidentified group with a mass of 104. This moiety was suggested to be CH$_2$N$_3$O$_3$, C$_2$H$_6$N$_3$O$_2$, or C$_3$H$_6$NO$_3$. The third possibility would be empirically and chemically consonant with a —CH$_2$OCOCH$_2$NHOH (N-hydroxyglycyl) or —CH$_2$OCOCH$_2$ONH$_2$ moiety.

9.1.3. Zetekitoxin (Atelopidtoxin).

The discovery of tetrodotoxin in atelopid frogs had its origin in investigation of the diurnal "golden frog" of El Valle, Panama. This frog, referred to previously as *Atelopus zeteki* but later considered only as a subspecies of *A. varius*, had not been previously investigated. The toxin was water-soluble and was isolated by dialysis, followed by gel filtration, cellulose column chromatography, and precipitation from water with acetone [531,536]. Thin-layer chromatography indicated that the toxin, termed in these studies atelopidtoxin, was not pure. The toxin had an R_f value of 0.27 on cellulose plates in the solvent mixture *t*-butanol:acetic acid:water (2:1:1). Tetrodotoxin has an R_f of 0.4 in this solvent system. The infrared spectrum was stated to show strong absorption peaks at 3000–3600 cm^{-1} (NH, OH), multiple absorption peaks from 1570–1680 cm^{-1}, a band at 1540 cm^{-1}, and broad general absorption from 1000–1250 cm^{-1} (carbon-oxygen bonds). The proton magnetic resonance spectrum (3 mg, D$_2$O) was stated to contain a methyl singlet at 1.6 ppm, a major peak at 5.3 ppm (OH, NH), and broad peaks at 4–5 ppm. No signals beyond 5.4 ppm were detected. The toxin was stable to chymotrypsin, negative to ninhydrin and the Reindel-Hoppe reagent for polypeptides, and positive to Weber's reagent, the latter considered diagnostic for a guanidinium moiety. The toxin was unstable in 0.1 N NaOH.

In the early 1970s the frog was becoming rare in the resort area of El Valle in Panama, and the Panamanian government declared it an endangered species, although it remains very common in the mountain ranges to the east of the volcano crater in which El Valle is located [Myers and Daly, unpublished observations]. Because of the embargo against further studies on *Atelopus zeteki*, Mosher and co-workers turned their attention to other Central American atelopid frogs. These studies resulted in the discovery of tetrodotoxin and the closely related chiriquitoxin in atelopid frogs (see above).

The name atelopidtoxin was now changed to zetekitoxin to reference the "species" name [537]. Further purification by electrophoresis indicated the presence of one major and one minor component. The major, more toxic component was designated zetekitoxin AB (LD$_{50}$ 11 µg/kg in mice), since it may consist of two closely related toxins or two forms. The minor, less toxic component was designated zetekitoxin C (LD$_{50}$ 80 µg/kg). The major toxin had the same electrophoretic mobility as tetrodotoxin, while the minor had a higher mobility. Zetekitoxin AB was stated as probably zwitterionic at pH 7. An empirical formula of (C$_2$H$_5$NO$_3$)$_n$ was derived from C, H, N analysis and a molecular weight from gel exclusion chromatography of 500 ± 100. Californium plasma desorption mass spectrometry gave positive ions at 496, 411, 407, 393, 323, 281. Tetrodotoxin affords a positive ion (M + 1) at 320 and chiriquitoxin a positive ion (M + 1) at 392. The infrared spectrum was apparently similar to that reported previously (see above). A triple peak was now reported for the 1600–1700 cm^{-1} region and a shoulder at 1760 cm^{-1}, while the 1590 cm^{-1} peak was not mentioned. Peaks in the 1400–600 cm^{-1} range were mentioned. The proton magnetic resonance spectrum was also apparently similar to that reported previously (see above). The spectrum did not show the doublets at 5.91 ppm and 2.76 ppm of the 4 and 4a protons of tetrodotoxin (Fig. 16). Most if not all of the CH groups were attached to a heteroatom. No derivatives could be formed and the nature of the zetekitoxins remains obscure. Although the pharmacology of zetekitoxins are different from tetrodotoxin, their chemical and physical properties suggest that they are probably related in structure to tetrodotoxin/chiriquitoxin from other atelopid frogs.

9.2.   The Synthesis and Chemistry of Tetrodotoxin

The synthesis of (±)-tetrodotoxin was accomplished in 1972 by Kishi and colleagues in Goto's laboratory in Nagoya [538]. Their 31-step synthesis has the remarkable distinction of being stereospecific in every step involving a chiral center. Furthermore it features, as a key step, an imaginative, quantitative rearrangement of a 7-membered ring lactone ether to a 6-membered ring lactone intermediate, incorporating all but one of the chiral centers of tetrodotoxin.

Their synthesis utilized a novel Lewis-acid-catalyzed Diels-Alder reaction between butadiene and the oximinoquinone 1 to provide the adduct 2 having the requisite cis-C(8a), C(4a) stereochemistry of tetrodotoxin (Scheme 108) [539]. Conversion of the oxime to the oxime mesylate and Beckmann rearrangement provided the 8a-acetamidoquinone 3 [540]. In many of the ensuing steps, application is made of steric control of reagent approach by the cage-like structure of many of the intermediates, approach being restricted exclusively to the convex face of the molecule. Thus the next step (d), a borohydride reduction, provided solely the 5β-alcohol 4. Epoxidation, again from the convex face, with neighboring group participation yields the tricyclic ether alcohol 5 [540], which after oxidation (f) and ketalization (g), then another stereospecific reduction (this time under thermodynamic control) and acetylation, affords 6 [541]. Selenium dioxide oxidation of the 6-methyl group to a formyl group, reduction, epoxidation, and acetylation, give, in excellent yield, the α-oxide 7. Deblocking of the C(10)-ketone, reacetylation, then diethyl ketalization of the C(10)-ketone (p) and pyrolysis (q) generate the enol ether 8. Another α-epoxidation was followed by acetolysis (s) to yield the ketol acetate 9 in good overall yield. Next, a quantitative, regiospecific Baeyer-Villiger oxidation provided the 7-membered ring lactone 11, which undergoes a dramatic rearrangement during acetolysis (v). Presum-

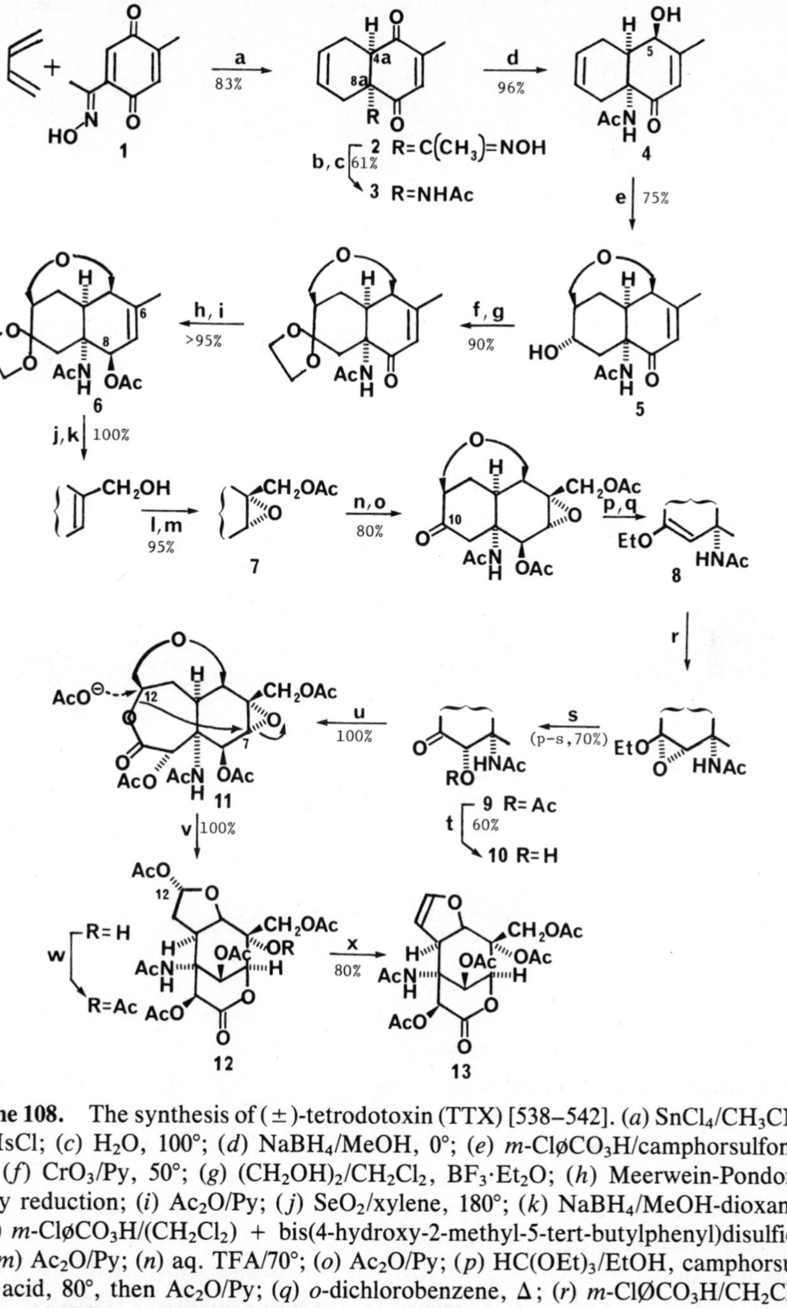

Scheme 108. The synthesis of ($\pm$)-tetrodotoxin (TTX) [538–542]. (*a*) SnCl$_4$/CH$_3$CN; (*b*) MsCl; (*c*) H$_2$O, 100°; (*d*) NaBH$_4$/MeOH, 0°; (*e*) *m*-ClØCO$_3$H/camphorsulfonic acid; (*f*) CrO$_3$/Py, 50°; (*g*) (CH$_2$OH)$_2$/CH$_2$Cl$_2$, BF$_3$·Et$_2$O; (*h*) Meerwein-Pondorf-Verley reduction; (*i*) Ac$_2$O/Py; (*j*) SeO$_2$/xylene, 180°; (*k*) NaBH$_4$/MeOH-dioxane, 0°; (*l*) *m*-ClØCO$_3$H/(CH$_2$Cl$_2$) + bis(4-hydroxy-2-methyl-5-tert-butylphenyl)disulfide 90°; (*m*) Ac$_2$O/Py; (*n*) aq. TFA/70°; (*o*) Ac$_2$O/Py; (*p*) HC(OEt)$_3$/EtOH, camphorsulfonic acid, 80°, then Ac$_2$O/Py; (*q*) *o*-dichlorobenzene, Δ; (*r*) *m*-ClØCO$_3$H/CH$_2$Cl$_2$, K$_2$CO$_3$; (*s*) HOAc; (*t*) 1*N* HCl/CH$_2$Cl$_2$; (*u*) *m*-ClØCO$_3$H/CH$_2$Cl$_2$; (*v*) KOAc/HOAc, 90°; (*w*) Ac$_2$O, camphorsulfonic acid, 100°; (*x*) 290–300°, vacuum; (*y*) OsO$_4$/THF-Py, $-20°$ then (CH$_3$)$_2$CO, H$^+$; (*z*) Et$_3$O$^{\oplus}$BF$_4{}^{\ominus}$/CH$_2$Cl$_2$, NaHCO$_3$; (*a'*) CNBr/Na-HCO$_3$, 60°; (*b'*)H$_2$S, 100°; (*c'*) Et$_3$O$^{\oplus}$BF$_4{}^{\ominus}$/CH$_2$Cl$_2$; (*d'*) Ac$_2$O/Py; (*e'*) AcNH$_2$, 150°; (*f'*) BF$_3$·Et$_2$O/CH$_2$Cl$_2$-TFA; (*g'*) MeOH, 100°; (*h'*) aq. TFA, 60°; (*i'*) HIO$_4$/aq. MeOH, 0°; (*j'*) NaIO$_4$; (*k'*) NH$_4$OH/aq. MeOH.

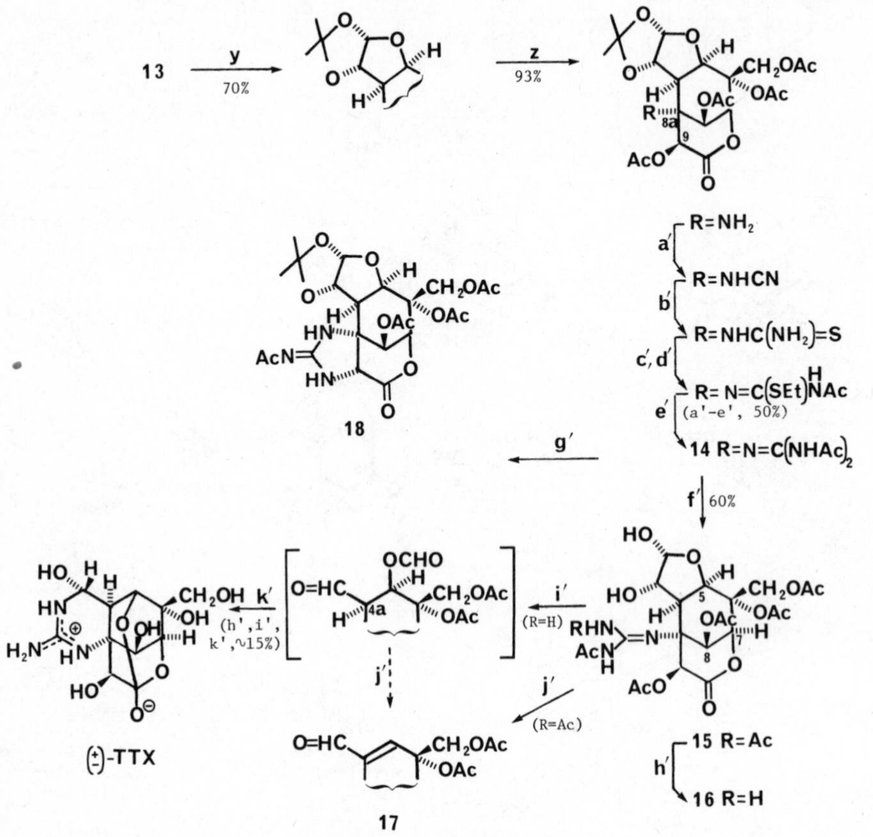

Scheme 108. (*Continued*)

ably, an S_N2 displacement of carboxylate by acetate at the least hindered side of C(12) is followed by a second displacement at C(7) as the liberated carboxylate opens the epoxide to yield the key intermediate **12** (R = H), in which all but one of the chiral centers of tetrodotoxin are now in place. Acetylation and pyrolysis produce the dihydrofuran **13** [541], which could be converted in 11 steps (Scheme 108, *y–k'*) to tetrodotoxin (3% overall yield from **13**) [538] or more preferably, in 8 steps (4% overall yield from **13**) by the route of Scheme 109 [538].

In the first route, after glycolization and acetonide protection (*y*), the C(8a) acetamido group is deacetylated (*z*) and converted stepwise to the diacetylguanidine moiety of **14** (*a'–e'*) [538]. The glycol is next deblocked, providing **15**, which is converted to the key mono-*N*-acetylguanidine **16** under acidic conditions (*h'*). Periodate cleavage of the glycol (*i'*), with concomitant addition of the nucleophilic monoacetyl guanidine group to the C(4a)-aldehyde group, so generated, followed by weakly alkaline hydrolysis

(k') of all acyl groups, produces tetrodotoxin. During this last step the novel hemiorthoester anion of tetrodotoxin is produced by intramolecular ring closure. Surprisingly, an attempt (g') to convert the acetonide 14 to the monoacetylguanidine led, instead, to the useless dialkylguanidine 18 by intramolecular displacement of the C(9) acetate.

In their alternative route from the key intermediate 13 to tetrodotoxin (Scheme 109) [538], formation of the diol is deferred until the monoacetylguanidine group has been created. It was found expedient to employ the reagent of Wheeler and Merriam (c) in preparing the intermediate N-acetylethylisothiourea 19 used for the preparation of the diacetylguanidine 20.

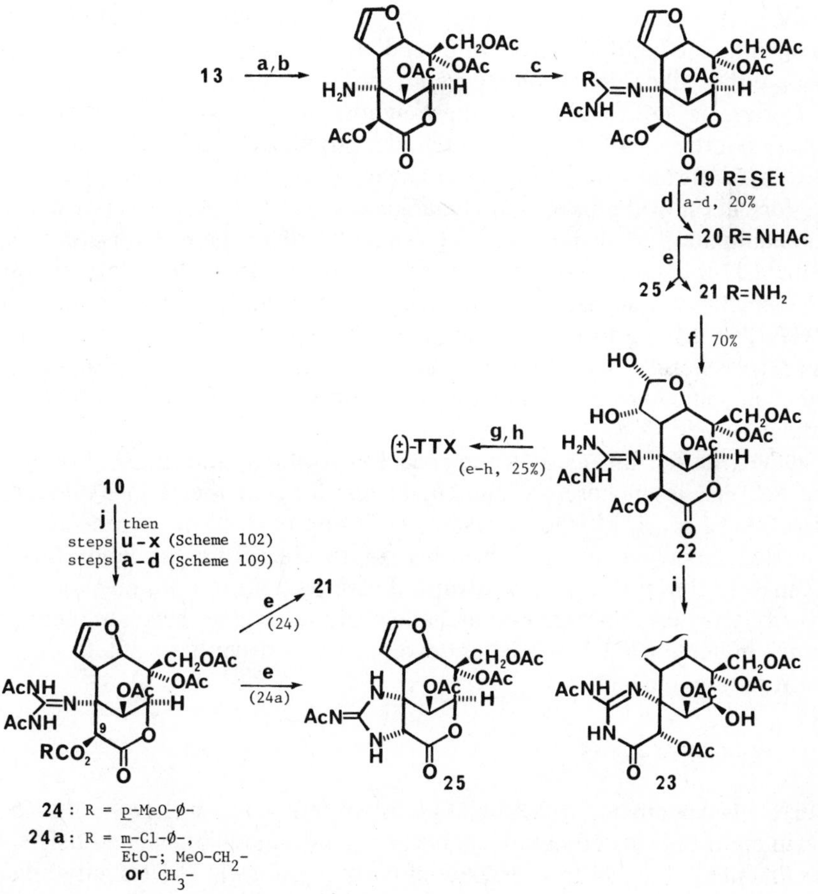

Scheme 109. An alternative and preferable synthesis of ($\pm$)-tetrodotoxin (TTX) [538,542]. (a) Et$_3$O$^\oplus$ BF$_4^\ominus$/CH$_2$Cl$_2$, NaHCO$_3$ (b) aq. HOAc/CH$_2$Cl$_2$; (c) AcN=C(SEt)$_2$), 120°; (d) AcNH$_2$, 150°; (e) NH$_4$OH/CH$_2$Cl$_2$-MeOH; (f) OsO$_4$/THF, -20°; (g) NaIO$_4$/aq. THF; (h) NH$_4$OH/aq. MeOH; (i) silica gel or base; (j) RCOCl/THF-NaH.

During the conversion of **20** to the monoacetylguanidine **21** (*e*), some (~30%) dialkylguanidine **25** is inevitably produced by the side reaction mentioned earlier in connection with intermediate **14**. Subsequent work [542] demonstrated that this side reaction could be prevented by replacement of the C(9) acetate group by an anisoate group. Models for this study (**20, 24, 24a**) were prepared from ketol **10** (Scheme 108) derived by hydrolysis of the ketol acetate **9** (*t*), then the acylation (*j*) and subsequent eight steps as indicated (Scheme 109). The acyl derivatives **24a** all gave solely the dialkylguanidine **25** on exposure to deacetylation conditions (*e*), while the anisoate **24** was converted cleanly to the desired uncyclized monoacetyl derivative.

Osmium tetroxide glycolization of **21** provided **22**, which was cleaved with periodate, then hydrolyzed with dilute ammonia to provide (±)-tetrodotoxin by a more convenient, reproducible procedure than that of Scheme 108 [538]. The highly reactive intermediate **22** rearranges to the lactam triol **23** on either silica gel chromatography or basification.

During the development of these syntheses, it was found imperative to delay the creation of the C(4a)-aldehyde, produced by oxidative cleavage of the dihydrofuran moiety (Scheme 108, step *i'*; Scheme 109, step *g*) until *after* the formation of the monoacetylguanidine group at C(8a), otherwise a ruinous elimination of formic acid intervenes yielding the α,β unsaturated aldehyde **17** (Scheme 108). To illustrate, compound **15** on periodate treatment gave solely **17**, and ozonolysis of **13** gave only the C(8a)-acetamido analog of **17**. The driving force in these eliminations is thought to be the relief of the severe steric compression caused by substituents at C(5), C(7), and C(8), which accompanies β-elimination of the *trans*-diaxial C(4a)-proton and C(5) formate group [538].

Other, as yet unsuccessful approaches to the synthesis of tetrodotoxin are not discussed here. Nachman, in his doctoral thesis [543] under the direction of H. S. Mosher, presents their approach to the synthesis of this complex molecule, along with other approaches from the laboratories of Keana, Yoshimura, and Woodward. There have been a number of derivatives of tetrodotoxin prepared as biological probes and these are mentioned briefly in Section 9.3.1. Kishi has developed a third synthesis of tetrodotoxin, but it is unpublished.

9.3. Biological Activity

9.3.1. Tetrodotoxin. The high toxicity of tetrodotoxin ($LD_{50} \sim$ 10–15 µg/kg) in many species of mammals, birds, reptiles, amphibians, and fish is well documented [see 544 for a review of early literature]. The toxicity appears to result from blockade of neuronal conduction. This results in paresthesia, muscular weakness and paralysis, emesis, hypotension, respiratory failure, and finally death. Convulsions occur in some species. In very early studies primarily on vertebrates [545], only two species were found resistant to tetrodotoxin. One was the puffer fish (*Spheroides vermicularis*), ovaries of which were used as the source of the crude toxin. The other was a Japanese salamander, now classified as *Cynops pyrrhogaster*, and now known to contain tetrodotoxin [525]. Tetrodotoxin (1000 μg/kg) interperitoneally did not paralyze or kill *Taricha* species [519]. Far smaller dosages killed frogs (*Rana pipiens*), gold fish, and tiger salamanders (*Ambystoma*). While tetrodotoxin blocks action potentials at 30–300 nM in desheathed frog nerves, 100 μM produced only a partial blockade in sciatic nerves of *Taricha granulosa* [519]. The red spotted newt (*Notophthalamus viridescens*) was resistant to the toxic effects of tetrodotoxin [529], and is known to contain tetrodotoxin [525,530].

Structural modification of tetrodotoxin results in marked reduction in activity [546]. Most structural modifications have involved the facile conversions possible at the 4-position of the carbinolguanidine function (see Fig. 16), such as conversion to 4-deoxytetrodotoxin by catalytic reduction, 4-*O*-methylation, and conversions to 4-amino-4-deoxytetrodotoxin. The 4-deoxy derivative is the most active, but is still tenfold less active than tetrodotoxin. Anhydrotetrodotoxin and derivatives, all of which contain a 4,9-oxygen bridge, are virtually inactive [546,547]. Indeed the activity of derivatives that are 200-fold or more less active than tetrodotoxin might involve traces of unmodified tetrodotoxin. Oxidation of tetrodotoxin leads to an inactive 6-keto derivative, nortetrodotoxin, which on reaction with methoxylamine yields an active compound [548]. These results have been questioned by another group [549] who found nortetrodotoxin to exist almost completely as the hydrated ketone and to be 20–40% as active as tetrodotoxin. Reduction to a 6-ol resulted in a further reduction in activity. Reductive condensation reactions of nortetrodotoxin with amines, hydrazines, or amino acids yield active derivatives [82,550,551].

The pharmacological basis for the toxicity of tetrodotoxin has been well characterized as due to the blockade of neuronal conductances [see 544 for a review]. Blockade occurs without depolarization in nerve and muscle [552–554], and initially the action of tetrodotoxin appeared reminiscent of the action of local anesthetics. However, tetrodotoxin, unlike local anesthetics, specifically blocked the sodium current without any effect on potassium currents [554,555]. This discovery was fundamental to the development of the concept of separate voltage-dependent sodium and potassium channels. Tetrodotoxin also had no effect on acetylcholine receptor mediated conductances. Tetrodotoxin acted at a site on the outer surface [556,557] of the sodium channel and is now thought merely to block the channel pore by high affinity interactions at the external opening. Both the guanidinium moiety and the 4-hydroxy group are considered to be important to interaction with the channel. Tarichatoxin, prior to its identification as tetrodotoxin, was reported to have no depolarizing action and to be specific to sodium currents [558,559]. A large number of reviews on the biological activity of tetrodotoxin are available including those by Kao [544,560], Narahashi [561–563], and Ritchie and Robard [564].

9.3.2. Chiriquitoxin. This closely related toxin shares some properties with tetrodotoxin but is not as specific [535,565]. Chiriquitoxin is as potent as tetrodotoxin in blocking sodium channels in frog muscles, but in addition, apparently affects potassium channels, as evident in delayed repolarization and reductions in steady state currents and delayed rectification in voltage clamped preparations. The effects of chiriquitoxin on potassium currents can be prevented by tetrodotoxin. Chiriquitoxin has been proposed to bind to the sodium channel, but because of close proximity of sodium and potassium channels, it thereby interferes with function of potassium channels.

9.3.3. Zetekitoxin. The pharmacology of zetekitoxin (atelopidtoxin) differs from that of tetrodotoxin [531,566]. In mammals, zetekitoxin causes hypotension, followed by "bizarre" cardiac rhythms, atrioventricular block, and terminal ventricular fibrillation. Neuromuscular transmission appeared unimpaired. Locomotor difficulties, convulsions, and respiratory failure precede death in mice. In frogs intravenous zetekitoxin causes atrioventricular block in ventricular diastole. Unlike tetrodotoxin, zetekitoxin had little effect on sciatic nerve and conduction even at about 20–30 μM. Zetekitoxin had no effect on beating of chick heart cells [196]. Zetekitoxin did block accumulations of cyclic AMP elicited by the depolarizing agent veratridine in brain slices, a property it shares with tetrodotoxin [228]. In later studies with pure zetekitoxin AB, hypotension occurred in cats and dogs [537]. Cardiac effects were similar to those reported previously. Constrictions of ductus deferens elicited by norepinephrine via α_1-adrenergic receptors and positive tropic responses of atria to the β-adrenergic agonist isoproterenol were not blocked by zetekitoxin, nor were contractions of ileum elicited by acetylcholine or norepinephrine. Vasoconstrictor responses of rabbit ear artery to nerve stimulation were blocked by zetekitoxin. Norepinephrine-elicited vasoconstriction was unaffected by zetekitoxin, and a norepinephrine-releasing agent D-amphetamine reversed blockade by zetekitoxin. The effects of zetekitoxin in this artery preparation were reminiscent of the adrenergic blocking agent, guanethidine. The results suggest that, while zetekitoxin has little effect on function of sodium channels in axons or striated muscles, it may cause blockade of sodium channels in certain cells or cellular entities. It is unfortunate that further supplies of the frog that produces this toxin are no longer available, in spite of the frog's abundance in certain mountain areas of Panama.

10. OTHER ALKALOIDS

10.1. Piperidine Alkaloids

Over 100 alkaloids from dendrobatid frogs cannot as yet be rigorously assigned to the histrionicotoxin, gephyrotoxin, indolizidine, pumiliotoxin-A, or decahydroquinoline classes of alkaloids. Some were previously tentatively assigned to decahydroquinoline or pumiliotoxin-A classes [23], but were later merely tabulated under "other alkaloids" [24]. It seems likely that the majority of these compounds will, like most other alkaloids from frogs of the genus *Dendrobates*, prove to contain a piperidine ring.

10.1.1. Piperidines. Although simple 2,6-disubstituted piperidines seem logical precursors to histrionicotoxin, gephyrotoxins, decahydroquinolines, and indolizidines, such piperidines have not been detected in almost all of the dendrobatid extracts thus far examined. One exception is a 2,6-dialkyl-piperidine detected in an extract from skin of a single specimen of *Dendrobates trivittatus* obtained at Pebas, Peru in 1978. Gas chromatographic mass spectral characterization identified its gross structure as a 2-heptyl-6-butyl-piperidine (**239I**) of unknown stereochemistry. Recently, a 2,5-dipentylpiperidine (**223B**) was detected in extracts of two populations of *Dendrobates histrionicus* [251]. The properties of the two dendrobatid alkaloids are as follows:

Piperidines

   225B. $C_{15}H_{31}N$,—, 174°, m/z 225 (1), 224 (1), 154 (100). H_0 derivative. 1D. Tentative structure: a 2,6-dipentylpiperidine.

   239I. $C_{16}H_{33}N$,—, 170°, m/z 239 (3), 162 (40), 140 (100). H_0 derivative. 1D. Tentative structure: a 2-heptyl-6-butylpiperidine.

10.1.2. Pyridyl piperidines. One piperidine alkaloid is remarkable even for dendrobatid frogs. This is noranabasamine (Fig. 17), which was isolated as a minor constituent (15 mg) from extracts of 426 skins of *Phyllobates terribilis* by high-pressure liquid chromatography [22]. The mass spectrum exhibited a parent ion at 239 ($C_{15}H_{17}N_3$) and only two fragment ions at m/z 157 ($C_{10}H_9N_2$) and m/z 84 ($C_5H_{10}N$). Proton and ^{13}C magnetic resonance spectral analysis indicated a 2,3'-dipyridyl structure with a 2-piperidyl group on the 5-position of the 2-substituted pyridine. The ultraviolet absorption spectrum (244 nM, ϵ 11,000; 275 nm, ϵ 10,000, methanol) also supported a 2,3'-dipyridyl structure. Such an alkaloid corresponds to an N-desmethyl analog of anabasamine, a plant alkaloid from the family Chenopodiaceae [567]. Both noranabasamine with an $[\alpha]_D^{25}$ $-14.4°$ and anabasamine

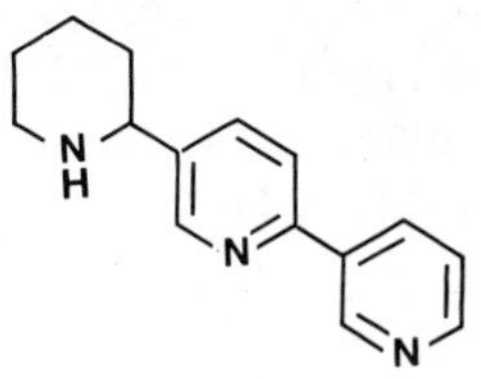

Figure 17. Structure of noranabasamine (absolute configuration unknown).

with an $[\alpha]_D^{20}$ $-82°$ are levorotatory, but whether noranabasamine has the same absolute configuration (2S) is unknown. Noranabasamine also occurs as a trace constituent in *Phyllobates aurotaenia* and *P. bicolor*, but not in various other dendrobatid frogs, based on the standard gas chromatographic mass spectral analysis (emergent temperature 204°, see Section 3.1 for conditions). The R_f for anabasamine on silica gel thin-layer chromatography is 0.21 ($CHCl_3$: methanol, 9:1).

Anabaseine, and various bipyridyl and tetrapyridyl alkaloids have been isolated from marine hoplonemertine worms [568]. Extracts from such worms were observed originally by Bacg [569] to have nicotinelike effects on vertebrate skeletal muscle and autonomic ganglia. Anabaseine was more toxic to mice than nicotine [568]. A bipyridyl marine worm alkaloid was, like nicotine, a potent convulsant agent in crayfish, while being twentyfold less toxic to mice than nicotine. The toxicity and pharmacological activity of noranabasamine from the dendrobatid frog *Phyllobates terribilis* have not been investigated.

10.1.3. Piperidine-Based Alkaloids.

For the present review most of the remaining dendrobatid alkaloids are treated in this section on piperidine alkaloids, with the term piperidine-based alkaloids referring to the probable presence of a piperidine ring, either as such or as part of a fused ring system. Three 2,5-disubstituted pyrrolidines have now been detected in dendrobatid frogs [251] and are treated in Section 10.2. Most of the following piperidine-based alkaloids are minor or trace constituents in extracts and isolation of sufficient quantities for establishment of definitive structures will be difficult. Certain of these alkaloids do occur in significant amounts, but in dendrobatid species that will be difficult or impossible to collect in large numbers.

The piperidine-based dendrobatid alkaloids, whose mass spectral characterization does not allow ready assignment into the histrionicotoxin, gephyrotoxin, indolizidine, pumiliotoxin-A, decahydroquinoline, or pyrrolidine classes are reported below in the standard format (see Section 3.1). Perhydrogenation results are not reported where results were ambiguous. These "other" alkaloids have been in part subdivided into groups according to their major fragment ions. Where no data on exchangeable hydrogen are reported, this signifies that the alkaloids could no longer be detected in extracts, which were frequently many years old at the time when the deuteroammonia chemical ionization mass spectral analysis was introduced.

The first group are tertiary amines with no exchangeable hydrogen and with a single major fragment ion at m/z 152 ($C_{10}H_{18}N$). Probably some are either 4- or 6-substituted-9-methylquinolizidines or 3- or 5-substituted-8-ethylindolizidines (see indolizidine **207A**, Section 5.1). The properties are as follows:

I. Tertiary Amines (Base Peak 152)

153B. "$C_{10}H_{19}N$,"—, 151°, m/z 153 (45), 152 (100). H_0 derivative. 0D.

181A. "$C_{12}H_{23}N$,"—, 152°, m/z 181 (2), 180 (1), 152 (100). H_0 derivative. 0D.

195C. "$C_{13}H_{25}N$,"—, 153°, m/z 195 (65), 152 (100). H_0 derivative. 0D.

207C. "$C_{14}H_{25}N$,"—, 155°, m/z 207 (16), 152 (100). H_2 derivative, m/z 209, 152. 0D.

209C. $C_{14}H_{27}N$,—, 164°, m/z 209 (11), 152 (100). H_0 derivative. 0D.

217. $C_{15}H_{23}N$, 0.40, 166°, m/z 217 (2), 216 (3), 152 ($C_{10}H_{18}N$, 100). H_6 derivative, m/z 223, 152. 0D.

219B. "$C_{15}H_{25}N$,"—, 162°, m/z 219 (1), 218 (2), 152 (100). H_4 derivative, m/z 223, 152. 0D.

223C. "$C_{15}H_{29}N$," 0.3, 159°, m/z 223 (1), 222 (2), 152 (100). H_0 derivative. 0D.

231B. $C_{16}H_{25}N$, 0.30, 166°, m/z 231 (2), 230 (1), 152 (100). H_6 derivative, m/z 237, 152. 0D. An isomer **231B'** has been detected at 168°.

235E. "$C_{16}H_{29}N$,"—, 168°, m/z 235 (3), 152 (100). H_2 derivative, m/z 237, 152. 0D.

263A. "$C_{18}H_{33}N$,"—, 174°, m/z 263 (2), 152 (100). 0D.

275A. $C_{19}H_{33}N$, 0.28, 198°, m/z 275 (3), 274 (2), 260 (5), 152 ($C_{10}H_{18}N$, 100). H_4 derivative, m/z 279, 278, 152. 0D.

277. "$C_{19}H_{35}N$,"—, 176°, m/z 277 (5), 152 (100). H_2 derivative, m/z 279, 152. 0D.

289A. "$C_{20}H_{35}N$,"—, 216°, m/z 289 (2), 287 (2), 274 (3), 152 (100). H_4 derivative, m/z 293, 152. 0D.

The second group are tertiary amines with no exchangeable hydrogen and with a single major fragment ion at m/z 166 ($C_{11}H_{20}N$). Probably some represent higher homologs of group I, namely 4- or 6-substituted-9-ethylquinolizidines or 3- or 5-substituted-8-propylindolizidines. Other structures are clearly possible. The properties are as follows:

II. Tertiary Amines (Base Peak 166)

167C. $C_{11}H_{21}N$,—, 152°, m/z 167 (100), 166 (55). H_0 derivative. 0D.

195D. "$C_{13}H_{25}N$,"—, 158°, m/z 195 (2), 166 (100). H_0 derivative. 0D.

203B. "$C_{14}H_{21}N$,"—, 159°, m/z 203 (<1), 166 (100). 0D.

209E. "$C_{14}H_{27}N$,"—, 158°, m/z 209 (17), 166 (100). H_0 derivative. 0D.

219E. "$C_{15}H_{25}N$,"—, 161°, m/z 219 (1), 218 (3), 166 (100). 0D.

223B. "$C_{15}H_{29}N$," 0.3, 159°, m/z 223 (1), 222 (2), 166 (100). H_0 derivative. 0D.

231A. $C_{16}H_{25}N$, 0.30, 166°, m/z 231 (2), 230 (1), 166 (100). H_6 derivative, m/z 237, 166. 0D. An isomer, **231A'**, has been detected at 171°.

233A. "$C_{16}H_{27}N$,"—, 167°, m/z 233 (2), 232 (2), 166 (100). H_4 derivative, m/z 237, 166. 0D.

The third group are tertiary amines with no exchangeable hydrogen and with a single major fragment ion at m/z 180 ($C_{12}H_{22}N$). Probably some represent the next higher homologs of groups I and II, namely 4- or 6-substituted-9-propylquinolizidines or 3- or 5-substituted-8-butylindolizidines. Other structures are clearly possible.

III. Tertiary Amines (Base Peak 180)

181C. "$C_{12}H_{23}N$,"—, 153 °, m/z 181 (100), 180 (68). H_0 derivative. 0D.

207D. "$C_{14}H_{25}N$,"—, 159°, m/z 207 (58), 180 (100). H_2 derivative, m/z 209, 180. 0D.

223A. "$C_{15}H_{29}N$," 0.3, 159°, m/z 223 (10), 222 (2), 180 (100). H_0 derivative. 0D. An isomer **223A'** has been detected at 163°.

231F. "$C_{16}H_{25}N$,"—, 164°, m/z 231 (1), 180 (100). H_6 derivative, m/z 237, 180. 0D.

237C. "$C_{16}H_{31}N$,"—, 168°, m/z 237 (2), 236 (1), 180 (100). H_0 derivative. 0D.

Certain of the remaining "other" dendrobatid alkaloids have properties distinctive enough to warrant separate comment. Three have anomalous base peaks at odd integer masses. Others have a major fragment ion at m/z 70 like the pumiliotoxin-A class, but, unlike that class of alkaloids, they do not have a major ion at either m/z 166 or 182. Another pair of closely related alkaloids exhibited a mass fragmentation pattern, which led to an early assignment to the pumiliotoxin-A class. These isomers are major alkaloids in two extracts and are distinctive enough to warrant separate comment. A third alkaloid is tentatively included with them. Finally, there are four dendrobatid alkaloids that contain two nitrogens. These alkaloids, three of which have been shown to be amidine in nature, are treated in Section 10.5.

The three alkaloids with anomalous base peaks at odd integer masses are all from a single dendrobatid species. *Dendrobates bombetes* [254]: Alkaloid **251F** ($C_{16}H_{29}NO$) shows a base peak at m/z 111 ($C_7H_{13}N$), while a higher homolog **265B** ($C_{17}H_{31}NO$) shows a base peak at m/z 125 ($C_8H_{15}N$). The third alkaloid **247** is a trace constituent in *Dendrobates bombetes*. It shows a base peak at m/z 109 and is probably an unsaturated analog of **251F**. All of this group of alkaloids show a major fragment ion at m/z 70 (C_4H_8N), which suggests a similarity to the pumiliotoxin-A class of dendrobatid alkaloids. They are tricyclic tertiary amines containing one hydroxy group. The properties of these three alkaloids are as follows:

IV. Tertiary Amines (Odd Integer Base Peak)

247. "$C_{16}H_{25}NO$,"—, 175°, m/z 247 (5), 246 (3), 230 (5), 170 (25), 166 (25), 150 (40), 109 (100), 70 (50). H_4 derivative, m/z 251, 232, 111, 70. 1D.

251F. $C_{16}H_{29}NO$, 0.25, 184°, m/z 251 (20), 250 (23), 236 (10), 234 (3), 222 ($C_{14}H_{24}NO$, 13), 220 ($C_{15}H_{26}N$, 25), 194 ($C_{12}H_{20}NO$, 30), 166 (10), 164 ($C_{11}H_{18}N$, 14), 152 ($C_{10}H_{18}N$, 32), 150 ($C_{10}H_{16}N$, 16), 112 (40), 111 ($C_7H_{13}N$, 100), 98 ($C_6H_{12}N$, 52), 70 (15). H_0 derivative. 1D. *O*-acetyl derivative.

265B. $C_{17}H_{31}NO$, 0.28, 193°, m/z 265 (18), 264 (22), 250 (12), 236 (17), 234 (20), 194 (16), 166 (25), 152 (10), 126 (30), 125 (100), 112 (28), 70 (20). H_0 derivative. 1D.

Three of the remaining dendrobatid alkaloids also may prove to be closely related to the pumiliotoxin A class. All have a major fragment at m/z 70 (C_4H_8N), but the other major fragments are not typical of the pumiliotoxin-A or allopumiliotoxin classes. The properties of these alkaloids are as follows:

V. Piperidine-Based Alkaloids (Base Peak 70)

251E. "$C_{15}H_{25}NO_2$,"—, 175°, m/z 251 (3), 250 (1), 234 (2), 168 (30), 84 (18), 70 (100).

281B. "$C_{18}H_{35}NO$,"—, 200°, m/z 281 (4), 264 (12), 208 (25), 206 (20), 150 (65), 98 (5), 96 (20), 70 (100). H_0 derivative.

351. "$C_{21}H_{37}NO_3$," 0.15, 230°, m/z 351 (6), 350 (2), 336 (4), 152 (38), 138 (65), 70 (100). 4D.

Two isomeric alkaloids, **283B** and **283C**, from one population of *Dendrobates histrionicus* and a sibling species, *D. occulator*, have not been detected in other dendrobatid frogs: Their empirical formulas and complex fragmentation led to an early tentative assignment to the pumiliotoxin-A class [23]. It now appears more likely that they represent a different structural class. Alkaloid **283B** loses a C_5H_{11} side chain, while **283C** loses a C_4H_9 side chain. A base peak at m/z 140 is present in one other piperidine-based alkaloid (**285D**), which, therefore, has been tentatively included in this grouping. The properties of these alkaloids are as follows:

VI. Tertiary Amines (Base Peak 140 or 126)

283B. $C_{17}H_{33}NO_2$, 0.36, 197°, m/z 283 (<1), 282 (1), 254 (2), 212 ($C_{12}H_{22}NO_2$, 40), 152 ($C_{10}H_{18}N$, 23), 140 ($C_9H_{18}N$, 100). H_0 derivative. 1D.

283C. $C_{17}H_{33}NO_2$. 0.40, 195°, m/z 283 (<1), 282 (1), 240 (5), 226 ($C_{13}H_{24}NO_2$, 28), 224 ($C_{15}H_{30}N$, 10), 166 ($C_{11}H_{20}N$, 60), 126 ($C_8H_{16}N$, 100). H_0 derivative. 1D.

285D. "$C_{19}H_{27}NO$,"—, 190°, m/z 285 (3), 270 (2), 256 (2), 180 (35), 140 (100).

Whether **283B** and **283C**, which are major alkaloids in the one population of *Dendrobates histrionicus*, are responsible for the clonic convulsions, sustained penile erections, and thrusting movements evoked by skin extracts from this frog [23] are unknown. None of the other major alkaloids (e.g., the histrionicotoxins) from this frog have these effects, nor do extracts from any other populations of *D. histrionicus*.

The remaining dendrobatid alkaloids are too diverse and poorly defined in properties to warrant speculation as to their structures. Most have relatively simple mass spectra with only one or two major fragment ions. Some of those with exchangeable hydrogens are probably hydroxy congeners of the indolizidine or decahydroquinoline classes or of groups I, II or III of other piperidine-based alkaloids (see above). The properties are as follows:

VII. Other Piperidine-Based Alkaloids

151. "$C_{10}H_{17}N$,"—, 152°, m/z 151 (100), 150 (25). 0D.

183A. $C_{12}H_{25}N$,—, 154°, m/z 183 (3), 154 (100). H_0 derivative. 1D. A possible structure would be a 2-ethyl-3, 4, or 5-pentyl piperidine.

185. "$C_{11}H_{23}NO$," 0.2, 153°, m/z 185 (1), 170 (100). H_0 derivative.

193A. $C_{13}H_{23}N$,—, 152°, m/z 193 (100), 192 (65). H_0 derivative. 0D.

193B. "$C_{13}H_{23}N$,"—, 158°, m/z 193 (23), 192 (13), 150 (100). H_0 derivative. 0D.

195E. "$C_{13}H_{25}N$,"—, 156°, m/z 195 (45), 194 (100). H_0 derivative. 0D.

197A. "$C_{12}H_{23}NO$,"—, 160°, m/z 197 (1), 180 (100), 126 (35). H_0 derivative. 1D.

201. "$C_{14}H_{19}N$,"—, 167°, m/z 201 (<1), 200 (2), 136 (100). 0D.

207E. "$C_{14}H_{25}N$,"—, 158°, m/z 207 (12), 164 (100), 84 (35). 0D.

207F. "$C_{14}H_{25}N$,"—, 157°, m/z 207 (7), 192 (100). H_2 derivative, m/z 209, 194. 0D.

209A. "$C_{13}H_{25}NO$,"—, 162°, m/z 209 (5), 168 (100). H_2 derivative, 211, 168.

211A. "$C_{13}H_{25}NO$,"—, 169°, m/z 211 (3), 210 (2), 168 (100). H_0 derivative. 2D.

211B "$C_{14}H_{25}NO$,"—, 168°, m/z 211 (4), 160 (100). 1D.

221B. "$C_{14}H_{25}NO$,"—, 173°, m/z 221 (2), 192 (100). H_2 derivative. 1D.

223E. "$C_{14}H_{27}NO$,"—, 163°, m/z 223 (2), 222 (3), 168 (100). H_2 derivative m/z 225, 168.

225A. "$C_{14}H_{29}NO$,"—, 164°, m/z 225 (3), 224 (6), 208 (2), 168 (100), 152 (25). H_0 derivative.

233B. "$C_{15}H_{23}NO$,"—, 180°, m/z 233 (<1), 168 (100).

233C. "$C_{16}H_{27}N$,"—, 173°, m/z 233 (1), 192 (60), 96 (100). 1D.

235D. "$C_{15}H_{25}NO$,"—, 182°, m/z 235 (<1), 196 (20), 170 (100). H_6 derivative. 1D.

235F. "$C_{16}H_{29}N$,"—, 170°, m/z 235 (5), 234 (3), 166 (36), 138 (100). 0D.

235G. "$C_{16}H_{29}N$,"—, 172°, m/z 235 (2), 206 (100), 194 (65).

237E. "$C_{15}H_{27}NO$," 0.25, 180°, m/z 237 (1), 236 (3), 208 (70), 152 (100). H_2 derivative, m/z 239, 152.

239A. "$C_{15}H_{29}NO$,"—, 178°, m/z 239 (2), 238 (3), 182 (100). H_0 derivative. An isomer **239A′** has been detected at 174°.

239B. "$C_{15}H_{29}NO$,"—, 178°, m/z 239 (2), 238 (3), 180 (100). H_0 derivative. 1D. An isomer **239B′** has been detected at 174°.

239C. "$C_{15}H_{29}NO$,"—, 179°, m/z 239 (2), 238 (3), 196 (100). H_0 derivative. ?D. An isomer **239C′** has been detected at 174°.

239D. "$C_{15}H_{29}NO$,"—, 179°, m/z 239 (2), 238 (3), 166 (100). H_0 derivative. An isomer **239D′** has been detected at 174°.

239E. "$C_{15}H_{29}NO$,"—, 176°, m/z 239 (2), 238 (3), 210 (40), 152 (100). H_0 derivative 1D.

239F. "$C_{15}H_{29}NO$," 0.30, 176°, m/z 239 (1), 168 (100). H_0 derivative. 1D. O-acetyl derivative.

241A. "$C_{14}H_{27}NO$," 180°, m/z 241 (2), 240 (3), 166 (100), 126 (48).

241C. "$C_{14}H_{27}NO_2$,"—, 177°, m/z 241 (1), 152 (100).

241D. "$C_{15}H_{31}NO$,"—, 188°, m/z 241 (2), 240 (3), 224 (5), 154 (15), 114 (100), 98 (25), 70 (32). H_0 derivative. 2D.

241E. "$C_{14}H_{27}NO_2$,"—, 174°, m/z 241 (3), 222 (62), 154 (100). 2D.

243B. "$C_{17}H_{25}N$,"—, 178°, m/z 243 (<1), 202 (28), 176 (100). H_6 derivative. 0D.

249. "$C_{16}H_{27}NO$,"—, 172°, m/z 249 (3), 192 (100). H_2 derivative. 0D.

251C. "$C_{16}H_{29}NO$,"—, 190°, m/z 251 (2), 234 (4), 154 (100). H_2 derivative, m/z 253, 154. 1D.

251H. "$C_{16}H_{29}NO$,"—, 170°, m/z 251 (1), 178 (30), 150 (100).

253C. "$C_{16}H_{31}NO$,"—, 175°, m/z 253 (3), 192 (100). 1D.

255. $C_{15}H_{29}NO_2$,—, 195°, m/z 255 (3), 238 (6), 198 (14), 152 (12), 114 (100), 70 (20). H_0 derivative. 2D.

257A. "$C_{18}H_{27}N$," 0.30, 188°, m/z 257 (1), 256 (2), 216 (100). H_8 derivative, m/z 265, 222.

257B. "$C_{18}H_{27}N$," 0.35, 192°, m/z 257 (60), 256 (100), 152 (20). 1D. The mass spectrum, which was considered unusual for a dendrobatid alkaloid, led to the suggestion that it might represent that of a degradation product [23].

257D. "$C_{18}H_{27}N$,"—, 190°, m/z 257 (5), 256 (3), 190 (100). H_8 derivative. 0D.

263B. "$C_{18}H_{33}N$,"—, 170°, m/z 263 (3), 198 (100). 0D.

265A. — 0.35, 198°, m/z 265 (50), 264 (100), 222 (58), 180 (72). The spectrum, which was considered unusual for a dendrobatid alkaloid, led to the suggestion that it might represent that of a degradation product [23]. Found to be an octadecenoic acid methyl ester—presumably an artifact.

265C. $C_{17}H_{31}NO$,—, 182°, m/z 265 (14), 236 (10), 210 ($C_{13}H_{24}NO$, 100), 192 (10), 138 (21), 84 ($C_5H_{10}N$, 45). H_2 derivative. 1D.

267B. "$C_{16}H_{29}NO_2$,"—, 208°, m/z 267 (7), 266 (4), 250 (1), 170 (100), 152 (4), 112 (13). 2D.

267E. "$C_{17}H_{33}NO$,"—, 182°, m/z 267 (18), 266 (11), 196 (100), 96 (58). H_0 derivative. 0D.

267F. "$C_{16}H_{29}NO_2$,"—, 198°, m/z 267 (12), 250 (20), 178 (100).

267G. "$C_{16}H_{29}NO_2$,"—, 190°, m/z 267 (4), 152 (100). H_2 derivative. 2D.

269A. "$C_{19}H_{27}N$,"—, 207°, m/z 269 (4), 204 (100). H_{10} derivative, m/z 279, 208. 1D.

269B. "$C_{19}H_{27}N$,"—, 207°, m/z 269 (4), 202 (100). H_{10} derivative, m/z 279, 208. 1D.

269C. "$C_{16}H_{31}NO_2$,"—, 207°, m/z 269 (2), 98 (100). 2D.

269D. "$C_{19}H_{27}N$,"—, 199°, m/z 269 (2), 176 (100). H_{10} derivative. 0D.

271. "$C_{19}H_{29}N$,"—, 198°, m/z 271 (3), 178 (100). H_8 derivative. 0D.

275B. "$C_{19}H_{33}N$,"—, 195°, m/z 275 (11), 222 (12), 206 (100). H_4 derivative, 279, 208, 1D.

279. "$C_{18}H_{33}NO$,"—, 190°, m/z 279 (35), 210 (90), 190 (75), 84 (100). H_0 derivative. 1D.

281C. "$C_{17}H_{31}NO_2$,"—, 195°, m/z 281 (25), 208 (100). H_0 derivative. 2D.

281D. "$C_{17}H_{31}NO_2$,"—, 196°, m/z 281 (16), 210 (100). H_0 derivative. 2D.

291B. "$C_{19}H_{33}NO$," 0.12, 221°, m/z 291 (2), 290 (3), 276 (6), 168 (100). H_4 derivative, m/z 295, 168. 1D.

291C. "$C_{19}H_{33}NO$," 0.20, 220°, m/z 291 (1), 290 (2), 276 (4), 210 (10), 152 (100). H_4 derivative, m/z 295, 152. 1D.

291D. $C_{19}H_{33}NO$,—, 201°, m/z 291 (<1), 168 (100). H_4 derivative. 1D.

295. "$C_{19}H_{37}NO$," 0.09, 224°, m/z 295 (3), 278 (4), 138 (100).

301. "$C_{21}H_{35}N$,"—, 213°, m/z 301 (<1), 260 (100).

309B. "$C_{20}H_{39}NO$," 0.09, 220°, m/z 309 (1), 152 (100).

309E. "$C_{18}H_{33}NO_3$,"—, 235°, m/z 309 (32), 266 (13), 240 (100), 205 (22), 124 (35), 114 (25). H_2 derivative. 3D.

309F. "$C_{19}H_{39}NO_2$,"—, 206°, m/z 309 (2), 152 (100). H_0 derivative. 2D.

Biological Activity.   The biological activity of these "other" piperidine-based alkaloids from dendrobatid frogs is unknown. However, alkaloid fractions from many dendrobatid species contain compounds with marked activities in mice [23,250,253,254,259], and some of these activities may be due to "other" alkaloids.

10.2.   Pyrrolidine Alkaloids

The range of structural classes of alkaloids, elaborated by dendrobatid frogs, has currently been extended by the identification of a *trans*-2-*n*-butyl-6-*n*-pentylpyrrolidine [251]. This monocyclic saturated alkaloid (**197B**, $C_{13}H_{27}N$) was a secondary amine and exhibited two major mass spectral fragments, one at m/z 140 corresponding to lose of C_4H_9 and one at m/z 126 corresponding to loss of C_5H_{11}. This pyrrolidine **197B** (Fig. 18) co-eluted on gas chromatography with synthetic *trans*-2-*n*-butyl-5-*n*-pentylpyrrolidine, an alkaloid previously identified from ants of the genus *Monomarium* [570]. Two other 2,6-disubstituted pyrrolidines have been identified from dendrobatid frogs, based on mass spectral analyses. The properties of the three dendro-

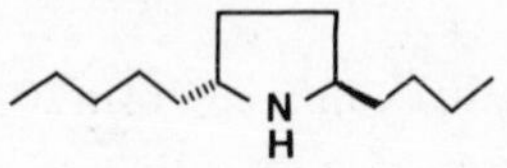

197B

Figure 18.   Structure of pyrrolidine **197B** (absolute configuration unknown).

batid alkaloids belonging to the pyrrolidine class are presented below in the standard format:

Pyrrolidines
   183B. "$C_{12}H_{25}N$,"—, 151°, m/z 183 (1), 126 (100). H_0 derivative. 1D. Tentative structure: a 2,5-dibutylpyrrolidine.
   197B. $C_{13}H_{27}N$,—, 163°, m/z 197 (1), 196 (2), 140 (78), 126 (100). H_0 derivative. 1D. Structure: Fig. 18.
   225C. $C_{15}H_{31}N$,—, 172°, m/z 225 (1), 224 (2), 168 (70), 126 (100). H_0 derivative. 1D. Tentative structure: a 2-butyl-5-heptylpyrrolidine.

10.3. Indole Alkaloids

10.3.1. Dihydroindoles. Two isomeric indole alkaloids, calycanthine and chimonanthine, have recently been isolated from a dendrobatid frog, *Phyllobates terribilis* [22]. From 426 skins, 11 mg of calycanthine and 8 mg of chimonanthine were obtained by high-pressure liquid chromatography. The alkaloids exhibited a parent ion at m/z 346 ($C_{22}H_{46}N_4$) and a major fragment ion at m/z 173 ($C_{11}H_{23}N_2$). The proton and ^{13}C magnetic resonance spectra indicated the structures of these "dimeric" indole alkaloids. Proton magnetic resonance spectra are depicted in [22]. The properties of the two dendrobatid alkaloids proved to be identical with those of calycanthine and chimonanthine from calycanthaceaid plants except for optical rotations. The frog calycanthine ($[\alpha]_D^{25}$ $-570°$, methanol) and chimonanthine ($[\alpha]_D^{25}$ $+$ 280°, methanol) had rotations opposite in sign to those of the *d*-calycanthine and *l*-chimonanthine from plants and were thus their enantiomers (Fig. 19). *l*-Chimonanthine in plants has been proposed as the biological precursor of *d*-calycanthine [571].

10.3.2. Biogenic Amines. Many amphibians contain high levels of serotonin and its *N*- and *O*-methylation products [see 571–577 for reviews). These biogenic amines include serotonin, *N*-methylserotonin, bufotenine, bufotenine *O*-sulfate, bufotenidine, bufoviridine, *N*-methyl-5-*O*-methylsero-

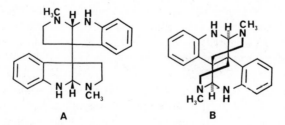

Figure 19. Structure of (*A*) *d*-chimonanthine, and (*B*) *l*-calycanthine from a dendrobatid frog, *Phyllobates terribilis*.

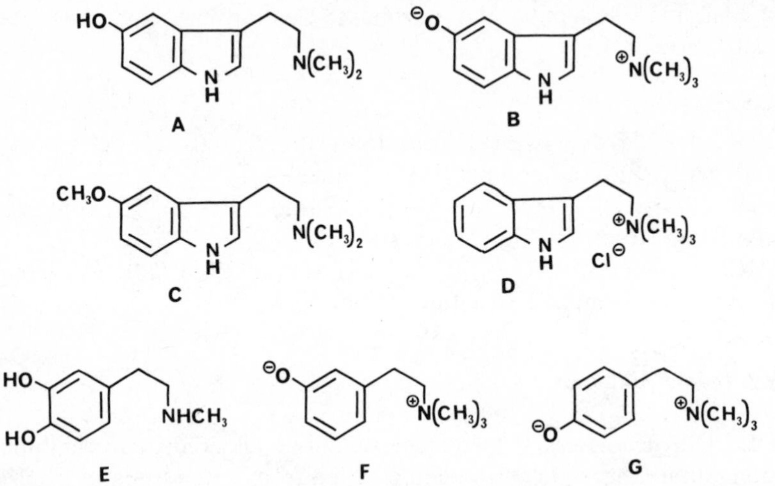

Figure 20. Structures of the methylated amines occurring in amphibians, which are derived from serotonin, tryptamine, dopamine, and tyramines. (*A*) Bufotenine; (*B*) bufotenidine; (*C*) *O*-methylbufotenine, (*d*) *N,N,N*-trimethyltryptamine chloride; (*E*) epinine; (*F*) leptodactyline; (*G*) candicine. The *O*-sulfate of bufotenidine (bufoviridine) also occurs in amphibians.

tonin, and 5-*O*-methylbufotenine (Fig. 20). Certain amphibians contain tryptamine and others, quaternary *N,N,N*-trimethyltryptamine. For the purpose of the present review on amphibian alkaloids, such simple methylated derivatives of serotonin and tryptamine are not treated except to tabulate their occurrence (Section 11). Similarly, although *N*-methylated congeners of catecholamines, such as epinine, and of phenolic amines, such as candicine and leptodactyline (Fig. 20), occur in skin or skin glands of certain amphibians, those methylation products of biogenic amines are not treated except to tabulate their occurrence (Section 11).

10.3.3. Dehydrobufotenine. Although originally suggested as being a side chain unsaturated congener of *N,N*-dimethylserotonin, based on its hydrogenation to bufotenine [578], dehydrobufotenine actually contains an additional heterocyclic ring (Fig. 21). Dehydrobufotenine and its *O*-sulfate (bufothionine) occur in large amounts in the parotid glands of the toad *Bufo marinus* and were originally isolated by the Wielands. The correct structure was proposed by two laboratories in 1961 based on the proton magnetic resonance spectrum [579,580].

The synthesis of dehydrobufotenine was reported some six years later [581] (Scheme 110). Nitration of 5-benzyloxygramine **1** afforded the 4-nitro derivative **2**, which was converted to the nitrile **3** via the quaternary methosulfate. Reductive cyclization of the nitrile was unsuccessful, perhaps due to hydrogenolysis of the *O*-benzyl group. The nitrile was converted to the

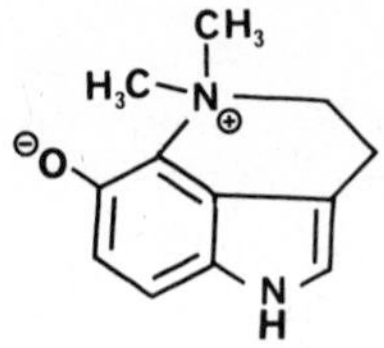

Figure 21. Structure of dehydrobufotenine. The *O*-sulfate (bufothionine) also occurs in amphibians.

ethyl ester by treatment with ethanolic hydrogen chloride. Reduction of the nitro group with dithionite was accompanied by hydrolysis of the ester. The resulting 4-amino-5-benzyloxyindole-3-acetic acid **4** was cyclized to the lactam with dicyclohexylcarbodiimide, then reduced to the secondary amine with diborane and quaternized with methyl iodide to **5**. Catalytic debenzylation with palladium-on-carbon yielded dehydrobufotenine hydroiodide **6**. Ion exchange provided the natural, zwitterionic, product.

Dehydrobufotenine is moderately toxic but its pharmacology has not been investigated. The lethal dose in mice on subcutaneous injection is 6 mg/kg [577]. Death occurs with clonic convulsions.

10.3.4. Tetrahydrocarbolines. Recently, a tetrahydro-β-carboline, trypargine, has been isolated from extracts of skin of the hyperolid frog, *Kassina*

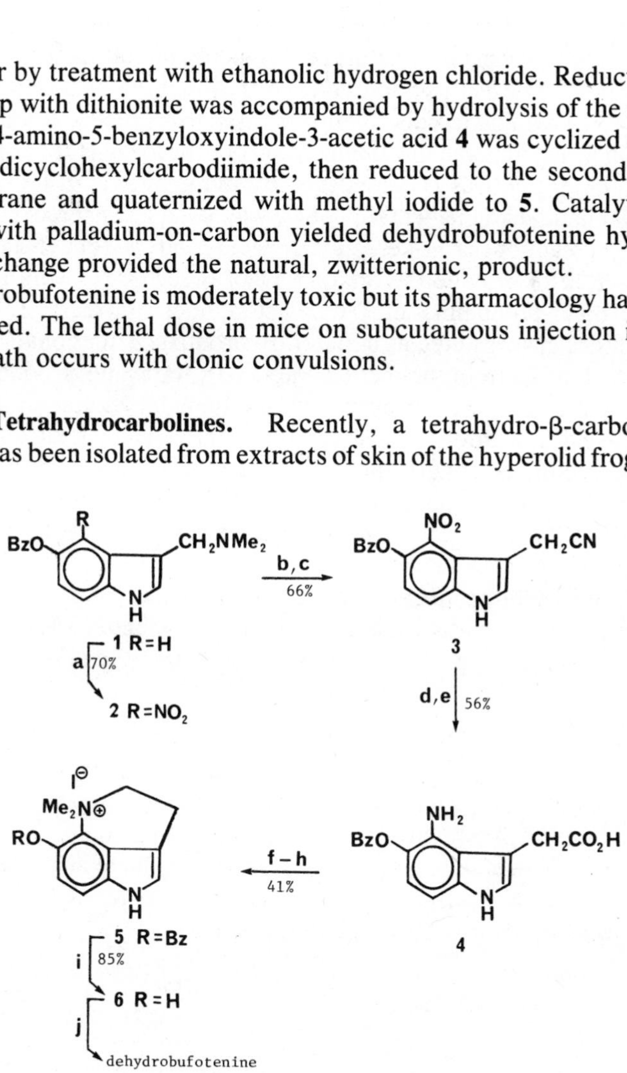

Scheme 110. The synthesis of dehydrobufotenine [581]. (*a*) HNO$_3$; (*b*) Me$_2$SO$_4$; (*c*) CN$^-$; (*d*) HCl/EtOH; (*e*) Na$_2$S$_2$O$_4$/OH$^-$; (*f*) DCC/THF; (*g*) B$_2$H$_6$; (*h*) MeI/KOH/aq. EtOH; (*i*) H$_2$/Pd-C/MeOH, 60°; (*j*) ion exchange chromatography.

senegalensis [582]. Trypargine contains a guanidine group and its spectral properties, including proton and ^{13}C magnetic resonance, led to definition of its structure (Fig. 22). The alkaloid appears likely to originate biosynthetically from condensation of tryptamine with the aldehyde derived from arginine by oxidation decarboxylation or by condensation with the α-keto acid followed by decarboxylation. Trypargine or a related compound was stated also to occur in large amounts in another hyperolid frog, *Hylambates maculatus*.

The proposed gross structure of trypargine **7** was confirmed by the Pictet-Spengler synthesis using tryptamine and γ-guanidinobutyraldehyde ethylene acetal [583] (Scheme 111). Resolution (*b*) provided the natural levorotatory enantiomer **8**, having a circular dichroism spectrum similar to the C(1) α-H tetrahydroharman alkaloid **9**. Conclusive proof for the α-H (**S**) configuration at C(1) of trypargine was accomplished by the synthesis of (−)-trypargine by Ishikawa and Nakajima's group [584], who used a diastereoselective Pictet-Spengler reaction of α-ketoglutaric acid and *N*-benzyl-D-tryptophan methyl ester **10** to provide the mixture **11** of β-carbolines with the desired C(1) α-H isomer predominating. The diastereomeric mixture, **12** and **13**, of 6-oxocanthines also resulted as a minor by-product. After conversion of the carboline acid mixture to the methyl esters (*d*), the major product was separated (*e*), subjected to a slow ammonolysis, then dehydrated to the dinitrile **14**. Reductive decyanation (*h*), conversion of the remaining nitrile to a methyleneamino group, debenzylation, and guanidation with S-methylisothiourea provided 1(**S**)-(−)-trypargine, identical with the natural material. Conclusive proof for the 1-**S** configuration was provided by X-ray analysis of the major Pictet-Spengler product **11**.

Intravenous administration of a high dose (10 mg/kg) of trypargine to mice caused paralysis, respiratory failure, and within 2 minutes, death [582]. The pharmacology of trypargine was stated to be under investigation.

10.4. Imidazole Alkaloids

Many species of frogs produce and store large quantities of histamines in skin [see 572,573,574,576 for reviews]. *N*-Methylhistamine, *N*,*N*-dimethylhistamine, and *N*-acetylhistamine (Fig. 23*A* and *B*) have also been found in frog skins. These *N*-methylated compounds are not treated in this review except to tabulate their occurrence (Section 11).

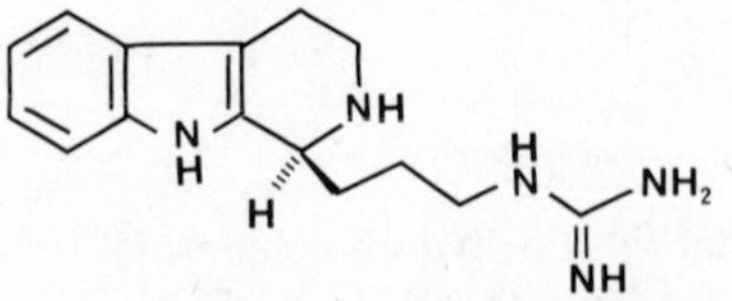

Figure 22.   Structure of trypargine.

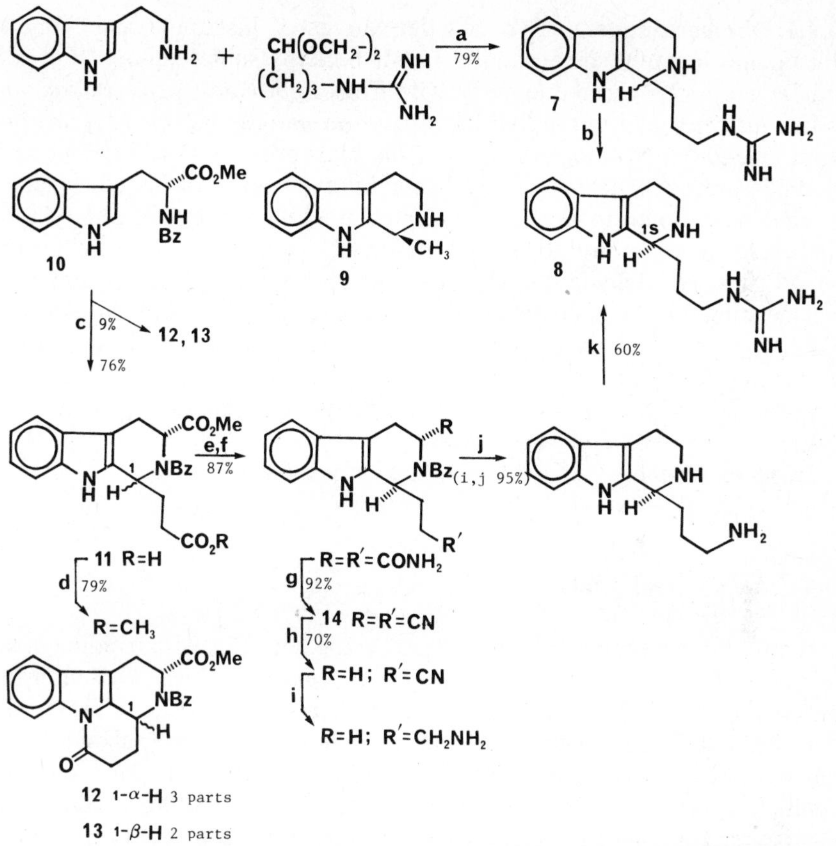

Scheme 111. Synthesis of (−)-trypargine [583,584]. (a) 80% HOAc, 50°; (b) resolution using D-(+)-10-camphorsulfonic acid; (c) α-ketoglutaric acid/øH-dioxane, Δ; (d) CH₂N₂/Et₂O; (e) chromatographic separation on SiO₂; (f) MeOH-NH₃, 20 d; (g) POCl₃/Py-DMF, 0°; (h) NaBH₄/EtOH, 60°; (i) LiAlH₄; (j) H₂/Pd-C/EtOH-HCl; (k) NH₂C(SMe)═NH·H₂SO₄/H₂O, 50°.

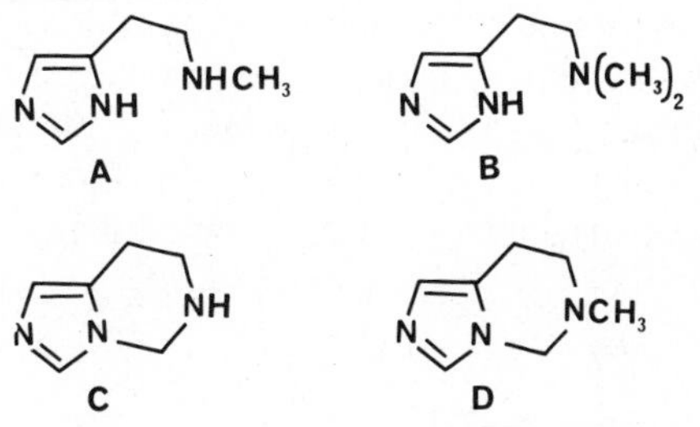

Figure 23. Structures of methylated and cyclized amines occurring in amphibians and derived from histamine. (A) N-Methylhistamine; (B) N,N-dimethylhistamine; (C) spinaceamine; (D) 6-methylspinaceamine. N-Acetylhistamine also occurs in amphibians.

207

10.4.1. Spinaceamines.   Cyclized derivatives of histamine and *N*-methylhistamine, namely spinaceamine and/or 6-methylspinaceamine (Fig. 23*C* and *D*) were found in the leptodactylid frogs *Leptodactylus pentadactylus* and *L. laticeps*, and in the hylid frogs *Litoria moorei* and *Nictimystes disrupta* [585–588]. Nothing is known of the biosynthetic pathways, but such alkaloids are likely to be formed either by cyclization of histidine with a formaldehyde donor to yield the acid spinacine followed by decarboxylation to spinaceamine, or directly by cyclization of histamine or *N*-methylhistamine with a formaldehyde donor. It has been stated [572] that 6,7-dimethylspinaceamine has been detected in extracts of *Leptodactylus pentadactylus* and *L. laticeps*, but that it was uncertain as to whether it was "an artifact of acetone extraction." Synthesis of the bicyclic alkaloid spinaceamine by condensation of histamine with formaldehyde was reported in 1920 [589] long before its detection in amphibians.

Spinaceamine has no histamine-like activity in a variety of systems [590,591].

10.5. Amidine Alkaloids

Recently, three dendrobatid alkaloids, **222**, **236**, and **252**, with amidine moieties were detected in extracts of *Dendrobates pumilio*: These had empirical formulas of $C_{13}H_{22}N_2O$, $C_{14}H_{24}N_2O$, and $C_{14}H_{24}N_2O_2$ [412]. Interestingly, in earlier extracts from the same populations of *D. pumilio*, alkaloids **236** and **252** were trace constituents when detected at all, but now represent significant, albeit minor, components of the alkaloid fraction. Alkaloid **252** was isolated in sufficient quantity for proton and ^{13}C magnetic resonance spectral analysis. The results suggested a tentative structure for **252** (Fig. 24), which was reported to the Japanese Chemical Society by T. Tokuyama in 1983.

A trace dendrobatid with potent analgesic activity may also be an amidine [unpublished results]. The presence of this compound in extracts from the frog *Dendrobates tricolor* was suggested by a marked Straub-tail reaction after injection of amounts of crude alkaloid fraction [23]. Small amounts of such an alkaloid also appeared to be present in extracts from *D. espinosai* and *D. pictus*. A Straub-tail reaction is typical of the opiate class of analgesics. The Straub-tail reaction was used as a bioassay to isolate small amounts of the active principle. It was a trace alkaloid in *D. tricolor* and

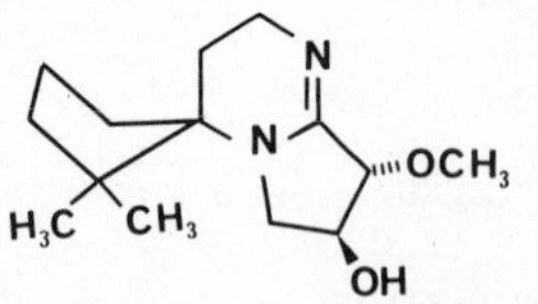

Figure 24.   Tentative structure for the dendrobatid amidine alkaloid **252**.

appeared to be present in quantities of less than 1 μg per frog skin. The true parent ion of this analgesic alkaloid has not been defined by mass spectrometry, apparently because of facile pyrolysis both on attempts at direct probe analysis and during gas chromatography. Instead, an apparent parent ion at m/z 208/210 ($C_{11}H_{13}N_2Cl$) has been detected with a base peak at m/z 69 (C_4H_7N). Analogies to the amidine alkaloids 236 and 252 suggest that this analgesic alkaloid is also an amidine. The chlorine is not introduced during exposure to hydrochloric acid during isolation of alkaloids since replacing chloroform with methylene bromide and using hydrobromic acid for isolation still yielded the same chlorine containing compound [unpublished results]. The properties of the dendrobatid amidine alkaloids are presented in the standard format:

Amidines

222. $C_{13}H_{22}N_2O$,—, 180°, m/z 222 (1), 221 (2), 112 ($C_5H_8N_2O$, 100). 1D.

236. $C_{14}H_{24}N_2O$,—, 172°, m/z 236 (12), 126 ($C_6H_{10}N_2O$, 100). 0D.

252. $C_{14}H_{24}N_2O_2$,—, 179°, m/z 252 (4), 251 (3), 142 ($C_6H_{10}N_2O_2$, 100). 1D.

"**208/210.**" $C_{11}H_{13}N_2Cl$, 177°, m/z (parent ion not detected) 210 (4, $C_{11}H_{13}N_2{}^{37}Cl$), 208 (12, $C_{11}H_{13}N_2{}^{35}Cl$), 181 (1), 179 (3), 173 (5, $C_{11}H_{13}N_2$), 69 (100, C_4H_7N). 1D.

10.6. Other Alkaloids

During the past two decades a large number of nondendrobatid amphibians have been screened by combined gas chromatography–mass spectrometry for alkaloids. Most of these species have not contained detectable amounts of alkaloids (see Section 11). But recently alkaloids have been detected with this technique from a Brazilian bufonid toad, *Melanophryniscus moreirae*, an Australian myobatrachid frog, *Pseudophryne semimarimorata*, and two Madagascan mantellid frogs, *Mantella aurantiaca* and *M. madagascariensis* [261]. All contained (allo)pumiliotoxin-A class alkaloids. The Mantellid species contained, in addition, three alkaloids of types not detected previously in dendrobatid frogs. The properties of these alkaloids of unknown structure are as follows:

235C. $C_{15}H_{25}NO$,—, 166°, m/z 235 (28), 234 (53), 220 (20), 162 (100, $C_{11}H_{16}N$). H_2 derivative. 1D. Occurrence: minor in *M. aurantiaca* and *M. madagascariensis*.

241B. $C_{10}H_{35}N$,—, 167°, m/z 241 (15), 125 (45), 58 (100). H_0 derivative. 0D. Occurrence: major alkaloid in *M. madagascariensis*; trace in *M. aurantiaca*.

251G. $C_{15}H_{25}NO_2$,—, 178°, m/z 251 (26), 250 (45), 162 (100). H_2 derivative. 2D. Occurrence: trace in *M. aurantiaca*.

The *Pseudophryne* species contained a mixture of alkaloids with apparent molecular weights of 311, 313, and 315. Further material will be required for characterization of these and other alkaloids present in frogs of the genus *Pseudophryne*.

11. BIOLOGICAL SIGNIFICANCE, FORMATION, AND DISTRIBUTION OF AMPHIBIAN ALKALOIDS

11.1. Biological Role

Amphibians have elaborated a remarkable diversity of secondary metabolites that are present in skin—probably stored in cutaneous glands—and often released as a chemical defense against predators. Such compounds include peptides, biogenic amines, bufodienolides, proteins, and the many different classes of alkaloids covered in the present review. The distribution of these various classes of biologically active secondary metabolites in amphibians is presented in Table 13. Obviously only a limited sampling of the various amphibian genera have been analyzed for toxic or noxious substances. The overview of the current knowledge, presented in Table 13 and surely incomplete, should serve as a guide to future research. The first column lists examples where the taste of skin exudates indicates the presence (+) or absence (−) of a noxious or toxic agent. Such data are, of course, quite subjective. The following three columns list three categories of alkaloids: (1) those "derived" from the biogenic amines or their amino acid precursors; (2) water-soluble alkaloids, such as tetrodotoxin; and (3) the lipophilic dendrobatid and samandarine alkaloids. The table then lists biogenic amines, peptides, and bufodienolides.

The first category of alkaloids includes the N-methylated and/or cyclized congeners of the various biogenic amines (see below). The indole alkaloids calycanthine and chimonanthine are also entered in this category. The second category covers the water-soluble alkaloids, primarily the tetrodotoxins (TTX). The third category is comprised of the various lipophilic alkaloids, including the batrachotoxins (BTX), histrionicotoxins (HTX), gephyrotoxins (GTX), indolizidines (I), pumiliotoxin-A class (PTX-A) alkaloids, decahydroquinolines (DHQ), piperidines (PIP), pyrrolidines (PYR), amidines (AMI), and samandarines (SAM). Where no lipophilic alkaloids were detected using the techniques of thin-layer chromatography and gas chromatography–mass spectrometry, developed for dendrobatid alkaloids, the abbreviation n.d. for not detected is entered in this column and the number of species examined within each genus follows in parentheses.

A variety of biogenic amines, such as serotonin (S) and histamine (H), tryptamine (Tr), tyramine (Ty), epinephrine (Epi), and dopamine (DA) are found in amphibian skin, sometimes in relatively high concentrations (Table 13). These are listed and the abbreviation n.d. for not detected is entered as appropriate.

Table 13. Occurrence of Alkaloids and Other Toxic or Noxious Compounds in Amphibian Skin

Order/Family/Genus[a,b]	Noxious Secretion[c]	Alkaloids			Biogenic Amines[g]	Peptides[h]	Bufodienolides[i]
		Biogenic Amine Derived[d]	Water-Soluble[e]	Lipophilic[f]			
Caudata (Salamanders)							
Ambystomatidae							
Ambystoma	+		–				
Dicamptodon	+						
Rhyacosiredon	+						
Rhyacotriton	+						
Amphiumidae							
Amphiuma			–				
Cryptobranchidae							
Cryptobranchus	+		–				
Hynobiidae[j]							
Hynobius	+						
Plethodontidae[k]							
Aneides	+		–				
Batrochoseps	–		–				
Bolitoglossa							
Desmognathus	–		–				
Ensatina	+		–				
Eurycea	+						
Gyrinophilis	–						
Hemidactylium	+						
Hydromantes	+						
Plethodon	+						
Pseudotriton	+			n.d.(1)		TOX	
Typhlotriton	+						

Table 13. (*continued*)

Order/Family/Genus[a,b]	Noxious Secretion[c]	Biogenic Amine Derived[d]	Alkaloids Water-Soluble[e]	Alkaloids Lipophilic[f]	Biogenic Amines[g]	Peptides[h]	Bufodienolides[i]
Proteidae[j]							
Necturus			—				
Salamandridae[m]							
Cynops	+		+				
Euproctus	+		+	n.d.(1)		TOX	
Notophthalmus	+		+				
Paramesotriton	+		+				
Pleurodeles	+				S		
Salamandra	+		+	SAM	S,Tr	HE	
Taricha	+		+			TOX	
Triturus	+		+	n.d.[x]	Tr	HE	
Sirenidae[l]							
Siren			—				
Anura (Frogs-Toads)							
Ascaphidae							
Ascaphus						B	
Brachycephalidae[l]							
Brachycephalus			+				
Bufonidae[n]		Bufotenine					
Atelopus	+	Bufotenidine	+	n.d.(4)			Bu
Bufo	+	Dehydrobufotenine		n.d.(5)	S,Epi,DA		Bu
		O-Methylbufotenine					
		Epinine					

Taxon		Indole / other compounds		Alkaloids			
Dendrophryniscus	+	Bufotenine		n.d.(1)			+
Melanophryniscus	+	Phenolic amine		PTX			+
Dendrobatidae							
Colustethus	−		+[y]	n.d.(7)	CAR	n.d.	
Dendrobates	+		−	HTX,GTX,I, PTX,DHQ, PIP,PYR, AMI		±	±
Phyllobates	+	Calycanthine	−	BTX,HTX,I, DHQ,PIP	n.d.	n.d.	±
		Chimonanthine					
Discoglossidae[l]							
Alytes						Bo	
Bombina	+				S	B,Bo,TRH, HE,ENZ	
Discoglossus							
Heleophrynidae							
Heleophryne					S	B	
Hylidae[o]							
Acris		Dehydrobufotenine		n.d.(1)			
Agalychnis				n.d.(1)			
Hemiphractus							±
Hyla	+	Bufotenine		n.d.(1)	S,H	C,HE	
Litoria	+	Bufotenidine		n.d.(4)	S,H	A,Bo,C	−
		Bufotenidine					
		N,N,N-Trimethyltryptamine					
		Imidazole					
		Indoles					

Table 13. *(continued)*

Order/Family/Genus[a,b]	Noxious Secretion[c]	Alkaloids Biogenic Amine Derived[d]	Alkaloids Water-Soluble[e]	Alkaloids Lipophilic[f]	Biogenic Amines[g]	Peptides[h]	Bufodienolides[i]
Nyctimystes		Bufotenine Bufotenidene *N,N,N*-Trimethyltryptamine *N*-Methylhistamine Spinceamines Indole			S,H	n.d.	—
Phrynohyas	+			n.d.(1)			—
Phyllomedusa	+			n.d.(1)		B,Bo,C,D, P,S,HE	±
Smilisca							
Trachycephalus	+			n.d.(1)		HE	±
Hyperolidae[p]							
Hyperolius				n.d.(1)			±
Kassina		Trypargine		n.d.(1)		C,P	±
Leptopelis				n.d.(1)			—
Leptodactylidae[q]							—
Adenomera				n.d.(1)			
Barycholus				n.d.(1)			
Batrachophrynus					n.d.	n.d.	
Caudiverbera		Leptodactyline			n.d.	n.d.	
Ceratophrys		Leptodactyline			n.d.		
Crossodactylus				n.d.(2)			±
Cycloramphus				n.d.(2)	n.d.	n.d.	±

Genus		Compound				
Edalorhina	+		n.d.(1)			±
Eleutherodactylus	−		n.d.(5)	S	CAR	±
Euparkerella			n.d.(1)			
Eupsophus				n.d.	n.d.	±
Holoaden			n.d.(1)	n.d.	n.d.	±
Hylodes			n.d.(4)	H		
Kyarranus				n.d.		
Lepidobatrachus						
Leptodactylus	+	Leptodactyline	n.d.(2)	S,Ty,H	C,TRH,CAR	±
		Bufotenine				
		N-Methylhistamines				
		Spinceamines				
		Leptodactyline				
		Candicine				
Lithodytes			n.d.(1)	n.d.		±
Megaelosia			n.d.(1)	S		
Odontophrynus		Leptodactyline	n.d.(1)			±
Paratelmatobius						±
Physalaemus	+	Leptodactyline	n.d.(2)	n.d.	P	−
Pleurodema		Leptodactyline	n.d.(1)	S	n.d.	±
Proceratophrys			n.d.(1)			
Telmatobius						±
Thoropa		Leptodactyline	n.d.(1)	S	P	±
Microhylidae					n.d.	
Asterophrys				H		
Breviceps	+		n.d.(1)	S	n.d.	±
Cophixalus						
Microhyla	+		n.d.(1)	S	n.d.	
Phrynomerus			n.d.(1)	n.d.		
Sphenophryne				n.d.		
Uperodon						−

Table 13. (continued)

Order/Family/Genus[a,b]	Noxious Secretion[c]	Alkaloids			Biogenic Amines[g]	Peptides[h]	Bufodienolides[i]
		Biogenic Amine Derived[d]	Water-Soluble[e]	Lipophilic[f]			
Myobatrachidae[s]							
Adelotus		Amine					
Crinia		Imidazoles			S	A	
Cyclorana		N-Methylserotonin		n.d.(1)	S	n.d.	
		Bufotenidine					
Heleioporus	+	Indole			S		
Lechriodus					S	n.d.	
Limnodynastes				n.d.(1)	S,H	n.d.	
Metacrinia					S	A	
Mixophyes		Indole		n.d.(1)	n.d.	n.d.	
Myobatrachus					S		
Neobatrachus					n.d.		
Notaden	+	Bufotenidine			n.d.		
Pseudophryne	+	Indole		PTX Alkaloids SAM?	S	P	
Taudactylus		Indole			S,H	P	
Uperoleia		Imidazole			S	Bo,P	
Pelobatidae[t]							
Leptobrachium	+			n.d.(1)		HE	
Pelobates	+					TRH	
Scaphiopus					n.d.		

Taxon						
Pipidae[a]						
Xenopus		Bufotenidine	n.d.(1)	S	C,X	±
Ranidae[v]						
Arthroleptis				n.d.		
Hemisus			n.d.(1)			
Hylarana	+		n.d.(1)			
Mantella			HTX,I,PTX, DHQ Alkaloid			
Phrynobatrachus			n.d.(1)			±
Ptychadena			n.d.(1)	n.d.		±
Pyxicephalus			n.d.(1)			±
Rana	+	Bufotenine	n.d.(5)	S	B,Bo,TRH, VIP,G, HE	±
Rhacophoridae[w]						
Chiromantis	+		n.d.(1)	n.d.		
Rhacophorus	+		n.d.(1)	n.d.	D	±
Rhinodermatidae						
Rhinoderma			n.d.(1)	S	n.d.	

[a] The order Gymnophiona (Caecilians) has apparently not been investigated for noxious substances. The anuran families Centrolenidae (3 genera), Leiopelmatidae (1 genus), Pelotydidae (1 genus), Pseudidae (2 genera), Rhinophrynidae (1 genus), and Sooglossidae (1 genus) have apparently not been investigated. An updated version of this table is in preparation [259].

[b] Gorham [592] was the source for the various taxa, but certain subsequent revisions (Centrolinidae, Hyperolidae, Leptodactylidae, Myobatrachidae, Rhacophoridae) have been included.

[c] Noxious secretions by taste [for examples see 574,575,593; also unpublished results].

[d] Indole, imidazole, catecholic, and phenolic alkaloids [574–577,585,589,594–596]. N-Methylserotonin, which usually occurs with bufotenine, is not listed in most cases. Amines, indoles, and imidazoles of unknown structure have also been detected.

[e] Unless otherwise designated, tetrodotoxin (TTX) or a similar alkaloid was detected (see Section 9).

Table 13. *(continued)*

f Lipophilic alkaloids include batrachotoxins (BTX), histrionicotoxins (HTX), gephyrotoxins (GTX), indolizidines (I), pumiliotoxin-A class, and allo-pumiliotoxins (PTX), decahydroquinolines (DHQ), piperidines (PIP), pyrrolidines (PYR), amidines (AMI), samandarines (SAM), and alkaloid(s) of unknown structure. Unless otherwise noted, not detected (n.d.) refers to standard gas chromatographic–mass spectral analysis used for dendrobatid alkaloids [23,24,261].

g Biogenic amines include serotonin (S), histamine (H), tryptamine (Tr), tyramine (Ty), epinephrine (Epi), dopamine (DA) [see 572–577,596–599].

h Peptides include the following classes: angiotensins (A), bradykinins (B), bombesins (Bo), caeruleins (C), dermorphins (D), and physalaemins (P) and also sauvagine (S), xenopsin (X), thyroid-releasing factor (TRH), vasoactive intestinal peptide (VIP), granuliberin (G), hemolytic (HE) or toxic (TOX) peptides or proteins, enzymes (ENZ), and carnosine (CAR) [see 600–611]. Amino acids, including γ-aminobutyric acid, have also been detected [612,613].

i Bufodienolides (Bu) or related compounds (+). The designation ± indicates very low levels [532].

j Four genera uninvestigated.

k Eleven genera uninvestigated.

l One genus uninvestigated.

m Seven genera uninvestigated.

n More than 14 genera uninvestigated.

o Twenty-seven genera uninvestigated.

p Ten genera uninvestigated.

q Twenty-three genera uninvestigated.

r Forty-seven genera uninvestigated.

s Eight genera uninvestigated.

t Five genera uninvestigated.

u Three genera uninvestigated.

v Forty-two genera uninvestigated.

w Six genera uninvestigated.

x Samandarines were not detected [500].

y A water-soluble toxin was detected in one (C. inquinalis) of seven species examined [unpublished results]. It seems unlikely that this is the same compound as that which inhibits ouabain binding (see last column) since the latter is present in very low amounts.

218

A variety of biologically active peptides are found in amphibian skin and are briefly presented in Table 13. These include: bradykinins (B) with some eight members including bradykinin, phyllokinin, bombinakinin, and ranakinin; physaelaemins (P), also termed tachykinins, with some seven members including physaelaemin, phyllamedusin, kassinin, hylambatin, and uperolein; caeruleins (C) with three members including caerulein and phyllocaerulein; bombesins (Bo) with some ten members including bombesin, litorin, phyllolitorin, ranatensin, and alytesin; dermorphins (D) with some four members; xenopsin (X); thyroid releasing hormone (TRH); vasoactive intestinal peptide (VIP); sauvagine (S); angiotensins (A) with one member, crinia-angiotensin; hemolytic (HE) or toxic (TOX) peptides or proteins; enzymes (ENZ); and finally a mast cell degranulating protein, granuliberin (G); and carnosine (CAR). These are listed and the abbreviation n.d. for not detected is entered as appropriate.

The presence of bufodienolides (Bu) or other compounds that inhibit Na^+-K^+-ATPase is indicated in the last column of Table 13. The abbreviation n.d. for not detected is entered where compounds that inhibit radioactive ouabain binding could not be detected [532]. The nitrogen-containing derivatives of bufodienolides such as bufotoxin (bufotalin 3β-O suberyl arginine ester) [614] and other bufotalin conjugates containing suberyl histidine [615], suberyl-3-methyl histidine [615], and suberyl glutamine [616], although formally deserving of classification as animal alkaloids [1], are not treated in this review [for reviews of the bufodienolides see 574,617].

Secondary metabolites, including alkaloids, undoubtedly serve certain amphibians as noxious or toxic substances in "passive" defense against predators. The aposematic coloration, the ready exudation of a milky and unpleasant secretion by such amphibians upon injury or attack, the adverse reactions of predators to such secretions, and often the bold or at least nonsecretive and even diurnal life style of some such amphibians provide strong evidence for such a defensive role [2,250,261,574]. Within the diurnal dendrobatid frogs, it is the some 50 brightly colored species that make up the genera *Phyllobates* and *Dendrobates* that produce biologically active alkaloids, while the cryptically colored species that make up the genus *Colustethus* do not produce alkaloids, at least in the seven species that have been investigated [2,261]. The levels of alkaloids and the unpalatability of various populations of Panamanian *Dendrobates pumilio* showed no obvious correlations with color or life style [424], suggesting that various interacting factors may be involved. Similar extraordinary variation in alkaloids, coloration, and escape behavior have been noted in various populations of *D. histrionicus* [250]. In some amphibians, most notably toads of the genus *Bufo* and salamanders of the genus *Salamander*, the microscopic cutaneous glands that in most amphibians appear to store the active principles are supplemented by macro (parotoid) glands whose contents are expressed by pressure with some force and with unpleasant sequalae to any predator [577,617]. Indeed, toxic principles from these species have often been isolated from gland exudates rather than whole skin.

On the other hand, even amphibians that do not produce noxious secondary metabolites have the cutaneous granular or "poison" glands that undoubtedly serve as storage and secretatory loci for noxious substances in other amphibians [see 618 and references therein]. The physiological roles of such glands in nonnoxious amphibians are unknown, but control of water and salt balance through some peptide or other factor is an attractive hypothesis. Compounds of unknown structure affecting ouabain binding to Na^+-K^+-ATPase have been detected at low levels in skin extracts from a wide range of amphibians [532]. Many of these amphibians do not appear to be noxious. Such unknown substances have been suggested as possible chemical ancestors of the bufodienolides [261], whose toxicity is due to inhibition of Na^+-K^+-ATPase.

Another role for the granular cutaneous glands and their contents has been suggested, namely that amphibians may elaborate cutaneous factors with antibacterial or antimycotic activity and that the toxins may, in some cases, serve to prevent skin infections [612]. Such a dual role is attractive, since a toxin might not only repel a predator, but prevent infections in abraded or otherwise injured skin. However, a systematic study of antibacterial and antimycotic activity of skin extracts or exudates from amphibians has not been conducted. In 1952, gland exudate from the fire salamander (*Salamandra maculosa*) had been found to have a broad spectrum of antibiotic activity [620]. Later studies [621,622] confirmed this, and the samandarine alkaloids were shown to have bacteriostatic activity. Secretion or extracts from skin or glands of various other amphibians from diverse genera, including *Bufo* [620], *Bombina* [623,624], *Rana* [625], and *Leptodactylus* [622], exhibited bacteriostatic activity. Spinceamine and bufotenine, present in the leptodactylid skin secretions, had bacteriostatic activity, while histamine did not [622]. Tetrodotoxin has no antibacterial effect on *Staphyllococcus* or *Streptococcus* [cited in 530]. Indeed, bacteria were stated to be found often growing in concentrated solutions of toxins from atelopid frogs.

Accounts of unpalatability (noxious) or toxicity of the amphibians that produce biologically active substances are quite common (Table 13) and suggest that at least for many amphibians, those substances do serve in defense against predators. Antipredator postures, and the avoidance of certain noxious salamanders by avian and mammalian predators has been discussed in detail [593]. Profuse skin secretions of brightly colored European toads of the genus *Bombina* are reported [626] to afford protection against natural predators such as snakes. Adverse effects in a human from handling an African microhylid *Phrynomerus bifasciatus* have been reported [627]. Snakes from the eastern United States were cited as dying after ingestion of the Californian newt *Taricha torosa*, while western snakes would not accept this newt [523]. Numerous dead catfish were stated to be observed in one Californian pond during the spawning season of *T. torosa*. At least one human death has been reported after ingestion of a *T. granulosa* [628].

Electrical shocks caused newts of the genus *Taricha* to release significant amounts of tetrodotoxin into water [629]. The eastern newt (*Notophthalmus*) was cited as being avoided by predators [529]. Neotropical snakes of the genus *Rhadinaea* and an opossum were observed to react to dendrobatid frogs immediately and with evident aversion [23]. Silverstone [630,631] has also reported adversive or toxic effects for predators after contact with various species of dendrobatid frogs. Observations that the bufonid toad *Dendrophryniscus stelzneri* is avoided by snake and avian predators have been cited [531]. There are many other observations of certain amphibians being avoided by predators, and the above examples are meant only to illustrate the phenomena. However, some predators apparently have developed mechanisms to cope with skin secretions from poisonous amphibians. Thus neotropical snakes of the genus *Leimadophis* feed on a variety of poisonous neotropical frogs [23]. Large spiders have been noted to eat small dendrobatid frogs [cited in 2]. A captive bullfrog (*Rana catesbeiana*) swallowed proferred *Dendrobates auratus* with no ill-effects [personal observation].

The distasteful quality of skin secretions has been suggested as a field assay for the presence and quantity or "quality" of alkaloids from dendrobatid frogs [23,618] and, of course, for the presence of other biologically active compounds in other amphibians. Field tasting of certain dendrobatids, such as those species of *Phyllobates* used to poison blow darts, would be ill-advised [9]. Even the less toxic alkaloids present in skin exudate from frogs of the genus *Dendrobates* cause burning, numbing, and sometimes "an unpleasant feeling of tightening of the throat." The noxious quality of a wide range of salamanders has been reported [593].

Thus while other roles are possible, the primary role for biologically active secondary metabolites, including alkaloids, in some amphibians appears to be as a defense against predation. The evolutionary development of amphibian defensive compounds has been suggested to fall into at least two categories [261]:

1. A widespread naturally occurring active compound, such as a peptide (bradykinin, bombesin) or a biogenic amine (serotonin, histamine), is elaborated in excessive quantities for a secondary function of defense.

2. A new biologically active substance, frequently with a complex structure, is elaborated for the primary function of defense. The amphibian alkaloids are obvious examples of this latter evolutionary strategy and occur with only a limited distribution in nature and have biological activities admirably suited as adversive or toxic agents.

11.2. Biosynthesis

The complex structures of many of the amphibian alkaloids suggest the development of whole complexes of biosynthetic enzymes. Little, however, is known of the biosynthetic pathways, let alone the enzymes involved. Radioactive cholesterol upon shaking for three days at room temperature with gland exudate from 35 salamanders was converted to a slight extent (~3%) into samandarine alkaloids [632]. An *in vivo* study on alkaloid biosynthesis in dendrobatid frogs was not successful. There was no significant incorporation over a period of up to six weeks of radioactive, acetate, mevalonate, or cholesterol into alkaloid fractions from *Dendrobates pumilio* or *Phyllobates aurotaenia* [633]. Both radioactive acetate and mevalonate were converted in high yield to skin cholesterol. No incorporation of radioactive serine into batrachotoxin alkaloids of *P. aurotaenia* was detected [633]. Similar negative results were obtained in an *in vivo* study on tetrodotoxin biosynthesis in Californian newts: There was no significant incorporation of radioactive acetate, arginine, citrulline, or glucose into tetrodotoxin using various routes of administration [629]. There was, however, significant incorporations into cholesterol, amines, and amino acids. Such negative results suggest the presence of factors or precursors of dietary origin, necessary for the synthesis of skin alkaloids in some amphibians. However, the relative constancy of alkaloid profiles in some dendrobatids after maintenance of frogs in captivity for several years [23] suggests that diet has little influence on the pathways in a particular species. No obvious alkaloid precursors were detected in stomach contents of field-collected *D. histrionicus* [unpublished results].

The lack of accumulation of toxic alkaloids in certain captive-raised amphibians is a remarkable finding. Thus second generation captive-raised *Phyllobates terribilis* do not have detectable levels of batrachotoxins, whereas wild-caught parental frogs that have been maintained in terraria still have batrachotoxins, albeit at reduced levels, after living in captivity for more than a decade on a diet of fruitflies and crickets [66]. Batrachotoxin was not identified in tadpoles of *P. aurotaenia* hatched in captivity [cited in 633]. It should be noted that tetrodotoxin is absent in hatchery-raised puffer fish [634]. Hatchery-raised puffer fish do not accumulate oral tetrodotoxin unless administered with liver extracts of wild-caught puffer fish.

Obviously the situation, at least with respect to the biosynthesis of dendrobatid alkaloids and tetrodotoxin, is complex. Genetic control of the expression of the necessary biosynthetic pathways must be involved in the case of the 200 dendrobatid alkaloids, since different species of dendrobatid frogs coexisting in neotropical forest exhibit quite different and fairly consistent profiles of alkaloids. Possible roles of diet and/or symbiotic organisms at present cannot be excluded.

There has been speculation as to a possible pivotal role of a 2,6-disubstituted (dehydro)piperidine in the biosynthesis of histrionicotoxins, gephyrotoxins, indolizidines, and decahydroquinolines [23,24,38]. "Biomimetic" pathways from a diketoamine or a piperidine for chemical synthesis of such alkaloids have been proposed [296–298,460]. Further speculation on the relevance of such intermediates ought to be deferred until evidence for their

existence or metabolism is obtained. It should be noted that such intermediates do not account for the formation of alkaloids of the pumiliotoxin-
A class or for recently discovered 3,8-disubstituted indolizidines. In view
of the recent discovery of 2,5-disubstituted pyrrolidines [25], a biosynthetic
intermediate of this class should also be considered a potential intermediate
in the formation of dendrobatid alkaloids that contain an indolizidine moiety,
such as the gephyrotoxins, indolizidines, and pumiliotoxin-A class. Some
possible biosynthetic pathways are depicted in Scheme 112.

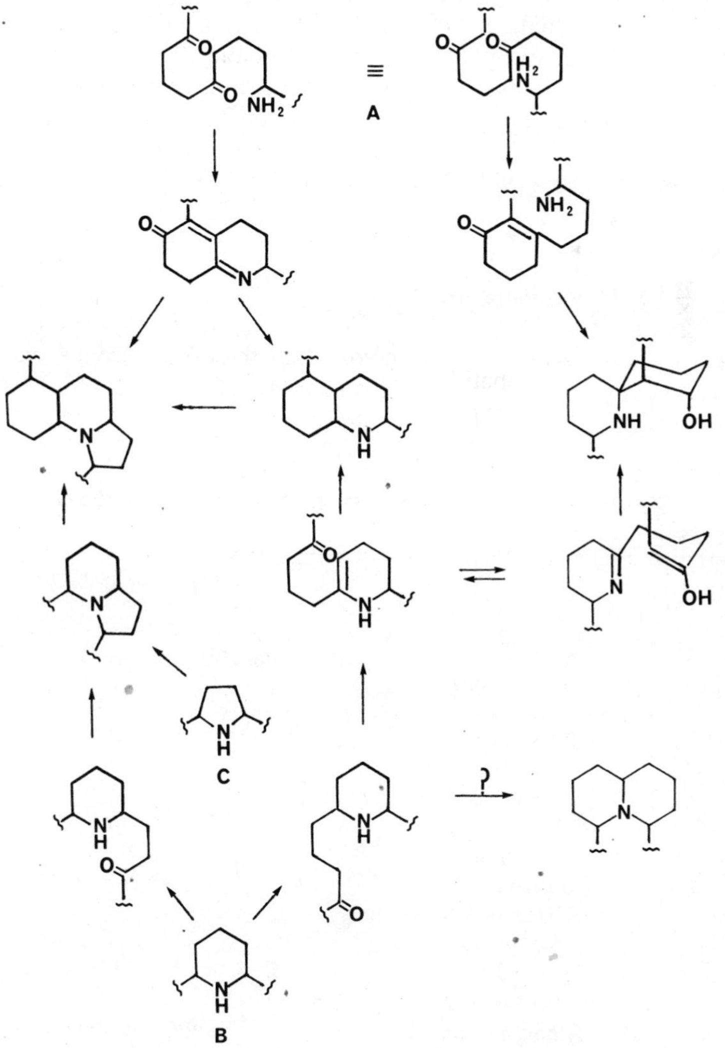

Scheme 112. Hypothetical biosynthetic pathways to dendrobatid alkaloids from:
(A) straight chain diketoamines; (B) 2,6-disubstituted piperidines; (C) 2,5-disubsti-
tuted pyrrolidines.

11.3. Distribution and Taxonomic Significance

11.3.1. Batrachotoxins. This unique class of highly toxic steroidal alkaloids, as yet, has not been identified in nature except in all five species of a monophyletic genus (*Phyllobates*) of dendrobatid frogs [9]. The relative levels of batrachotoxin differ considerably in these five species, ranging from not detectable levels in some populations of Panamanian *P. lugubris* to about 1 mg per frog in Colombian *P. terribilis* (Table 14). The presence of batrachotoxin served as one of several characters considered to be uniquely derived, which thereby indicated *Phyllobates* to represent a monophyletic lineage. Frogs of this genus also contain simpler dendrobatid alkaloids including histrionicotoxins, indolizidines, the pumiliotoxin-A class, and decahydroquinolines. The occurrence of these alkaloids in the five species of *Phyllobates* is presented in Table 15. The occurrence of noranabasamine, chimonanthine, and calycanthine along with the batrachotoxins in extracts of *P. terribilis* [22] defies explanation.

11.3.2. Other Dendrobatid Alkaloids.

The rich array of dendrobatid alkaloids other than the steroidal batrachotoxins would appear to provide an invaluable set of characters for definition

Table 14. **Distribution of Batrachotoxins in Dendrobatid Frogs of the Genus** *Phyllobates* [9]

		Micrograms per frog		
Species	Locale	Batracho-toxin	Homo-batracho-toxinin	Batracho-toxinin A
P. terribilis	Rio Saija, Cauca, Colombia	500	300	200
P. aurotaenia	Rio San Juan, Choco, Colombia	20	10	50[a]
P. bicolor	Rio San Juan, Risaralda, Colombia	24	12	60
P. vittatus	Palmar Norte, Costa Rica	0.2	0.2	2
P. lugubris	Almirante, Panama	not detected		
	Cope, Panama	0.2	0.1	0.5

[a] Including pseudobatrachotoxin initially present at 20 μg per frog. This alkaloid undergoes very facile conversion to batrachotoxinin A and has been detected only in *P. aurotaenia*.

of relationships among the frogs of the large and heterophyletic genus *Dendrobates*. Instead, their diversity and the variability of alkaloid profiles within a single species have confounded their simple use as taxonomic characters.

In an early study, the profiles of alkaloids in nine populations of the variable species *Dendrobates histrionicus* appeared to form a rather coherent group with a set of several histrionicotoxins appearing in all populations [250]. The presence of two alkaloids **219A** and **243A**, now classified as decahydroquinolines (Section 7), varied considerably in these populations from absence to presence as a major constituent. Indolizidines **223AB**, **239AB**, and **239CD** were also variable in their occurrence and quantity in various populations, as was the allopumiliotoxin **267A**. The number of shared alkaloids was lowest with geographically remote populations. Two "sibling" species of *D. histrionicus* were also first described in this study [250]. One, *D. occultator*, had a very similar profile of alkaloids to a macrosympatric population of *D. histrionicus*. The other, *D. lehmanni*, shared no alkaloids with any of the eight populations of *D. histrionicus*. This was a significant consideration in defining *D. lehmanni* as a distinct species rather than as yet another population of *D. histrionicus*, as previously defined [630]. After these pioneering studies, alkaloid profiles became one taxonomic character in the description of additional new species of *Dendrobates* [250,252–257] and *Phyllobates* [9], and in ongoing investigations of the evolutionary relationships among dendrobatid frogs. Over 200 of these alkaloids have been detected and partially characterized through the use of the gas chromatographic mass spectral protocols described in Section 3.1. These alkaloids are listed in Table 16 along with their classification as: histrionicotoxin (HTX); gephyrotoxin (GTX); indolizidine (I); pumiliotoxin A (PTX), allopumiliotoxin (aPTX), or homopumiliotoxin (HPTX); decahydroquinoline (DHQ); piperidine (PIP); pyrrolidine (PYR); amidine (AMI); or unknown (—). Key properties such as gas chromatographic emergent temperature, major mass spectral fragment ions, perhydrogenation result, and exchangeable hydrogens are also listed in the table. The occurrence of these alkaloids in some 35 species of *Dendrobates* are presented in Table 17 in a format where alkaloids are listed as major, minor, and trace and according to the above classes. Such data, upon detailed analysis, may provide insights into relationships both among alkaloids and among dendrobatid frogs. For example, the 20 odd populations of *D. pumilio* appear at first to be remarkably diverse, but tend to group into two categories, one of which has histrionicotoxins as a major grouping, while the other has the pumiliotoxins [259]. A *D. histrionicus* species group of eight related dendrobatid frogs has recently been proposed [255]. These species, namely *D. arboreus*, *D. granuliferus*, *D. histrionicus*, *D. lehmanni*, *D. occultator*, *D. pumilio*, *D. speciosus*, and *D. species* (undescribed), contain very different profiles of dendrobatid alkaloids, including all of the major classes of dendrobatid alkaloids yet described, except, of course, for the batrachotoxins [259].

Table 15. Occurrence of Dendrobatid Alkaloids Other than Batrachotoxins in Frogs of the Genus Phyllobates[a]

Species	Locale and Date	Number of skins	Number and Amount[b]	Major[c] Minor Trace	Alkaloids[a]							
					HTX	GTX	I	PTX	DHQ	Other	PYR or PIP	AMI
P. aurotaenia	Playa de Oro, Rio San Juan, Choco, Colombia, Aug. 1977	100	10 +	Major	—	—	223AB	—	—	—	—	—
				Minor	287A	—	—	243A	—	—	—	—
				Trace	283A, 285A	—	—	251D	219A, 219A', 243A', 269AB	—	—	—
P. bicolor	Santa Cecilia, Rio San Juan, Risaralda, Colombia, Sept. 1977	10	11 +++	Major	—	—	243A	—	—	—	—	—
				Minor	235A, 259, 283A, 285A, 285C	—	—	—	219A, 219A', 243A', 269AB	—	—	—
				Trace	—	—	—	195B	—	—	—	—
P. lugubris	Almirante, Bocas, Panama, Nov. 1968	38	1 +	Major	—	—	—	—	—	—	—	—
				Minor	—	—	—	223A	—	—	—	—
				Trace	—	—	—	—	—	—	—	—
	El Cope, Panama, Oct. 1977	2	2 +	Major	—	—	—	—	—	—	—	—
				Minor	—	—	—	—	—	—	—	—
				Trace	285C	—	—	—	269AB	—	—	—

Species[a]	Locale, date	n[b]		HTX	GTX	I	PTX	DHQ	other	PYR/PIP	AMI
	Isla Colon, Bocas, Panama, May 1977	6	0	—	—	—	—	—	—	—	—
				—	—	—	—	—	—	—	—
				—	—	—	—	—	—	—	—
P. terribilis	Rio Saija, Cauca, Colombia, Feb. 1973	1	2 +	—	—	—	—	—	—	—	—
				—	—	—	—	—	—	—	—
				—	—	195B	—	—	251E	—	—
P. vittatus	Palmar Norte, Puntarenas, Costa Rica, July 1967	26	1 +	—	—	—	—	—	—	—	—
				—	—	—	—	—	—	—	—
				—	—	—	223D	—	—	—	—

[a] Species, locale, and date of collection, number of skins are reported.

[b] The number of alkaloids detected and the approximate total amount of alkaloids per 100 mg skins are reported: + + + = >150 µg/mg skin; + + = 50–150 µg; + = <50 µg.

[c] Major alkaloids presented in the first row, minor alkaloids in the second row, and trace alkaloids in the third row. The format for each row is as follows: histrionicotoxin (HTX), vertical line; gephyrotoxins (GTX), vertical line; indolizidine (I), vertical line; pumiliotoxin-A class and allo- and homopumiliotoxins (PTX), vertical line; decahydroquinolines (DHQ), vertical line; other alkaloids, vertical line; pyrrolidines (PYR) or piperidines (PIP), vertical line; amidines (AMI). The alkaloids are listed by code number in Table 16. The designation "major" does not indicate absolute amount, but only that in that particular species it was a predominant alkaloid. In extracts where only trace amounts of alkaloids were detected; these are entered as trace alkaloids.

227

Table 16. Summary of Dendrobatid Alkaloids: Properties and Occurrence in Frogs of the Genus *Dendrobates*

Alkaloid[a]	Class[b]	Temperature[c]	Fragment Ions[d]	Perhydro[e]	Exchange[f]	Occurrence[g] (Major|Minor|Trace)
151	—	152°	150	0	0	—|—|25A'
153A	DHQ	154°	152	0	1	—|—|25A'
153B	—	151°	152	0	0	—|—|25D
167A*	I	151°	138	0	0	29C|4, 5A, H, I, 17B, 25A', D, G, 29A, B
167B	I	150°	124	0	0	—|17A, 29B'
167C	—	152°	166	0	0	25O|4, 16B, 25D, F, G, N
167D	DHQ	154°	166	0	1	—|5H, I
181A	—	152°	152	0	0	—|5A
181B	I	153°	138	0	0	29C|5A, B, 6, 11, 12, 17A, B, 18C, 22B, 29B, B', 33B
181C	—	153°	180	0	0	—|5E, G, H, 16B, 25D, F, O
181D	DHQ	156°	152	0	1	25O|25P
181E	DHQ	156°	180	0	1	35A'|—
183A*	—	154°	154	0	1	29B, B'|8B
183B	PYR	151°	126	0	1	25N|25A', D
185	—	153°	170	0	1	—|5A
193A*	—	152°	192	2	0	—|8A, B
193B	—	158°	150	0	0	25B|5F|—
195A*	DHQ	157°	152	0	1	5A, B, 25A, A', H-J, L, R, S|5G, H, I, 25M, N, Q, 29B, B'|5E, 18E, 25B, D, F, K, O, P, 29C
195B*	I	156°	138	0	0	—|5F, G, 14C, 25B, E|4, 14A, B, D', G'-J, 25G, K, Q, 29A-C
195C	—	153°	152	0	0	29C|16B, 25D, L, 27, 29B, B', 30|5D, H, I, 15, 17A, B, 25B, C, E, J, M, 26B

Sample	Type	Temp	Value	n1	n2	Notes
195D	—	158°	166	0	0	—|25F|—
195E	—	156°	194	0	0	—|—|5E
197A	PYR	160°	180, 126	1	0	—|—|5B
197B*	—	158°	140, 126	1	0	14C|14B||14A, D'G', 25K
197C	—	157°	152	—	1	—|25J, R
201	—	167°	136	—	0	—|25A'
203A*	I	158°	138	6	0	4, 12, 25G, N, 29A|5C, E, H, 21, 25O, 29B', 32|5A, F, 8B, 10, 14J, 17B, 18B, 25A, A', C, F, H, Q, 29B
203A'	I	157°	138	6	0	—|—|25G
203B	—	159°	166	—	0	—|5F, 25O
205*	I	158°	138	4	0	4, 5H, 17B, 25D, E, L, N, 28|14C, 17A, 25A', B, I, K, M|5D, F, G, 8B, 14A, B, G', J, 25H, J, R
205'	I	156°	138	4	0	—|25A'
207A*	I	158°	138	2	0	5D|4, 5E, H, 8B, 17B, 25B, C, H, M, N, P, 29B–C|5G, 8A, 14G'I, K, 25A', D, S, 29A
207A'	I	156°	138	2	0	—|25B|—
207B*	PTX	161°	166, 70	2	1	—|4|—
207C	—	155°	152	2	0	—|25N|5E–I, 25G, M
207D	—	159°	180	2	0	—|5H, 25C
207E	—	158°	164, 84	2	0	—|25E, M, 29A
207F	—	157°	192	2	0	—|5I, 25D, F, S
"208/10"'*	AMI	175°	69	—	1	—|32
209A	—	162°	168	2	—	—|17A, B
209B	I	162°	138	0	0	—|25P, S|14G', 17A, B, 25Q
209C	—	164°	152	0	0	—|14A, B, 25D, S
209D	I	159°	124	0	0	—|25L
209E	—	158°	166	0	0	—|5E, 25E|5F, G, 25F
211A	—	169°	168	0	2	—|25A'

Table 16. (*continued*)

Alkaloid[a]	Class[b]	Temperature[c]	Fragment Ions[d]	Perhydro[e]	Exchange[f]	Occurrence[g] (Major|Minor|Trace)
211B	—	168°	160	—	1	—|—|25R, 29B, B'
217*	—	166°	152	0	0	7A, 21|5E, F, 14A–C, 18C, 32|10, 27
219A*	DHQ	165°	178	4	1	5A, B, 6, 13, 14D'–G', 22A, 34|5I, 14A, B, D, 22B, 24B, 26A, B|5H, 14C, H, 33C
219A'	DHQ	164°	178	4	1	—|—|14D', G'
219B	—	162°	152	4	0	2, 3, 7B, 14I, K, L, 16B|6, 7A, 10, 14A–C, 25I, J, 30
219C*	DHQ	170°	152	4	1	25F|25S
219D*	DHQ	170°	180	4	1	17B|—
219E	—	161°	166	—	0	—|—|25S
221A	I	162°	138	2	0	5H, 15
221B	—	173°	192	2	1	25D, E, G, M, P
221C*	DHQ	166°	152	2	1	14A, B
221D*	DHQ	168°	180	2	1	25A'
222*	AMI	180°	112	0	1	25A'
223AB*	I	160°	180, 166	0	0	5E–G, 14A, B, 29A–B'|14C, D', H, J, 29C|14D, 26A, B, 27
223A	—	159°	180	0	0	25C|25G, O, P, 29B'|5A, D, 7B, 14D, I, 15, 18E, 20, 25A', B, E, F, I, J, M, R, S, 34
223A'	—	163°	180	0	0	—|—|25E, G
223B	—	159°	166	0	0	5H, I, 16B, 26B|5A, 7B, 14A, B, D, G, I, J, 18B, E, 26A, 34
223C	—	159°	152	0	0	7B, 14K, L|5A, C, 27, 34

ID	Reagent	Temp	Mass			Fragments
223D	I	159°	138	0	0	—|14K, L|14I, J, 17A, B, 18B, 25A, A', H, 34
223E	DHQ	163°	168	2	1	—|—|15, 20, 25A', N, 31
223F	—	163°	180	0	1	—|—|25O
223G	HPTX	163°	180, 84	2	1	—|—|25A', 29A
225A	—	164°	168	0	1	—|29B|30, 31
225B*	PIP	174°	154, 140	0	1	—|17B|14A, B
225C*	PYR	172°	168, 126	0	1	—|14A, B|14C
225D*	I	164°	138	0	1	—|25D, E, K, P, S
231A*	—	166°	166	6	0	—|5C, 14I, L, 21, 27|5H, 7A, 8A, B, 18B, 25I, 32
231A'*	—	171°	166	6	0	—|8A, B|—
231B*	—	166°	152	6	0	8A, 25P|2, 4, 8B, 10, 14G', I, 20, 25B–G, 29A, C, 31|5D, F, 6, 7A, B, 14A–C, G, K, 15, 25I, N, 32
231B'	—	168°	152	6	0	—|25P
231C*	I	171°	138	6	0	—|4, 8A, 14C|8B
231D	—	168°	154	6	1	—|35A'
231E	DHQ	160°	152	—	1	—|25N
231F	—	165°	180	6	0	—|25C
233A	—	167°	166	4	0	—|7A, B, 25I|35A'
233B	—	180°	168	—	1	—|21|—
233C	—	173°	192	—	1	—|22B, 33C|—
235A*	HTX	176°	194, 96	4	2	16B, 26A, 33C|5A, B, D', F, G', 22B, 26B, 27, 33A, B, 34|5I, 13, 22A', 23, 29A
235B*	I	166°	138	2	0	29B, B'|14C, 25B, J, K, L, 29C|5D, 14A, B, H, 25A, A', E, M, S, 29A
235C*	—	166°	162	2	1	—|— Nondendrobatid (Section 10.6)
235D	—	182°	170	6	1	—|17B|25D
235E	—	168°	152	—	0	—|26B

Table 16. (continued)

Alkaloid[a]	Class[b]	Temperature[c]	Fragment Ions[d]	Perhydro[e]	Exchange[f]	Occurrence[g] (Major|Minor|Trace)
235F	—	170°	166, 138	—	1	—|—|25D
235G	—	172°	206, 194	—	0	—|—|33C
236*	AMI	172°	126	0	1	—|25A', R|—
237A*	PTX	167°	166, 70	2	1	1, 10|22B, 26A, B|12, 14K, 30, 32
237B*	aPTX	168°	182, 70	—	2	—|1
237C	—	168°	180	0	0	—|14K|25B
237D	I	163°	138	0	0	29B, B'|5B
237E	—	180°	208, 152	2	—	—|14H
239AB*	I	178°	182, 180	0	1	14A, B, D, D', H, 20|14C, E, G|14G'
239CD*	I	179°	196, 166	0	1	14H, 20|—|14A–C, D'
239A	—	178°	182	0	—	—|20
239A'	—	174°	182	0	—	—|7B, 20
239B	—	178°	180	—	1	7A, B|14A, 20
239B'	—	174°	180	0	—	—|7B, 20
239C	—	179°	196	0	—	—|20
239C'	—	174°	196	0	—	—|7B, 14H, 20
239D	—	179°	166	0	—	—|20, 25B
239D'	—	174°	166	0	—	—|14H, 20
239E	—	176°	152	0	—	—|14H
239F	—	176°	168	0	1	—|14K
239G*	I	178°	138	0	1	—|14A–C, H
239H	HTX	182°	196, 96	0	2	—|14J
239I*	PIP	171°	182, 140	—	1	33C|—
241A	—	180°	166, 126	—	—	—|20
241B*	—	167°	124, 58	2	1	—|—|Nondendrobatid (Section 10.6)

Code	Class					Occurrence
241C	—	177°	152	—	—	—|25G|3
241D	—	188°	114	2	0	29C|25D, 29A, B|25B, C, M, 29B'
241E	—	176°	222, 154	2	8	—|—|25J, K
243A*	DHQ	182°	202	1	8	5A, B, 6, 13, 14D', E–G'|7A, 23, 24B, 26B, 31, 33A, B, 34|14D, 24A, 27
243A'	DHQ	179°	202	1	6	—|24B|14D', G'
243B	—	178°	176	0	4	—|—|7A
247	—	175°	109, 70	1	2	—|—|7A
249	—	172°	192	0	0	—|—|5H
251A	DHQ	170°	152	1	2	—|—|14I
251B	I	184°	138	1	2	9|—|15, 25A, A', K
251C	—	190°	154	1	2	—|18B|18E, 29B', 35A'
251D*	PTX	172°	166, 70	1	2	7A, B, 8A, 18C, 25L, 28, 29B', C, 32|4, 5A, 8B, 10, 14A, B, 17A, 25D, F, I, Q, S, 26B, 29A, B, 35A'|5D, 14E, 15, 16B, 18A, 21, 25B, G, H, K, M, N, 30
251E	—	175°	168, 70	—	—	—|18E|15
251F*	—	184°	111, 70	1	0	7A|—|7B
251G*	—	178°	162	2	2	—|—| Nondendrobatid (Section 10.6)
251H	—	170°	178, 152	—	—	—|—|32
251I	aPTX	180°	182, 70	2	2	26B|—|—
252*	AMI	179°	142	1	0	—|25A', 25R|—
253A*	aPTX	179°	182, 70	2	2	1, 26A|8B, 14G', 15, 26B, 30|5A, 8A, 14A, B, 17A, B, 25J, K
253B	I	192°	138	1	0	—|17A, B, 25K
253C	—	177°	192	1	—	—|5E
255*	—	195°	114	2	0	29A, B|29B'
257A	—	188°	216	—	8	33A, B|—
257B	—	192°	256, 152	1	—	—|7A, 15
257C	I	190°	138	0	8	—|5F
257D	—	190°	190	0	8	—|25C, E|—

Table 16. (continued)

Alkaloid[a]	Class[b]	Temperature[c]	Fragment Ions[d]	Perhydro[e]	Exchange[f]	Occurrence[g] (Major|Minor|Trace)
259*	HTX	190°	218, 96	8	2	26B, 31, 33A, B, 34|5A, 6, 13, 14B, D'-G', 20, 24B, 27, 33A, B|5H, I, 14A, C, 23
261*	HTX	190°	220, 96	6	2	—|33C|26B
263A	—	174°	152	0	0	—|25O|—
263B	—	170°	198	—	0	—|25C, D
263C*	HTX	192°	222, 96	4	2	—|26A|—
265A	—	198°	264, 222, 180	—	—	—|1, 2, 5A, B, 14L, 17A, B, 20, 22A, B, 25K, 28, 31
265B*	—	193°	125, 70	0	1	—|7A|7B
265C*	—	182°	210, 84	2	1	—|8A, B|25K, N
265D	PTX	191°	222, 166, 70	4	2	—|29C
267A*	aPTX	186°	182, 70	2	2	2, 5A, D, G, 6, 8A, 9, 12, 13, 14A, B, 15, 16A, 18A–E, 25F, I, L, M, Q, S, 26B, 27, 29B–C, 30, 31, 35A'|4, 5B, C, I, 8B, 14C, 16B, 17A, 25D, G, O, P, 26A, 29A|10, 14D–E, G', J, 17B, 25A–B, H, K, 32
267B	—	208°	170	—	2	—|5A, 18B
267C*	PTX	190°	166, 70	2	2	—|—| Nondendrobatid (Section 6.1)
267D*	PTX	178°	166, 70	2	2	—|—| Nondendrobatid (Section 6.1)
267E	—	182°	196, 96	0	0	—|5D
267F	—	198°	250, 178	—	—	23|5D
267G	—	190°	152	2	2	—|25O|—
269AB*	DHQ	207°	204, 202	10	1	—|25C, N|5D, E, H, I, 13, 14A–D', G', J, K, 20, 25D–F, J, 33A, B, 34

234

269A	—	207°	204	10	1	—|14A|13, 14B
269B	—	207°	202	10	1	—|5I
269C	—	208°	98	—	2	—|25E|25C, D
269D	—	199°	176	8	0	—|25O
271	—	197°	178	4	0	—|25O
275A*	—	198°	152	4	0	15—|5A, 25D, 29B'
275B	—	195°	206	2	1	—|5E|—
277	—	176°	152	0	0	—|25O|—
279	—	190°	210, 190, 84	2	1	—|25N
281A	PTX	205°	166, 70	2	2	17A, 18A|—|1, 25P
281B*	—	200°	150, 70	0	2	—|8A|8B, 13
281C	—	195°	208	0	2	—|25E|—
281D	—	196°	210	0	2	—|9, 10
283A*	HTX	210°	218, 96	12	2	5C, E, H, I, 6, 13, 14B, D, E–L, 20, 22A, 23, 24A, B, 25C–F, 26B, 31, 33A, B, 34|5D, 14A, C, D', G', 25J, N, 27|—
283A'*	HTX	210°	218, 96	12	2	—|14J
283B*	—	197°	152, 140	0	1	14H, I, 20|—|—
283C*	HTX	195°	166, 126	0	1	14H, I, 20|—|—
285A*	HTX	215°	96	10	2	5C, 6, 13, 14B, D, E–L, 20, 22A, 24A, B, 25C–F, 29B, C, 31|5D–F, H, I, 14A, C, D', G', 23, 25J, N|—
285B*	HTX	211°	220, 96	10	2	24A|14E, H, J, K, 33A, B, 34|14D'
285C*	HTX	211°	96	10	2	5C, E, 14F, K, 24A, 25C, 26B, 33A, B, 34|5D, H, I, 13, 14A, B, D–E, G–J, L, 23, 24B, 25D–F, J, N, 27, 31|14C
285D	—	190°	180, 140	—	—	—|7A, B
285E*	HTX	212°	218, 200, 96	10	2	—|14J
287A*	HTX	216°	162, 96	8	2	—|6, 14D–J, 24A|5C–F, H, I, 14A–C, D', G', 25C–F, J, N

Table 16. (continued)

Alkaloid[a]	Class[b]	Temperature[c]	Fragment Ions[d]	Perhydro[e]	Exchange[f]	Occurrence[g] (Major|Minor|Trace)
287B*	HTX	213°	220, 96	8	2	—|14D, F, G, J–L, 22A|14A–C, D', G', 24B, 25D–F, J, N
287C*	GTX	218°	242, 222	6	1	—|14D, E, I, J|14C, D'
287D*	HTX	215°	162, 96	8	2	—|14G'|14A–C, D', J
289A	—	216°	152	4	0	—|15
289B*	GTX	217°	244, 222	4	1	—|14J
291A*	HTX	212°	254, 96	4	2	14K, 22B, 26A, 33C|10, 13, 14I, J, L, 22A, 31, 34|23, 24B
291B	—	221°	168	4	1	—|15, 25D, 29B'
291C	—	220°	152	0	1	—|15
291D*	—	201°	168	4	1	—|8A|—
293	DHQ	194°	152	0	1	—|25C, F, O, Q, R
295	—	224°	138	—	—	—|5A, B
297A*	aPTX	225°	182, 70	2	3	18A|17A|—
297B*	PTX	222°	166, 70	4	—	—|5A
301	—	213°	260	—	—	—|6
305	aPTX	214°	182, 70	6	2	—|25O|—
307A* (A', A")	PTX	216°	166, 70	4	2	4, 8B, 25A, A', K, 29B, B'|2, 5A, 13, 15, 18C, E, 25J, L, 27, 35A|5D, 7B, 8A, 14A, 17A, 20, 21, 25H, M, O, 32, 35A'
307B*	PTX	211°	166, 70	4	1	—|4, 5A, 8B, 13, 15, 25A', K, 35A'|5D, 8A, 18A, C, E, 25A, 27, 29B, B'
307C*	aPTX	214°	182, 70	4	2	8A|15, 21|8B, 25A, 28
307D	PTX	234°	166, 70	4	2	—|17A
309A	PTX	218°	166, 70	2	2	18C, E, 35A|2, 35A'|—
309B	—	220°	152	—	—	—|5A, 13
309C	PTX	210°	166, 70	2	—	—|17A, B|25F, R

309D	aPTX	182, 70	208°	2	2	—|25C, D|—
309E	—	240	235°	2	2	5E|—|—
309F	—	152	206°	0	2	—|—|25O
321*	PTX	166, 70	223°	4	1	—|8B|8A, 25A'
323A*	PTX	166, 70	230°	4	3	4, 5A, D, 8A, 13, 15, 16A, 20, 25A, A', 29B–C, 35A|1, 5I, 7A, B, 8B, 12, 14A, B, 16B, 18E, 25D, J, K, M, 35A'|5E, 10, 14H, 17A, B, 21, 23, 25C, 27, 29A
323B* (**B'**, **B''**)	aPTX	182, 70	228°	4	3	25O, 28|1, 7A, B, 8A, 16A, 25C, H, L, M, R, 32|5A, E, 8B, 16B, 18C, 25A, A', F, G, K, S
325A	aPTX	182, 70	232°	2	3	2, 12, 18B, C, E|16B, 25E, J|25B
325B	PTX	166, 70	228°	2	3	—|25C, 35A'|—
339A*	aPTX	182, 70	243°	4	4	—|5E|5A, I, 25A, A'
339B*	aPTX	182, 70	243°	4	4	—|—|5A
341A*	aPTX	182, 70	222°	2	3	5B|13, 15, 18C, 32, 35A, A'|5A, 10
341B	aPTX	182, 70	223°	—	—	—|—|15
351	—	152, 138, 70	230°	—	—	—|—|5B
353	PTX	166, 70	240°	—	3	—|5D, 29B, B'
357	aPTX	182, 70	240°	—	4	—|5B, 10|5A, 32

[a] Batrachotoxins, chimonanthine, calycanthine, and noranabasamine are not listed. Certain alkaloids from nondendrobatid frogs are listed. Alkaloids designated as described in text (see Section 3.1). An asterisk indicates that the empirical formula has been determined by high resolution mass spectrometry.

[b] Histrionicotoxin, HTX; gephyrotoxin, GTX; indolizidine, I; pumiliotoxin-A class, PTX; allopumiliotoxin subclass, aPTX; homopumiliotoxin subclass, HPTX; decahydroquinoline, DHQ; piperidine, PIP; pyrrolidine, PYR; amidine, AMI; unknown class, —.

[c] Temperature (°C) at which alkaloid emerges from a 1.5% OV-1 gas chromatographic column programmed from 150–280° (see legend of Fig. 2 for conditions).

[d] Major mass spectral fragment ions.

[e] Number of hydrogens incorporated in perhydro derivative.

[f] Number of exchangeable hydrogens incorporated in deuteroammonia chemical ionization mass spectrometry.

[g] Occurrence in various species and populations of dendrobatid frogs (see Table 17 for listing of frogs). Occurrences as a major alkaloid are followed by a vertical line and then occurrences as a minor alkaloid, a vertical line, and finally occurrences as a trace alkaloid.

Table 17. Occurrence of Dendrobatid Alkaloids in Frogs of the Genus *Dendrobates*[a]

Species	Locale and Date[b]	Number of Skins	Number[c] and Amount	Alkaloids — Major[d] / Minor / Trace — HTX|GTX|I|PTX|DHQ|Other|PYR or PIP|AMI
1. *D. abditus*	Volcan Reventador, Napo, Ecuador, Feb. 1974	3	7 +++	—|—|237A, 253A|—|—|—|— —|—|323A, 323B|—|—|—|— —|—|237B, 281A|—|265A|—|—
2. *D. altobueyensis*	Altos de Buey, Colombia, Oct. 1978	1	7 +++	—|—|267A, 325A|—|—|—|— —|—|307A, 309A|219B|231B|—|— —|—|265A|—|—|—|—
3. *D. anthonyi*	Pasaje, Ecuador, Nov. 1979	5	2 +	—|—|—|—|—|—|— —|—|219B|—|—|—|— —|—|241C|—|—|—|—
4. *D. arboreus*	Border Chiriqui/ Bocas, Panama, Jan. 1983	10,10	14 +++	—|—|203A, 205|307A, 323A|—|—|— —|—|207A, 231C|207B, 251D, 267A, 307B|—|231B|— —|167A, 195B|—|—|167C|—|—
5A. *D. auratus*	Isla Taboga, Panama, 1968–1975	10 and others	30 +++	—|—|267A, 323A|195A, 219A, 243A|—|—|— 235A, 259|—|—|251D, 307A, 307B|—|—|— —|167A, 181B, 203A|253A, 297B, 323B, 339A, 339B, 341A, 357|—|181A, 185, 223A, 223C, 265A, 267B, 275A, 295, 309B|—|—

	Species	Locality				Allele data
5B.	*D. auratus*	Rio Campana, Panama, Mar. 1972	2	13	+ +	—|—|341A|195A, 219A, 243A|—|—| 235A|—|267A, 357|—| |181B, 237D|—|—|197A, 265A, 295, 351|—|—|
5C.	*D. auratus*	El Llano-Carti Rd., Panama, May 1979	1	8	+ + +	283A, 285A, 285C|—|—|—|—|—| —|203A|267A|—|231A|—|—| 287A|—|—| |223C|—|—|
5D.	*D. auratus*	El Cope, Panama, Oct. 1977	10	19	+ + +	—|—|207A|267A, 323A|—|—|—| 283A, 285A, 285C|—|—|—|—|—| 287A|—|205, 235B|251D, 307A, 307B, 353|269AB|195C, 223A, 231B, 267E, 267F|—|—|
5E.	*D. auratus*	Changuinola, Panama, Apr. 1980	1	20	+ + +	283A, 285C|—|—| |223AB, 309E|—|—| 285A|—|203A, 207A|339A|275B|209E, 217|—|—| 287A|—|323A, 323B|195A, 269AB|181C, 195E, 207C, 253C|—|—|
5F.	*D. auratus*	El Valle, Colombia, Oct. 1978	1	13	+	—|—|223AB|—|—|—| 285A|—|195B|—|—|193B, 217|—|—| 287A|—|203A, 205, 209E, 257C|—|—|203B, 207C, 231B|—|—|
5G.	*D. auratus*	Rampala, Bocas, Panama, Oct. 1981	1	9	+ + +	—|—|223AB|267A|—|—|—| |195B|—|195A|—|—| 205, 207A|—|—|181C, 207C, 209E|—|—|
5H.	*D. auratus*	Miramar, Bocas, Panama, Oct. 1982	13	21	+ + +	283A|—|205|—|—|—| 285A, 285C|—|203A, 207A|323A|195A|223B|—|—| 259, 287A|—|167A|—|167D, 219A, 221A, 269AB|181C, 195C, 207C, 207D, 231A, 249|—|—|

Table 17. (continued)

Species	Locale and Date[b]	Number of Skins	Number[c] and Amount	Alkaloids
				Major[d] / Minor / Trace — HTX|GTX|I|PTX|DHQ|Other|PYR or PIP|AMI
51. *D. auratus*	Isla Pastores, Bocas, Panama, Oct. 1982	45	19 ++	283A|—|—|—|—|—|—| 285A, 285C|—|—|267A, 323A|195A, 219A|223B|—| 235A, 259, 287A|—|167A|339A|167D, 269AB|167D, 269AB|195C, 207C, 207F, 269B|—
6. *D. azureus*	Sipaliwini Savanna, Surinam, July 1972	5	11 ++	283A, 285A|—|—|267A|219A, 243A|—|—| 259, 287A|—|—|—|—|—|—| —|181B|—|—|219B, 231B, 301|—|—
7A. *D. bombetes*	Lago de Calima, Valle, Colombia, Nov. 1976	10,10	16 ++	—|—|251D|—|217, 251F|—|—| 323A, 323B|243A|233A, 239B, 265B|—| —|219B, 231A, 231B, 243B, 247, 257B, 285D|—|
7B. *D. bombetes*	Quebrada de la Chapa, Valle, Colombia, Feb. 1974	7	17 ++	—|—|251D|—|—|—| 323A, 323B|—|219B, 223C, 233A, 239B|—| 307A|—|223A, 223B, 231B, 239A', 239B', 239C', 251F, 265B, 285D|—|
8A. *D. species*[e]	Cerro Caracoral, Panama, 1982	2	18 ++	—|—|251D|—|—|—| 231C|323B|—|231A', 265C, 281B, 291D|—| 207A|253A, 307A, 307B, 321|—|193A, 231A|—

240

8B.	D. species	El Copé, Panama, Oct. 1977	5	20 ++	—|—|307A|—|—|— —|207A|251D, 253A, 267A, 307B, 321, 323A|— 231A′, 231B, 265C|—|— —|203A, 205, 231C|307C, 323B|—|183A, 193A, 231A, 281B|—|—
9.	D. erythromos	Río Palenque Biol. Station, Pichincha, Ecuador Nov. 1979	5	3 +	—|—|251B|267A|—|—|—|— —|—|—|— —|—|281D|—|—
10.	D. espinosai	Rio Palenque, Ecuador, Nov. 1979	21	12 +	—|—|237A|—|—|— 291A|—|251D, 357|—|231B|—|—|217, 219B, 281D|— —|203A|267A, 323A, 341A|—|217, 219B, 281D|— |—|
11.	D. femoralis[f]	Santa Cecilia, Ecuador	1	1 +	—|—|—|—|— —|—|—|— 181B|—|—
12.	D. fulquritus	El Llano-Carti Rd., Panama, Mar. 1974	6	6 +	—|203A|267A, 325A|—|—|— —|—|323A|—|— 181B|237A|—|—|—
13.	D. granuliferus	Palmar Norte, Puntarenas, Costa Rica, July 1967	9	17 ++	283A, 285A|—|—|267A, 323A|219A, 243A|—|—|— 259, 285C, 291A|—|307A, 307B, 341A|—|—|— 235A|—|—|269AB|269A, 281B, 309B|—|—
14A.	D. histrionicus	Altos de Buey, Colombia (higher elevations), Oct. 1978	5	32 +++	—|—|223AB, 239AB|267A|—|—|— 283A, 285A, 285C|—|251D, 323A|219A|217, 269A|225C|— 259, 287A, 287B, 287D|—|195B, 205, 235B, 239CD, 239G|253A, 307A|221C, 269AB|209C, 219B, 223B, 231B, 239B|197B, 225B|—

Table 17. (*continued*)

Species	Locale and Date[b]	Number of Skins	Number[c] and Amount	Alkaloids — Major[a] / Minor / Trace HTX|GTX|I|PTX|DHQ|Other|PYR or PIP|AMI
14B. *D. histrionicus*	Altos de Buey, Colombia (lower elevations), Oct. 1978	5	30 +++	283A, 285A|—|223AB, 239AB|267A|—|—|—|— 259, 285C|—|—|251D, 323A|219A|217|197B, 225C|— 287A, 287B, 287D|—|195B, 205, 235B, 239CD, 239G|253A|221C, 269AB|209C, 219B, 223B, 231B, 269A|225B|—
14C. *D. histrionicus*	El Valle, Colombia, Oct. 1978	10	24 ++	—|—|—|—|197B|— 283A, 285A|—|195B, 205, 223AB, 231C, 235B, 239AB|267A|—|217|— 259, 285C, 287A, 287B, 287D|287C|239CD, 239G|—|219A, 269AB|219B, 231B|225C|—
14D. *D. histrionicus*	Santa Cecilia, Río San Juan, Risaralda, Colombia, Feb. 1970	12	14 ++	283A, 285A|—|239AB|—|—|—|— 285C, 287A, 287B|287C|—|—|219A|—|— —|223AB|267A|243A, 269AB|223A, 223B|—|—
14D′. *D. histrionicus*	Feb. 1983	3	21 ++	—|—|239AB|—|219A, 243A|—|—|— 235A, 259, 283A, 285A, 285C|—|223AB|— 285B, 287A, 287B, 287D|287C|195B, 239CD|267A|219A′, 243A′, 269AB|—|197B|—
14E. *D. histrionicus*	Playa de Oro, Río San Juan, Choco, Colombia, Feb. 1971	8	12 +++	283A, 285A|—|—|219A, 243A|—|—|— 259, 285B, 285C, 285C, 287A|287C|239AB|— —|—|251D, 267A|—|—

14F. *D. histrionicus*	Quebrada Vicordo, Río San Juan, Choco, Colombia, Feb. 1971	3	9 ++	283A, 285A, 285C|—|—|219A, 243A|—|—| 235A, 259, 287A, 287B|—|—|—|—|—|
14G. *D. histrionicus*	Quebrada Docordo, Río San Juan, Choco, Colombia, Feb. 1971	8	10 +++	283A, 285A|—|—|219A, 243A|—|—| 259, 285C, 287A, 287B|—|239AB|—|—|—|—| —|—|—|—|231B|—|
14G'. *D. histrionicus*	Quebrada Docordo, Río San Juan, Choco, Colombia, June 1983	5	23 +++	—|—|219A, 243A|—|—| 235A, 259, 283A, 285A, 285C, 287D|—|—|253A| |231B|—| 287A, 287B|—|195B, 205, 207A, 209B, 239AB|267A|219A', 243A', 269AB|223B|197B|—
14H. *D. histrionicus*	Río Saija, Cauca, Colombia, Feb. 1973	10	19 ++	283A, 285A|—|239AB, 239CD|—|—|283B, 283C|—| 285B, 285C, 287A|—|223AB|—|—| —|195B, 235B, 239G|323A|219A|237E, 239C', 239D', 239E|—|—
14I. *D. histrionicus*	Río Guapi, Cauca, Colombia, Feb. 1973	5	15 ++	283A, 285A|—|—|—|—| 285C, 287A, 291A|287C|—|—|219B, 231A, 231B|— —|—|195B, 207A, 223D|—|251A|223A, 223B|—|—
14J. *D. histrionicus*	Guayacana, Narino, Colombia, 1972–1978	10 and others	21 +++	283A, 285A|—|—|—|—|—| 285B, 285C, 287A, 287B, 291A|287C|223AB|—| —| 239H, 283A', 285E, 287D|289B|195B, 203A, 205, 223D|267A|269AB|223B|—|
14K. *D. histrionicus*	Río Baba, Pichincha, Ecuador, Feb. 1974	5	15 +++	283A, 285A, 285C, 291A|—|—|—|—| 285B, 287B|—|223D|—|219B, 223C, 237C|—|— —|—|207A|237A|269AB|231B, 239F|—|

Table 17. (continued)

Species	Locale and Date[b]	Number of Skins	Number[c] and Amount	Major[d] / Minor / Trace	HTX	GTX	I	PTX	DHQ	Other	PYR or PIP	AMI
14L. *D. histrionicus*	Río Palenque, Biol. Station, Pichincha, Ecuador, Feb. 1974	5	10 +++	Major	283A, 285A	—	—	—	—	—	—	—
				Minor	285C, 287B, 291A	—	223D	—	—	219B, 223C, 231A	—	—
				Trace	—	—	—	—	265A	—	—	—
15. *D. lehmanni*	Anchicaya Valley, Valle, Colombia, Jan. 1973	8	21 ++	Major	—	—	267A, 323A	—	275A	—	—	—
				Minor	—	—	253A, 307A, 307B, 307C, 341A	—	—	—	—	—
				Trace	251B	251D, 341B	—	195C, 221A, 223A, 223E, 231B, 251E, 257B, 289A, 291B, 291C	—	—		
16A. *D. leucomelas*[g]	Guri Dam, Boliva, Venezuela, Nov. 1968	6	3 +	Major	—	—	267A, 323A	—	—	—	—	—
				Minor	—	—	323B	—	—	—	—	—
				Trace	—	—	—	—	—	—	—	—
16B. *D. leucomelas*	Lowland at Cerro, Yapacana, Venezuela, Feb. 1978	1	11 +	Major	235A	—	—	—	—	—	—	—
				Minor	—	—	267A, 323A, 325A	—	195C, 219B, 223B	—	—	
				Trace	—	—	251D, 323B	—	167C, 181C	—	—	
17A. *D. species*[h]	Isla Colon, Bocas, Panama, May 1977	10,10	18 ++	Major	—	—	281A	—	—	—	—	—
				Minor	205	251D, 267A, 297A, 309C	—	—	—	—	—	
				Trace	167B, 181B, 209B, 223D, 253B	253A, 307A, 307D, 323A	—	195C, 209A, 265A	—	—	

				Mass spectral data
17B. *D. species*[i]	Isla Colon, Bocas, Panama, Nov. 1983	10	18 +	—|—|205|—|—|—|— —|207A|309C|235D|219D|225B|— —|167A, 181B, 203A, 209B, 223D, 253B|253A, 267A, 323A|—|195C, 209A, 265A|—|—
18A. *D. minutus*	Cerro Campana, Panama, Mar. 1972	36	5 +	—|—|267A, 281A, 297A|—|—|—|— —|—|251D, 307B|—|—|—|—
18B. *D. minutus*	El Llano-Carti Rd., Panama, Mar. 1974	10	8 +	—|—|267A, 325A|—|—|—|— —|—|251C|—|—|— 203A, 223D|—|223B, 231A, 267B|—|—
18C. *D. minutus*	Altos de Buey, Colombia, Oct. 1978	10	10 ++	—|—|251D, 267A, 309A, 325A|—|—|—|— —|307A, 341A|217|—|—|— —|181B|307B, 323B|—|—|—
18D. *D. minutus*[g]	Playa de Oro, Río San Juan, Choco, Colombia, Feb. 1970	6	1 +	—|—|267A|—|—|— —|—|—|— —|—|—|—
18E. *D. minutus*	Río Saija, Cauca, Colombia, Feb. 1973	9	11 +	—|—|267A, 309A, 325A|—|—|— —|307A, 323A|—|251E|—|— —|307B|195A|223A, 223B, 251C|—|—
19. *D. myersi*	Río Vaupes, Colombia, April, 1979	1	0	None detected
20. *D. occultator*	Río Saija, Cauca, Colombia, Feb. 1973	4	23 ++	283A, 285A|—|239AB, 239CD|323A|—|283B, 283C|— 259|—|—|231B|— —|307A|269AB|223A, 223E, 239A, 239A', 239B, 239B', 239C, 239C', 239D, 239D', 241A, 265A|—|—

245

Table 17. (*continued*)

Species	Locale and Date[b]	Number of Skins	Number[c] and Amount	Alkaloids HTX|GTX|I||PTX|DHQ|Other|PYR or PIP|AMI (Major[d] / Minor / Trace)
21. *D. opistomelas*	Santa Rita, Colombia, Nov. 1976	10	8 +	—|—|—||217|—|— —|—|203A|307C|231A, 233B|—|— —|—|—||251D, 307A, 323A|—|—|—
22A. *D. parvulus*[g]	Mocoa, Putamayo, Colombia, Feb. 1970	25	7 +	283A, 285A|—|—||219A|—|— 287B, 291A|—|—||265A|—|— 235A|—|—||—|—
22B. *D. parvulus*	Pebas, Peru, Apr. 1977	5	7 ++	291A|—|—||—|—|— 235A|—|237A|219A|233C|—|— —|181B|—||265A|—|—
23. *D. petersi*	Río Llullapichis, Peru, Feb. 1975	1	9 +	283A|—|—||—|—|— 285A, 285C|—|—|243A|267F|—|— 235A, 259, 291A|—|—||323A|—
24A. *D. pictus*	Boquerón de Padre, Abad, Río Aquaytia, Loreto, Peru, Nov. 1974	10	6 ++	283A, 285A, 285B, 285C|—|—|—|—|—|— 287A|—|—||243A|— —|—|—||243A|—
24B. *D. pictus*	Santa Cecilia, Ecuador, Jan. 1977	5	9 ++	283A, 285A|—|—||—|—|— 259, 285C|—|—||219A, 243A, 243A'|—|—|— 287B, 291A|—|—||—|—

25A. *D. pumilio*	Isla Bastimentos (North), Bocas, Panama, Oct. 1972	10	12 +++	—|—|307A, 323A|195A|—|—|— —|—|—|—|—|—|— —|—|203A, 223D, 235B, 251B|267A, 307B, 307C, 323B, 339A|—|—|—
25A'. *D. pumilio*	Isla Bastimentos (North) Bocas, Panama, Oct. 1981	10 and others	28 +++	—|—|307A, 323A|195A|—|—|— 205|307B|—|—|—|236, 252 167A, 203A, 205', 207A, 223D, 235B, 251B|267A, 321, 323B, 339A|153A|151, 183B, 201, 211A, 221D, 223A, 223E, 223G|—|222
25B. *D. pumilio*	Isla Bastimentos (South) Bocas, Panama, Oct. 1981	10	16 ++	—|—|—|193B|—|—|— —|—|195B, 205, 207A, 207A', 235B|—|—|231B|—|— —|—|251D, 267A, 325A|195A|195C, 223A, 237C, 239D, 241D|—|—
25C. *D. pumilio*	Dos Bocas, Bocas, Panama, Apr. 1980	5	21 +++	283A, 285A, 285C|—|—|—|223A|—|— —|207A|309D, 323B, 325B|269AB|231B, 257D|—|— 287A|—|203A|323A|293|195C, 207D, 231F, 241D, 263B, 269C|—|—
25D. *D. pumilio*	Changuinola, Bocas, Panama, Apr. 1980	10	31 +++	283A, 285A|—|205|—|—|—|— 285C|—|—|251D, 267A, 309D, 323A|—|195C, 231B, 241D|—|— 287A, 287B|—|167A, 207A, 225D|—|153B, 195A, 269AB|167C, 181C, 183B, 207F, 209C, 221B, 235D, 235F, 263B, 269C, 275A, 291B|—|—
25E. *D. pumilio*	Isla Colon, Bocas, Panama, May 1977	10	20 +++	283A, 285A|—|205|—|—|—|— 285C|—|195B|325A|—|209E, 257D, 269C, 281C|—|— 287A, 287B|—|225D, 235B|—|269AB|195C, 207E, 221B, 223A, 223A'|—|—

Table 17. *(continued)*

Species	Locale and Date[b]	Number of Skins	Number[c] and Amount	Alkaloids Major[d] / Minor / Trace	HTX	GTX	I	PTX	DHQ	Other	PYR or PIP	AMI
25F. *D. pumilio*	Isla Pastores, Bocas, Panama, Oct. 1981	10,10	21 ++	Major	283A, 285A	—	—	267A	—	—	—	—
				Minor	285C	—	—	251D	219C	195D, 231B	—	—
				Trace	287A, 287B	—	203A	309C, 323B	195A, 269AB, 293	167C, 181C, 207F, 209E, 223A	—	—
25G. *D. pumilio*	Isla Cristóbal, Bocas, Panama, Oct. 1981	10,10	14 ++	Major	—	203A	—	—	—	—	—	—
				Minor	—	—	267A	—	223A, 231B, 241C	—	—	—
				Trace	—	167A, 195B, 203A'	251D, 323B	—	167C, 207C, 221B, 223A'	—	—	—
25H. *D. pumilio*	Cayo Nancy, Bocas, Panama, Dec. 1978	1	9 ++	Major	—	—	195A	—	—	—	—	—
				Minor	—	207A	323B	—	—	—	—	—
				Trace	—	203A, 205, 223D	251D, 267A, 307A	—	—	—	—	—
25I. *D. pumilio*	Isla Popa (North) Bocas, Panama, Oct. 1981	10	9 +	Major	—	—	267A	195A	—	—	—	—
				Minor	—	205	251D	—	233A	—	—	—
				Trace	—	—	219B, 223A, 231A, 231B	—	—	—	—	—
25J. *D. pumilio*	Isla Split Hill, Bocas, Panama, Oct. 1981	10	18 +	Major	—	—	195A	—	—	—	—	—
				Minor	283A, 285A, 285C	—	235B	307A, 323A, 325A	—	—	—	—
				Trace	287A, 287B	—	205	253A	269AB	195C, 197C, 219B, 223A, 241E	—	—

					Allele data
25K. *D. pumilio*	Mainland, Split Hill, Bocas, Panama, Oct. 1981	10	18 +++		—|—|307A|—|—|— —|—|205, 235B|307B, 323A|—|—|— —|—|195B, 225D, 251B, 253B|251D, 253A, 267A, 323B|195A|241E, 265A, 265C|197B|—
25L. *D. pumilio*	Isla Cayo Agua, Bocas, Panama, Oct. 1981.	10	9 +++		—|—|205|251D, 267A|195A|—|—|— —|—|307A, 323B|—|195C, 235B|—|— —|—|—|209D|—|—
25M. *D. pumilio*	Bahia Azule, Bocas, Panama, Oct. 1981	10,10	14 +		—|—|267A|—|—|— —|—|205, 207A|323B|195A —|—|251D, 307A|—|195C, 207C, 207E, 221B, 223A, 235B, 241D|—|—
25N. *D. pumilio*	Miramar, Bocas, Panama, Oct. 1981	5	19 ++		—|—|203A, 205|—|—|— 283A, 285A, 285C|—|207A|—|195A, 269AB|183B, 207C|—|— 287A, 287B|—|—|251D|231E|167C, 223E, 231B, 265C, 279|—
25O. *D. pumilio*	Chirigui Grande, Bocas, Panama, Oct. 1981	10 (red)	19 ++		—|—|323B|—|—|— —|—|203A|267A, 305|181D|167C, 223A, 263A, 267G, 277|—|— —|—|307A|195A, 223F, 293|181C, 203B, 269D, 271, 309F|—
25P. *D. pumilio*	Chirigui Grande, Bocas, Panama, Oct. 1981	10 (green)	11 ++		—|—|231B|—|—| —|—|207A, 209B|267A|—|223A|—|— —|—|225D|281A|181D, 195A|221B, 231B'|—|—
25Q. *D. pumilio*	Rampala, Bocas, Panama, Oct. 1981	10 (red)	7 +++		—|—|267A|—|—|— —|—|251D|195A|—|—|— —|—|195B, 203A, 209B|—|293|—|—|—

Table 17. *(continued)*

Species	Locale and Date[b]	Number of Skins	Number[c] and Amount	Major[d] / Minor / Trace — HTX|GTX||PTX|DHQ|Other|PYR or PIP|AMI
25R. *D. pumilio*	Rampala, Bocas, Panama, Oct. 1981	10,10 (green)	10 +	—|—|—|195A|—|—|— —|—|—|323B|—|—|236, 252 205|—|293|211B, 223A, 309C|197C|—
25S. *D. pumilio*	Rampala, Bocas, Panama, Oct. 1981	10 (yellow)	13 +++	—|—|267A|195A|—|—|— —|—|209B|251D|—|—|— —|207A, 225D, 235B|323B|219C|207F, 209C, 219E, 223A|—|—
26A. *D. quinquevittatus*	Pebas, Peru, Apr. 1977	4	9 ++	235A, 291A|—|253A|—|—|—|— 263C|—|237A, 267A|219A|—|—|— —|223AB|—|223B|—|—
26B. *D. quinquevittatus*	Mishana, Peru, Apr. 1977	4	16 ++	259, 283A, 285C|—|—|267A|—|—|— 235A|—|237A, 251D, 251I, 253A|219A, 243A|223B|—|— 261|—|223AB|195C, 235E|—|—|—
27. *D. reticulatus*	Mishana, Peru, Apr. 1977	10,10	14 +++	—|—|267A|—|—|—|— 235A, 259, 283A, 285C|—|307A|—|195C|231A|—|— —|223AB|307B, 323A|243A|217, 223C|—|—
28. *D. silverstonei*	Loreto, Peru, Nov. 1974	4	5 ++	—|251D, 323B|—|—|—|— —|—|—|—|—|— 307C|—|265A|—|—

250

29A. *D. speciosus*	Highland Panama, Feb./Mar. 1976	5,5	15 ++	—|—|203A, 223AB|—|—|—|— —|—|251D, 267A|—|231B, 241D, 255|—|— 235A|—|167A, 195B, 207A, 223G, 235B|323A|— |207E|—|—
29B. *D. speciosus*	Continental Divide, Bocas/Changuinola, Panama, Jan. 1983	10	21 +++	—|—|223AB, 235B|267A, 307A, 323A|—|—|— —|—207A, 237D|251D|195A|183A, 195C, 225A, 241D, 255|—|— —|—|167A, 181B, 195B, 203A, 211B|307B, 353]—|— |—|
29B.' *D. speciosus*	Continental Divide, Bocas/Changuinola, Panama, Apr. 1984	10,10	24 +++	—|—|223AB, 235B|251D, 267A, 307A, 323A|—|—|— —|—| —|—|203A, 207A, 237D|—|195A|183A, 195C, 223A|— |—| —|167B, 181B, 195B, 211B|307B, 353]—|241D, 251C, 255, 275A, 291B|—|—
29C. *D. speciosus*	Continental Divide, Changuinola, Panama, Jan. 1983	10	14 +++	—|—|251D, 267A, 323A|—|195C, 241D|—|— —|—|167A, 181B, 207A, 223AB, 235B|—|—|231B|— |—|195B|265D|195A|—|—|—
30. *D. steyermarki*	Cerro Yapacana, Venezuela, Feb. 1978	10	7 +	—|—|267A|—|—|—|— —|—|253A|—|195C|—|— —|—|237A, 251D|—|219B, 225A|—|—
31. *D. tinctorius*	Raleigh Cataracts, Coppename River, Surinam, Feb. 1973, 1974	3	11 ++	259, 283A, 285A|—|—|267A|—|—|— 285C, 291A|—|—|243A|231B|—|— —|—|—|223E, 225A, 265A|—|—

251

Table 17. (continued)

Species	Locale and Date[b]	Number of Skins	Number[c] and Amount	Alkaloids — Major[d] / Minor / Trace — HTX|GTX|I|PTX|DHQ|Other|PYR or PIP|AMI
32. *D. tricolor*	Santa Isabel, Ecuador, Feb. 1974–Nov. 1979	10 and others	13 +++	—|—|251D|—|—|—|—|— —|—|203A, 323B, 341A|—|217|—|—|— —|—|237A, 267A, 307A, 357|—|231A, 231B, 251H|—|"208/10"|—
33A. *D. trivittatus*	Coppename River, Surinam, Feb. 1973	3	9 +++	259, 283A, 285C|—|—|—|—|—|—|— 235A, 259, 285B|—|243A|257A|—|— —|269AB|—|—
33B. *D. trivittatus*	Tingo Maria, Huanuco, Peru, Nov. 1974	2	10 +++	259, 283A, 285C|—|—|—|—|—|—|— 235A, 259, 285B|—|243A|257A|—|— —|181B|—|269AB|—|—
33C. *D. trivittatus*	Pebas, Peru, Apr. 1977	1	7 ++	235A, 291A|—|—|—|—|—|—|— 261|—|—|233C|239I|— —|—|219A|235G|—|—
34. *D. truncatus*	Mariquita, Tolima, Colombia, Jan. 1970	24	13 +++	259, 283A, 285C|—|—|219A|—|—|—|— 235A, 285B, 291A|—|243A|—|— —|223D|—|269AB|223A, 223B, 223C|—|—

| 35A. *D. viridis*[g] | Anchicaya Valley, Valle, Colombia, Jan. 1973 | 2 | 4 | + | —\|—\|—\|**309A, 323A**\|—\|—\|—\|—\|
—\|—\|—\|**307A, 341A**\|—\|—\|—\|—\|
—\|—\|—\|—\|—\|—\|—\|—\| |
| 35A′. *D. viridis* | Anchicaya Valley, Valle, Colombia, Jan. 1983 | 10 | 12 | + | —\|—\|—\|**267A**\|—\|—\|—\|—\|
—\|—\|—\|**251D, 307B, 309A, 323A, 325B, 341A\|181E**\|—\|
—\|—\|
—\|—\|—\|**307A**\|—\|**231D, 233A, 251C**\|—\|—\| |

[a] Species, locale, and date of collection, number of skins are reported.

[b] Numbers of assigned to each species and an identifying letter to each locale. Where there were two samples from the same locale of different times, prime designations (A, A′, etc.) have been used. In three cases (5A, 14J, 32) data from samples from the same locale collected at several different times have been combined: Some of the trace alkaloids were identified only in large samples of these populations. In one case (25A′), some of the trace alkaloids were identified only in a large sample obtained at the same time as the 10-skin sample.

[c] The number of alkaloids detected and the approximate total amount of alkaloids per 100 mg skins are reported: + + + = > 150 μg/100 mg skin; + + = 50–150 μg/ + = < 50 μg.

[d] Major alkaloids presented in the first row, minor alkaloids in the second row, and trace alkaloids in the third row. The format for each row is as follows: histrionicotoxins (HTX), vertical line; gephyrotoxins (GTX), vertical line; indolizidine (I), vertical line; pumiliotoxin-A class and allo- and homopumiliotoxins (PTX), vertical line; decahydroquinolines (DHQ), vertical line; other alkaloids, vertical line; pyrrolidine (PYR) or piperidine (PIP), vertical line; amidines (AMI). The alkaloids are listed by code number in Table 16. The designation "major" does not indicate absolute amount, but only that in that particular species it was a predominant alkaloid. In extracts where only trace amounts of alkaloids were detected, these are entered as trace alkaloids.

[e] A new species to be described [256].

[f] No alkaloids were detected in skins from Pepino, Putamayo, Colombia; Lago Agrio, Ecuador; Pebas and Rio Llullapichis, Peru; and Raleigh Cataracts, Surinam.

[g] Not analyzed by gas chromatography–mass spectrometry.

[h] A new species to be described [257].

[i] Not the same population as 17A.

The consideration of dendrobatid alkaloids as a taxonomic character is now further complicated by the discovery of such alkaloids in three additional families of anurans, namely Bufonidae (genus *Melanophryniscus*); Ranidae (subfamily Matellinae, genus *Mantella*), and Myobatrachidae (genus *Pseudophryne*) [261]. Several proposals for the occurrence of shared dendrobatid alkaloid in such diverse genera of frogs have been put forward. The most likely is that of an independent evolutionary occurrence in the several lineages. Whether this represents parallelism (the *de novo* expression of a shared-primitive capacity for the alkaloid production) or convergence (the *de novo* development of evolutionarily new biosynthetic pathways) cannot be decided at present. Delineation of the biosynthetic pathways in detail would be one approach to answering this question, since the occurrence of identical pathways would be more likely in the case of parallelism. Dendrobatid or other alkaloids have not been detected in a wide range of other amphibians (Table 13).

11.3.3. Samandarines. The samandarine alkaloids had appeared, like the batrachotoxins, to be unique to a single genus of amphibians, namely *Salamandra* of the family Salamandridae. The tentative chromatographic identification of samandarines in the myobatrachid frog *Pseudophryne corroboree* [501] requires more rigorous proof, especially in view of recent identification of other classes of alkaloids in frogs of this genus [261].

11.3.4. Tetrodotoxins. The tetrodotoxins represent one of the most striking examples of the random phylogenetic occurrence of a secondary metabolite in animals. Such complex alkaloids are found in an octopus [635], in certain marine snails [636–638], in a starfish [639], in a xanthid crab [640], in puffer and goby fishes [641 and references therein], in several genera of salamanders and newts [525, see Section 9.1.1] all of the family Salamandridae, in frogs of the bufonid genus *Atelopus* [530], and in a frog of the brachycephalid genus *Brachycephalus* [533]. The explanation for such independent occurrence is uncertain. Mosher and Fuhrman [642] have discussed various possibilities, including biosynthesis by the organism, exogenous origin from diet or symbiotic organisms, and multiple origins. Recently, the liberation of tetrodotoxin from liver of puffer fish after treatment with RNAase was reported [643]. Chiriquitoxin appears to be present in eggs of atelopid frogs in a form requiring acid treatment for extraction [534].

11.3.5. Conclusion. The investigation of amphibian alkaloids has played a significant role in the fields of chemistry, pharmacology, and biology. This review has attempted to provide not only an overview, but a sense of the importance of the field. The first animal alkaloid, namely samandarine, was isolated from a European salamander in 1866. The last 20 years (1964–1984) have resulted in the discovery of over 200 unique new alkaloids from the family Dendrobatidae. The structures of the various amphibian alkaloids have provided a challenge to chemists, who have introduced new synthetic strategies and reactions in meeting this challenge. The remarkable biological activities of amphibian alkaloids have established them as invaluable tools for the investigation of ion transport in many biological systems. The past will undoubtedly prove to be only a prologue to the future as chemists, pharmacologists, and biologists continue to investigate all of the fascinating aspects of amphibian alkaloids.

REFERENCES

1. S. W. Pelletier, in *Alkaloids: Chemical and Biological Perspectives*, Vol. 1, S. W. Pelletier, Ed., Wiley, New York, 1983, p. 1.
2. C. W. Myers and J. W. Daly, *Sci. Amer.*, **248,** 120 (1983).
3. C. S. Cochrane, *Journal of a Residence and Travels in Colombia During the Years 1823 and 1824*, Henry Colburn, London, 1825.
4. Le Docteur Saffray, *Le Tour du Monde, Noveau Journal des Voyages* (Paris) **26,** 97 (1873).
5. A. Posada Arango, *Estudios Cientificos del Doctor Andres Posado con Algunos Otros Escritos Suyos Sobre Diversos Temas*, C. A. Molina, Ed., Imprenta Oficial, Medellin, Colombia, 1909, p. 78.
6. S. H. Wassen, *Ethnografiska Museum, Göteborg Ethnologiska Studier*, **1935,** 35.
7. S. H. Wassen, *Ethnografiska Museum, Göteborg, Arstryck 1955–1956*, p. 73 (1957).
8. F. Märki and B. Witkop, *Experientia* **19,** 329 (1963).
9. C. W. Myers, J. W. Daly, and B. Malkin, *Bull. Amer. Mus. Nat. History* **161,** 307 (1978).
10. A. Posada Arango, *Ann. Acad. med. Medellin* **1,** 69 (1888).
11. C. G. Santesson, in *Comparative Ethnographic Studies*, Vol. 9, E. Nordenskiold, Ed., Elanders Boktryckeri, Göteborg, 1931, p. 157.
12. C. G. Santesson, *Naunyn-Schmiedeberg's Arch. Pharmacol.* **181,** 180 (1936).
13. K. Mezey, *Rev. Acad. Colombiana Ciencias Exactas fisicas y naturales, Bogotá* **7,** 319 (1947).
14. M. Latham, *Nat. Geographic Magazine* **129,** 682 (1966).
15. J. W. Daly, B. Witkop, P. Bommer, and K. Biemann, *J. Am. Chem. Soc.* **87,** 124 (1965).
16. K. Biemann, *Pure Appl. Chem.* **9,** 95 (1964).
17. T. Tokuyama, J. Daly, and B. Witkop, *J. Am. Chem. Soc.* **91,** 3931 (1969).
18. T. Tokuyama, J. Daly, B. Witkop, I. L. Karle, and J. Karle, *J. Am. Chem. Soc.* **90,** 1917 (1968).
19. I. L. Karle and J. Karle, *Acta Crystallogr.* **B25,** 428 (1969).
20. R. D. Gilardi, *Acta Crystallogr* **B26,** 440 (1970).
21. I. L. Karle, *Proc. Nat. Acad. Sci. USA* **69,** 2932 (1972).
22. T. Tokuyama and J. W. Daly, *Tetrahedron* **39,** 41 (1983).
23. J. W. Daly, G. B. Brown, M. Mensah-Dwumah, and C. W. Myers, *Toxicon* **16,** 163 (1978).
24. J. W. Daly, *Prog. Chem. Org. Nat. Prods.*, Vol. 41, W. Herz, H. Grisebach, and G. W. Kirby, Eds., Springer-Verlag, Vienna, 1982, p. 205.

25. R. Imhof, E. Gössinger, W. Graf, L. Berner-Fenz, H. Berner, R. Schaufelberger, and H. Wehrli, *Helv. Chim Acta* **56**, 139 (1973).

26. G. B. Brown and J. W. Daly, *Cell. Mol. Neurobiol.* **1**, 361 (1981).

27. H. Berner, L. Berner-Fenz, R. Binder, W. Graf, T. Grütter, C. Pascual, and H. Wehrli, *Helv. Chim. Acta* **53**, 2252 (1970).

28. L. Berner-Fenz, H. Berner, W. Graf, and H. Wehrli, *Helv. Chim. Acta* **53**, 2258 (1970).

29. W. Graf, H. Berner, L. Berner-Fenz, E. Gössinger, R. Imhof, and H. Wehrli, *Helv. Chim. Acta* **53**, 2267 (1970).

30. E. Gössinger, W. Graf, R. Imhof, and H. Wehrli, *Helv. Chim. Acta* **54**, 2785 (1971).

31. W. Graf, E. Gössinger, R. Imhof, and H. Wehrli, *Helv. Chim. Acta* **54**, 2789 (1971).

32. R. Imhof, E. Gössinger, W. Graf, W. Schnüringer, and H. Wehrli, *Helv. Chim. Acta* **54**, 2775 (1971).

33. R. Imhof, E. Gössinger, W. Graf, H. Berner, L. Berner-Fenz, and H. Wehrli, *Helv. Chim. Acta* **55**, 1151 (1972).

34. W. Graf, E. Gössinger, R. Imhof, and H. Wehrli, *Helv. Chim. Acta* **55**, 1545 (1972).

35. U. Kerb, H.-D. Berndt, U. Eder, R. Wiechert, P. Buchschacher, A. Furlenmeier, A. Fürst, and M. Müller, *Experientia* **27**, 759 (1971).

36. R. R. Schumaker and J. F. W. Keana, *J. Chem. Soc. Chem. Commun.* **1972**, 622.

37. J. F. W. Keana and R. R. Schumaker, *J. Org. Chem* **41**, 3840 (1976).

38. B. Witkop and E. Gössinger, in *The Alkaloids*, Vol. 21, A. Brossi, Ed., Academic Press, New York, 1982, p. 139.

39. E. A. Yelin, E. V. Grishin, V. N. Leonov, T. K. Prosolova, N. M., Soldatov, I. V. Torgov, N. F., Myasoyedov, and V. P. Shevchenko, *Biorganicheskaia Khimiia* **9**, 990 (1983).

40. E. Yelin, V. Leonov, O. Tikhomirova, and I. Torgov, in *Toxins as Tools in Neurochemistry*, F. Hucho and Y. A. Ovchinnikov, Eds., Walter de Gruyter, New York, 1983, p. 25.

41. N. K. Levchenko, A. P. Sviridova, G. P. Segal, and I. V. Torgov, *Bioorganicheskaii Khimiia* **4**, 1651 (1978).

42. G. B. Brown, S. C. Tieszen, J. W. Daly, J. E. Warnick, and E. X. Albuquerque, *Cell. Mol. Neurobiol.* **1**, 19 (1981).

43. K. J. Angelides and G. B. Brown, *J. Biol. Chem.* **259**, 6117 (1984).

44. J. E. Warnick, E. X. Albuquerque, and F. M. Sansone, *J. Pharmacol. Exp. Therap.* **176**, 497 (1971).

45. E. X. Albuquerque, J. E. Warnick, and F. M. Sansone, *J. Pharmacol. Exp. Therap.* **176**, 511 (1971).

46. P. M. Hogan and E. X. Albuquerque, *J. Pharmacol. Exp. Therap.* **176**, 529 (1971).

47. E. X. Albuquerque and J. E. Warnick, *J. Pharmacol. Exp. Therap.* **180**, 683 (1972).

48. S. O. Kayaalp, E. X. Albuquerque, and J. E. Warnick, *Eur. J. Pharmacol.* **12**, 10 (1970).

49. E. X. Albuquerque, M. Sasa, B. P. Avner, and J. W. Daly, *Nature New Biology* **234**, 93 (1971).

50. E. X. Albuquerque, M. Sasa, and J. M. Sarvey, *Life Sci.* **11**, 357 (1972).

51. E. X. Albuquerque and J. W. Daly, in *Receptors and Recognition*, Vol. 1, series B, P. Cuatracasas, Ed., Chapman and Hall, London, 1977, p. 297.

52. E. Bartels de Bernal, M. I. Llano, and E. Diaz, *Acta Med. Valle* **6**, 74 (1975).

53. E. Bartels-Bernal, T. L. Rosenberry, and J. W. Daly, *Proc. Natl. Acad. Sci. USA* **74**, 951 (1977).

54. S. V. Revenko, *Neirofiziologiya* **9**, 546 (1977).

55. L.-Y. M. Huang, N. Moran, and G. Ehrenstein, *Proc. Nat. Acad. Sci. USA* **79**, 2082 (1982).

56. J.-M. Dubois and A. Coulombe, *J. Gen. Physiol.* **84**, 25 (1984).

57. F. N. Quandt and T. Narahashi, *Proc. Nat. Acad. Sci. USA* **79**, 6732 (1982).

58. C. N. Allen and E. X. Albuquerque, *J. Physiol. (Paris)* **79**, 338 (1984).

59. W. A. Catterall, *J. Biol Chem.* **250**, 4053 (1975).

60. W. A. Catterall, *Proc. Nat. Acad. Sci. USA* **72**, 1782 (1975).

61. M. R. Costa and W. A. Catterall, *Mol. Pharmacol.* **22**, 196 (1982).

62. M. H. Davis, C. N. Pato, and E. Gruenstein, *J. Biol. Chem.* **257**, 4356 (1982).

63. C. Frelin, A. Lombet, P. Vigne, G. Romey, and M. Lazdunski, *Biochem. Biophys. Res. Commun.* **107**, 202 (1982).

64. W. A. Catterall, *Ann. Rev. Pharmacol. Toxicol.* **20**, 15 (1980).

65. E. X. Albuquerque, J. E. Warnick, F. M. Sansone, and J. Daly, *J. Pharmacol. Exp. Therap.* **184**, 315 (1973).

66. J. W. Daly, C. W. Myers, J. E. Warnick, and E. X. Albuquerque, *Science* **208**, 1383 (1980).

67. D. Colquhoun, R. Henderson, and J. M. Ritchie, *J. Physiol. (London)* **227**, 95 (1972).

68. C. Frelin, A. Lombet, P. Vigne, G. Romey, and M. Lazdunski, *J. Biol. Chem.* **256**, 12,355 (1981).

69. J. Baumgold, J. B. Parent, and I. Spector, *J. Neurosci.* **3**, 1004 (1983).

70. J. F. Renaud, G. Romey, A. Lombet, and M. Lazdunski, *Proc. Nat. Acad. Sci. USA* **78**, 5348 (1981).

71. J. C. Lawrence and W. A. Catterall, *J. Biol. Chem.* **256**, 6223 (1981); *ibid* **256**, 6213 (1981).

72. S. J. Sherman, J. C. Lawrence, D. J. Messner, K. Jacoby, and W. A. Catterall, *J. Biol. Chem.* **258**, 2488 (1983).

73. S.-E. Jansson, E. X. Albuquerque, and J. Daly, *J. Pharmacol. Exp. Therap.* **189**, 525 (1974).

74. L. L. Simpson, *J. Pharmacol. Exp. Therap.* **206**, 661 (1978).

75. R. G. Sharkey, D. A. Beneski, and W. A. Catterall, *Biochemistry* **23**, 6078 (1984).

76. W. A. Catterall and M. Risk, *Mol. Pharmacol.* **19**, 345 (1981).

77. S. J. Hong and C. C. Chang, *Toxicon* **21**, 503 (1983).

78. W. A. Catterall, C. S. Morrow, and R. P. Hartshorne, *J. Biol. Chem.* **254**, 11,379 (1979).

79. E. Jaimovich, M. Ildefonse, J. Barhanin, O. Rougier, and M. Lazdunski, *Proc. Nat. Acad. Sci. USA* **79**, 3896 (1982).

80. J. Barhanin, J. R. Giglio, P. Leopold, A. Schmid, S. V. Sampaio, and M. Lazdunski, *J. Biol. Chem.* **257**, 12,553 (1982).

81. Y. Jacques, G. Romey, M. T. Cavey, B. Kartalovski, and M. Lazdunski, *Biochim. Biophys. Acta* **600**, 882 (1980).

82. M. Lazdunski, M. Balerna, J. Barhanin, R. Chicheportiche, M. Fosset, C. Frelin, Y. Jacques, A. Lomet, J. Pouyssegur, J. F. Renaud, G. Romey, H. Schweitz, and J. P. Vincent, in *Neurotransmitters and Their Receptors*, U. Z. Littauer, Y. Dudai, I. Silman, V. I. Teichberg, and Z. Vogel, Eds., Wiley, New York, 1980, p. 511.

83. E. X. Albuquerque, I. Seyama, and T. Narahashi, *J. Pharmacol. Exp. Therap.* **184**, 308 (1973).

84. E. X. Albuquerque, N. Brookes, R. Onur, and J. E. Warnick, *Mol. Pharmacol.* **12**, 82 (1976).

85. B. I. Khodorov, E. M. Peganov, S. V. Revenko, and L. D. Shishkova, *Brain Res.* **84**, 541 (1975).

86. B. I. Khodorov, *Prog. Biophys. Molec. Biol.* **45**, 57 (1985).

87. L. D. Zaborovskaya and B. I. Khodorov, *Neurophysiology* **14**, 468 (1982).

88. L. D. Zaborovskaya and B. I. Khodorov, *Gen. Physiol. Biophys.* **1**, 283 (1982).

89. B. I. Khodorov and L. D. Zaborovskaya, *Gen. Physiol Biophys.* **3**, 233 (1983).

90. A. N. Zubov, A. P. Naumov, and B. I. Khodorov, *Gen. Physiol. Biophys.* **2**, 125 (1983).

91. S. V. Revenko, B. I. Khodorov, and L. M. Shapovalova, *Neuroscience* **7**, 1377 (1982).

92. C. Frelin, P. Vigne, and M. Lazdunski, *Eur. J. Biochem.* **119**, 437 (1981).

93. C. Frelin, P. Vigne, G. Ponzio, G. Romey, Y. Tourneur, H. P. Husson, and M. Lazdunski, *Mol. Pharmacol.* **20**, 107 (1981).

94. J. C. Matthews, J. E. Warnick, E. X. Albuquerque, and M. E. Eldefrawi, *Membrane Biochem.* **4**, 71 (1981).

95. C. R. Creveling, E. McNeal, J. W. Daly, and G. B. Brown, *Mol Pharmacol* **23**, 350 (1983).

96. M. Willow, E. A. Kuenzel, and W. A. Catterall, *Mol. Pharmacol.* **25**, 228 (1984).

97. W. A. Catterall, *Mol. Pharmacol.* **20**, 356 (1981).

98. L.-Y. M. Huang, G. Ehrenstein, and W. A. Catterall, *Biophys. J.* **23**, 219 (1978).

99. L.-Y. M. Huang and G. Ehrenstein, *J. Physiol.* **77**, 137 (1981).

100. B. I. Khodorov, B. Neumcke, W. Schwarz, and R. Stampfli, *Biochim. Biophys. Acta* **648**, 93 (1981).

101. S. V. Revenko and B. I. Khodorov, *Neirofiziologiya* **9**, 313 (1977).

102. A. N. Zubov, A. P. Naumov, and B. I. Khodorov, *Gen. Physiol. Biophys.* **2**, 75 (1983).

103. J. M. Dubois, M. F. Schneider, and B. I. Khodorov, *J. Gen. Physiol.* **81**, 829 (1983).

104. B. Khodorov, *Biochem. Pharmacol.* **28**, 1451 (1979).

105. B. I. Khodorov and S. V. Revenko, *Neuroscience* **4**, 1315 (1979).

106. B. I. Khodorov, *Neirofiziologiya* **12**, 317 (1980).

107. G. N. Mozhayeva, A. P. Naumov, and B. I. Khodorov, *Gen. Physiol. Biophys.* **1**, 453 (1982).

108. G. N. Mozhayeva, A. P. Naumov, and B. I. Khodorov, *Gen. Physiol. Biophys.* **1**, 281 (1982).

109. G. N. Mozhayeva, A. Naumov, and B. I. Khodorov, *Gen. Physiol. Biophys.* **1**, 463 (1982).

110. G. N. Mozhayeva, A. P. Naumov, and B. I. Khodorov, *Gen. Physiol Biophys.* **1**, 221 (1982).

111. G. N. Mozhayeva, A. P. Naumov, and B. I. Khodorov, *Neurophysiology* **15**, 357 (1983).

112. G. N. Mozhayeva, A. P. Naumov, and B. I. Khodorov, *Neurophysiology* **15**, 349 (1983).

113. G. N. Mozhayeva, A. P. Naumov, and B. I. Khodorov, *Neurophysiology* **16**, 14 (1984).

114. L.-Y. M. Huang, W. A. Catterall, and G. Ehrenstein, *J. Gen. Physiol* **73**, 839 (1979).

115. L.-Y. Huang, N. Moran, and G. Ehrenstein, *Biophys. J.* **45**, 313 (1984).

116. Y. Jacques, G. Romey, and M. Lazdunski, *Eur. J. Biochem.* **111**, 265 (1980).

117. M. Lazdunski, M. Balerna, J. Barhanin, R. Chicheportiche, M. Fosset, C. Frelin, Y. Jacques, A. Lombet, J. Pouyssequr, J. F. Renaud, G. Romey, H. Schweitz, and J. P. Vincent, *Ann. NY Acad. Sci* **358**, 169 (1980).

118. B. I. Khodorov, in *Membrane Transport Processes*, Vol. II, D. C. Tosteson, Y. A. Duchinnikov, and R. Latorre, Eds., Raven Press, New York, 1978, p. 153.

119. B. Khodorov, in *Toxins as Tools in Neurochemistry*, F. Hucho and Y. Ovchinnikov, Eds., Walter de Gruyter, Berlin, 1983, p. 35.

120. B. I. Khodorov, in *Structure and Function in Excitable Cells*, D. C. Chang, I. Tasaki, W. J. Adelman, Jr., and H. R. Leuchtag, Eds., Plenum, New York, 1983, p. 281, Chapter 14.

121. J. M. Dubois and B. I. Khodorov, *Pflugers Archiv* **395**, 55 (1982).

122. B. I. Khodorov and V. M. Bolotina, *Gen. Physiol. Biophys.* **3**, 85 (1984).

123. L. D. Zaborovskaya and B. I. Khodorov, *Gen. Physiol. Biophys.* **4**, 101 (1985).

124. P. Honerjaeger and M. Reiter, *Brit. J. Pharmacol.* **58**, 415P (1976).

125. P. Honerjaeger and M. Reiter, *Naunyn-Schmiedeberg's Arch. Pharmacol.* **299**, 239 (1977).

126. P. Honerjaeger, *Rev. Physiol. Biochem. Pharamcol.* **92**, 1 (1982).

127. G. S. Shotzberger, E. X. Albuquerque, and J. W. Daly, *J. Pharmacol. Exp. Therap.* **196**, 433 (1976).

128. M. Mensah-Dwumah and J. W. Daly, *Toxicon* **16**, 189 (1978).

129. J. Daly, E. X. Albuquerque, F. C. Kauffman, and F. Oesch. *J. Neurochem.* **19**, 2829 (1972).

130. S. Ochs and R. Worth, *Science* **187**, 1087 (1975).

131. R. M. Worth and S. Ochs, *J. Neurobiol.* **13**, 537 (1982).

132. M. H. Kumara-Siri, *J. Neurobiol.* **10**, 509 (1979).

133. D. S. Forman and W. G. Shain, *Brain Res.* **211**, 242 (1981).

134. R. J. Boegman and R. J. Riopelle, *J. Neurobiol.* **11**, 497 (1980).

135. R. J. Boegman, S. S. Deshpande, and E. X. Albuquerque, *Brain Res.* **187**, 183 (1980).

136. S. R. Max, S. S. Deshpande, and E. X. Albuquerque, *Brain Res.* **130**, 101 (1977).

137. S. R. Max, S. S. Deshpande, and E. X. Albuquerque, *J. Neurochem.* **38**, 386 (1982).

138. R. J. Boegman and B. Scarth, *Neurosci. Lett.* **24**, 261 (1981).

139. R. J. Boegman and T. W. Oliver, *Life Sci.* **27**, 1339 (1980).

140. K. K. Wan and R. J. Boegman. *Exp. Neurol.* **70**, 475 (1980).

141. K. K. Wan and R. J. Boegman, *FEBS Lett.* **112**, 163 (1980).

142. G. R. W. Moore, R. J. Boegman, D. M. Robertson, and R. J. Riopelle, *Brain Res.* **207**, 481 (1981).

143. J. H. Garcia, S. S. Deshpande, R. S. Pence, and E. X. Albuquerque, *Brain Res.* **140**, 75 (1978).

144. G. R. W. Moore, D. M. Robertson, and R. J. Boegman, *Brain Res.* **279**, 246 (1983).

145. C. S. Hudson, S. S. Deshpande, and E. X. Albuquerque, *Brain Res.* **296**, 319 (1984).

146. T. Narahashi, E. X. Albuquerque, and T. Deguchi, *J. Gen. Physiol.* **58**, 54 (1971).

147. Y. Jacques, C. Prelin, P. Vigne, G. Romey, M. Prajari, and M. Lazdunski, *Biochemistry* **20**, 6219 (1981).

148. G. Romey and M. Lazdunski, *Nature* **297**, 79 (1982).

149. M. Fosset, E. Jaimovich. E. Delpont, and M. Lazdunski, *J. Biol. Chem.* **258**, 6086 (1982).

150. D. L. Gill, E. F. Grollman, and L. D. Kohn, *J. Biol. Chem* **256**, 184 (1981).

151. D. L. Garrison, E. X. Albuquerque, J. E. Warnick, J. W. Daly, and B. Witkop, *Mol. Pharmacol.* **14**, 111 (1978).

152. D. L. Kilpatrick, R. Slepetis, and N. Kirshner, *J. Neurochem.* **37**, 125 (1981).

153. D. L. Kilpatrick, R. Slepetis, and N. Kirshner, *J. Neurochem.* **36**, 1245 (1981).

154. C. Amy and N. Kirshner, *J. Neurochem.* **39**, 132 (1982).

155. R. L. Rosenberg, S. A. Tomiko, and W. S. Agnew, *Proc. Nat. Acad. Sci. USA* **81**, 1239 (1984).

156. J. B. Weigele and R. L. Barchi, *Proc. Nat. Acad. Sci. USA* **79**, 3651 (1982).

157. E. Moczydlowski, S. S. Garber, and C. Miller, *J. Gen. Physiol.* **84**, 665 (1984).

158. E. Moczydlowksi, S. Hall, S. S. Garber, G. S. Strichartz, and C. Miller, *J. Gen. Physiol.* **84**, 687 (1984).

159. J. S. Weiner and B. Rudy, *Biophys. J.* **45**, 288a (1984).

160. F. V. Barnola and R. Villegas, *J. Gen. Physiol,* **67**, 81 (1976).

161. R. Villegas, G. M. Villegas, M. Condrescu-Guidi, and Z. Suarez-Mata, *Ann. N.Y. Acad. Sci.* **358**, 183 (1980).

162. R. Villegas, G. M. Villegas, Z. Suarez-Mata, and F. Rodriques, in *Structure and Function in Excitable Cells*, D. C. Chang, I. Tasaki, W. J. Adelman, Jr., and H. R. Leuchtag, Eds., Plenum Press, New York, 1983, p. 453, Chapter 23.

163. B. K. Krueger, J. F. Worley, III, and R. J. French, *Nature* **303**, 172 (1983).

164. R. J. French, J. F. Worley, III, and B. F. Krueger, *Biophys. J.* **45**, 301 (1984).

165. S. Seiler and S. Fleischer, *J. Biol. Chem.* **257**, 13,862 (1982).

166. J. E. Warnick, E. X. Albuquerque, R. Onur, S.-E. Jansson, J. Daly, T. Tokuyama, and B. Witkop, *J. Pharmacol. Exp. Therap.* **193**, 232 (1975).

167. J. A. Waters, C. R. Creveling, and B. Witkop, *J. Med. Chem.* **17**, 488 (1974).

168. G. B. Brown and R. J. Bradley, *J. Neurosci. Methods,* **13**, 119 (1985).

169. D. Bar-Sagi and J. Prives, *J. Cell, Physiol.* **114**, 77 (1983).

170. T. Masutani, I. Seyama, T. Narahashi, and J. Iwasa, *J. Pharmacol. Exp. Ther.* **217**, 812 (1981).

171. E. M. Kosower, *FEBS Lett.* **163**, 161 (1983).

172. J. R. Smythies, F. Benington, and R. D. Morin, *Nature* **231**, 188 (1971).

173. J. R. Smythies, F. Benington, R. J. Bradley, W. F. Bridgers, and R. D. Morin, *J. Theor. Biol.* **43**, 29 (1974).

174. J. R. Smythies, *Adv. Cytopharmacol.* **3**, 317 (1979).

175. J. R. Smythies, *Ala. J. Med. Sci.* **15**, 372 (1978).

176. M. Noda, S. Shimizu, T. Tanabe, T. Takai, T. Kayano. T. Ikeda, H. Takashi, H. Na-kayama, Y. Kanaoka, N. Minamino, K. Kangawa, H. Matsuo, M. A. Raftery, T. Hirose, S. Inayama, H. Hayashida. T. Miyata, and S. Numa, *Nature* **312**, 121 (1984).

177. W. A. Catterall, C. S. Morrow, G. B. Brown, and J. W. Daly, *J. Biol. Chem.* **256**, 8922 (1981).

178. G. B. Brown, J. A. Johnston, and L. C. Tollbert, *Toxicon* **21**, 699 (1983).

179. C. R. Creveling, E. T. McNeal, G. A. Lewandowski, M. Rafferty, E. H. Harrison, A. E. Jacobson, K. C. Rice, and J. W. Daly, *J. Neuropeptides*, **5**, 353 (1985).

180. E. T. McNeal, G. A. Lewandowski, J. W. Daly, and C. R. Creveling, *J. Med. Chem.*, **28**, 381 (1985).

181. S. W. Postma and W. A. Catterall, *Mol. Pharmacol.* **25**, 219 (1984).

182. M. Willow and W. A. Catterall, *Mol. Pharmacol.* **22**, 627 (1982).

183. J. W. Daly, *Proceedings of the Naito Symposium on Natural Products and Biological Activities*, The Naito Foundation, Univ. Tokyo Press, 1984, p. 121.

184. C. R. Creveling, G. A. Lewandowski, and J. W. Daly, in preparation (1985).

185. N. Soldatov, T. Prosolova, V. Kovalenko, A. Petrenko, E. Grishin, and Y. Ovchinnikov, in *Toxins as Tools in Neurochemistry*, F. Hucho and Y. A. Ovchinnikov, Eds., Walter de Gruyter, New York, 1983, p. 47.

186. G. B. Brown and R. W. Olsen, *Neurosci. Abst.* **1984**, 865.

187. K. J. Angelides, T. J. Nutter, L. W. Elmer, and E. S. Kempner, *J. Biol. Chem.* **260**, 3431 (1985).

188. E. X. Albuquerque, J. W. Daly, and B. Witkop, *Science* **172**, 995 (1971).

189. E. X. Albuquerque, *Fed. Proc.* **31**, 1133 (1972).

189. J. Daly and B. Witkop, *Clinical Toxicology* **4**, 331 (1971).

191. T. Narahashi, *Adv. Exp. Med. Biol.* **84**, 407 (1977).

192. T. Narahashi, in *Advances in Pharmacology and Toxicology*, Vol. 3, H. Yoshida, Y. Hagihara, and S. Ebashi, Eds., Pergamon, New York, 1982, p. 3.

193. J. W. Daly, *J. Toxicol.-Toxin Rev.* **1**, 33 (1982).

194. H. J. Adams, A. R. Mastri, D. Doherty, Jr., and D. Charron, *Pharmacol. Res. Comm.* **10**, 719 (1978).

195. J. H. Brown, *Mol. Pharmacol.* **20**, 113 (1981).

196. J. S. Roseen and F. A. Fuhrman, *Toxicon* **9**, 411 (1971).

197. W. A. Catterall and J. Coppersmith, *Mol. Pharmacol,* **20**, 533 (1981).

198. W. A. Catterall, *Biochem. Biophys. Res. Comm.* **68**, 136 (1976).

199. W. A. Catterall, *J. Biol. Chem.* **251**, 5528 (1976).

200. W. A. Catterall, *Dev. Biol.* **78**, 222 (1980).

201. C. M. Tang, G. R. Strichartz, and R. K. Orkand, *J. Gen. Physiol,* **74**, 629 (1979).

202. T. Narahashi, T. Deguchi, and E. X. Albuquerque, *Nature New Biology* **229**, 221 (1971).

203. R. Henderson and G. Strichartz, *J. Physiol. (London)* **238**, 329 (1974).

204. J. E. Warnick and E. X. Albuquerque, *Fed. Proc.* **37**, 2811 (1978).

205. E. X. Albuquerque, J. E. Warnick, and L. Guth, *Prog. Clin. Biol. Res.* **39**, 41 (1980).

206. R. J. Boegman and E. X. Albuquerque, *J. Neurobiology* **11**, 283 (1980).

207. R. J. Boegman and R. J. Riopelle, *Neurosci. Lett.* **18**, 143 (1980).

208. U. Otten and H. Thoenen, *Neurosci. Lett.* **2**, 93 (1976).

209. E. Bartels de Bernal, R. Cadena, and E. Diaz, *Colombia Medica* **13**, 2 (1982).

210. E. Bartels-Bernal, E. Diaz, R. Cadena, J. Ramos, and J. W. Daly, *Cell. Mol. Neurobiol.* **3**, 203 (1984).

211. H. A. Lester, *J. Gen. Physiol.* **72**, 847 (1978).

212. M. E. Eldefrawi, N. Shaker, N. A. Mansour, J. E. Warnick, and E. X. Albuquerque, *Life Sci.* **29**, 1033 (1981).

213. W. A. Catterall, *J. Biol. Chem.* **252**, 8669 (1971).

214. W. A. Catterall, *J. Biol. Chem.* **252**, 8660 (1977).

215. W. A. Catterall, *Adv. Cytopharmacol,* **3**, 305 (1979).

216. W. A. Catterall and R. Ray, *J. Supramol. Struct.* **5**, 397 (1976).

217. W. A. Catterall and L. Beress, *J. Biol. Chem.* **253**, 7393 (1978).

218. W. A. Catterall and D. A. Beneski, *J. Supramol. Struct.* **14**, 295 (1980).

219. W. A. Catterall, R. Ray, and C. S. Morrow, *Proc. Nat. Acad. Sci. USA* **73**, 2682 (1976).

220. R. Ray and W. A. Catterall, *J. Neurochem.* **31**, 397 (1978).

221. G. J. West and W. A. Catterall, *Proc. Nat. Acad. Sci. USA* **76**, 4136 (1979).

222. P. J. Conn and D. C. Rogers, *Endocrinology* **107**, 2133 (1980).

223. H. Shimizu and J. W. Daly, *Eur. J. Pharmacol.* **17**, 240 (1972).

224. H. Shimizu, C. R. Creveling, and J. Daly, *Proc. Nat. Acad. Sci. USA* **65**, 1033 (1970).

225. H. Shimizu, C. R. Creveling, and J. W. Daly, *Mol. Pharmacol.* **6**, 184 (1970).

226. H. Shimizu, C. R. Creveling, and J. W. Daly, *Adv. Biochem. Psychopharmacol.* **3**, 135 (1970).

227. M. Huang, H. Shimizu, and J. W. Daly, *J. Med. Chem.* **15**, 462 (1972).

228. M. Huang and J. W. Daly, *J. Neurochem.* **23**, 393 (1974).

229. J. W. Daly, E. McNeal, C. Partington, M. Neuwirth, and C. R. Creveling, *J. Neurochem.* **35**, 326 (1980).

230. H. R. Wagner and J. N. Davis, *Proc. Nat. Acad. Sci. USA* **76**, 2057 (1979).

231. E. T. McNeal, C. R. Creveling, and J. W. Daly, *J. Neurochem.* **35**, 338 (1980).

232. M. J. Mullin, and W. A. Hunt, *J. Pharmacol. Exp. Therap.* **232**, 413 (1985).

233. R. Ray, C. S. Morrow, and W. A. Catterall, *J. Biol. Chem.* **253**, 7307 (1978).

234. S. Ramos, E. F. Grollman, P. S. Lazo, S. A. Dyer, W. H. Habig, M. C. Hardegree, H. R. Kaback, and L. D. Kohn, *Proc. Nat. Acad. Sci. USA* **76**, 4783 (1979).

235. J. C. Matthews, E. X. Albuquerque, and M. E. Eldefrawi, *Life Sci.* **25**, 1651 (1979).

236. B. K. Krueger and M. P. Blaustein, *J. Gen. Physiol.* **76**, 287 (1980).

237. C. R. Creveling, E. T. McNeal, D. H. McCulloh, and J. W. Daly, *J. Neurochem.* **35**, 922 (1980).

238. G. Ponzio, Y. Jacques, C. Frelin, R. Chicheportiche, and M. Lazundski, *FEBS Lett.* **121**, 265 (1980).

239. J. P. Vincent, M. Balerna, J. Barhanin, M. Fosset, and M. Lazdunski, *Proc. Nat. Acad. Sci. USA* **77**, 1646 (1980).

240. K. D. McCarthy and T. K. Harden, *J. Pharmacol. Exp. Therap.* **216**, 183 (1981).

241. M. M. Tamkun and W. A. Catterall, *Mol. Pharmacol.* **19**, 78 (1981).

242. S. M. Ghiasuddin and D. M. Soderlund, *Comparative Biochem. and Physiol.* **77**, 267 (1984).

243. R. W. Holz and J. T. Coyle, *Mol. Pharmacol.* **10**, 746 (1974).

244. F. Hery, G. Simonnet, S. Bourgoin, P. Soubrie, F. Artaud, M. Hamon, and J. Glowinski, *Brain Res.* **169**, 317 (1979).

245. J. W. Daly, I. Karle, C. W. Myers, T. Tokuyama, J. A. Waters, and B. Witkop, *Proc. Nat. Acad. Sci. USA* **68**, 1870 (1971).

246. I. L. Karle, *J. Am. Chem. Soc.* **95**, 4036 (1973).

247. T. Tokuyama, K. Uenoyama, G. Brown, J. W. Daly, and B. Witkop, *Helv. Chim. Acta* **57**, 2597 (1974).

248. J. W. Daly, B. Witkop, T. Tokuyama, T. Nishikawa, and I. L. Karle, *Helv. Chim Acta* **60**, 1128 (1977).

249. T. Tokuyama, J. Yamamoto, J. W. Daly, and R. J. Highet, *Tetrahedron* **39**, 49 (1983).

250. C. W. Myers and J. W. Daly, *Bull. Am. Mus. Nat. History* **157**, 173 (1976).

251. J. W. Daly, N. Whittaker, T. F. Spande, R. J. Highet, D. Feigl, N. Nishimori, T. Tokuyama, and C. W. Myers, *J. Nat. Products Chem. (Lloydia)*, in press (1986).

252. C. W. Myers, and J. W. Daly, *Occasional Papers Mus. Nat. History, Univ. Kansas*, Number 59, p. 1 (1976).

253. C. W. Myers and J. W. Daly, *Am. Mus. Novitates*, Number 2674, p. 1 (1979).

254. C. W. Myers and J. W. Daly, *Am. Mus. Novitates*, Number 2692, p. 1 (1980).

255. C. W. Myers, J. W. Daly, and V. Martinez, *Am. Mus. Novitates*, Number 2783, p. 1 (1984).

256. C. W. Myers and J. W. Daly, *Am. Mus. Novitates*, in preparation (1985).

257. C. W. Myers and J. W. Daly, *Am. Mus. Novitates*, in preparation (1985).

258. M. Edwards, Ph.D. Thesis, University of Maryland, College Park, MD, October 1984.

259. J. W. Daly, N. Whittaker, and C. W. Myers, *Toxicon*, in preparation (1986).

260. J. W. Daly, *Proc. 4th Asian Symp. Med. Plants Spices*, Akorn Charoen-Tat Publishing House, Bangkok, 1980, p. 49.

261. J. W. Daly, R. J. Highet and C. W. Myers, *Toxicon* **22**, 905 (1984).

262. K. Takahashi, B. Witkop, A. Brossi, A. C. Maleque, and E. X. Albuquerque, *Helv. Chim Acta* **65**, 252 (1982).

263. T. Fukuyama, L. V. Dunkerton, M. Aratani, and Y. Kishi, *J. Org. Chem.* **40**, 2011 (1975).

264. M. Aratani, L. V. Dunkerton, T. Fukuyama, Y. Kishi, H. Kakoi, S. Sugiura, and S. Inoue, *J. Org. Chem.* **40**, 2009 (1975).

265. S. C. Carey, M. Aratani and Y. Kishi, *Tetrahedron Lett.* **26,** 5887 (1985).

266. Y. Inubushi and T. Ibuka, *Heterocycles* **17,** 507 (1982).

267. E. J. Corey, J. F. Arnett, and G. N. Widiger, *J. Am. Chem. Soc.* **97,** 430 (1975).

268. P. Duhamel and M. Kotera, *J. Org. Chem.* **47,** 1688 (1982).

269. E. J. Corey, M. Petrizilka, and Y. Ueda, *Helv. Chim. Acta* **60,** 2294 (1977); *Tetrahedron Lett.*, 4343 (1975).

270. T. Ibuka, Y. Mitsui, K. Hayashi, H. Minakata, and Y. Inubushi, *Tetrahedron Lett.* **22,** 4425 (1981).

271. T. Ibuka, H. Minakata, Y. Mitsui, E. Tabushi, T. Taga, and Y. Inubushi, *Chem. Lett.*, 1409 (1981).

272. T. Ibuka, H. Minakata, Y. Mitsui, K. Hayashi, T. Taga, and Y. Inubushi, *Chem. Pharm. Bull.* **30,** 2840 (1982).

273. S. A. Godleski, D. J. Heacock, J. D. Meinhart, and S. van Wallendael, *J. Org. Chem.* **48,** 2101 (1983).

274. W. Kissing and B. Witkop, *Chem. Ber.* **108,** 1623 (1975).

275. F. T. Bond, J. E. Stemke, and D. W. Powell, *Synthetic Commun.* **5,** 427 (1975).

276. W. Carruthers and S. A. Cumming, *J. Chem. Soc. Perkin Trans. I* **10,** 2383 (1983); *J. Chem. Soc. Chem. Commun.*, 360 (1983).

277. S. A. Godleski, J. D. Meinhart, D. J. Miller, and S. van Wallendael, *Tetrahedron Lett.* **22,** 2247 (1981).

278. S. A. Godleski and D. J. Heacock, *J. Org. Chem.* **47,** 4820 (1982).

279. A. J. Pearson and P. Ham, *J. Chem. Soc. Perkin Trans, I,* 1421 (1983); see also A. J. Pearson, P. Ham, and D. C. Rees, *Tetrahedron Lett.* **21,** 4639 (1980); *J. Chem. Soc. Perkin I,* 489 (1982).

280. A. B. Holmes, K. Russell, E. S. Stern, M. E. Stubbs and N. K. Wellard, *Tetrahedron Lett.* **25,** 4163 (1984).

281. L. E. Overman, *Tetrahedron Lett.* 1149 (1975).

282. E. J. Corey and R. D. Balanson, *Heterocycles* **5,** 445 (1976).

283. E. Gössinger, R. Imhof, and H. Wehrli, *Helv. Chim. Acta* **58,** 96 (1975).

284. J. J. Tufariello and E. J. Trybulski, *J. Org. Chem.* **39,** 3378 (1974).

285. E. R. Koft and A. B. Smith, III, *J. Org. Chem.* **49,** 832 (1984).

286. G. E. Keck and J. B. Yates, *J. Org. Chem.* **47,** 3590 (1982).

287. G. E. Keck and J. B. Yates, *J. Am. Chem. Soc.* **104,** 5829 (1982).

288. H. E. Schoemaker and W. N. Speckamp, *Tetrahedron Lett.*, 1515 (1978).

289. H. E. Schoemaker and W. N. Speckamp, *Tetrahedron Lett.*, 4841 (1978).

290. H. E. Schoemaker and W. N. Speckamp, *Tetrahedron* **36,** 951 (1980).

291. D. A. Evans and E. W. Thomas, *Tetrahedron Lett.* 411 (1979).

292. D. A. Evans, E. W. Thomas, and R. E. Cherpeck, *J. Am. Chem. Soc.* **104,** 3695 (1982).

293. K. Takahashi, A. E. Jacobson, C.-P. Mak, B. Witkop, A. Brossi, E. X. Albuquerque, J. E. Warnick, M. A. Maleque, A. Bavoso, and J. W. Silverton, *J. Med. Chem.* **25,** 919 (1982).

294. E. J. Corey, Y. Ueda, and R. A. Ruden, *Tetrahedron Lett.* 4347 (1975).

295. J. J. Venit and P. Magnus, *Tetrahedron Lett.* 4815 (1980).

296. M. Glanzmann, C. Karalai, B. Ostersehlt, U. Schön, C. Frese, and E. Winterfeldt, *Tetrahedron,* **38,** 2805 (1982); see also E. Winterfeldt, *Heterocycles,* **12,** 1631 (1979).

297. D. H. Gnecco Medina, D. S. Grierson, and H.-P. Husson, *Tetrahedron Lett.* **24,** 2099 (1983).

298. H. Harris, D.-S. Grierson, and H.-P. Husson, *Tetrahedron Lett.* **22,** 1511 (1981).

299. E. J. Corey and R. A. Ruden, *Tetrahedron Lett.* 1495 (1973).
300. Y. Yamakado, M. Ishiguro, N. Ikeda, and H. Yamamoto, *J. Am. Chem. Soc.* **103**, 5568 (1981).
301. E. J. Corey and C. Rücker, *Tetrahedron Lett.* **23**, 719 (1982).
302. A. B. Holmes, R. A. Raphael, and N. K. Wellard, *Tetrahedron Lett.* 1539 (1976).
303. A. O. King, N. Okukado, and E. Negishi, *Chem. Comm.* **1977**, 683.
304. R. S. Aronstam, C. T. King, E. X. Albuquerque, J. W. Daly, and D. M. Feigl, *Biochem. Pharmacol.*, **34**, 3037 (1985).
305. C. E. Spivak, M. A. Maleque, A. C. Oliveria, L. Masukawa, T. Tokuyama, J. W. Daly, and E. X. Albuquerque, *Mol. Pharmacol.* **21**, 351 (1982).
306. A. J. Lapa, E. X. Albuquerque, J. M. Sarvey, J. Daly, and B. Witkop, *Exp. Neurol.* **47**, 558 (1975).
307. E. X. Albuquerque, E. A. Barnard, T. H. Chiu, A. J. Lapa, J. O. Dolly, S.-E. Jansson, J. Daly, and B. Witkop, *Proc. Nat. Acad. Sci. USA* **70**, 949 (1973).
308. E. X. Albuquerque, K. Kuba, and J. Daly, *J. Pharmacol. Exp. Therap.* **189**, 513 (1974).
309. E. X. Albuquerque, K. Kuba, A. J. Lapa, J. W. Daly, and B. Witkop, in *Exploratory Concepts in Muscular Dystrophy*, Vol. II, A. T. Molhorat, Ed., Excerpta Medica, Amsterdam, 1974, p. 585.
310. L. M. Masukawa and E. X. Albuquerque, *J. Gen. Physiol.* **72**, 351 (1978).
311. E. X. Albuquerque and P. W. Gage, *Proc. Nat. Acad. Sci. USA* **75**, 1596 (1978).
312. E. X. Albuquerque, P. W. Gage, and A. C. Oliveira, *J. Physiol. (London)* **297**, 423 (1979).
313. W. C.-S. Wu and M. A. Raftery, *Biochem. Biophys. Res. Commun.* **99**, 436 (1981).
314. W. Burgermeister, W. A. Catterall, and B. Witkop, *Proc. Nat. Acad. Sci. USA* **74**, 5754 (1977).
315. G. Kato, M. Glavinovic, J. Henry, K. Krnjevic, E. Puil, and B. Tattrie, *Croatica Chemica Acta* **47**, 439 (1975); G. Kato and J.-P. Changeux, *Mol. Pharmacol.* **12**, 92 (1976).
316. N. D. Boyd and J. B. Cohen, *Biochemistry* **23**, 4023 (1984).
317. J. Elliott and M. A. Raftery, *Biochem. Biophys. Res. Comm.* **77**, 1347 (1977).
318. S. M. Sine and P. Taylor, *J. Biol. Chem.* **257**, 8106 (1982).
319. T. Heidmann, R. E. Oswald, and J.-P. Changeux, *Biochemistry* **22**, 3112 (1983).
320. T. Heidmann and J.-P. Changeux, *FEBS Lett.* **131**, 239 (1981).
321. B. M. Conti-Tronconi and M. A. Raftery, *Ann. Rev. Biochem.* **51**, 491 (1982).
322. T. Heidmann, R. Oswald, and J.-P. Changeux, *Comptes Rendus Sci. (Paris)* **295**, Serie III, 345 (1982).
323. R. Oswald and J.-P. Changeux, *Proc. Nat. Acad. Sci. USA* **78**, 3925 (1981); *Biochemistry* **20**, 7166 (1981).
324. L. P. Wennogle, R. Oswald, T. Saitoh, and J.-P. Changeux, *Biochemistry* **20**, 2492 (1981).
325. R.-R. J. Kaldany and A. Karlin, *J. Biol. Chem.* **258**, 6232 (1983).
326. R. N. Cox, R.-R. J. Kaldany, M. DiPaola, and A. Karlin, *J. Biol. Chem.*, in press (1985).
327. V. Witzemann and M. Raftery, *Biochemistry* **17**, 3598 (1978).
328. M. I. Schimerlik, U. Quast, and M. A. Raftery, *Biochemistry* **18**, 1902 (1979).
329. S. G. Blanchard and M. A. Raftery, *Proc. Nat. Acad. Sci. USA* **76**, 81 (1979).
330. S. M. J. Dunn, S. G. Blanchard, and M. A. Raftery, *Biochemistry* **20**, 5617 (1981).
331. S. M. J. Dunn and M. A. Raftery, *Proc. Nat. Acad. Sci USA* **79**, 6757 (1982).
332. M. Glavinovic, J. L. Henry, G. Kato, K. Krnjevic, and E. Puil, *Canad. J. Physiol. Pharmacol.* **52**, 1220 (1974).

333. D. B. Sattelle and J. A. David, *Neurosci. Lett.* **43**, 37 (1983).

334. H. Betz, *Neurosci. Lett.* **33**, 152 (1982).

335. W. Burgermeister, W. L. Klein, M. Nirenberg, and B. Witkop, *Mol. Pharmacol.* **14**, 751 (1978).

336. R. S. Aronstam, A. T. Eldefrawi, and M. E. Eldefrawi, *Biochem. Pharmacol.* **29**, 1311 (1980).

337. C. Cremo and M. I. Schimerlik, *Arch. Biochem. Biophys.* **224**, 506 (1983).

338. E. X. Albuquerque and A. C. Oliveira, *Adv. Cytopharmacol.* **3**, 197 (1979).

339. C. E. Spivak, M. A. Maleque, K. Takahashi, A. Brossi, and E. X. Albuquerque, *FEBS Lett.* **163**, 189 (1983).

340. M. A. Maleque, A. Brossi, B. Witkop, S. A. Godleski, and E. X. Albuquerque, *J. Pharmacol. Exp. Therap.* **229**, 72 (1984).

341. M. A. Maleque, K. Takahashi, B. Witkop, A. Brossi, and E. X. Albuquerque, *J. Pharmacol. Exp. Therap.* **230**, 619 (1984).

342. R. Anwyl and T. Narahashi, *Brit. J. Pharmacol.* **68**, 611 (1980).

343. A. T. Eldefrawi, M. E. Eldefrawi, E. X. Albuquerque, A. C. Oliveira, N. Mansour, M. Adler, J. W. Daly, G. B. Brown, W. Burgermeister, and B. Witkop, *Proc. Nat. Acad. Sci. USA* **74**, 2172 (1977).

344. M. E. Eldefrawi, A. T. Eldefrawi, N. A. Mansour, J. W. Daly, B. Witkop, and E. X. Albuquerque, *Biochemistry* **17**, 5474 (1978).

345. E. X. Albuquerque, A. T. Eldefrawi, M. E. Eldefrawi, N. A. Mansour, and M.-C. Tsai, *Science* **199**, 788 (1978).

346. E. X. Albuquerque, M.-C. Tsai, R. S. Aronstam, A. T. Eldefrawi, and M. E. Eldefrawi, *Mol. Pharmacol.* **18**, 167 (1980).

347. E. X. Albuquerque, M.-C. Tsai, R. S. Aronstam, B. Witkop, A. T. Eldefrawi, and M. E. Eldefrawi, *Proc. Nat. Acad. Sci. USA* **77**, 1224 (1980).

348. C. T. King, Jr., and R. S. Aronstam, *Eur. J. Pharmacol.* **90**, 419 (1983).

349. E. X. Albuquerque, M. Adler, C. E. Spivak, and L. Aguayo, *Ann. NY Acad. Sci.* **358**, 204 (1980).

350. M. Adler, A. C. Oliveira, E. X. Albuquerque, N. A. Mansour, and A. T. Eldefrawi, *J. Gen. Physiol.* **74**, 129 (1979).

351. M. Adler, A. C. Oliveira, M. E. Eldefrawi, A. T. Eldefrawi, and E. X. Albuquerque, *Proc. Nat. Acad. Sci. USA* **76**, 531 (1979).

352. R. S. Aronstam, A. T. Eldefrawi, I. N. Pessah, J. W. Daly, E. X. Albuquerque, and M. E. Eldefrawi, *J. Biol. Chem.* **256**, (1981).

353. R. S. Aronstam, *Life Sci.* **28**, 59 (1981).

354. R. S. Aronstam and B. Witkop, *Proc. Nat. Acad. Sci. USA* **78**, 4639 (1981).

355. M. A. Abbassy, M. E. Eldefrawi, and A. T. Eldefrawi, *Life Sci.* **31**, 1547 (1982).

356. M. A. Abbassy, M. E. Eldefrawi, and A. T. Eldefrawi, *Pesticide Biochem. Physiol.* **19**, 299 (1983).

357. J. S. Carp, R. S. Aronstam, B. Witkop, and E. X. Albuquerque, *Proc. Nat. Acad. Sci. USA* **80**, 31 (1983).

358. E. F. El-Fakahany, A. T. Eldefrawi, and M. E. Eldefrawi, *J. Pharmacol. Exp. Therap.* **221**, 694 (1982).

359. E. F. El-Fakahany, E. R. Miller, M. A. Abbassy, A. T. Eldefrawi, and M. E. Eldefrawi, *J. Pharmacol. Exp. Therap.* **224**, 289 (1983).

360. M. E. Eldefrawi and A. T. Eldefrawi, *Adv. Cytopharmacol.* **3**, 213 (1979).

361. M. E. Eldefrawi and A. T. Eldefrawi, *Ann. NY Acad. Sci.* **358**, 239 (1980).

362. A. T. Eldefrawi, N. M. Bakry, M. E. Eldefrawi, M.-C. Tsai, and E. X. Albuquerque, *Mol. Pharmacol.* **17,** 172 (1980).

363. A. T. Eldefrawi, E. R. Miller, and M. E. Eldefrawi, *Biochem. Pharmacol.* **31,** 1819 (1982).

364. A. T. Eldefrawi, E. R. Miller, D. L. Murphy, and M. E. Eldefrawi, *Mol. Pharmacol.* **22,** 72 (1982).

365. M. E. Eldefrawi, R. S. Aronstam, N. M. Bakry, A. T. Eldefrawi, and E. X. Albuquerque, *Proc. Nat. Acad. Sci. USA* **77,** 2309 (1980).

366. M. E. Eldefrawi, D. S. Copia, C. S. Hudson, J. Rash, N. A. Mansour, A. T. Eldefrawi, and E. X. Albuquerque, *Exp. Neurol.* **64,** 428 (1979).

367. M. E. Eldefrawi, J. E. Warnick, G. G. Schofield, E. X. Albuquerque, and A. T. Eldefrawi, *Biochem. Pharmacol.* **30,** 1391 (1981).

368. M. E. Eldefrawi, A. T. Eldefrawi, R. S. Aronstam. M. A. Maleque, J. E. Warnick, and E. X. Albuquerque, *Proc. Nat. Acad. Sci. USA* **77,** 7458 (1980).

369. S. G. Blanchard, J. Elliott, and M. A. Raftery, *Biochemistry* **18,** 5880 (1979).

370. J. Elliott and M. A. Raftery, *Biochemistry* **18,** 1868 (1979).

371. J. Elliott, S. M. J. Dunn, S. G. Blanchard, and M. A. Raftery, *Proc. Nat. Acad. Sci. USA* **76,** 2576 (1979).

372. G. J. Pascuzzo, A. Akaike, M. A. Maleque, K. Shaw, R. S. Aronstam, D. L. Rickett, and E. X. Albuquerque, *Mol. Pharmacol.* **25,** 92 (1984).

373. A. Desouki, A. T. Eldefrawi, and M. E. Eldefrawi, *Exp. Neurol.* **73,** 440 (1981).

374. T. N. Tiedt, E. X. Albuquerque, N. M. Bakry, M. E. Eldefrawi, and A. T. Eldefrawi, *Mol. Pharmacol.* **16,** 909 (1979).

375. M.-C. Tsai, N. A. Mansour, A. T. Eldefrawi, M. E. Eldefrawi, and E. X. Albuquerque, *Mol. Pharmacol.* **14,** 787 (1978).

376. M.-C. Tsai, A. C. Oliveira, E. X. Albuquerque, M. E. Eldefrawi, and A. T. Eldefrawi, *Mol. Pharmacol.* **16,** 382 (1979).

377. N. Shaker, A. T. Eldefrawi, E. R. Miller, and M. E. Eldefrawi, *Mol. Pharmacol.* **20,** 511 (1981).

378. N. Shaker, A. T. Eldefrawi, L. G. Aguayo, J. E. Warnick, E. X. Albuquerque, and M. E. Eldefrawi, *J. Pharmacol. Exp. Therap.* **220,** 172 (1982).

379. S. R. Ikeda, R. S. Aronstam, J. W. Daly, Y. Aracava, and E. X. Albuquerque, *Mol. Pharmacol.* **26,** 293 (1984).

380. C. Souccar, W. A. Varanda, R. S. Aronstam. J. W. Daly, and E. X. Albuquerque, *Mol. Pharmacol.* **25,** 395 (1984).

381. J. E. Warnick, P. J. Jessup, L. E. Overman, M. E. Eldefrawi, Y. Nimit, J. W. Daly, and E. X. Albuquerque, *Mol. Pharmacol.* **22,** 565 (1982).

382. A. Sobel, T. Heidmann, and J.-P. Changeux, *Compt. Rendus Acad. Sci. (Paris)* **285D,** 1255 (1977).

383. A. Sobel, T. Heidmann, J. Hofler, and J.-P. Changeux, *Proc. Nat. Acad. Sci. USA* **75,** 510 (1978).

384. R. R. Neubig, E. K. Krodel, N. D. Boyd, and J. B. Cohen, *Proc. Nat. Acad. Sci. USA* **76,** 690 (1979).

385. E. K. Krodel, R. A. Beckman, and J. B. Cohen, *Mol. Pharmacol.* **15,** 294 (1979).

386. J. Vignon, J. P. Vincent, J. N. Bidard, J. M. Kamanka, P. Genetste, S. Monier, and M. Lazdunski, *Eur. J. Pharmacol.* **81,** 531 (1982).

387. E. X. Albuquerque, L. G. Aguayo, J. E. Warnick, H. Weinstein, S. D. Glick, S. Maayani, R. K. Ickowicz and M. P. Blaustein, *Proc. Nat. Acad. Sci. USA* **78,** 7792 (1981).

388. G. T. Bolger, M. F. Rafferty, and P. Skolnick, *J. Neurochem.,* in press (1985).

389. J. O. Dolly, E. X. Albuquerque, J. M. Sarvey, B. Mallick, and E. A. Barnard, *Mol. Pharmacol.* **13,** 1 (1977).

390. J. E. Warnick, E. X. Albuquerque, A. J. Lapa, J. Daly, and B. Witkop, *Proc. Sixth Int. Congr. Pharmacol.* **1,** 67 (1976).

391. J. L. Popot, J. Cartaud, and J.-P. Changeux, *Eur. J. Biochem.* **118,** 203 (1981).

392. E. X. Albuquerque and C. E. Spivak, in *Natural Products and Drug Development*, P. Krogsgaard-Larsen, S. Brogger-Christensen, and H. Koford, Eds., Alfred Benzon Symposium 20, Munksgaard, Copenhagen, 1984, p. 301.

393. M. E. Eldefrawi, in *Myasthenia Gravis*, E. A. Albuquerque and A. T. Eldefrawi, Eds., Chapman and Hall, London, 1983, p. 189.

394. T. Heidmann and J.-P. Changeux, *Ann. Rev. Biochem.* **47,** 317 (1978).

395. C. E. Spivak and E. X. Albuquerque, in *Progress in Cholinergic Biology: Model Cholinergic Synapses*, I. Hanin and A. M. Goldberg, Eds., Raven, New York, 1982, p. 323.

396. R. Fujimoto and Y. Kishi, *Tetrahedron Lett.* 4197 (1981).

397. L. E. Overman and R. L. Freerks, *J. Org. Chem.* **46,** 2833 (1981).

398. L. E. Overman and C. Fukaya, *J. Am. Chem. Soc.* **102,** 1454 (1980).

399. R. Fujimoto, Y. Kishi, and J. Blount, *J. Am. Chem. Soc.* **102,** 7154 (1980).

400. L. E. Overman, D. Lesuisse, and M. Hashimoto, *J. Am. Chem. Soc.* **105,** 5373 (1983).

401. Y. Ito, E. Nakajo, M. Nakatsuka, and T. Saegusa, *Tetrahedron Lett.* **24,** 2881 (1983).

402. T. Ibuka, G.-N. Chu, and F. Yoneda, *J. Chem. Soc. Chem. Commun.* **1984,** 597.

403. D. J. Hart, *J. Org. Chem.* **46,** 3576 (1981).

404. D. J. Hart and K. Kanai, *J. Am. Chem. Soc.* **105,** 1255 (1983).

405. D. J. Hart, *J. Am. Chem. Soc.* **102,** 397 (1980).

406. D. J. Hart, and Y.-M. Tsai, *J. Am. Chem. Soc.* **104,** 1430 (1982).

407. G. Habermehl and O. Thurau, *Naturwiss.* **67,** 193 (1980).

408. C. Souccar, W. A. Varanda, J. W. Daly, and E. X. Albuquerque, *Mol. Pharmacol.* **25,** 384 (1984).

409. T. F. Spande, J. W. Daly, D. J. Hart, Y.-M. Tsai, and T. L. Macdonald, *Experientia* **37,** 1242 (1981).

410. P. E. Sonnet, D. A. Netzel, and R. Mendoza, *J. Heterocyclic Chem.* **16,** 1041 (1979); J. E. Oliver and P. E. Sonnet, *J. Org. Chem.* **39,** 2662 (1974); P. E. Sonnet and J. E. Oliver, *J. Heterocylic Chem.* **12,** 289 (1975).

411. J. Royer and H.-P. Husson, *Tetrahedron Lett.*, **26,** 1515 (1985).

412. T. Tokuyama, N. Nishimori, I. L. Karle, M. W. Edwards, and J. W. Daly, *Tetrahedron*, Submitted (1985).

413. T. L. Macdonald, *J. Org. Chem.* **45,** 193 (1980).

414. M. Natsume, personal communication.

415. O. E. Edwards, A. Greves, and W. W. Sy, *Can. J. Chem.*, in preparation (1985).

416. T. H. Jones, R. J. Highet, M. S. Blum, and H. M. Fales, *J. Chem. Ecol.* **10,** 1233 (1984).

417. M. Przybylska and F. R. Ahmed, *Can. J. Chem.*, in preparation (1985).

418. M. Ogawa, J Nakajima, and M. Natsume, *Heterocycles* **19,** 1247 (1982).

419. D. J. Hart and Y.-M. Tsai, *J. Org. Chem.* **47,** 4403 (1982).

420. R. V. Stevens and A. W. M. Lee, *J. Chem. Soc. Chem. Commun.* **1982,** 103.

421. T. F. Spande, unpublished results.

422. R. S. Aronstam, J. W. Daly, and E. X. Albuquerque, unpublished results.

423. F. J. Ritter, I. E. M. Rotgans, E. Talman, P. E. J. Verwiel, and F. Stein, *Experientia* **29,** 530 (1973).

424. J. W. Daly and C. W. Myers, *Science* **156**, 970 (1967).

425. J. W. Daly, T. Tokuyama, T. Fujiwara, R. J. Highet, and I. L. Karle, *J. Am. Chem. Soc.* **102**, 830 (1980).

426. J. W. Daly, T. Tokuyama, G. Habermehl, I. L. Karle, and B. Witkop, *Justus Liebigs Ann. Chem.* **729**, 198 (1969).

427. T. Tokuyama, K. Shimada, M. Uemura, and J. W. Daly, *Tetrahedron Lett.* **23**, 2121 (1982).

428. L. E. Overman and R. J. McCready, *Tetrahedron Lett.* **23**, 2355 (1982).

429. M. Uemura, K. Shimada, T. Tokuyama, and J. W. Daly, *Tetrahedron Lett.* **23**, 4369 (1982).

430. T. Tokuyama, J. W. Daly, and R. J. Highet, *Tetrahedron* **40**, 1183 (1984).

431. R. J. Highet, J. W. Daly, T. Fujiwara, and T. Tokuyama, *Planta Medica* **39**, 260 (1980).

432. L. E. Overman and N.-H. Lin, *J. Org. Chem.* **50**, 3669 (1985).

433. L. E. Overman and K. L. Bell, *J. Am. Chem. Soc.* **103**, 1851 (1981).

434. L. E. Overman, L. K. Bell, and F. Ito, *J. Am. Chem. Soc.* **106**, 4192 (1984).

435. L. E. Overman and S. W. Goldstein, *J. Am. Chem. Soc.* **106**, 5360 (1984).

436. J. W. Daly, E. T. McNeal, L. E. Overman, and D. H. Ellison, *J. Med. Chem.*, **28**, 482 (1985).

437. E. X. Albuquerque, J. E. Warnick, M. A. Maleque, F. C. Kauffman, R. Tamburini, Y. Nimit, and J. W. Daly, *Mol. Pharmacol.* **19**, 411 (1981).

438. E. X. Albuquerque, J. E. Warnick, R. Tamburini, F. C. Kauffman, and J. W. Daly, in *Disorders of the Motor Unit*, D. L. Schotland, Ed., Wiley, New York, 1982, p. 611.

439. E. X. Albuquerque, J. E. Warnick, F. C. Kauffman, and J. W. Daly, in *Membranes and Transport*, Vol. 2, A. N. Martonosi, Ed., Plenum, New York, 1982, p. 355.

440. R. Tamburini, E. X. Albuquerque, J. W. Daly, and F. C. Kauffman, *J. Neurochem.* **37**, 775 (1981).

441. Anonymous, *Med. World News*, p. 20, July 16 (1965).

442. T. Tokuyama, N. Nishimori, M. W. Edwards, and J. W. Daly, *Tetrahedron*, in preparation (1985).

443. L. E. Overman and P. J. Jessup, *J. Am. Chem. Soc.* **100**, 5179 (1978).

444. G. Habermehl, H. Andres, K. Miyahara, B. Witkop, and J. W. Daly, *Justus Liebigs Ann. Chem.* **1976**, 1577.

445. W. Oppolzer and E. Flaskamp, *Helv. Chim. Acta* **60**, 204 (1977).

446. Y. Inubushi and T. Ibuka, *Heterocycles* **8**, 633 (1977).

447. W. Oppolzer and W. Frostl, *Helv. Chim. Acta* **58**, 590 (1975).

448. W. Oppolzer, W. Frostl, and H. P. Weber, *Helv. Chim. Acta* **58**, 593 (1975).

449. W. Oppolzer, C. Fehr, and J. Warneke, *Helv. Chim. Acta* **60**, 48 (1977).

450. T. Ibuka, Y. Mori, and Y. Inubushi, *Tetrahedron Lett.* 3196 (1976).

451. T. Ibuka, Y. Mori, and Y. Inubushi, *Chem. Pharm. Bull. (Japan)* **26**, 2442 (1978).

452. L. E. Overman and P. J. Jessup, *Tetrahedron Lett.* 1253 (1977).

453. S. Masamune, L. A. Reed, III, J. T. Davis, and W. Choy, *J. Org. Chem.* **48**, 4441 (1983).

454. T. Ibuka, N. Masaki, I. Saji, K. Tanaka, and Y. Inubushi, *Chem. Pharm. Bull. (Japan)* **23**, 2779 (1975).

455. T. Ibuka, Y. Inubushi, I. Saji, K. Tanaka, and N. Masaki, *Tetrahedron Lett.* 323 (1975).

456. K. Hattori, Y. Matsumura, T. Miyazaki, K. Maruoka, and H. Yamamoto, *J. Am. Chem. Soc.* **103**, 7368 (1981).

457. G. Habermehl, H. Andres, and B. Witkop, *Naturwiss.* **62**, 345 (1975).

458. W. Kissing, Ph.D. Thesis, University Darmstadt (1972), through [459].

459. J. L. Flippen, *Acta Crystallogr.* **B30**, 2906 (1974).

460. M. Bonin, R. Bellelievre, D. S. Grierson, and H.-P. Husson, *Tetrahedron Lett.* **24**, 1493 (1983).

461. G. Habermehl and H. Andres, *Justus Liebigs Ann. Chemie* **1977**, 800.

462. L. E. Overman and T. Yokomatsu, *J. Org. Chem.* **45**, 5229 (1980).

463. G. B. Bennett and H. Minor, *J. Het. Chem.* **16**, 633 (1979).

464. G. Habermehl and W. Kissing, *Chem. Ber.* **107**, 2326 (1974).

465. J. von Braun, W. Gmelin, and A. Petzold, *Ber. Deutschen Chemischen Gesellshaften* **57**, 382 (1924).

466. S. Zalesky, *Medizinisch-chemische Untersuchungen von Hoppe-Seyler Berlin* **1**, 85 (1866).

467. S. Faust, *Naunyn-Schmiedeberg's Arch. Exper. Path. Pharmak.* **41**, 229 (1898).

468. S. Faust, *Naunyn-Schmiedeberg's Arch. Exper. Path. Pharmak.* **43**, 84 (1900).

469. F. Netolitzky, *Naunyn-Schmiedberg's Arch. Exper. Path. Pharmak.* **51**, 118 (1904).

470. O. Gessner and K. Craemer, *Naunyn-Schmiedeberg's Arch. Exper. Path. Pharmak.* **152**, 229 (1930).

471. C. Schöpf and W. Braun, *Justus Liebigs Ann. Chem.* **514**, 129 (1934).

472. C. Schöpf and K. Koch, *Justus Liebigs Ann. Chem.* **552**, 37 (1942).

473. C. Schöpf and K. Koch, *Justus Liebigs Ann. Chem.* **552**, 62 (1942).

474. C. Schöpf, H.-K. Blödorn, D. Klein, and G. Seitz, *Chem. Ber.* **83**, 372 (1950).

475. C. Schöpf, D. Klein, and E. Hofmann, *Chem. Ber.* **87**, 1638 (1954).

476. C. Schöpf and O. W. Müller, *Justus Liebigs Ann. Chem.* **633**, 127 (1960).

477. E. Wölfel, C. Schöpf, G. Weitz, and G. Habermehl, *Chem. Ber.* **94**, 2361 (1961).

478. C. Schöpf, *Experientia,* **17**, 285 (1961).

479. G. Habermehl, *Chem. Ber.* **96**, 143 (1963).

480. G. Habermehl, *Chem. Ber.* **96**, 840 (1963).

481. C. W. Shoppee and G. Krueger, *J. Chem. Soc.* **1961**, 3641.

482. G. Habermehl and S. Göttlicher, *Angew. Chemie* **75**, 247 (1963).

483. G. Habermehl and S. Göttlicher, *Chem. Ber.* **98**, 1 (1965).

484. G. Habermehl and A. Haaf, *Zeit. Naturforschung,* **23**, 1551 (1968).

485. G. Eggart and H. Wehrli, *Helv. Chim. Acta* **49**, 2453 (1966).

486. G. Habermehl, *Justus Liebigs Ann. Chem.* **679**, 164 (1964).

487. G. Habermehl and A. Haaf, *Chem. Ber.* **98**, 3001 (1965).

488. K. Oka and S. Hara, *J. Am. Chem. Soc.* **99**, 3859 (1977); see also K. Oka, Y. Ike, and S. Hara, *Tetrahedron Lett.* 4543, 4547 (1968).

489. G. Habermehl, *Chem. Ber.* **99**, 1439 (1966).

490. G. Habermehl, *Chem. Ber.* **96**, 2029 (1963).

491. G. Habermehl and G. Vogel, *Toxicon* **7**, 163 (1969).

492. G. Habermehl and A. Haaf. *Justus Liebigs Ann. Chem.* **722**, 155 (1969).

493. S. Hara and K. Oka, *Japan Patent Number 7018*, 633 (1970); through *Chem. Abst.* **73**, 56,316g (1970).

494. S. Hara and K. Oka, *J. Am. Chem. Soc.* **89**, 1041 (1967).

495. G. Habermehl, *Naturwissenschaften* **53**, 123 (1966).

496. G. Habermehl, in *The Alkaloids*, Vol. 9, R. H. F. Manske, Ed., Academic, New York, 1967, p. 427.

497. G. Habermehl, *Progress in Organic Chemistry*, Vol. 7, Butterworth, London, 1968, p. 35.

498. G. Habermehl, in *Venomous Animals and Their Venoms*, Vol. 2, W. Bücherel and E. E. Buckley, Eds., Academic, New York, 1971, p. 569.

499. G. Habermehl and G. Spiteller, *Justus Liebigs Ann. Chem.* **706**, 213 (1967).

500. H. Bachmayer and H. Michl, *Monatsh. Chem.* **96**, 1166 (1965).

501. G. Habermehl, *Zeit. Naturforschung* **208**, 1129 (1965).

502. K. Oka and S. Hara, *Tetrahedron Lett.* 1193 (1969).

503. Y. Shimizu, *Tetrahedron Lett.* 2919 (1972).

504. Y. Shimizu. *J. Org. Chem.* **41**, 1930 (1976).

505. M. Benn and R. Shaw, *Can. J. Chem.* **52**, 2936 (1974).

506. M. H. Benn and R. Shaw, *Chem. Comm.* **1973**, 288.

507. G. Eggart, C. Pascual, and H. Wehrli, *Helv. Chim. Acta* **50**, 985 (1967).

508. K. Oka and S. Hara. *J. Org. Chem.* **43**, 4408 (1978).

509. R. B. Rao and L. Weiler, *Tetrahedron Lett.* 4971 (1973).

510. O. Gessner and W. Esser, *Naunyn-Schmiedeberg's Arch. Exper. Path. Pharmak.* **179**, 639 (1935).

511. O. Gessner and P. Möllenhoff, *Naunyn-Schmiedeberg's Arch. Exper. Path. Pharmak.* **167**, 638 (1932).

512. O. Gessner and G. Urban, *Naunyn-Schmiedeberg's Arch. Exper. Path. Pharmak.* **187**, 378 (1937).

513. V. C. Twitty and H. H. Johnson, *Science* **80**, 78 (1934).

514. V. C. Twitty and H. A. Elliott, *J. Exp. Zool.* **68**, 247 (1934).

515. V. C. Twitty, *J. Exp. Zool.* **76**, 67 (1937).

516. M. S. Brown and H. S. Mosher, *Science* **140**, 295 (1963).

517. D. B. Horsburgh, E. L. Tatum, and V. E. Hall, *J. Pharmacol. Exp. Therap.* **68**, 284 (1940).

518. W. J. van Wagtendonk, F. A. Fuhrman, E. L. Tatum, and J. Field, *Biol. Bull.* **83**, 137 (1942).

519. H. D. Buchwald, L. Durham. H. G. Fischer, R. Harada. H. S. Mosher, C. Y. Kao, and F. A. Fuhrman, *Science* **143**, 474 (1964); see also C. Y. Kao and F. A. Fuhrman, *Toxicon* **5**, 25 (1967).

520. T. Goto, Y. Kishi, S. Takahashi, and Y. Hirata, *Tetrahedron* **21**, 2059 (1965).

521. R. B. Woodward, *Pure Applied Chem.* **9**, 49 (1964).

522. K. Tsuda, *Naturwissenschaften* **53**, 171 (1966).

523. H. S. Mosher, F. A. Fuhrman, H. D. Buchwald, and H. C. Fischer, *Science* **144**, 1100 (1964).

524. Y. Tomiie, A. Furusaki, K. Kasami, N. Yasuoka, K. Miyake, M. Haisa, and I. Nitta. *Tetrahedron Lett.* 2101 (1963).

525. J. F. Wakely, G. J. Fuhrman, F. A. Fuhrman, H. G. Fischer, and H. S. Mosher, *Toxicon* **3**, 195 (1966).

526. E. D. Brodie, Jr., *Copeia* **1968**, 307.

527. E. D. Brodie, Jr., *Am. Midland Naturalist* **80**, 276 (1968).

528. E. D. Brodie, Jr., J. L. Hensel, Jr., and J. A. Johnson, *Copeia* **1974**, 506.

529. C. H. Levenson and A. M. Woodhull, *Toxicon* **17**, 184 (1979).

530. Y. H. Kim, G. B. Brown, H. S. Mosher, and F. A. Fuhrman, *Science* **189**, 151 (1975).

531. F. A. Fuhrman, G. J. Fuhrman, and H. S. Mosher, *Science* **26**, 1376 (1969).

532. J. Flier, M. W. Edwards, J. W. Daly, and C. W. Myers, *Science* **208**, 503 (1980).

533. A. Sebben, *Anais Acad. Bras. Cienc.* **54**, 767 (1982).

534. L. A. Pavelka, Y. H. Kim, and H. S. Mosher, *Toxicon* **15**, 135 (1977).

535. C. Y. Kao, P. N. Yeoh, M. D. Goldfinger, F. A. Fuhrman, and H. S. Mosher, *J. Pharmacol. Exp. Therap.* **217**, 416 (1981).

536. J. Shindelman and H. S. Mosher, *Toxicon* **7**, 315 (1969).

537. G. B. Brown, Y. H. Kim, H. Küntzel, H. S. Mosher, G. J. Fuhrman, and F. A. Fuhrman, *Toxicon* **15**, 115 (1977).

538. Y. Kishi, T. Fukuyama, M. Aratani, F. Nakatsubo, T. Goto, S. Inoue, H. Tanino, S. Sugiura, and H. Kakoi, *J. Am. Chem. Soc.* **94**, 9219 (1972).

539. Y. Kishi, F. Nakatsubo, M. Aratani, T. Goto, S. Inoue, H. Kakoi, and S. Sugiura, *Tetrahedron Lett.* 5127 (1970).

540. Y. Kishi, F. Nakatsubo, M. Aratani, T. Goto, S. Inoue, and H. Kakoi, *Tetrahedron Lett.* 5129 (1970).

541. Y. Kishi, M. Aratani, T. Fukuyama, F. Nakatsubo, T. Goto, S. Inoue, H. Tanino, S. Sugiura, and H. Kakoi, *J. Am. Chem. Soc.* **94**, 9217 (1972).

542. H. Tanino, S. Inoue, M. Aratani, and Y. Kishi, *Tetrahedron Lett.* 335 (1974).

543. R. J. Nachman, Ph.D. Thesis, Stanford University, Palo Alto, CA, 1981.

544. C. Y. Kao, *Pharmacol. Revs.* **18**, 997 (1966).

545. F. Ishihara, *Mittheil. Med. Fak. Tokio. Univ.* **20**, 375 (1918), through [544].

546. K. Tsuda, S. Ikuma, M. Kawamura, R. Tachikawa, K. Sakai, C. Tamura, and O. Amakasu, *Chem. Pharm. Bull.* **12**, 1357 (1964).

547. P. N. Strong and J. F. W. Keana, *Bioorg. Chem.* **5**, 255 (1976).

548. R. Y. Tsien, D. P. L. Green, S. R. Levinson, B. Rudy and J. K. M. Sanders, *Proc. Royal Soc. B.* **191**, 555 (1975).

549. L. A. Pavelka, F. A. Fuhrman, and H. S. Mosher, *Heterocycles* **17**, 225 (1982); see also C. Y. Kao, *Toxicon* **20**, 1043 (1982).

550. R. Chicheportiche, M. Balerna, A. Lombet, G. Romey, and M. Lazdunski, *J. Biol. Chem.* **254**, 1552 (1979).

551. R. Chicheportiche, M. Balerna, A. Lombet, G. Romey, and M. Lazdunski, *Eur. J. Biochem.* **104**, 617 (1980).

552. T. Furukawa, T. Sasaoka, and Y. Hasoya, *Jap. J. Physiol.* **9**, 143 (1959).

553. W. D. Dettbarn, H. W. Higman, P. Rosenberg, and D. Nachmansohn, *Science* **132**, 300 (1960).

554. T. Narahashi, T. Deguchi, N. Urakawa, and Y. Ohkubo, *Am. J. Physiol.* **198**, 934 (1960).

555. T. Narahashi, J. W. Moore, and W. Scott, *J. Gen. Physiol.* **47**, 965 (1964).

556. J. W. Moore, *J. Gen. Physiol.* **48**, Part 2, 11 (1965).

557. T. Narahashi, *Science* **153**, 765 (1966).

558. C. Y. Kao and F. A. Fuhrman, *J. Pharmacol. Exp. Therap.* **140**, 31 (1963).

559. M. Takata, J. W. Moore, C. Y. Kao, and F. A. Fuhrman, *J. Gen. Physiol.* **49**, 977 (1966).

560. C. Y. Kao, *Fed. Proc.* **31**, 1117 (1972).

561. T. Narahashi, *Fed. Proc.* **31**, 1124 (1972).

562. T. Narahashi, *Physiol. Rev.* **54**, 813 (1974).

563. T. Narahashi, in *Neurotoxins—Tools in Neurobiology,* B. Cecarelli and F. Clementi, Eds., Raven, New York, 1979, p. 293.

564. J. M. Ritchie, and R. B. Rogart, *Rev. Physiol. Biochem. Pharmacol.* **79**, 1 (1977).

565. C. Y. Kao, *Fed. Proc.* **40**, 30 (1981).

566. B. K. Ranney, G. J. Fuhrman, and F. A. Fuhrman, *J. Pharmacol. Exp. Therap.* **175**, 368 (1970).

567. H. G. Boit, *Ergehnisse der Alkaloid Chemie bis 1960,* Akademie Verlag, Berlin, 1961.

568. W. R. Kem, K. N. Scott, and J. H. Duncan, *Experientia* **32**, 684 (1976).

569. Z. M. Bacq, *Arch. Int. Physiol.* **44**, 109 (1937).

570. T. H. Jones, M. S. Blum, and H. M. Fales, *Tetrahedron* **38**, 1949 (1982).

571. E. S. Hall, F. McCapra, and A. I. Scott, *Tetrahedron* **23**, 4131 (1967).

572. V. Erspamer, *Ann. Rev. Pharmacol.* **11**, 327 (1971).

573. V. Deulofeu and E. A. Rúveda, in *Venomous Animals and Their Venoms,* Vol. 2, W. Bücherl and E. E. Buckley, Eds., Academic, New York, 1971, p. 475.

574. H. Michl and E. Kaiser, *Toxicon* **1**, 175 (1963).

575. G. Habermehl, *Naturwissenschaften* **56**, 615 (1969).

576. J. Daly and B. Witkop, in *Venomous Animals and Their Venoms,* Vol. 2, W. Bücherl and E. E. Buckley, Eds., Academic, New York 1971, p. 497.

577. G. Habermehl, in *Chemical Zoology,* Vol. 9, M. Florkin and B. T. Scheer, Eds., Academic, New York, 1974, p. 161.

578. H. Wieland and T. Wieland, *Justus Liebigs Ann.* **528**, 234 (1937).

579. F. Märki, A. V. Robertson, and B. Witkop, *J. Am. Chem. Soc.* **83**, 3341 (1961).

580. B. Robinson, G. F. Smith, A. H. Jackson, D. Shaw, B. Frydman, and V. Deulofeu, *Proc. Chem. Soc.* **1961,** 310.

581. W. F. Gannon, J. D. Benigni, J. Suzuki, and J. W. Daly, *Tetrahedron Lett.* 1531 (1967).

582. T. Akizawa, K. Yamazaki, T. Yasuhara, T. Nakajima, M. Roseghini, G. F. Erspamer, and V. Erspamer, *Biomed. Res.* **3**, 232 (1982).

583. M. Shimizu, M. Ishikawa, Y. Komoda, and T. Nakajima, *Chem. Pharm. Bull.* **30**, 909 (1982).

584. M. Shimizu. M. Ishikawa, Y. Komoda, and T. Nakajima, *Chem. Pharm. Bull.* **30**, 3453 (1982).

585. V. Erspamer, M. Roseghini, and J. M. Cei, *Biochem. Pharmacol.* **13**, 1083 (1964).

586. V. Erspamer, T. Vitali, M. Roseghini, and J. M. Cei, *Experientia* **19**, 346 (1963).

587. V. Erspamer, T. Vitali, M. Roseghini, and J. M. Cei, *Arch. Biochem. Biophys.* **105**, 620 (1964).

588. M. Roseghini, R. Endean, and A. Temperrili, *Zeitschrift Naturforschung* **31C**, 118 (1976).

589. S. Fränkel and K. Zeimer, *Biochem. Zeitschrift.* **110**, 234 (1920).

590. H. H. Dale and H. W. Dudley, *J. Pharmacol. Exp. Therap.* **18**, 103 (1920).

591. G. Bertaccini and T. Vitale, *J. Pharm. Pharmacol.* **16**, 441 (1964).

592. S. W. Gorham, *Checklist of World Amphibians,* Lingley, Saint John, New Brunswick, 1974.

593. E. O. Brodie, Jr., *Copeia* **1977**, 523.

594. V. Erspamer, T. Vitali, M. Roseghini, and J. M. Cei. *Biochem. Pharmacol.* **16**, 1149 (1967).

595. V. Erspamer, J. M. Cei, and M. Roseghini, *Life Sci.* **11**, 825 (1963).

596. F. Märki, J. Axelrod, and B. Witkop, *Biochim. Biophys. Acta* **58**, 367 (1962).

597. J. M. Cei, V. Erspamer, and M. Roseghini, *Syst. Zool.* **16**, 328 (1967); *ibid,* **17**, 232 (1968).

598. J. M. Cei and V. Erspamer, *Copeia* **1966**, 74.

599. M. Roseghini, V. Erspamer, and R. Endean, *Comp. Biochem. Physiol.* **54C**, 31 (1976).

600. V. Erspamer, and P. Melchiorri, *Trends Pharmacological Sci.* **1**, 391 (1980).

601. V. Erspamer, *Trends Neurosci.* **4**, 267 (1981).

602. T. Nakajima, *Trends Pharmacological Sci.* **2**, 202 (1981).

603. T. Yasuhara, T. Nakajima, K. Nokihara, C. Yanaihara, N. Yanaihara, V. Erspamer, and G. F. Erspamer, *Biomed. Res.* **4**, 407 (1983).

604. I. M. D. Jackson, and S. Reichlin, *Science* **198**, 414 (1977).

605. T. Yasuhara and T. Nakajima, *Chem. Pharm. Bull.* **23**, 3301 (1975).

606. H. Michl and H. Molzer, *Toxicon* **2**, 281 (1965).

607. H. Michl and A. Pastuszyn, *Monatsch. Chem.* **95**, 978 (1964).

608. L. Lábler, H. Keilova, F. Sorm, F. Kornalik, and Z. Styblova, *Toxicon* **5**, 247 (1968).

609. A. Csordás and H. Michl, *Monatsch. Chem.* **101**, 182 (1972).

610. R. A. Brandon and J. E. Huheey, *Toxicon* **19**, 25 (1981).

611. H. Heatwole and J. W. Daly, *Experientia* **22**, 764 (1966).

612. H. Michl and A. Pastuszyn, *Monatsch. Chem.* **95**, 480 (1964).

613. L. Bolognani and A. M. Bolognani-Fantin, *Ital. J. Biochem.* **14**, 81 (1965).

614. H. Wieland and R. Alles, *Ber. Deutsch. Chem. Gesellschaften* **55**, 1789 (1922).

615. K. Shimada, K. Ohishi, and T. Nambara, *Tetrahedron Lett.* **25**, 551 (1984).

616. K. Shimada and T. Nambara, *Chem. Pharm. Bull.* **28**, 1559 (1980).

617. K. Meyer and H. Linde, in *Venomous Animals and Their Venoms*, Vol. 2, W. Bücherl and E. E. Buckley, Eds., Academic, New York, 1971, p. 521.

618. M. Neuwirth, J. W. Daly, C. W. Myers, and L. W. Tice, *Tissue & Cell* **11**, 755 (1979).

619. H. Bachmayer, H. Michl, and B. Roos, in *Animal Toxins*, F. E. Russell and P. R. Saunders, Eds., Pergamon, New York, 1967, p. 395.

620. M. Pavan, *Zeitschrift Hygiene Infections-Krankheit* **134**, 136 (1952).

621. G. Habermehl and H. J. Preusser, *Zeit. Naturforsch.* **24B**, 1599 (1969).

622. H. J. Preusser, G. Habermehl, M. Sablofski, and D. Schmall-Haury, *Toxicon* **13**, 285 (1975).

623. A. Csordás and H. Michl, *Toxicon* **7**, 103 (1969).

624. G. Croce, N. Giglioli, and L. Bolognani, *Toxicon* **11**, 99 (1973).

625. A. Cevikbas, *Toxicon* **16**, 195 (1978).

626. G. Kiss and H. Michl, *Toxicon* **1**, 33 (1962).

627. R. G. Jaeger, *Copeia* **1971**, 160.

628. S. C. Bradley and L. J. Klika, *J. Am. Med. Assoc.* **246**, 247 (1981).

629. Y. Shimizu and M. Kobayashi, *Chem. Pharm. Bull.* **31**, 3625 (1983).

630. P. A. Silverstone, *Nat. Hist. Museum, Los Angeles County Sci. Bull.*, Number 21, p. 1 (1975).

631. P. A. Silverstone, *Nat. Hist. Museum. Los Angeles County Sci. Bull.*, Number 27, p. 1 (1976).

632. G. Habermehl and A. Haaf, *Chem. Ber.* **101**, 198 (1968).

633. D. F. Johnson and J. W. Daly, *Biochem. Pharmacol.* **20**, 2555 (1971).

634. T. Matsui, S. Hamada, and S. Konosu, *Bull. Jap. Soc. Sci. Fish.* **47**, 535 (1981).

635. D. D. Sheumack, M. E. H. Howden, I. Spence, and R. J. Quinn, *Science* **199**, 188 (1978).

636. H. Narita, T. Noguchi, J. Maruyama, Y. Ueda, K. Hashimoto, Y. Watanabe, and K Hida, *Bull. Japan. Soc. Sci. Fish.* **47**, 935 (1981).

637. T. Noguchi, J. Maruyama, Y. Ueda, K. Hashimoto, and T. Harada, *Bull. Japan. Soc. Sci. Fish.* **47**, 909 (1981).

638. T. Yasumoto, Y. Oshima, M. Hosaka, and S. Miyakoshi, *Bull. Japan Soc. Sci. Fish.* **47**, 929 (1981).

639. T. Noguchi, H. Narita, J. Maruyama, and K. Hashimoto, *Bull. Japan Soc. Sci. Fish.* **48**, 1173 (1982).

640. T. Noguchi, A. Uzu, K. Koyama, J. Maruyama, Y. Nagashima, and K. Hashimoto, *Bull. Japan Soc. Sci. Fish.* **49**, 1887 (1983).

641. K. S. Elam, F. A. Fuhrman, Y. H. Kim, and H. S. Mosher, *Toxicon* **15**, 45 (1977).

642. H. S. Mosher and F. A. Fuhrman, *ACS Symposium Series No. 262, Seafood Toxins*, E. P. Ragelis, Ed., American Chemical Society, 1984, p. 333.

643. M. Kodama, T. Noguchi, J. Maruyama, T. Ogata, and K. Hashimoto, *J. Biochem. (Jap.)* **93**, 243 (183).

Chapter Two

Marine Alkaloids
and Related Compounds

William Fenical
Institute of Marine Resources
Scripps Institution of Oceanography
La Jolla, California 92093

CONTENTS

1.	INTRODUCTION	276
2.	ALKALOIDS FROM MARINE PLANTS	277
	2.1. Alkaloids from Marine Phytoplankton	277
	2.2. Alkaloids from Marine Seaweeds	278
	2.2.1. Blue-green Algae (Cyanophyta)	278
	2.2.2. Green Algae (Chlorophyta)	282
	2.2.3. Red Algae (Rhodophyta)	283
3.	ALKALOIDS FROM MARINE INVERTEBRATES AND MICROORGANISMS	285
	3.1. Alkaloids from Marine Bacteria	285
	3.2. Alkaloids from Marine Invertebrates	287
	3.2.1. Sponges (Porifera)	287
	3.2.2. Bryozoans (Ectoprocta)	309
	3.2.3. Tunicates (Urochordata)	312
	3.2.4. Coelenterates (Cnidaria)	315
	3.2.5. Molluscs (Mollusca)	317
	3.2.6. Miscellaneous Marine Invertebrates	319
4.	ALKALOIDS FROM MARINE VERTEBRATES	320
5.	ALKALOIDS INVOLVED IN MARINE BIOLUMINESCENCE	322
	REFERENCES	323

1. INTRODUCTION

The past ten years have produced considerable chemical information defining the marine environment. Along with the many unique natural products that have been described are numerous reports of nitrogenous compounds that, according to modern definitions, fall within the nature of an "alkaloid." These substances are widely distributed in marine organisms, being found in both marine plants and animals, but they are not the familiar alkaloid structures found in the terrestrial environment. Rather, marine alkaloids tend to possess new carbon skeletons, or to be simple nitrogen-containing structures with added complexities such as halogen substituents. In reviewing this topic I have continually referred to the recent working definition of an "alkaloid" proposed by Pelletier [1].

To fully compare marine and terrestrial alkaloids, the fundamental differences between marine and terrestrial organisms should be defined. Marine plants, for example, do not compare with the mainly terrestrial angiosperms, which produce numerous alkaloids. Marine plants are mostly of the algae class (nonvascular), and hence they are totally unrelated to the majority of the terrestrial flora. The same point can be made about marine invertebrates. The abundant marine invertebrates, sponges (Porifera), coelenterates (Cnidaria), molluscs (Mollusca), and so on, have few, if any, close terrestrial relatives. It is not surprising, given the major biotic differences, that marine organisms produce very nontraditional alkaloidal compounds.

The marine environment also contains complexities with regard to the origin of alkaloids which are of less importance in the terrestrial environment. The major complexity lies in the interwoven marine food web. Alkaloids produced by microorganisms, for example, are readily accumulated by filter-feeding organisms such as clams, oysters, and mussels. Symbiosis between microorganisms and marine invertebrates creates additional insecurity in defining the origins of complex metabolites. Many sponges, for example, are recognized to be composed of as much as 50% symbiotic bacteria or blue-green algae. Hence, the question of true source arises when alkaloids are isolated from these complex associations.

This chapter is the first comprehensive review of marine alkaloids. There are, however, several reviews that provide partial coverage of this subject. The series edited by Scheuer [2] contains major reviews on dinoflagellate toxins [3], uncommon amino acids from marine algae [4], nitrogenous pigments from marine invertebrates [5], nitrogenous compounds from blue-green algae [6], guanidine derivatives from marine sources [7], and marine indoles [8]. The recent general reviews of the current literature by Faulkner [9,10] were also consulted in constructing this review.

In reviewing the marine alkaloids I have adopted a biological origin approach rather than collating similar structure types. While both approaches are used regularly in reviewing alkaloids, the biogenetic approach, which allows biological sources to be compared and contrasted, was selected here to highlight the nonclassical sources of marine alkaloids. Where appropriate, cross-references to similar structure types from diverse sources are made.

2. ALKALOIDS FROM MARINE PLANTS

The basis of primary productivity (photosynthesis) in the oceans lies in the 13 or more distinct Phyla of both unicellular and multicellular marine plants. With the exception of a few vascular plants (the seagrasses), the vast majority of marine plants are algae, and these may be either the unicellular, free-floating forms, that is, the "phytoplankton," or they may be multicellular and attached to hard substrates, as in the "seaweeds." Several genera from both algal groups are known to produce alkaloids; however, reports of alkaloids from marine angiosperms have not appeared.

2.1. Alkaloids from Marine Phytoplankton

Although there are at least nine diverse Phyla of marine phytoplankton, only one group, the dinoflagellates (Dinophyta), produces nitrogenous metabolites. Historically, the best known compounds are saxitoxin (**1**) and the related metabolites neosaxitoxin (**2**) and the gonyautoxins I–VIII (**3–10**). These compounds are collectively recognized as the neurotoxic substances associated with the fatal ingestion of toxic shellfish. The saxitoxin "complex" represents metabolites that are now known to be metabolic products of various *Gonyaulax* (= *Protogonyaulax*) species, such as the Pacific alga *G. excavata* and the Atlantic *G. tamarensis*. The structures of these interesting and important toxins are based upon the original X-ray structure determinations of saxitoxin (**1**) [11,12], upon a more recent X-ray study [13], and upon spectral analyses and chemical interconversions [14–20]. Saxitoxin and its relatives are not exclusively of marine origin and also occur in the freshwater blue-green alga *Amphanizomenon flosaquae* [21]. The perhydropurine skeleton of toxins **1–10** is unique among the marine alkaloids, and until recently it was of an unknown biosynthetic origin. Elegant labeling studies by the Shimizu group recently [22] showed neosaxitoxin (**2**) to have a complex origin apparently involving a Claisen-type condensation of an acetate precursor to the α carbon of arginine followed by imidazol ring formation after loss of the carboxyl carbon.

Although perhaps not strictly alkaloidal, the nitrogen- and phosphorus-containing metabolites **11** and **12** are two more toxins associated with marine dinoflagellates. Both were isolated from cultures of the tropical dinoflagellate *Ptychodiscus brevis* (= *Gymnodinium breve*), which is known as one of the producers of the "red tide" phenomenon. The structure of metabolite **11**, known as GB-4, was determined by X-ray methods [23], while the structure of the cyclooctyl phosphonamide **12** was confirmed by synthesis [24].

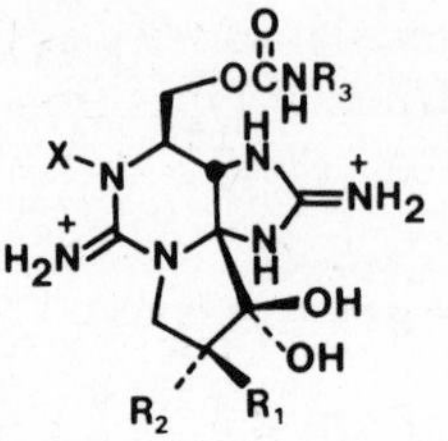

	X	R_1	R_2	R_3	
1	H	H	H	H	Saxitoxin
2	OH	H	H	H	Neosaxitoxin
3	OH	H	OSO_3^-	H	Gonyautoxin I
4	H	H	OSO_3^-	H	II
5	H	OSO_3^-	H	H	III
6	OH	OSO_3^-	H	H	IV
7	H	H	H	SO_3^-	V
8	OH	H	H	SO_3^-	VI
9	H	OSO_3^-	H	SO_3^-	VIII
10	H	H	OSO_3^-	SO_3^-	

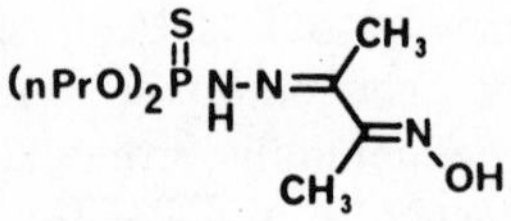

11 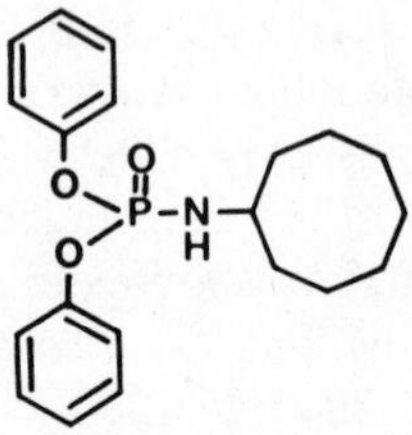   12

2.2. Alkaloids from Marine Seaweeds

The seaweeds consist of four major Phyla, the blue-green algae (Cyano-phyta), the green algae (Chlorophyta), the red algae (Rhodophyta), and the brown algae (Phaeophyta). Although seaweeds are found in all marine habitats and are particularly abundant in colder waters, it is mainly the tropical seaweeds that are recognized to produce alkaloids. Of the four major groups, alkaloids are most prevalent in the blue-green algae. Alkaloids have not been observed as metabolites of the brown seaweeds (Phaeophyta).

2.2.1. Blue-Green Algae (Cyanophyta). Considerable debate exists in the classification of the procaryotic blue-green algae as either "true" algal forms or as photosynthetic bacteria. The alkaloidal components of these plants are, in the majority, characterized by their syntheses via acetate and amino

acid biosynthetic pathways. Thus from a chemical perspective the blue-green algae would appear to be more closely related to the bacteria. My preference, however, is to classify them as algae based upon their photosynthetic activities and similarities in habitat with other algal groups. A concise discussion of this question, as well as a more thorough review of marine algae, is given by Moore [6].

The cosmopolitan blue-green alga *Lyngbya majuscula* seems to vary chemically depending upon its area of collection, and also perhaps the extraction and work-up methods involved. One collection of *L. majuscula*, for example, was found to contain the simple amidobutyrolactone **13**, as well as the complex group of bisbutyroimides known as the pukeleimides A–G (**14–20**) [25–27]. The structures of these latter compounds were determined by an X-ray investigation [26] of pukeleimide C (**19**) and by subsequent spectral comparisons. These metabolites contain structural similarities (4-methoxy-Δ^3-pyrrolin-2-one rings) with the known fungal metabolite penicillic acid [28].

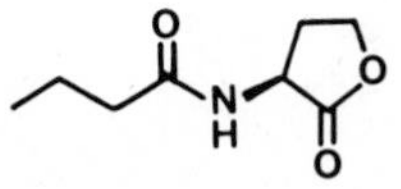

13

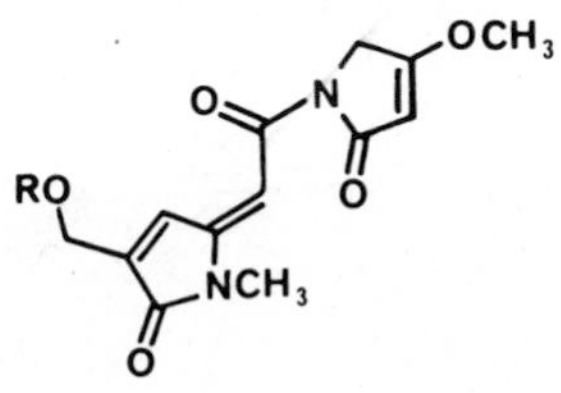

14 R = H pukeleimide A
15 R = CH$_3$ puk G

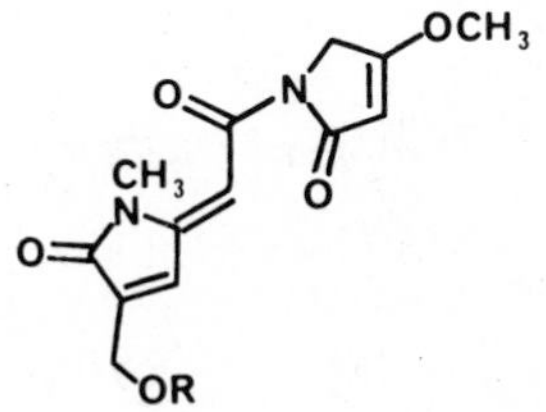

16 R = H puk B
17 R = CH$_3$ puk F

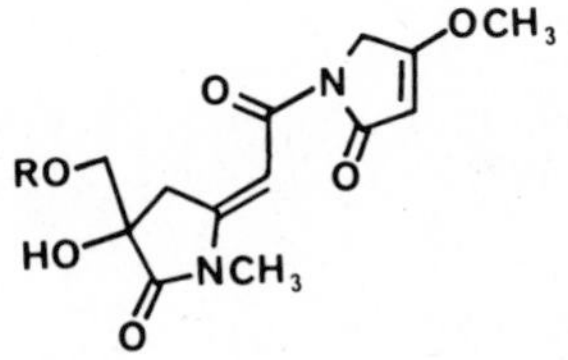

18 R = H puk D
19 R = CH$_3$ puk C

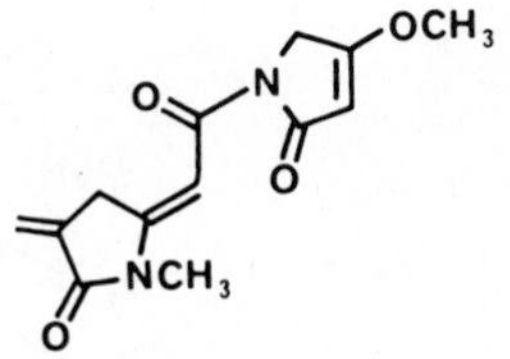

20   puk E

Other more complex cyclic amides related to the pukeleimides have also been isolated from geographically different collections of *Lyngbya majuscula*. Malyngamide A (**21**) contains the same 4-methoxy-Δ^3-pyrrolin-2-one constellation as found in the pukeleimides [29]. In addition to malyngamide A, several other amides referred to as ''malyngamides'' have been reported. The structures of malyngamides D and E (**22,23**) were determined by spectral analyses and by interconversion through dehydration [30]. Several assignments of stereochemistry remain to be defined in these molecules. Although not yet fully described, related molecules referred to as the malyngamides B, C, and C-acetate have been isolated and partially defined [6,31].

Several forms of *Lyngbya majuscula* are also known to produce peptides of unusual structure. The majusculamides A and B (**24,25**) are lipodipeptides that are epimers at the methyl-bearing carbon. The structure of majusculamide B was provided by X-ray methods, and A and B were conveniently interrelated by thermal epimerization at 140°C [32]. More recently, a related lipophilic depsipeptide, majusculamide C (**26**), was defined from this same algal source [33]. Unlike majusculamides A and B, majusculamide C possesses considerable antifungal activity toward the common plant pathogens *Phytophthora infestans* and *Plasmopora viticola*.

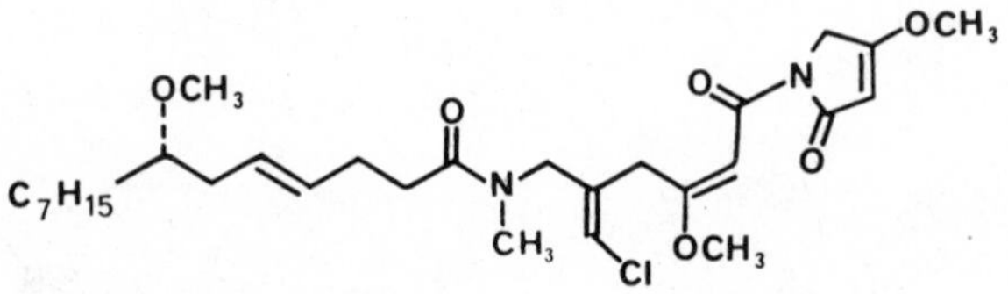

21

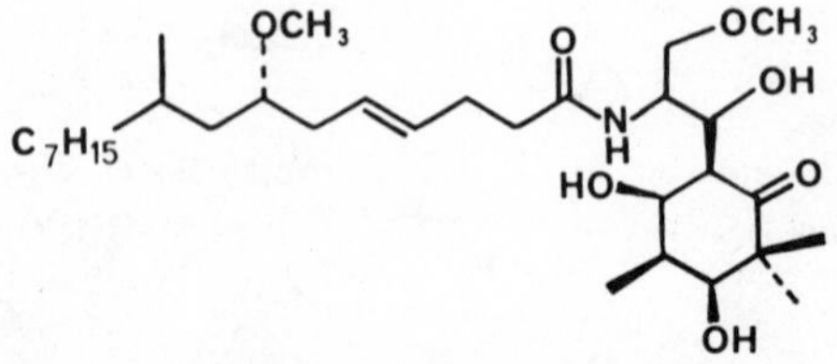

22

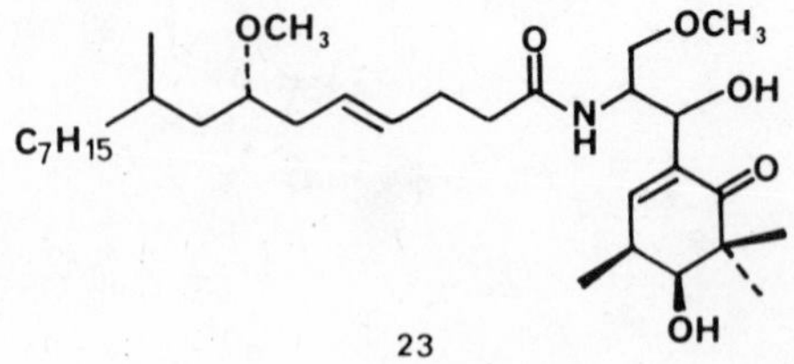

23

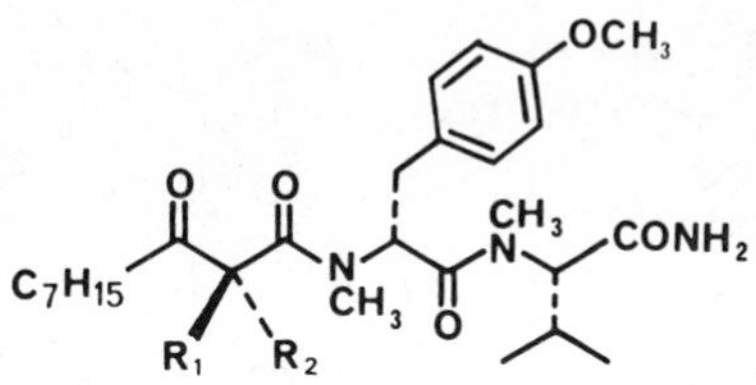

24 $R_1 = H$, $R_2 = CH_3$

25 $R_1 = CH_3$, $R_2 = H$

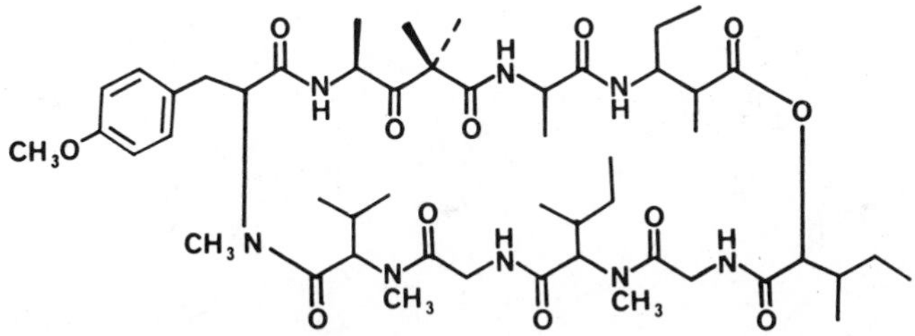

26

Alkaloids that possess the indole nucleus have also been found in blue-green algae, and these compounds clearly resemble the more "classical" indole alkaloids. Yet another collection of *Lyngbya mujuscula* was found to contain the unique alkaloid **27** referred to as lyngbyatoxin A [34]. This toxin is the long-sought-after causative substance that produces a contact dermatitis known as "swimmer's itch" in Hawaii. This highly inflammatory alkaloid is closely related to teleocidin B, a toxic indole alkaloid produced by several actinomycetes of the genus *Streptomyces* [35].

Additional indole alkaloids that have been described from blue-green algae include the carbazoles chlorohyellazole (**28**) and hyellazone (**29**) from *Hyella caespitosa* [36] and the series of indole dimers **30–35** isolated from the intertidal blue-green alga *Rivularia firma* [37]. Halogen substituents are prevalent in these latter simple indole derivatives. Halogenated indoles from marine sources are common and numerous examples of these compounds are presented in the following sections.

The freshwater blue-green algae are apparently closely related chemically to their marine counterparts. As mentioned earlier, saxitoxin has been reported as a constituent of *Amphizaminon flos-aquae*, the alkaloid anatoxin A is produced by *Anabena flos-aquae* [38], and unique chlorinated indole alkaloids that contain isonitrile and isothiocyanate functionalities have recently been isolated from *Hapalosiphon fontinalis* [39].

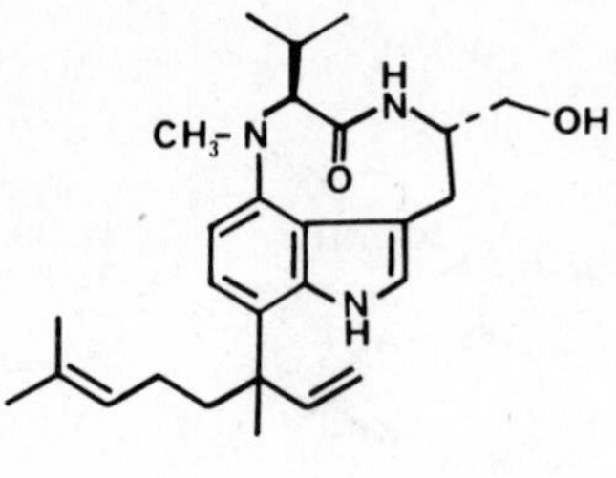

27

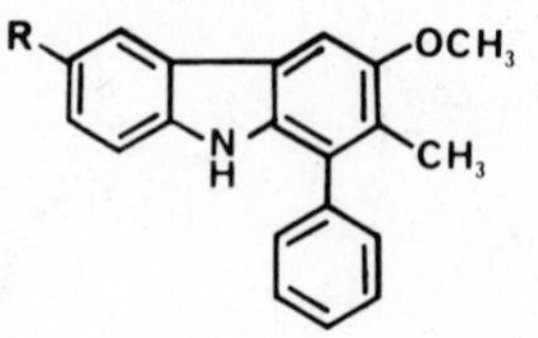

28 R = Cl
29 R = H

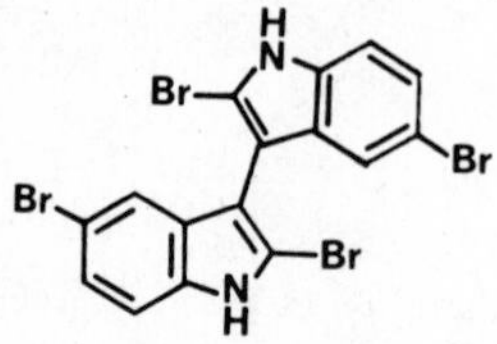

30

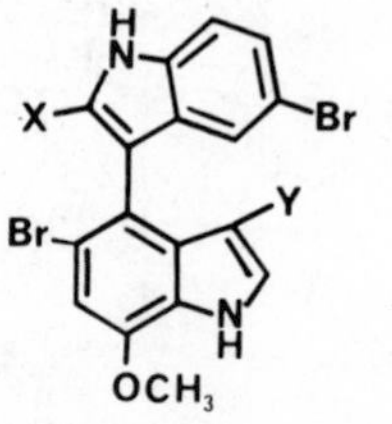

31   X = Y = Br
32   X = H , Y = Br
33   X = Br , Y = H

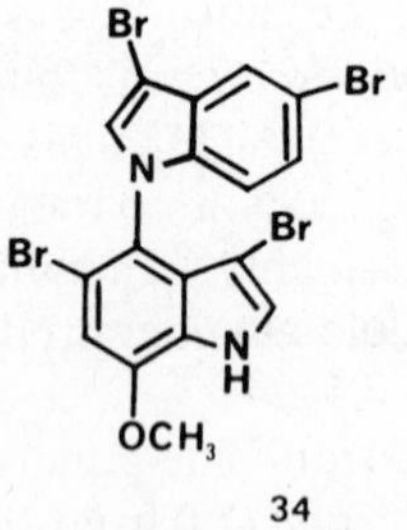

34

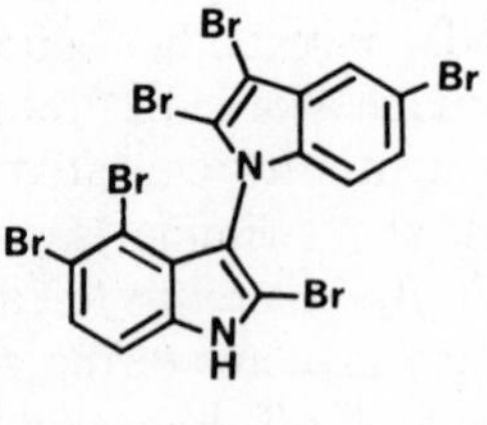

35

2.2.2. Green Algae (Chorophyta). Only a single cyclic nitrogenous compound has been isolated from the green seaweeds. Studies of numerous species of the tropical green alga *Caulerpa* have shown the almost ubiquitous presence of the orange pigment caulerpin (**36**) [40]. Caulerpin was originally isolated from *C. racemosa*, which is a common edible seaweed in the Philippines. An incorrect structure was initially assigned to this indole dimer [41,42], but a revision has subsequently been made based upon extensive spectral analysis and upon a total synthesis [43].

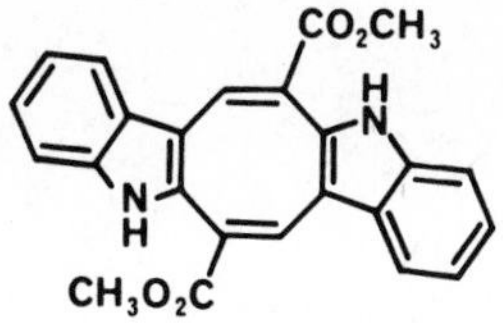

36

2.2.3. Red Algae (Rhodophyta).

Only four or five species of red seaweeds have been shown to produce alkaloids or simple indole derivatives. The extract of the red alga *Rhodophylis membranaceae* from New Zealand possesses significant antifungal activity, and from it very complex mixtures of brominated and chlorinated simple indole derivatives have been isolated. The indoles **37–46** were subsequently purified and identified on the basis of their high-resolution mass spectral properties, particularly their halogen isotope ratios, by NMR methods, and in several cases by comparison with synthetic samples [44,45]. Halogen substitution was observed only in the 2-, 3-, 4-, and 7-positions of the indole ring.

In an unrelated red alga, *Laurencia brongniartii*, significant antibacterial and antigungal activity was also observed. A series of four related bromoindoles and bromomethylindoles (**47–50**) were subsequently isolated and defined [46]. Of these isomers only 2,3,5,6-tetrabromoindole (**49**) showed antimicrobial activity (*Bacillus subtilis, Saccharomyces cerevisiae*).

More classical indole alkaloids have recently been reported from the red seaweed *Martensia fragilis* collected in Hawaii [47]. Isolated from this source were the indole alkaloids martensine A and B (**51,52**), along with the related benzyl amine, fragilamide (**53**). The identification of these com-

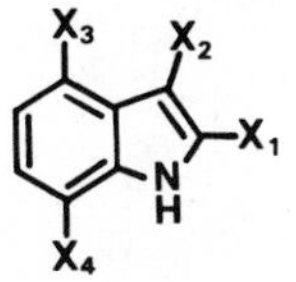

37 $X_1X_2X_4 = Cl$, $X_3 = H$	47 R=CH$_3$, X_1=H, X_2=Br
38 $X_1X_2 = Cl$, X_4=Br, X_3=H	48 R=CH$_3$, X_1=Br, X_2=H
39 X_1X_4=Br, X_2=Cl, X_3=H	49 R=H, X_1X_2=Br
40 $X_1X_2X_4$=Br, X_3=H$_4$	50 R=CH$_3$, X_1X_2=Br
41 $X_1X_2X_3$=Cl, X=H	
42 $X_1X_2X_3X_4$=Cl	
43 X_1X_2=Cl, X_3=Br(Cl), X_4=Cl(Br)	
44 X_1X_2=Cl, X_3X_4=Br	
45 X_3X_4=Br, X_1=Br(Cl), X_2=Cl(Br)	
46 $X_1X_2X_3X_4$=Br	

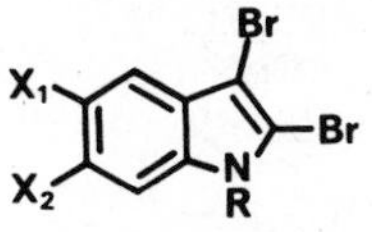

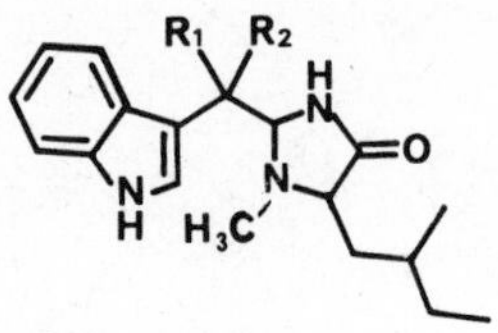

51 R$_1$=OH, R=H

52 R$_1$R$_2$=O

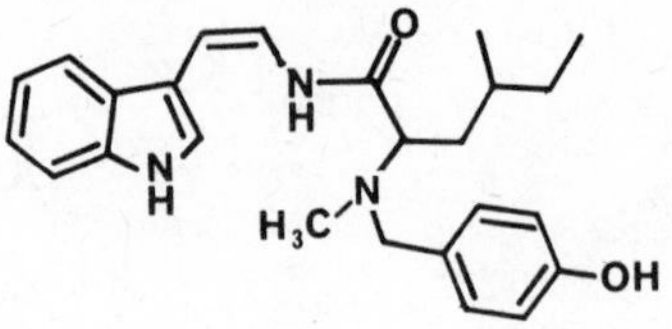

53

pounds was accomplished by spectral methods, by interconversion, and by chemical degradation. Fragilamide, for example, was cleaved by singlet oxygen photooxidation to yield indole-3-carboxaldehyde and *p*-(methoxymethyl) phenol. Martensine A (**51**) was reported to show antibacterial activity toward *Bacillus subtilis, Staphylococcus aureus*, and *Mycobacterium smegmatis*.

Although nucleosides are often excluded from reviews of alkaloids, the iodotubercidin derivative **54** must be considered novel and of very limited distribution. In their search for bioactive substances from marine sources, the Australian group at the Roche Research Institute of Marine Pharmacology discovered that the red alga *Hypnea valendiae* produced the iodinated nucleoside **54** (5-iodo-5′-deoxytubercidin) [48]. This interesting metabolite was found to produce unique neuromuscular effects in several bioassays. Nucleoside **54** is one of only a few naturally occurring iodinated metabolites.

Several structurally unique and historically important amino acids are also found in red algae. Recent studies of "unknown UV-absorbing substances" in several species have led to the isolation of the microsporine derivatives **55–57** [49–51]. In addition, the "terpenoid" amino acids, α-kainic acid (**58**)

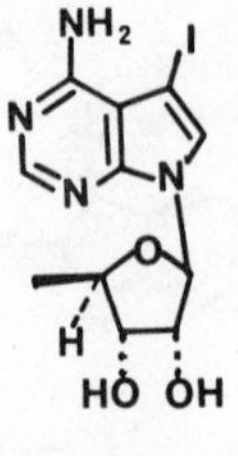

54

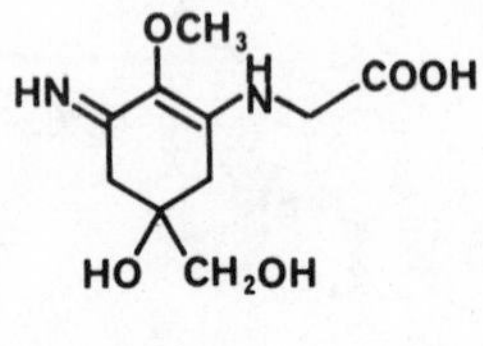

55

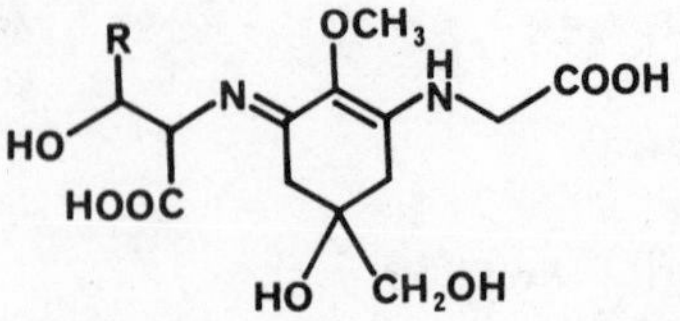

56 R=H

57 R=CH$_3$

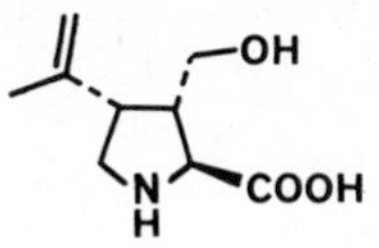

58 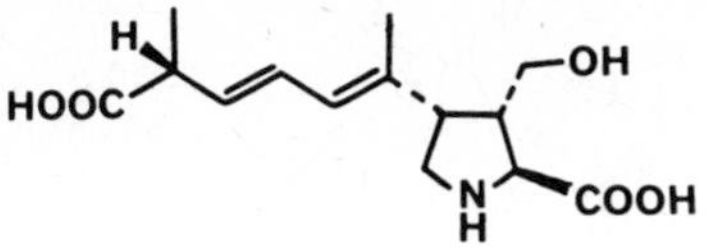   59

and domoid acid (**59**), from the red alga *Digenia simplex* [52] and from various *Chondria* species [53], are now recognized as the active anthelminthic components of these algae. *Digenia* and *Chondria* have had a long history in the Orient in traditional medicinal applications.

3.   ALKALOIDS FROM MARINE INVERTEBRATES AND MICROORGANISMS

By far, the majority of the marine alkaloidal compounds have been isolated from marine invertebrates. Sponges (Phylum Porifera) have yielded the largest number of nitrogenous compounds, but sponges also contain numerous nonnitrogenous substances, the majority of which are terpenoids [10,54]. Although sponges have been studied extensively because of their availability, other marine invertebrates, such as the bryozoans, appear to have a preference for the production of alkaloids. It should again be emphasized that the "classical" alkaloids do not occur in the vast majority of these sources.

3.1.   Alkaloids from Marine Bacteria

Terrestrial microorganisms have been extensively investigated for the derivation of new antibiotics. In contrast, marine microorganisms seem to have been systematically avoided for reasons that remain unclear. Of the few studies reported, only marine bacteria have been studied and several alkaloidal compounds have been described. Although the criteria to define a "true" marine microorganism are under constant debate, the production of a unique group of brominated pyrroles by several isolates serves to illustrate their distinct phylogeny. Several marine *Pseudomonas* and *Chromobacterium* species produce the bromopyrroles **60–62** [55,56]. The organism producing the phenol **60** was subsequently named *Pseudomonas bromoutilis* in recognition of the unexpected biosynthesis involving bromination [55]. The indole **63**, and two 4-hydroxyquinoline bases, **64** and **65**, were also described from a marine pseudomonad [57]. Derivatives of the bacterial pigment prodigiosin have also been found in marine bacteria. The structure of a cyclic prodigiosin derivative was recently revised to **66** [58]. Several interesting

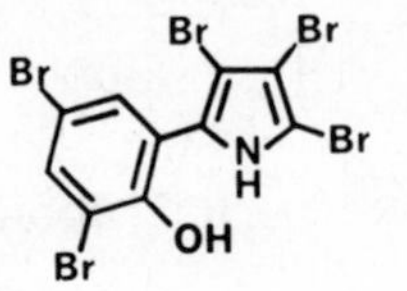

60 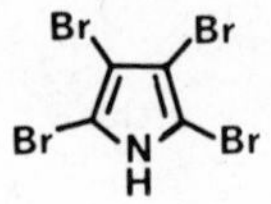   **61** 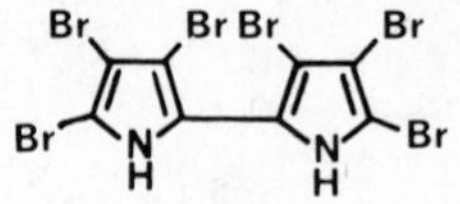   **62**

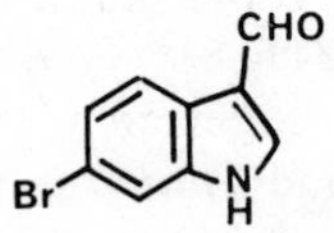

63

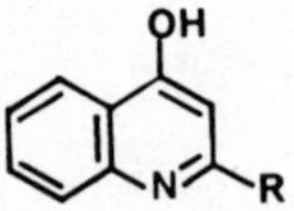

64 R=C_5H_{11}
65 R=C_7H_{13}

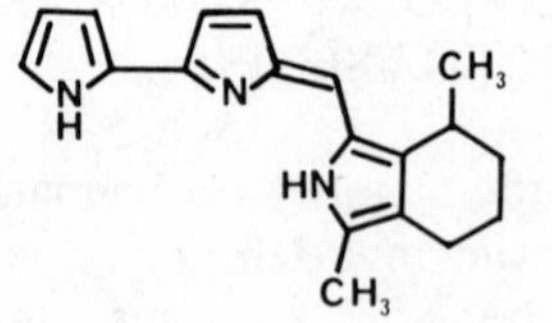

66

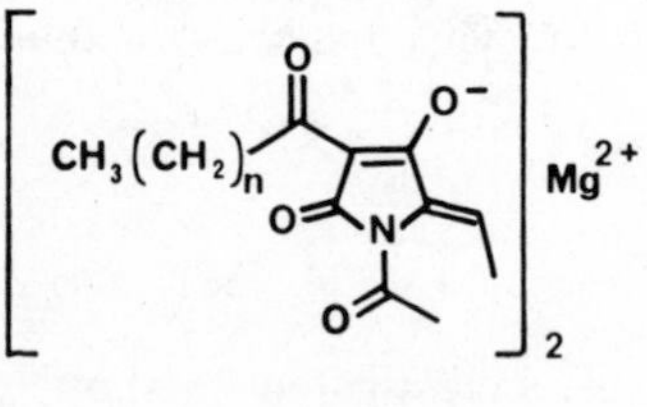

67 $(n=4,6)$

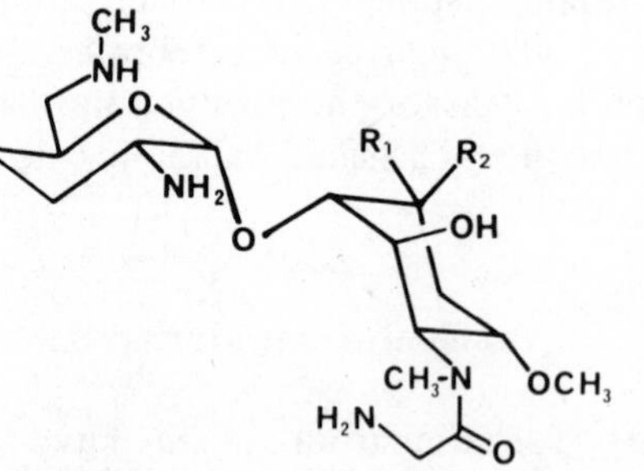

68 R_1=H, R_2=NH_2
69 R_1=NH_2, R_2=H

nitrogen-containing magnesium salts, the magnesidins (**67**), were also found in marine *Pseudomonas* species. These antibiotic compounds are thought to be degradation products of prodigiosin [59].

Perhaps the most interesting new group of antibiotics from marine microorganisms is the istamycins (**68,69**) isolated from the marine actinomycete *Streptomyces tenjimariensis* [60]. These interesting compounds are, in part, deoxyaminoglycosides, but they possess an unusal cyclohexane acetal constellation.

3.2. Alkaloids from Marine Invertebrates

3.2.1. Sponges (Porifera). A wide variety of nitrogenous compounds have been isolated from sponges. *The true sources of these metabolites remain to be fully defined.* Since sponges are filter-feeding animals, some metabolites could be obtained by food-chain accumulation. Most evidence suggests, however, that sponges produce the metabolites they contain. In many cases though, sponges are not discrete animals, but consist of as much as 50% by weight symbiotic bacteria and blue-green algae [61]. A possible source for the following nitrogenous metabolites may also be these symbiotic microorganisms.

Bromotyrosine Derivatives. Sponges of the genus *Verongia* (= *Aplysina*) are abundant, and along with some sponges of the related genus *Ianthella*, produce the simple halogenated metabolites **70–78** [62–68]. These

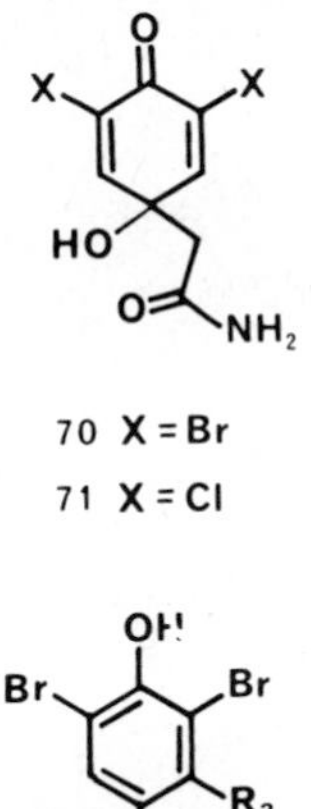

70 X = Br
71 X = Cl

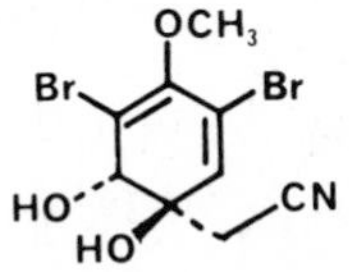

72

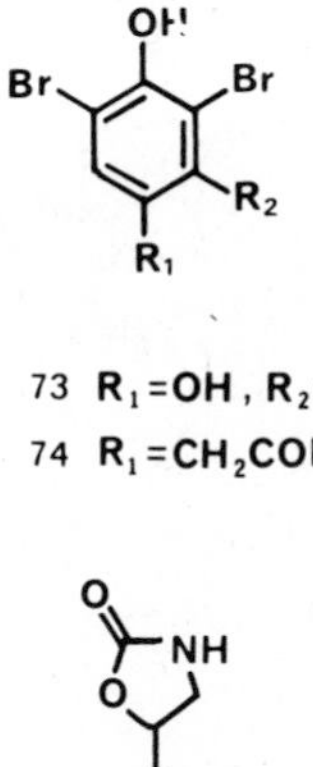

73 R$_1$ = OH, R$_2$ = CH$_2$CONH$_2$
74 R$_1$ = CH$_2$CONH$_2$, R$_2$ = OH

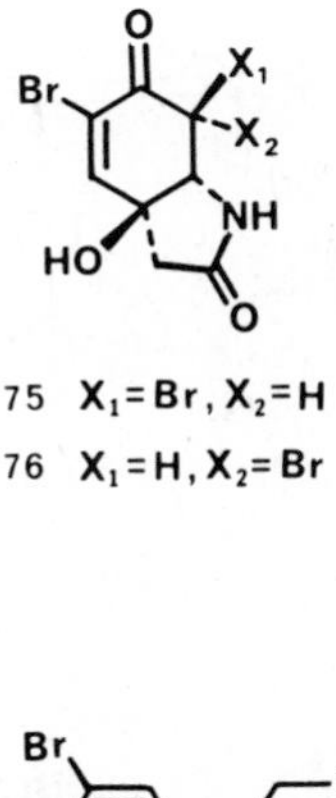

75 X$_1$ = Br, X$_2$ = H
76 X$_1$ = H, X$_2$ = Br

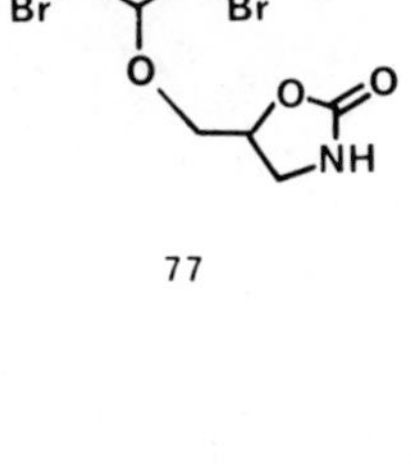

77

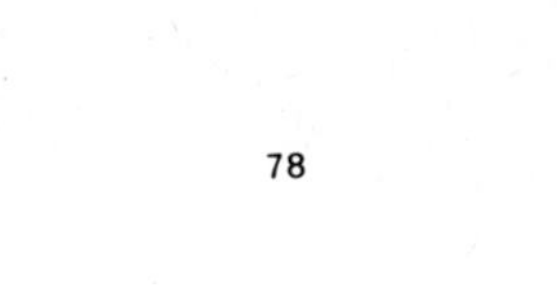

78

simple compounds often possess antibiotic activities that can be potentially attributed to their high levels of halogenation. The dibromophenethylammonium salt **78**, isolated from *V. fistularis* [68], was studied in detail and found to produce pharmacological effects similar to epinephrine and acetylcholine. Products resulting from the dimerization and further condensation of these cyclic amides with smaller molecules have also been observed. The interesting metabolites aerothionin and homoaerothionin (**79,80**) were described in 1972 as metabolites of *V. thiona* and *V. aerophoba*, respectively [69,70]. The conversion of a presumed monomeric precursor (such as **70** or **72**) to these compounds involves an addition to the amide carbonyl group with concomitant production of the spiro isoxazole ring. Other related metabolites, fistularin-3 and isofistularin-3 (**81,82**), also isolated from *Verongia* species, represent modifications on this theme [71,72]. Isofistularin-3 (**82**) is

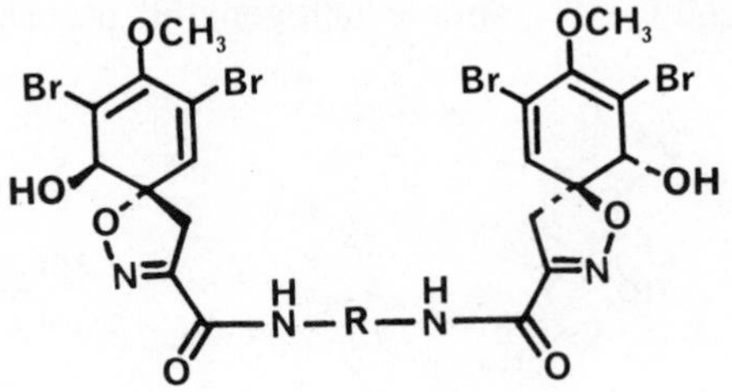

79 R = -C₄H₈-
80 R = -C₅H₁₀-

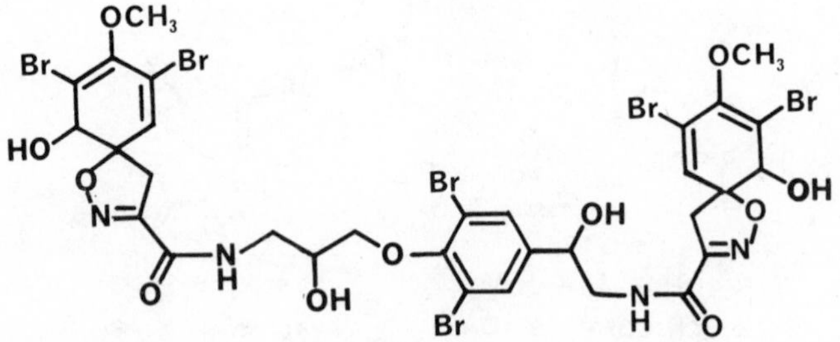

81
82 (2 isomers)

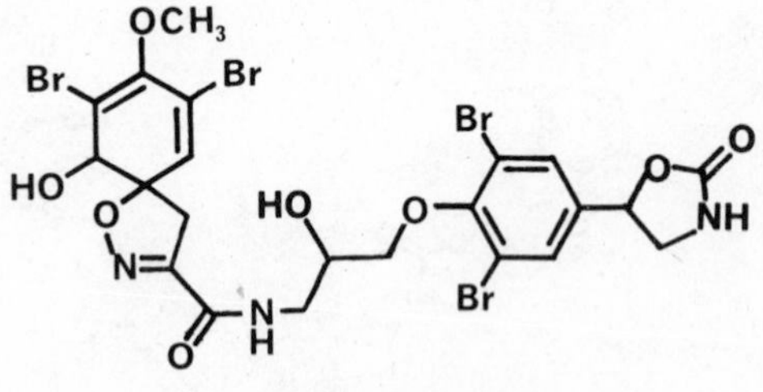

83

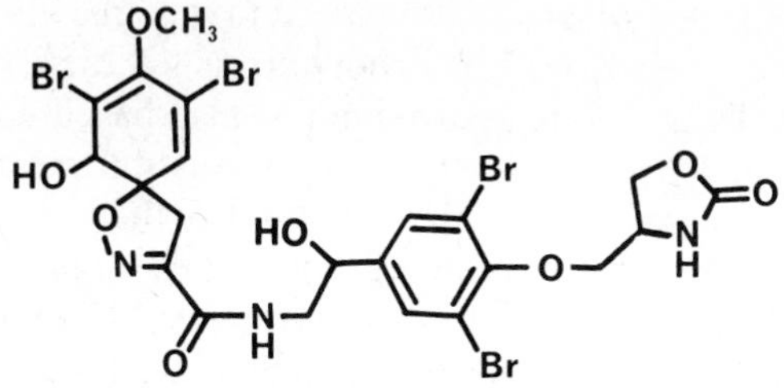

84

isomeric (epimeric) with **81**, but it is not known which of the six chiral centers are involved. The isomeric metabolites fistularin-1 (**83**) and fistularin-2 (**84**), which also bear the spiro isoxazole ring, are metabolites of the sponge *A. fistularis* [71]. The related sponge *Psammaplysilla purpurea* contains two metabolites that were first misidentified [73] and later accurately assigned as **85** and **86** on the basis of X-ray diffraction analysis [74]. The ketal structures of these latter compounds would appear to arise via rearrangement (ring enlargement) of a more typical dibromomethyoxycyclohexadiene precursor. Presumably this alteration proceeds by spiro ring opening and reclosure. Lastly, a more recent investigation of *V. aerophoba* provided the structures of two new imidazole derivatives aerophobins-1 and -2 (**87,88**) [72].

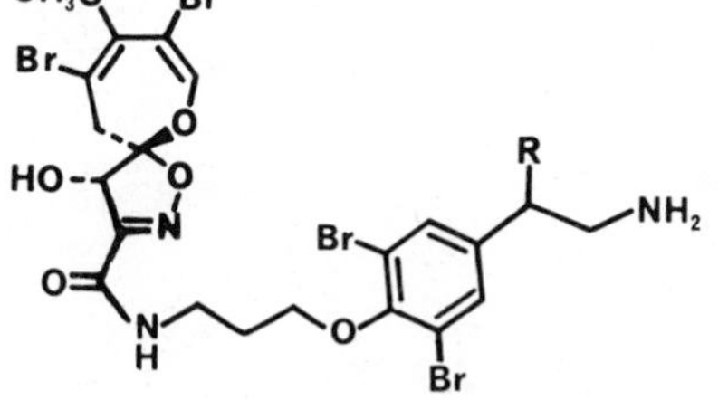

85 R=H

86 R=OH

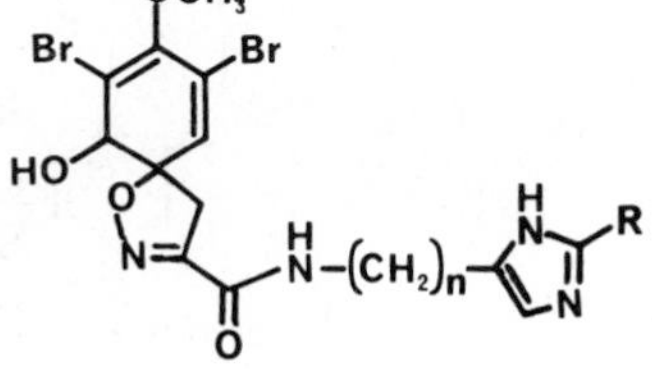

87 R=H, n=2

88 R=NH$_2$, n=3

Sponges of the genus *Ianthella* are placed taxonomically within the same order as *Verongia* (= *Aplysina*). It is not surprising, therefore, that a series of cyclic and acyclic aromatic hydroximines, also based upon tyrosine metabolism, have been isolated from this source. Australian samples of *I. basta* contained metabolites **89–95**, which are known as the bastadins 1–7 [75,76]. The structures of these metabolites were provided largely through spectral analyses, but subsequent syntheses of bastadins 1–3 [77] and bastadin 6 [78,79] have confirmed these spectral assignments. Like many brominated phenols, the bastadins generally show potent *in vitro* antibacterial activities.

Indole Derivatives. Indole derivatives are among the most common marine metabolites and they are also found in a diversity of sponges. Bro-

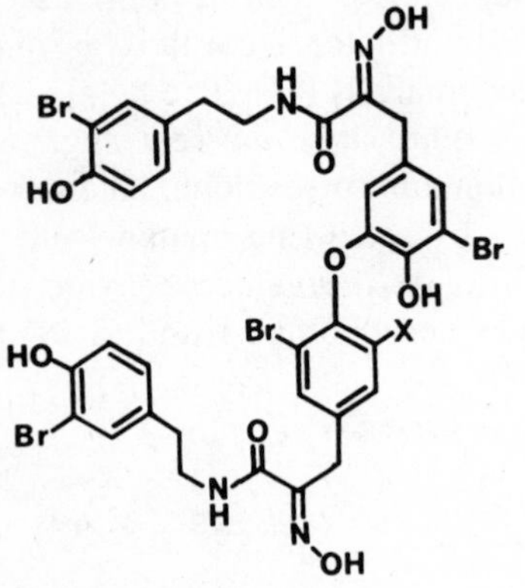

89   X=H
90   X=Br

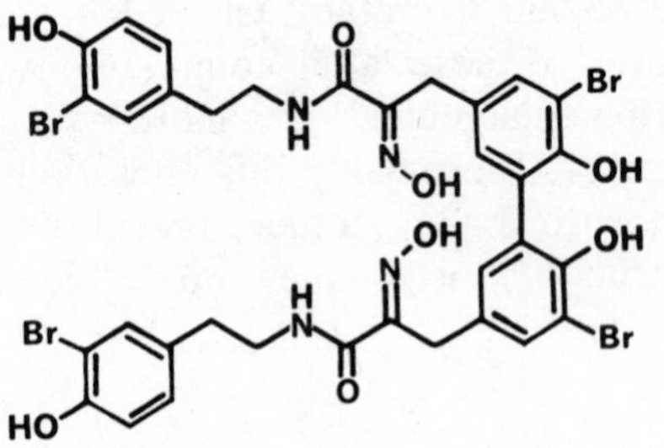

91

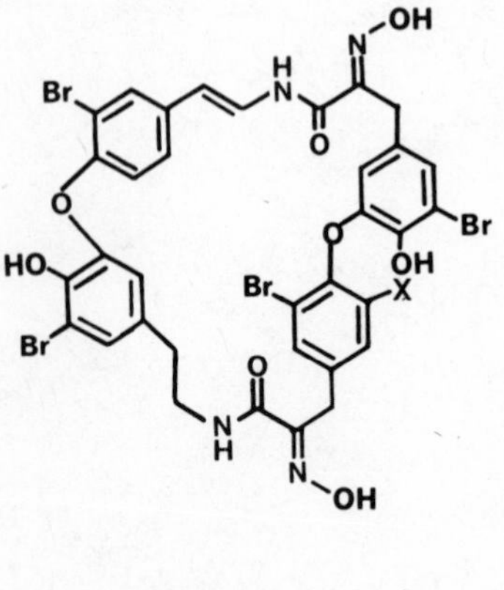

92   X=Br
95   X=H

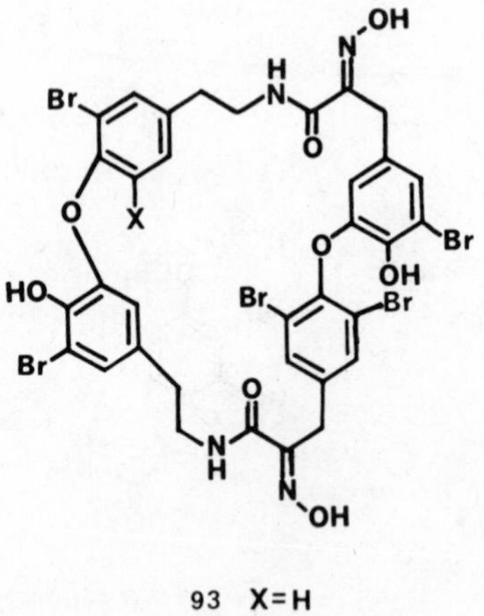

93   X=H
94   X=Br

moindoles, substituted in either the C(5) or C(6) position, are most common with alkyl substitution at the C(3) position, as predicted by their probable biosynthesis from tryptophane. The simplest examples are the bromo derivatives of tryptamine that have been isolated from various *Smenospongia* (= *Polyfibrospongia*) species. The dimethyltryptamines **96** and **97** were recently found in *S. echina* and *S. aurea* [80], while earlier studies had shown the Caribbean sponge *P. maynardii* to contain metabolites **98** and **99** [81]. The latter compounds showed *in vitro* antibacterial activity against both gram + and gram − bacteria. A simple 6-bromoindole **100** was isolated from an *Iotrochota* species [82].

The indole derivative aplysinopsin (**101**) was simultaneously isolated from an Australian Barrier Reef sponge initially identified as a *Thorecta* species [83], and from a Caribbean sponge identified as *Verongia spengelii* [84]. In a comprehensive chemotaxonomic study of this group of sponges, Bergquist has subsequently revised the naming of these organisms to *Fascaplysinopsis reticulara* and *Smenospongia echina*, respectively [85]. Aplysinopsin was shown to possess cytotoxicity against the P-388, KB, and L1210 cancer cell lines [84]. Demethylated derivatives of aplysinopsin (**102,103**) have also been isolated from a sponge of the genus *Dercitus* [86], while methylaplysinopsin (**104**) was also later reported as a natural product [87]. Methylaplysinopsin is a potent serotonergenic agent with significant antidepressant properties [87]. The bromoindole **105**, a distant relative of the aplysinopsin indoles,

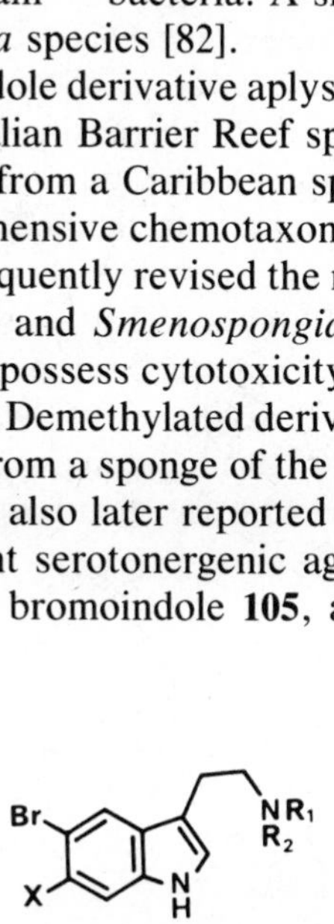

```
96   R₁R₂ = CH₃ , X = Br
97   R₁R₂ = CH₃ , X = H
98   R₁ = CH₃ , R₂ = H , X = Br
99   R₁R₂ = H , X = Br
```

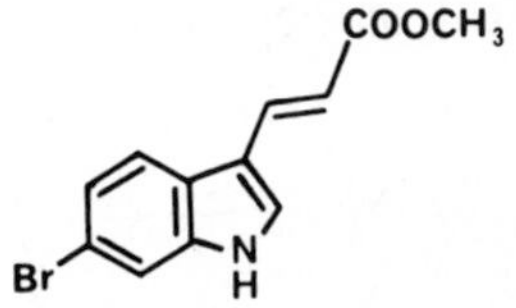

100

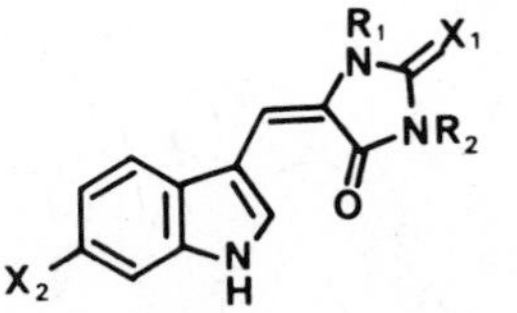

```
101   R₁R₂ = CH₃ ,  X₁ = NH , X₂ = H
102   R₁ = H, R₂ = CH₃ ,  X₁ = NH, X₂ = H
103   R₁ = H, R₂ = CH₃ ,  X₁ = NH, X₂ = Br
104   R₁R₂ = CH₃ ,  X₁ = NCH₃ , X₂ = H
105   R₁R₂ = H ,  X₁ = O , X₂ = Br
```

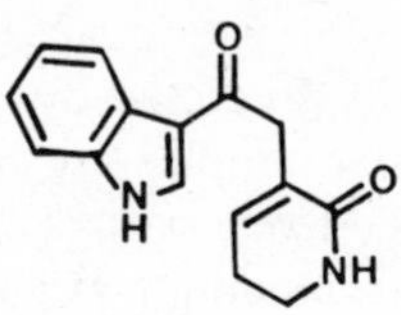

106 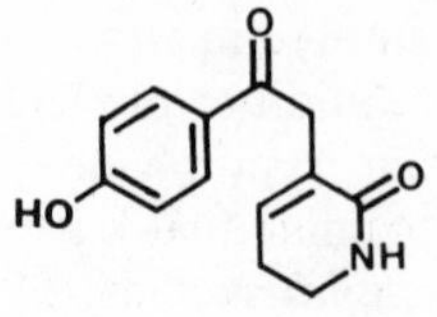   107

was reported to be a minor metabolite of *Smenospongia aurea* [80]. The indole lactam **106**, and the analogous tyrosine derivative **107**, are products of the sponge *Halichondria melanodocia* [88].

A series of interesting indole alkaloids, **108–112**, possessing antibiotic properties, were recently reported from the temperate sponge *Cliona celata* [89–92]. Difficulties in isolation were encountered due to the polarities and instabilities of these metabolites. Thus in several cases the fully acetylated derivatives were synthesized and subsequently defined. Tetraacetyl clionamide (**108**) was the major acetylated compound identified, along with acetylated celenamides A (**110**), B (**111**), and C (**112**). Evidence of the presence of the unacetylated clionamide (**109**) was given, but the compound was unstable in light and air [90]. The related alkaloid **113** was also found among these interesting compounds [92].

Pyrrole Derivatives.   Sponges contain a diversity of both simple and complex pyrrole derivatives. The Mediterranean sponge *Oscarella lobalaris*,

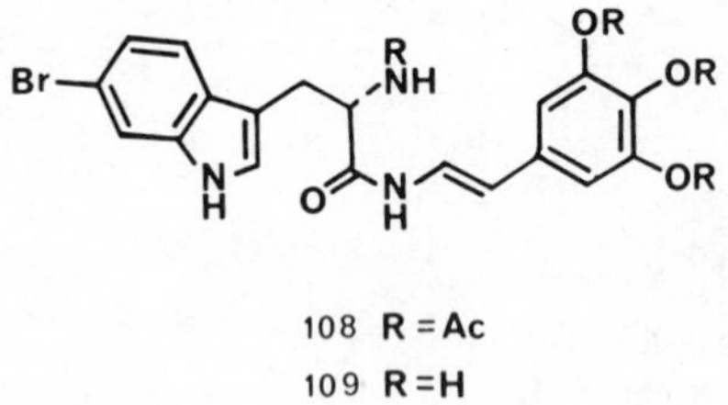

108 R = Ac
109 R = H

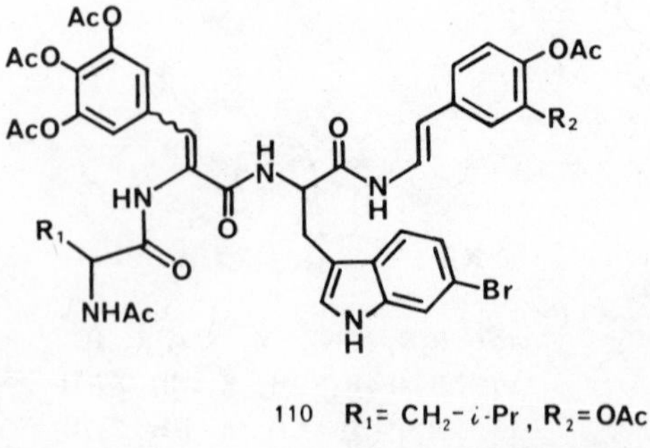

110   $R_1 = CH_2\text{-}i\text{-}Pr$, $R_2 = OAc$

111   $R_1 = i\text{-}Pr$, $R_2 = OAc$

112   $R_1 = CH_2\text{-}i\text{-}Pr$, $R_2 = H$

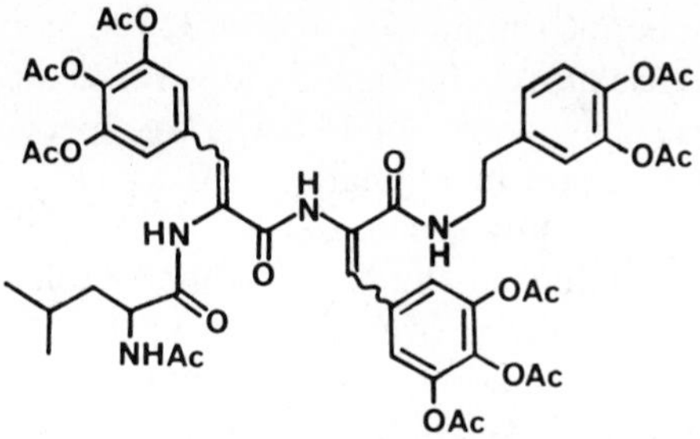

113

for example, contains complex mixtures of secondary metabolites composed of the alkylated pyrrole carboxaldehydes **114–120**, the methyl esters **121–126**, and the acids **127,128**. These latter compounds were assigned as 2-alkyl-1-carboxypyrroles on the basis of their respective ^{1}H NMR spectra in comparison with model compounds [93]. More recently, a closely related group of pyrroles has been obtained from a sponge of the genus *Laxosuberites* [94]. These compounds (**129–134**) were assigned as 5-alkylpyrrole-2-carboxal-

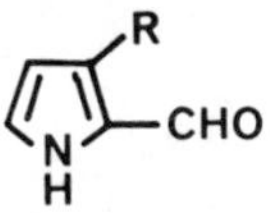

114 R = $n\text{-}C_{19}H_{39}$
115 R = $n\text{-}C_{20}H_{41}$
116 R = $n\text{-}C_{21}H_{43}$
117 R = $n\text{-}C_{22}H_{45}$
118 R = $(CH_2)_{15}$ ⌁ $(CH_2)_5CH_3$
119 R = $(CH_2)_{11}$ ⌁ $(CH_2)_7CH_3$
120 R = $(CH_2)_6$ ⌁ $(CH_2)_7$ ⌁ $(CH_2)_5CH_3$

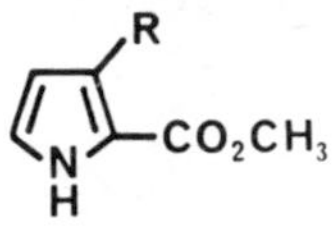

121 R = $n\text{-}C_{19}H_{39}$
122 R = $n\text{-}C_{20}H_{41}$
123 R = $n\text{-}C_{23}H_{47}$
124 R = $n\text{-}C_{23}H_{45}$
125 R = $n\text{-}C_{21}H_{41}$
126 R = $n\text{-}C_{23}H_{43}$

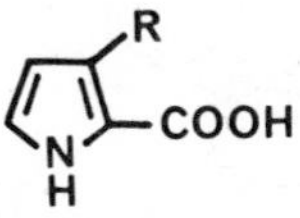

127 R = $n\text{-}C_{21}H_{40}$
128 R = $n\text{-}C_{23}H_{45}$

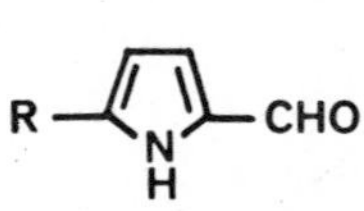

129 R = $n\text{-}C_{15}H_{31}$
130 R = $n\text{-}C_{16}H_{33}$
131 R = $n\text{-}C_{17}H_{35}$
132 R = $n\text{-}C_{19}H_{39}$
133 R = $(CH_2)_5$ ⌁ $(CH_2)_5CN$
134 R = $(CH_2)_5$ ⌁ $(CH_2)_{15}$ $\underset{OH}{CH}$ $-CN$

dehydes, and the spectral similarities of **129–134** to the aldehydes **114–120** may indicate some insecurities in the assignments of the earlier defined derivatives. The unusual cyanohydrin **134** was unexpectedly stable to chromatography and subsequent manipulation [94].

Simple bromopyrroles, potentially derived from proline, have also been isolated from marine sponges. The 4,5-dibromopyrrole carboxylic acid derivatives **135–138** were isolated from extracts of the Mediterranean sponge *Agelas oroides* [95]. The methyl ester **136** was recognized as an artifact formed during methanol extraction. The *N*-methyl analog of these latter compounds (**139**), known as midpacamide, was identified as a component of an unidentified sponge from the Marshall Islands [96]. This latter sponge also contained the secondary amide derivative **140**, which possesses an *N*-methyl hydantoin substituent.

A series of pentacyclic terpenoids, the millorins, in which pyrrole rings are incorporated via 1,4-dialdehyde cyclization, have been isolated from the Mediterranean sponge *Cacospongia mollior* [97–101]. Millorins A–E (**141–145**) are derivatives of the sesterterpenoid "spongiane" skeleton, found earlier in several scalarin derivatives [10], but with differing *N*-alkyl substituents. The millorins, with the exception of millorin C (**143**), were synthe-

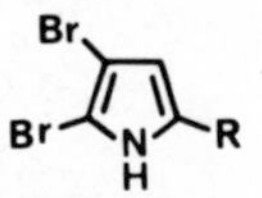

135 R = COOH
136 R = COOCH₃
137 R = CN
138 R = CONH₂

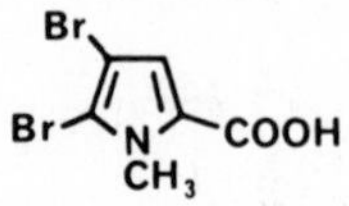

139

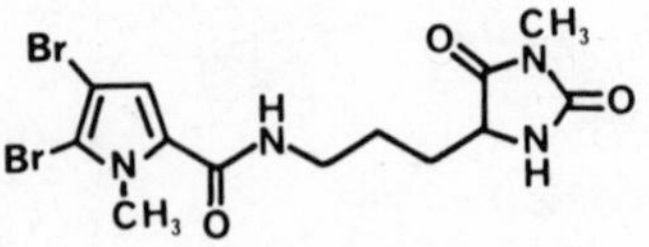

140

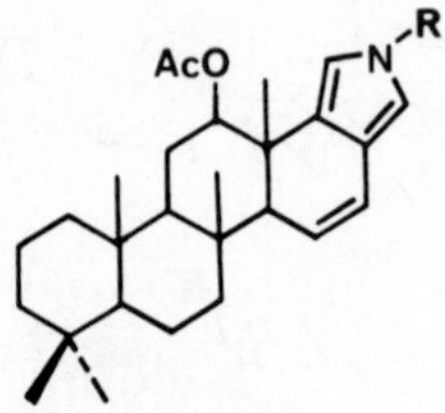

141 R =

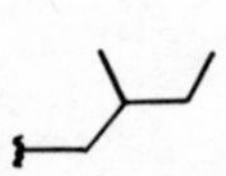

143 R =

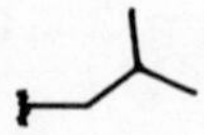

144 R =

145 R = CH₃

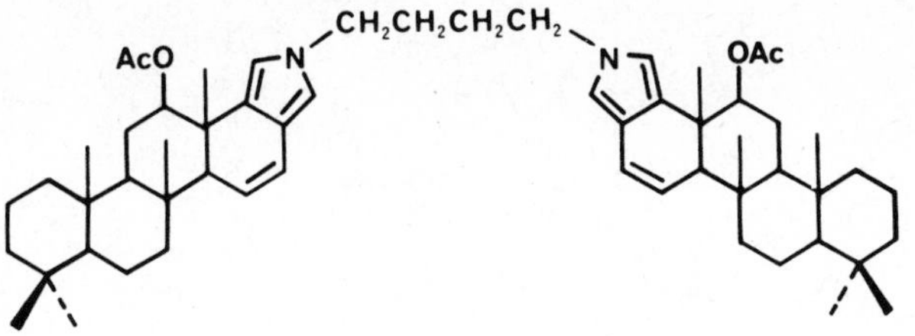

142

sized by the condensation of the appropriate amine with the corrresponding terpenoid dialdehyde, scalaradial, which was also isolated from a Mediterranean sponge [54].

Aminoimidazole and Guanidine Derivatives. One of the very first alkaloidal structures isolated from sponges was the 2-aminoimidazole derivative, oroidin (**146**), a salt isolated first in 1971 from the Mediterranean sponge *Agelas oroides* [95]. The structure of oroidin was originally incorrectly assigned and a revision, based upon the synthesis of dihydrooroidin, was later reported [102]. Complete confirmation of the structure of oroidin as a hydrochloride salt by X-ray analysis was recently mentioned in a footnote [108], but details of this work have not appeared. Oroidin has subsequently been isolated from sponges of the genus *Axinella* [103], and several related compounds have subsequently been described. Mono- and dibromophakellin (**147,148**) are examples of more complex compounds isolated from the Great Barrier Reef sponge *Phakellia flabellata*. These compounds were extensively studied for the period 1971–1977 [104,105]. The structures of the phakellins can be considered to be derived from a dihydro derivative of

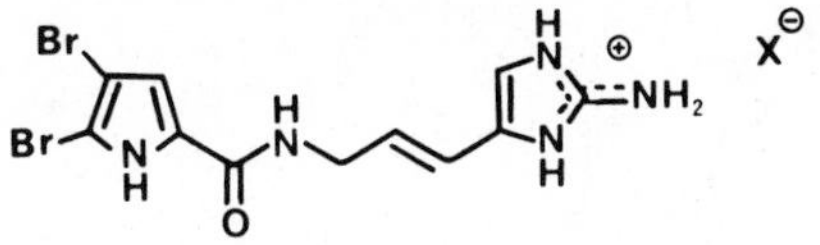

146

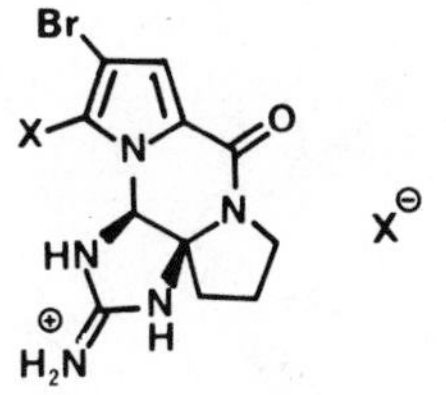

147 X = H
148 X = Br

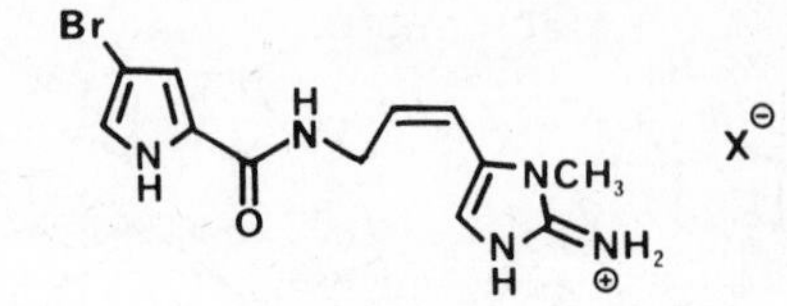

149

oroidin (146) via addition of the pyrrole and amide nitrogen atoms to the imidazole double bond. A monobromo-*N*-methyl derivative of oroidin, known as keramadine (149), has been found in an Okinawan sponge of the genus *Agelas* [106]. Keramidine possesses antagonistic activity on the serotonergic receptor sites in the rabbit aorta [107].

Compounds are also known that are related to oroidin by subsequent dimerization or intramolecular cyclization. The unique cyclobutane derivative sceptrin (150) is an antibiotic metabolite of the Caribbean sponge *Agelas sceptrum* [108]. This unusual substance is formally a [$2\pi + 2\pi$] "cycloaddition dimer" of 5-debromooroidin. Cyclization of the double bond in debromooroidin to the 3-position of the pyrrole ring has also been observed in the metabolites of several sponges. The sponge *Phakellia flabellata* was shown to contain the cyclization product 151, which was identified on the basis of its spectral properties [109]. Later, the same compound and the brominated analog 152, were isolated from the Mediterranean sponge *Axi-*

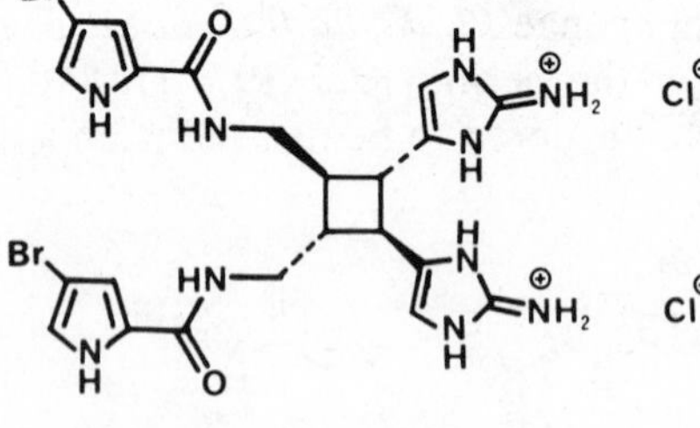

150

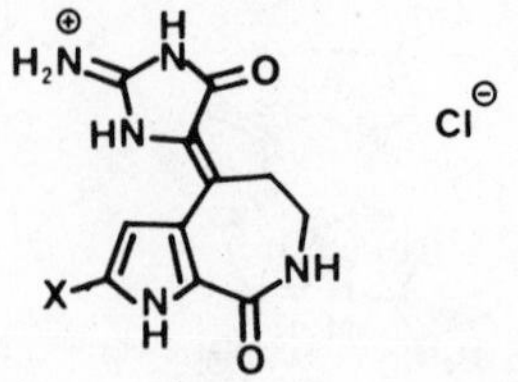

151 X = H
152 X = Br

nella verrucosa and the Red Sea sponge *Acanthella aurantiaca* [110]. In this latter investigation the structure of the bromo compound **152** was secured by X-ray crystallography. More recently, the Okinawan sponge *Hymenia-cidon aldis* was shown to also contain **151**, as well as the free bases corresponding to **151** and **152** [111].

Several miscellaneous guanidines are also found in sponges from various marine habitats. Ptilocaulin (**153**) and isoptilocaulin (**154**) are two examples of isomeric cyclic guanidines possessing both cytotoxic and antimicrobial properties. These compounds were isolated from the sponge *Ptilocaulis* cf. *spiculifera* [112], and are of an unknown biosynthetic origin. The structure of ptilocaulin was confirmed by X-ray crystallography and later by chiral synthesis [113]. The structure of isoptilocaulin, while not rigorously established, has been assumed to be isomeric on the basis of favorable spectral comparisons. Siphonodictidine (**155**) is a sesquiterpenoid metabolite of several sponges of the genus *Siphonodictyon* [114] that burrow into and destroy living coral heads. Laboratory bioassays showed that siphonodictidine was toxic to coral tissue, and thus this metabolite was proposed to play an offensive role in the infestation of coral by the sponge.

Pyridine, Quinoline, and Isoquinoline Derivatives. Alkaloidal structures in which nitrogen is part of a six-membered ring are less frequently found in marine organisms. Among the marine invertebrates known as nemertines, or ribbon worms, several species within the order Hoplonemertinae contain anabaseine (**156**), a close relative of nicotine that is the unsaturated (imine)

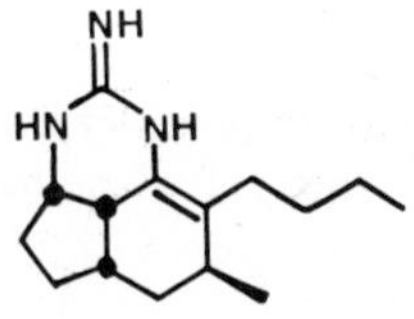

153

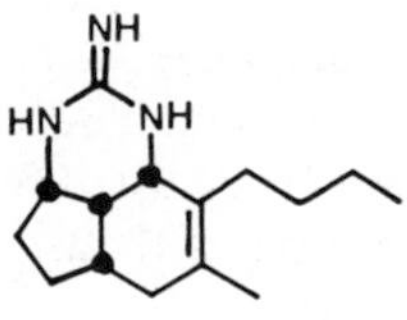

154

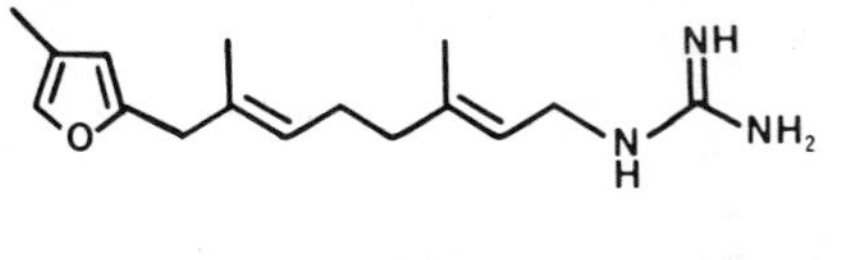

155

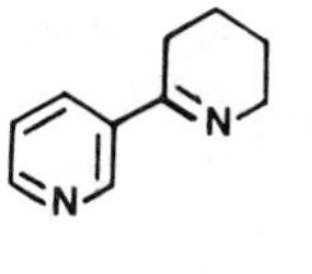

156

analog of the common plant alkaloid anabasine [115]. In more recent research, a toxic mixture of high molecular weight pyridinium salts, referred to as halitoxin (157), has been isolated from marine sponges of the genus *Haliclona* [116]. Halitoxin (157) was initialy isolated for structure determination from *H. rubens*, but this complex toxin was also detected in *H. viridis* and *H. erina*. It was concluded that halitoxin consists of mixtures of repeating units of 3-alkyl-*N*-alkylpyridinium salts within a wide molecular weight range (1,000–25,000 amu).

Although sponges of the genus *Aplysina* (= *Verongia*) characteristically contain brominated tyrosine derivatives (see the earlier discussion of bromotyrosine derivatives), the dihydroxyquinoline carboxylic acid 158 was among the first marine natural products to be described. Acid 158 was isolated from the aqueous extract of the Mediterranean sponge *Verongia aerophoba*, and although a simple structure, it had not been previously isolated as a natural product [117]. While this latter research was initiated in the early 1970s, it was not until very recently that similar nitrogenous substances have been reported from *Verongia aerophoba*. The species name "*aerophoba*," or "air-fearing," accurately describes these bright yellow sponges, which rapidly blacken upon exposure to air. Many *Verongia* species possess this unusual characteristic, which has been a subject of curiosity since the late 1800s. The precise structure of the "zoochrome" responsible for this behavior has very recently been shown, on the basis of its spectral and chemical properties [118], to be 3,5,8-trihydroxy-4-quinolinone (159). The *Verongia* zoochrome is obviously related to the dihydroxyquinoline 158, and it has been proposed as a logical precursor.

At least one sponge contains isoquinoline alkaloids. An unknown representative of the genus *Reniera*, from Pacific Mexico, contains renierone (160) as the major metabolite, along with minor amounts of the related compounds 161–168. The structure of renierone (160) was conclusively secured by X-ray diffraction methods [119], while the structures of the related com-

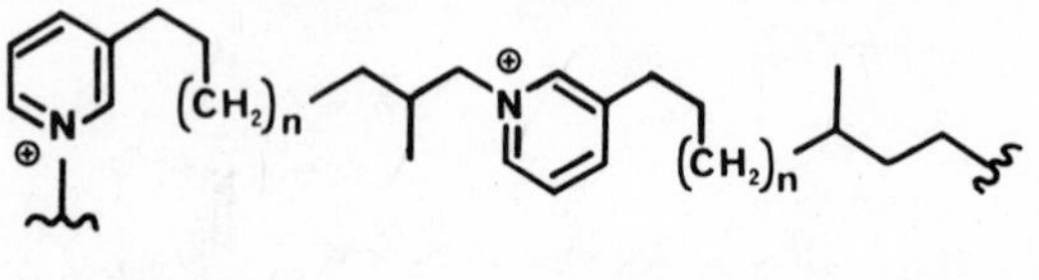

157

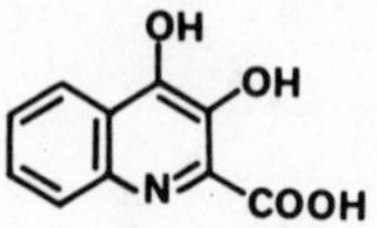

158

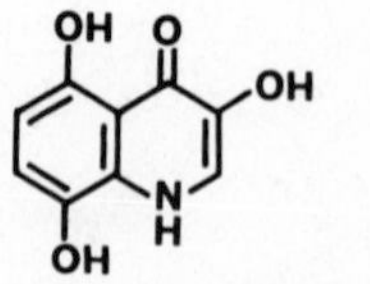

159

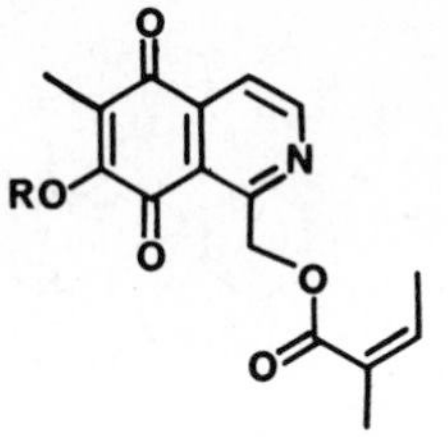

160   R = CH₃

161   R = H

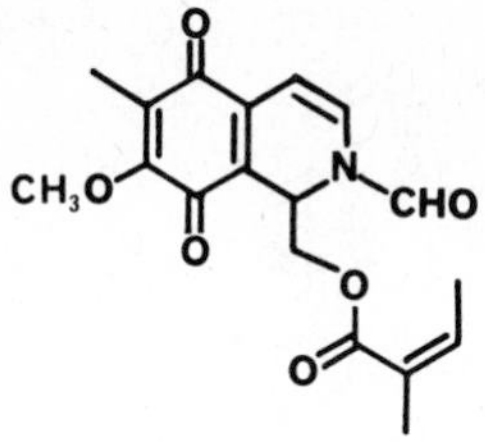

162

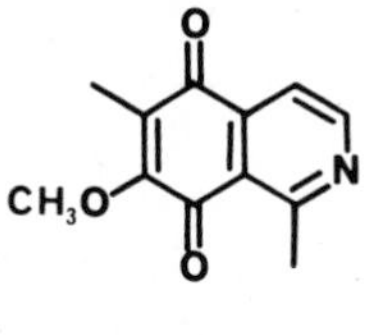

163

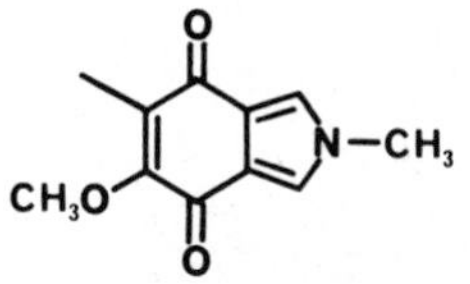

164

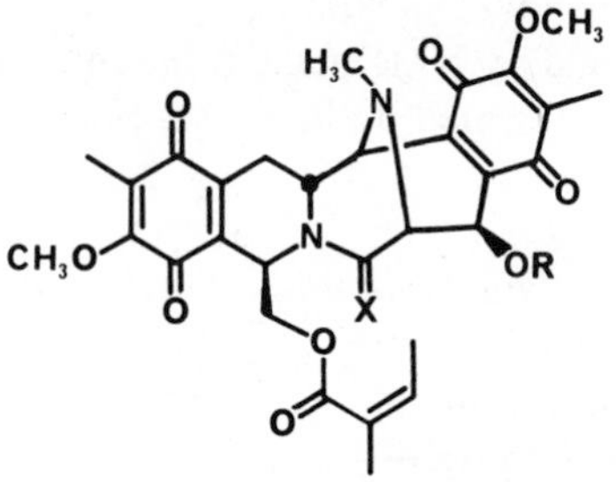

165   R = H , X = H₂
166   R = C₂H₅, X = H₂
167   R = H , X = O
168   R = C₂H₅, X = O

pounds **161–168** were assigned on the basis of their spectral properties [120]. In assigning the structures of the dimeric isoquinoline alkaloids reniera-mycins A–D (**165–168**), the power of ¹H NMR nuclear Overhauser enhance-ment (NOE) spectral methods in assigning overall stereochemistry was con-vincingly illustrated [120]. These isoquinoline alkaloids are closely related to several isoquinoline alkaloids produced by terrestrial actinomycetes. Sim-ilar structures include mimosamycin and mimosin, from strains of *Strep-tomyces lavendulae*, which also possess the A-ring *p*-quinone functionalities. It is interesting to note that the same terrestrial bacterium also produces dimeric alkaloids, the saframycins, which are closely related to the renier-amycins A–D (**165–168**). The question soon arises as to the true "source" of these antibiotics given the known symbiotic associations of sponges with marine microorganisms. Although unsuccessful to date, attempts have been made to isolate and culture sponge symbionts. It is also not unreasonable, however, that sponges themselves produce these isoquinoline alkaloids. Of

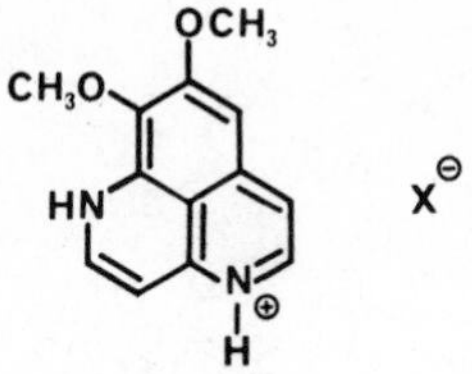

169 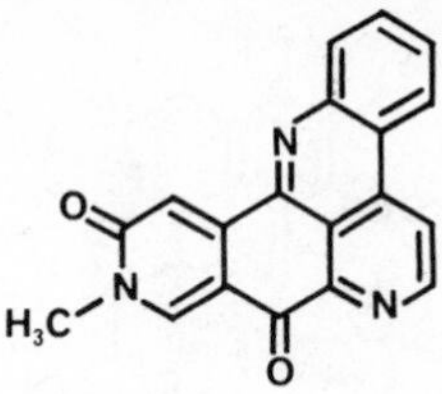   170

the marine invertebrates, sponges are considered the most primitive and, therefore, more closely related to the procaryotic microorganisms.

Quinoline and isoquinoline bases that are part of larger polycyclic systems have also been recently found in sponges. The Okinawan sponge *Aaptos aaptos*, for example, contains the unprecedented tricyclic alkaloid, aaptamine (**169**) [121]. This exceptional compound was isolated by monitoring α-adrenoacceptor blocking activity in the extract fractionation process, and the structure of **169** was assigned on the basis of chemical and spectral studies. In another recent investigation of a Pacific sponge of the genus *Amphimedon*, an unprecedented pentacyclic alkaloid, amphimedine (**170**) was discovered [122]. This novel substance showed cytotoxic properties and its structure was determined by NMR measurements, which emphasized the determination of ^{13}C-^{13}C natural abundance couplings that provided informative carbon connectivities.

Uncommon Nucleosides and Related Heteroaromatic Bases. Although numerous nucleic acid derivatives are ubiquitous in nature, several sponges have been recognized to contain modified derivatives of this class as nucleosides that often possess potent biological activities. Bergmann, in his pioneering work with marine natural products, was the first to recognize the presence of modified nucleosides in the Caribbean sponge *Crypotethia crypta*. In a series of papers in the 1950s, Bergmann and co-workers described the isolation of spongothymidine (**171**), spongouridine (**172**), and spongosine (**173**) from this latter source [123,124]. More recently, the nucleoside **174** (9-β-D-ribofuranosyl-1-methylisoguanine) was isolated from the Australian sponge *Tedania digitata* [125]. This latter nucleoside was sub-

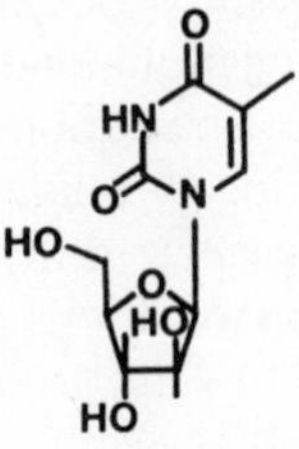

171 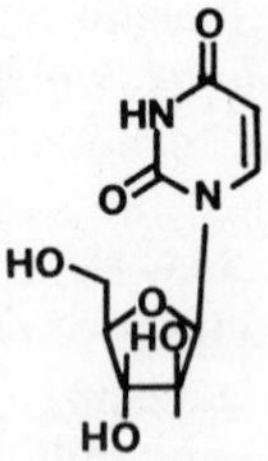   172

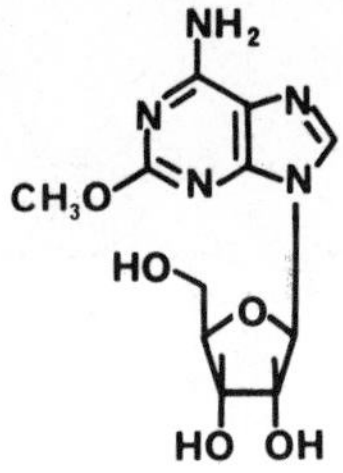

173 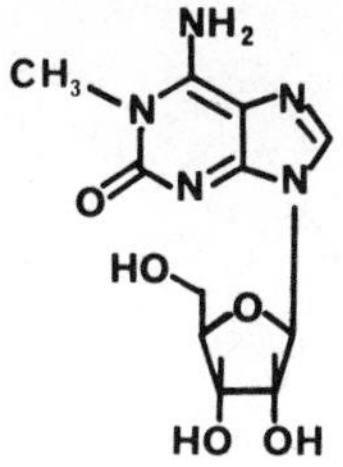   174

sequently shown to produce muscle relaxation and hypothermia, hypotension associated with bradycardia, and antiinflammatory and antiallergic activity in rats [125].

In 1975 the sponge *Agelas dispar* was reported to contain a 9-methyladenine derivative, agelasine, which was deduced to possess a $C_{20}H_{33}$ alkyl substituent in the 7-position [126]. Although the 9-methyladenine unit was conclusively identified, the nature of this latter alkyl component was not defined. Two recent investigations of Pacific sponges of the genus *Agelas* have probably illustrated the general composition of this fragment. The antimicrobial substitutents of an *Agelas* sponge from Palau were shown to be ageline A (175) and ageline B (177) [127]. In two papers, the structures of agelasines A (178), B (179), C (180), D (181), E (176), and F (= ageline A, 175) were subsequently provided from investigations of the Okinawan sponge *A. nakamurai* [128,129]. Unfortunately these latter papers suggested new names for at least one previously described compound. The agelasines, as a group, were reported to possess antimicrobial properties and to produce inhibitory effects against the enzymes Na- and K-ATPase.

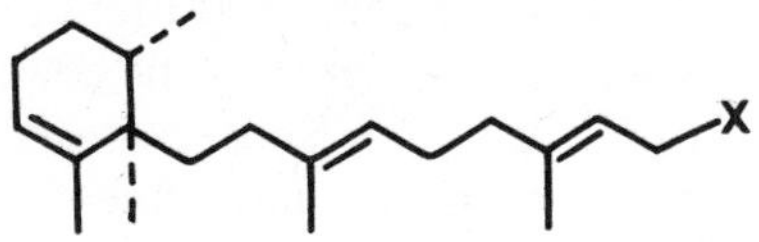

175

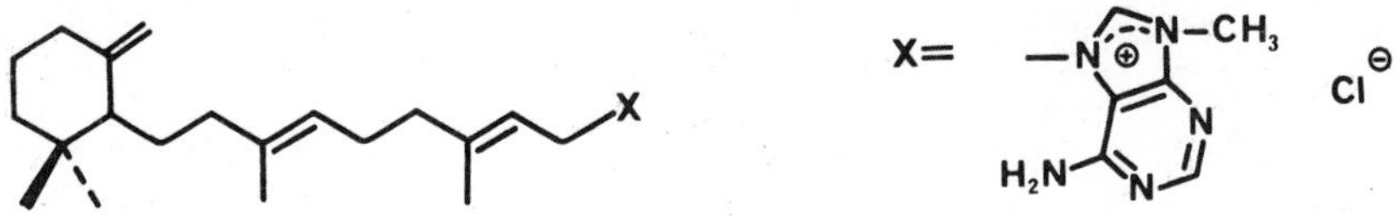

176

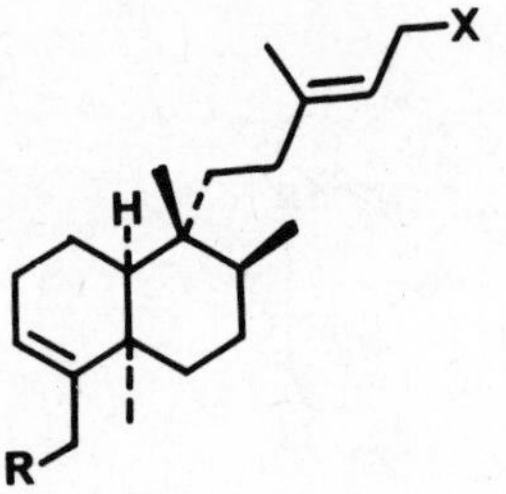

177 R = OCO-〈pyrrole〉

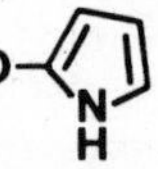

178 R = H

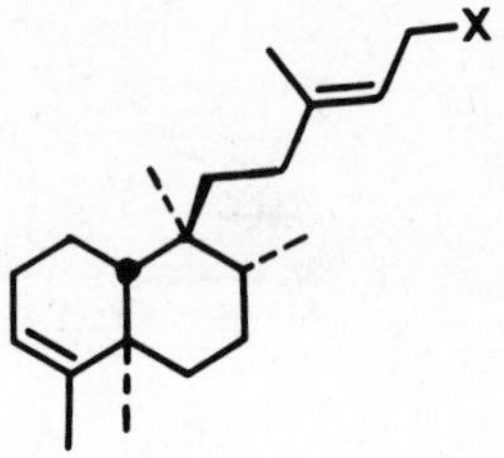

179

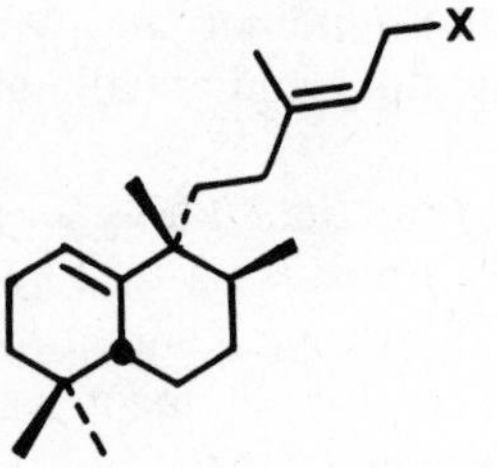

180

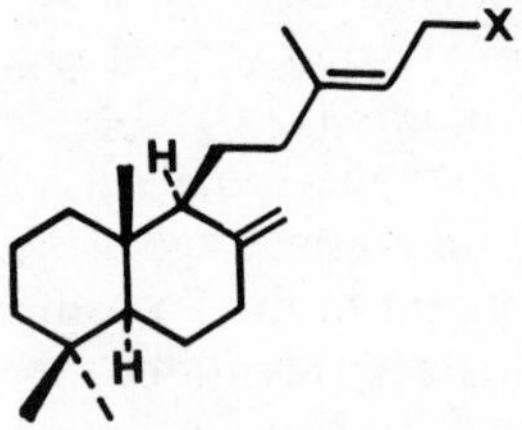

181

Although pyrrolo[2,3-*d*]pyrimidine bases are very rare in nature, the interesting bromo derivative **182** was recently isolated from an unidentified Australian sponge of the genus *Echinodictyum* [48]. This interesting base showed strong bronchodilating activity in the isolated guinea pig trachea.

Isonitriles and Related Nitrogen-Containing Metabolites. Naturally occurring isonitriles are uncommon in the terrestrial environment, found in only rare cases as metabolites of several *Penicillium* species. In contrast, marine sponges of several related genera have been recognized to contain both sesquiterpene- and diterpene-derived isonitriles. In most cases the isonitriles are accompanied by the corresponding isothiocyanates and formamides, which have been recently shown to be produced *in vivo* from the parent isonitrile [130,131]. The biosynthetic origin of the isonitrile carbon atom remains unknown, but incorporation experiments have ruled out formate as the precursor [131].

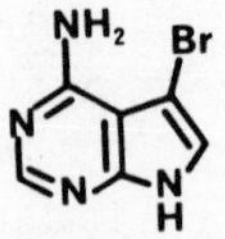

182

The first marine-derived isonitrile to be described was axisonitrile-1 (**183**) from the Mediterranean sponge *Axinella cannabina* [132]. Axisonitrile-1 was isolated, along with its corresponding isothiocyanate (**184**) and formamide (**185**) [132]. The related vinyl isonitrile, axisonitrile-4 (**186**) and the corresponding metabolites **187** and **188** were also described [132,133]. In extensive studies of the same sponge, sesquiterpenoid isonitriles of two additional ring systems, **189–191** and **192–194**, were described [134,135]. More recently the isonitrile **195** and the isothiocyanate **196**, which possess yet another sesquiterpene skeleton, were reported from an unidentified California *Axinella* species [136].

Earlier research with marine isonitriles involved several other sponge genera. Acanthellin-1 (**197**) was described in 1974 from the Mediterranean sponge *Acanthella acuta* [137] and the metabolites **198–200** were isolated from a Hawaiian *Halichondria* species [138]. Both of these isonitrile ring systems represent common terrestrial skeletons.

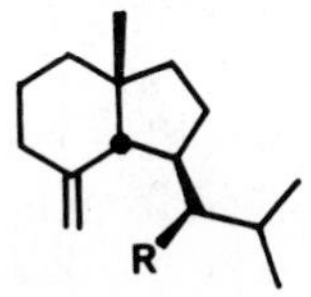

183 R=NC
184 R=NCS
185 R=NHCHO

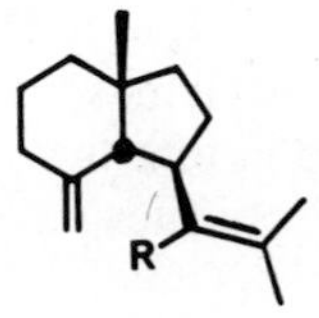

186 R=NC
187 R=NCS
188 R=NHCHO

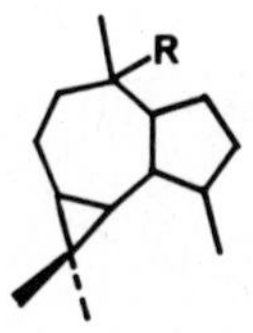

189 R=NC
190 R=NCS
191 R=NHCHO

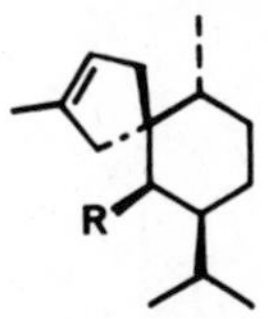

192 R=NC
193 R=NCS
194 R=NHCHO

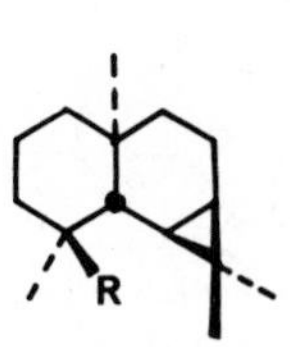

195 R=NC
196 R=NCS

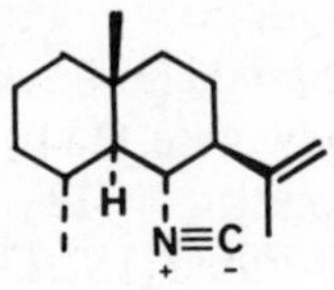

197

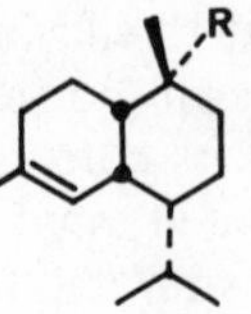

198 R=NC
199 R=NCS
200 R=NHCHO

One of the more interesting classes of sponge-derived isonitriles are the isocyanopupukeananes (**201,202**) isolated from both the nudibranch mollusc *Phyllidia varicosa* and its prey, a sponge of the genus *Hymeniacidon*. The structures of these tricyclic sesquiterpenoids were provided by X-ray diffraction, and it was shown that these metabolites are used as a defensive secretion by the nudibranch [139,140]. A monocyclic isothiocyanate (**203**) and a formamide (**204**) of the bisabolane ring system have also been recently described from the Okinawan sponge *Theonella* cf. *swinhoei* [141]. This latter case was unique in that the presumed parent isonitrile was not observed.

Diterpenoid isonitriles have also been observed from numerous sponges, and the ring systems involved are usually unprecedented. The first discovery of a diterpenoid isonitrile was in 1976, when the unprecedented tetracyclic diisonitrile **205** was isolated from an unidentified Great Barrier Reef sponge of the genus *Adocia* [142]. The structure of **205** was based on X-ray crystallography. In a subsequent paper, the same group provided the structures of six minor but equally interesting isonitriles **206–211**, from what is apparently the same sponge source [143]. In this latter case the structures of

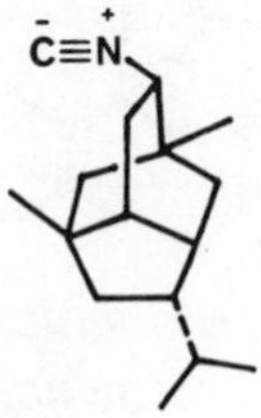

201

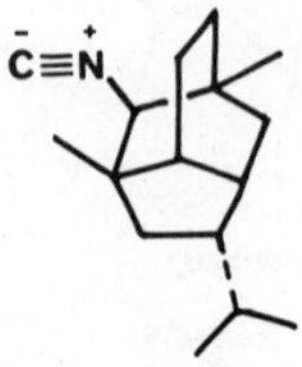

202

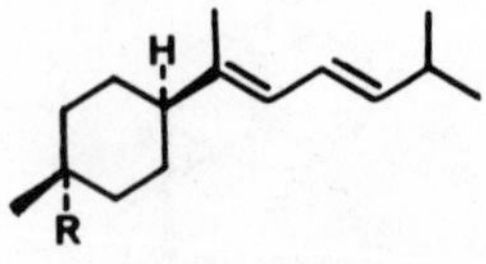

203 R=NCS
204 R=NHCHO

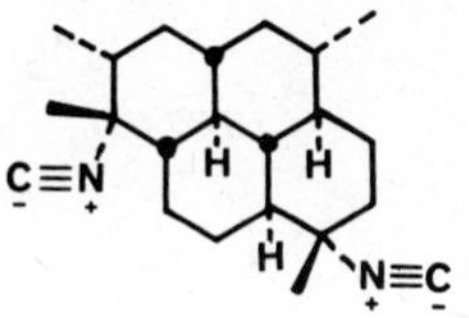

205

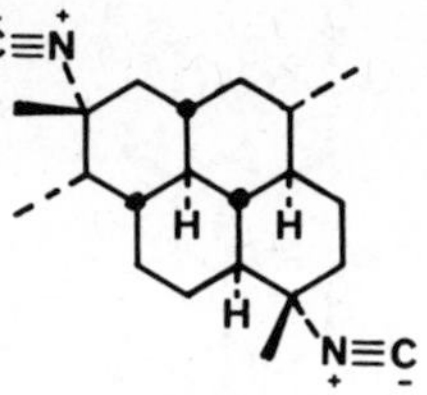

206

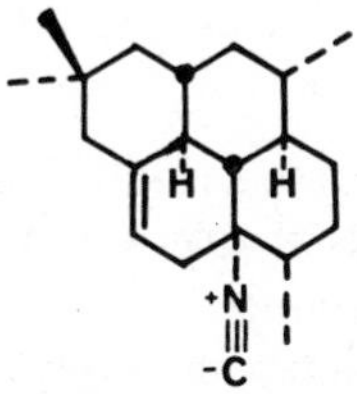

207

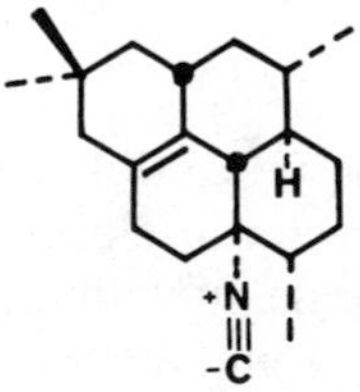

208

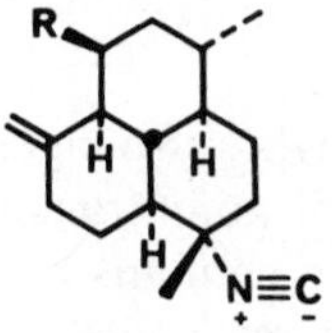

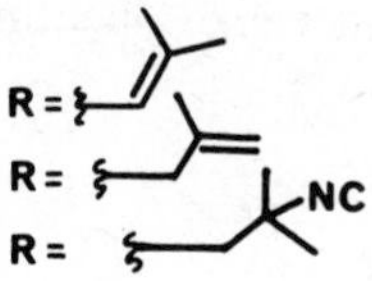

209 R =

210 R =

211 R =

206 and **207** were also provided by X-ray methods. These authors did not report the co-isolation of isothiocyanates nor formamides from these sponges.

The minor tricyclic metabolites of an unidentified *Adocia*, **209–211**, were very similar in structure to an isonitrile/formamide pair, **212** and **213**, reported earlier from the sponge *Hymeniacidon amphilecta* [144]. In this case as well, X-ray methods were selected to provide the structure of the diisonitrile **212**. Lastly, a recent paper has provided the X-ray structure of kalihinol A (**214**), a chlorodiisonitrile of an interesting diterpenoid skeleton from a unidentified Pacific sponge of the genus *Acanthella* [145]. Kalihinol A was reported to possess *in vitro* antimicrobial activity against human pathogenic bacteria and yeast.

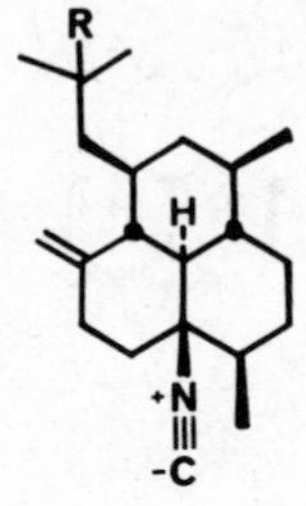

212 R=NC
213 R=NHCHO

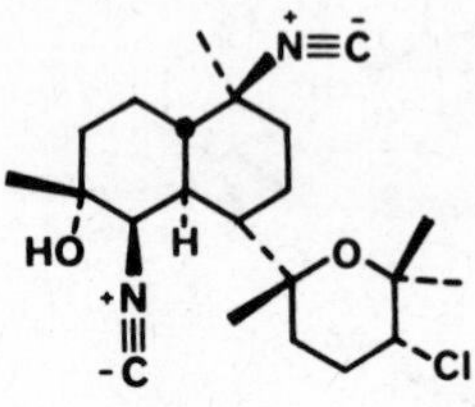

214

It would be impossible to discuss sponge-derived isonitriles without re-ferring to a group of closely related compounds that possess rare carbon-imidic dichloride functionalities. The Pacific sponge *Pseudaxinyssa pitys* contains the sesquiterpenoids **215–219**, all of which possess this unique func-tionality [146–148]. Carbonimidic dichlorides had not been observed in na-ture, but this functionality had been synthesized by the molecular chlori-nation of isonitriles. Given the precedence for natural halogenation in the marine environment, it follows that these interesting metabolites probably arise by a similar biosynthetic mechanism.

Unusual Amino Acid Derivatives.   Sponges of the genus *Dysidea* are well known for their production of nonnitrogenous terpenoids [54]. In one case, however, a *Dysidea* species contains a wide diversity of secondary metabolites including several unique amides apparently derived from un-common amino acids. Numerous collections of the Pacific sponge *D. her-*

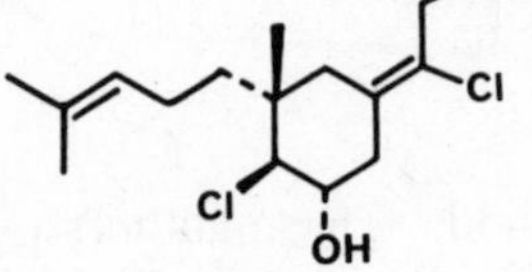

215

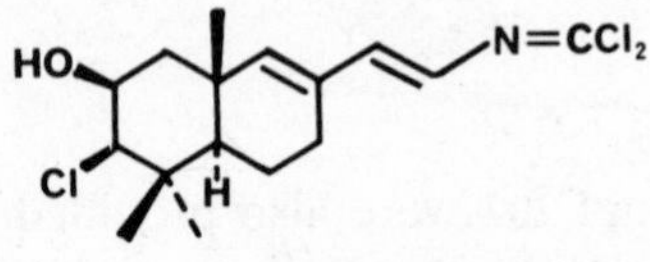

216

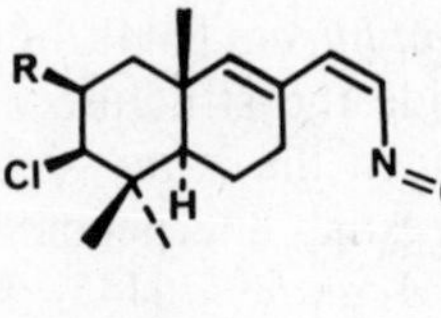

217 R = H
218 R = OH

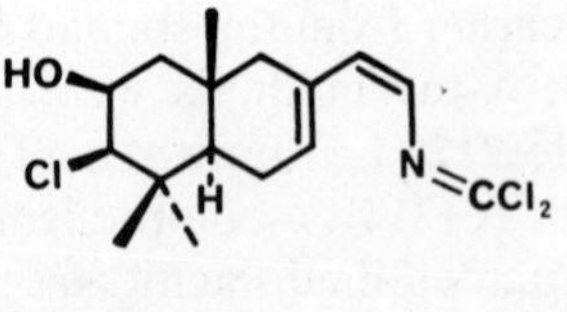

219

bacea contain the tetramic acid derivative, dysidin (**220**), the structure of which was provided by X-ray [149]. In several studies the structures of the epimeric thiazoles dysidenin (**221**) and isodysidenin (**223**) were described initially without full stereochemical details [150,151]. Subsequently, both the relative and absolute stereochemistries of these molecules have been fully defined [152,153]. Dysidenin and isodysidenin are the first examples of natural products possessing trichloromethyl groups. In subsequent investigations of the same sponge several demethyl analogs (**222,224–226**) were also described [154]. A diketopiperazine derivative (**227**) with an apparently similar origin has also been isolated from the same source [155]. Since sponges are known to contain symbiotic microorganisms, investigators have concluded that these latter compounds are of microbial origin. A favorable comparison exists between these metabolites and those produced by the blue-green algae (Section 2.2.1.), and these latter microorganisms are well known as sponge symbionts.

Macrocyclic and Steroidal Alkaloids. Several sponge alkaloids possessing 16- to 20-membered carbocyclic rings have recently been described. Petrosin (**228**), an ichthyotoxic bisquinolizidine alkaloid, was described in 1982 from the sponge *Petrosia seriata* [156]. Petrosin is a symmetrical dimer

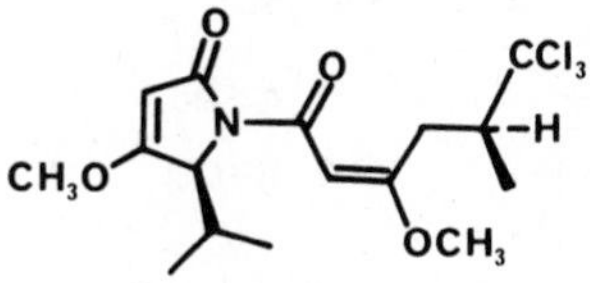

220

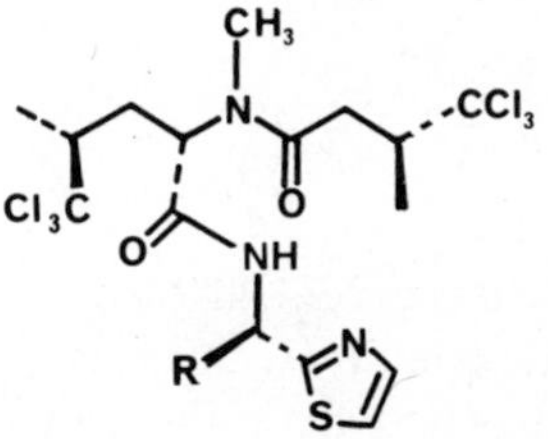

221 R=CH₃
222 R=H

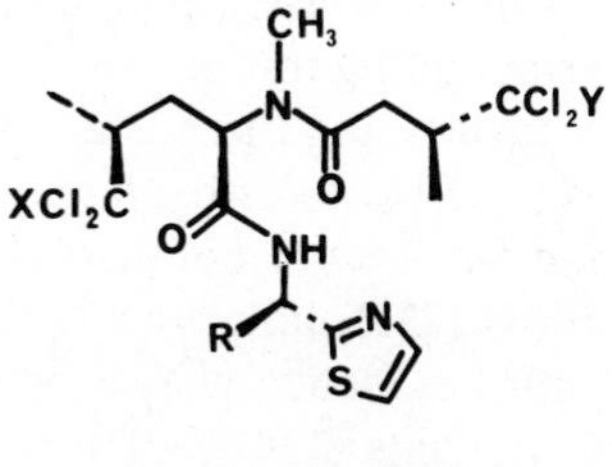

223   R=CH₃ X,Y=Cl
224   R=H, X,Y=Cl
225   R=H, X=Cl,Y=H
226   R=H, X=H,Y=Cl

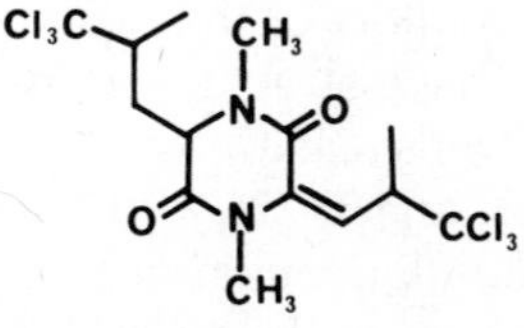

227

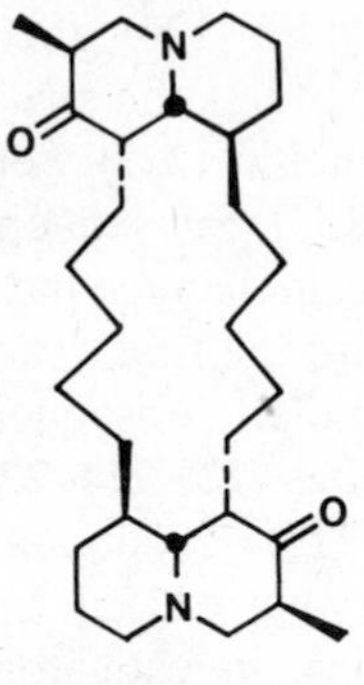

228

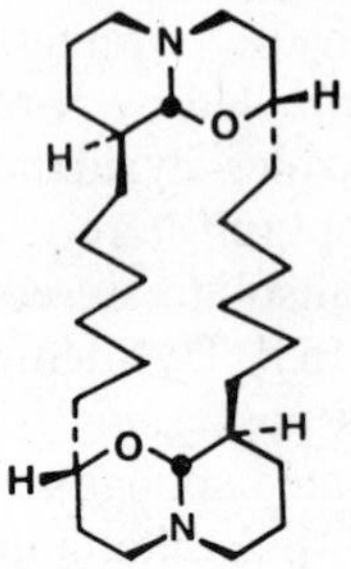

229

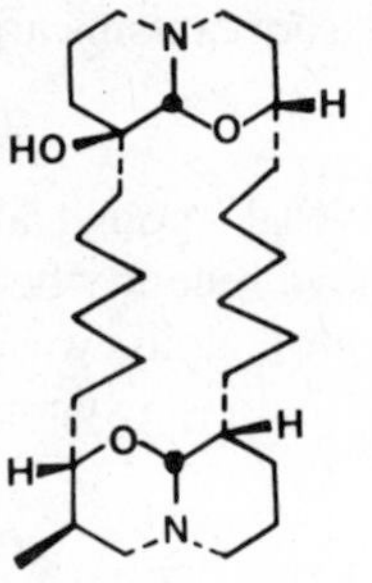

230

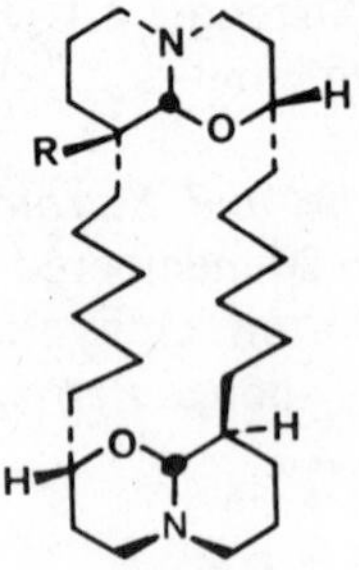

231 R=H
232 R=OH

possessing a central 16-membered ring. In a recent paper four apparently related structures, the xestospongins A (229), B (230), C (231), and D (232), were isolated from the Australian sponge *Xestospongia exigua* [157]. The structure of xestospongin C (231), determined by X-ray analysis, revealed this metabolite to possess a formally symmetrical bis-1-oxaquinolizidine structure, apparently also produced by an unknown mechanism of dimerization. The structure of the remaining metabolites, xestospongins A, B, and D, were then provided via spectral analyses. An important point emphasized by the authors of this work was the differing configurations of the bicyclic rings in each structure. Based upon X-ray results and the presence of Bohlmann bands (ca. 2760 cm^{-1}) in the infrared spectra of xestospongins A (229), C (231), and D (232), the oxaquinolizidine rings were confirmed to adopt *trans*-decalin configurations. Xestospongin B (230), however, lacked these bands, and on the basis of additional spectral evidence was assigned a *cis*-oxaquinolizidine configuration. The xestospongins (229–232) would appear to be related to petrosin (228) by nonobvious biosynthetic conversions.

Another class of macrocyclic alkaloidal structures are the potent ichthyotoxins, latrunculins A (233) and B (234), isolated from the Red Sea sponge *Latruncularia magnifica* [158,159]. The structures of these interesting new

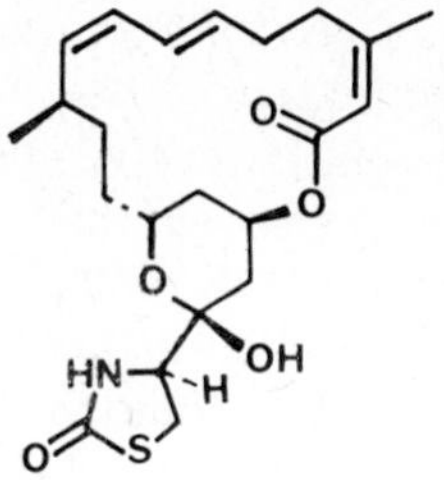

233

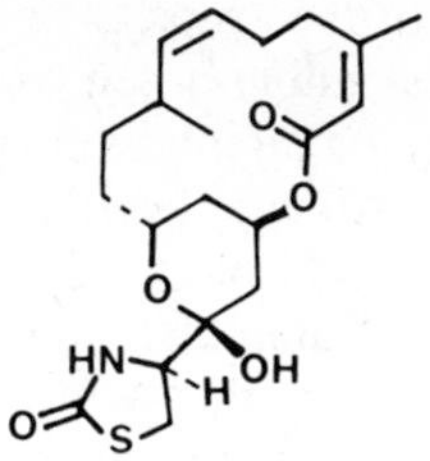

234

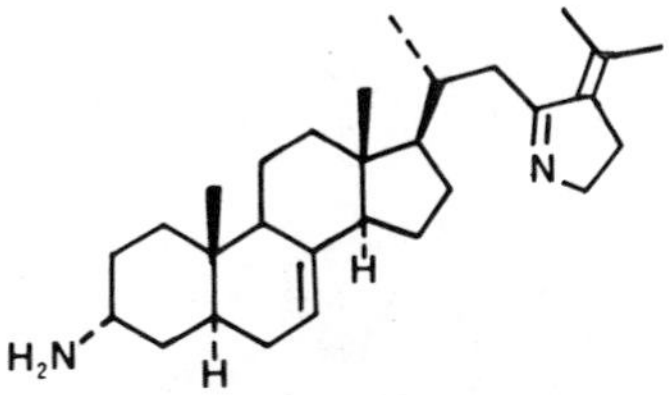

235

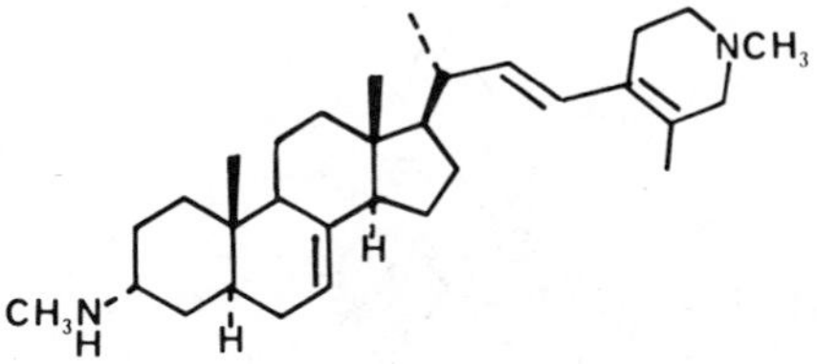

236

macrolides were based upon an X-ray diffraction study of latrunculin A (**233**), and subsequent biological studies have confirmed the potent toxicities of these metabolites apparently involving the deactivation of tubulin [160].

Although more common in terrestrial sources, two examples of steroidal alkaloids have recently been isolated from sponges. Two new antimicrobial alkaloids, plakinamines A (**235**) and B(**236**), were isolated from an unidentified Pacific sponge of the genus *Plakina* [161]. The structures of these new steroidal alkaloids were shown, by detailed spectral analyses and comparisons with model compounds, to be based on a 3-α-amino-5α-ergost-7-ene skeleton.

3.2.2. Bryozoans (Ectoprocta).

The "bryozoans" or "moss animals" (Phylum Ectoprocta) are represented by over 4000 species, most of which are marine. The bryozoans are found in both articulated and encrusting forms, and recent studies have shown these animals produce a wide spectrum of interesting alkaloids. The simple bromogramine derivatives, **237** and **238**,

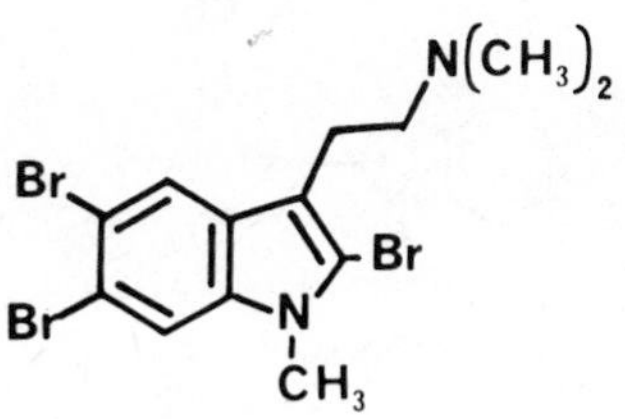

237

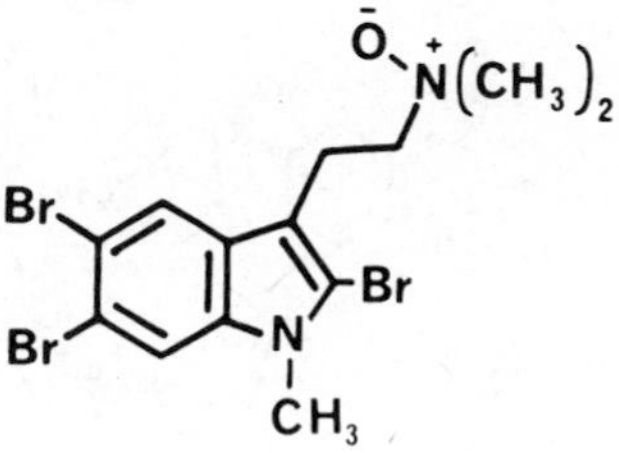

238

isolated from the tropical bryozoan *Zoobotryon verticillatum*, are characteristic of the indole-based alkaloids from this source [162]. The *N*-oxide **238** was found to be an inhibitor of mitotic cell division. Extensive investigation of the north Atlantic bryozoan *Flustra foliacea* has shown that this animal contains a variety of indole alkaloids. Flustramines A, B, and C (**239–241**), as well as frustraminols A and B (**242,243**) were the first bromophysostigmine alkaloids found in the marine environment [163–165]. In addition, *F. foliacea* was found to contain the indole derivatives flustrabromine (**244**) and flustramide A (**245**), and the trypamine derivative **246** [166,167]. The quinoline

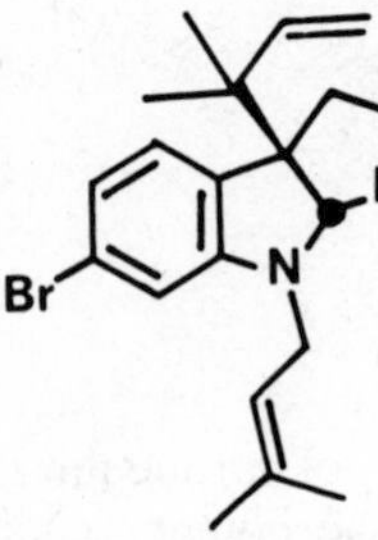

239

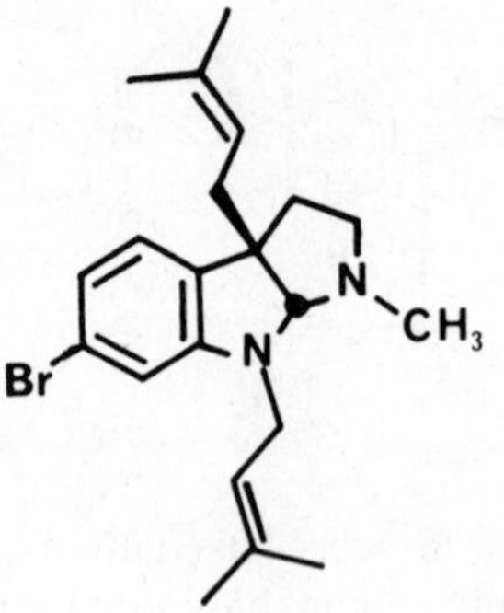

240

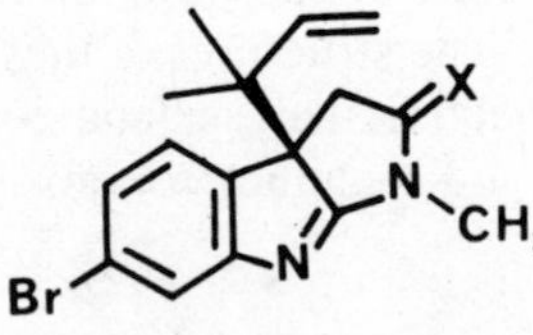

241 X=H₂
245 X=O

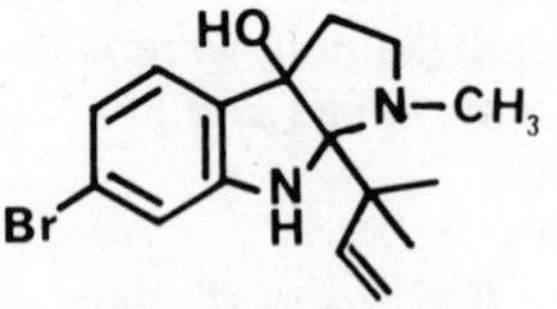

242

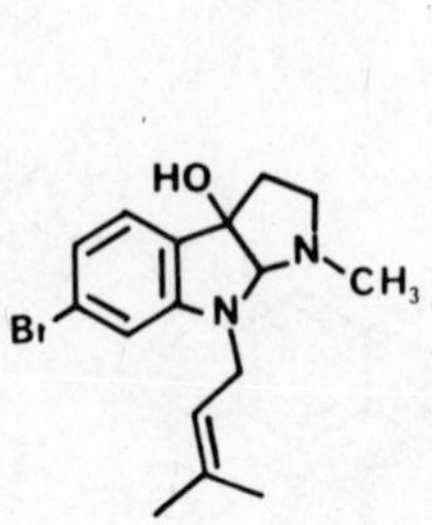

243

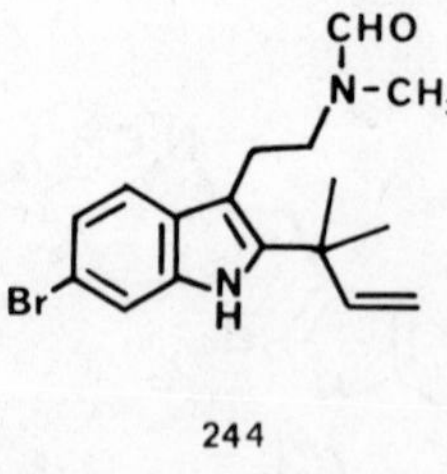

244

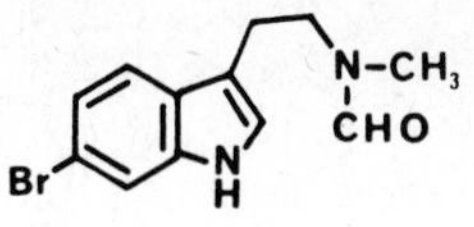

246

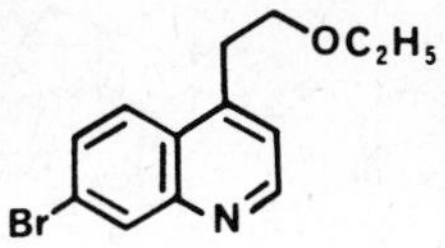

247

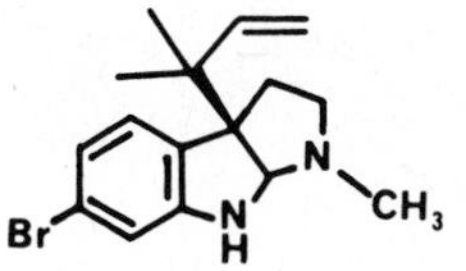

248

alkaloid **247** was also isolated from the same source [168]. A recent report has shown that dihydroflustramine C (**248**) is the major component of *F. foliacea* collected in Nova Scotia [169].

In a recent investigation of a nudibranch mollusc of the genus *Tambje*, it was recognized that this animal preys upon the Pacific bryozoan *Sessibugula transluscens*. Four interesting bipyrroles, the tambjamines A–D (**249–252**), were subsequently isolated from both the prey and its predator [170]. Like most nudibranchs, it appears that *Tambje* had concentrated these defensive substances from its food source.

Studies of the Tasmanian bryozoans *Costaticella hastata* and *Amathia wilsoni* yielded harman (**253**) and vinylharman (**254**), and the amathamides A–D (**255–258**) [171]. Lastly, a unique purine alkaloid, phidolopin (**259**), was recently isolated from the bryozoan *Phidolopora pacifica* from British Columbia [172].

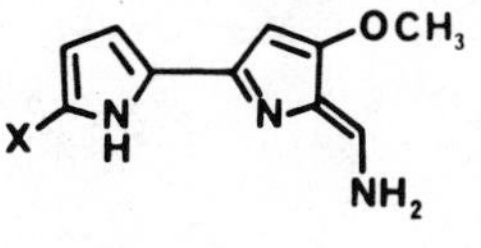

249 X = H
250 X = Br

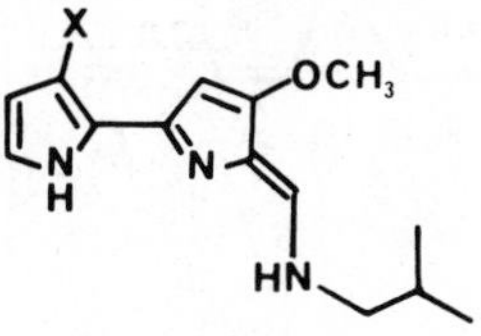

251 X = H
252 X = Br

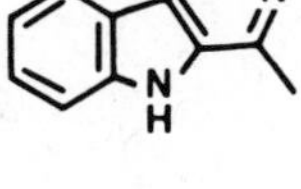

253

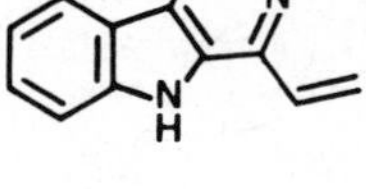

254

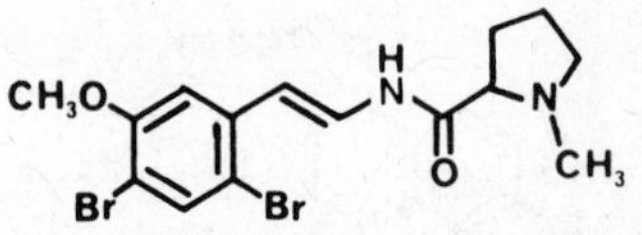

255

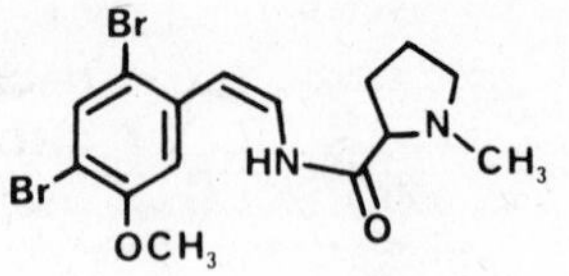

256

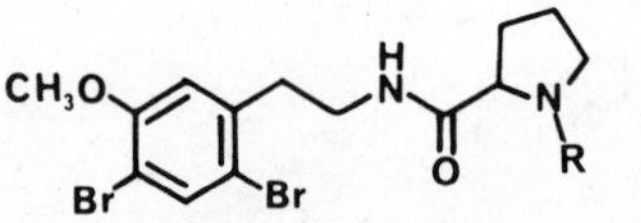

257 R=H
258 R=CH₃

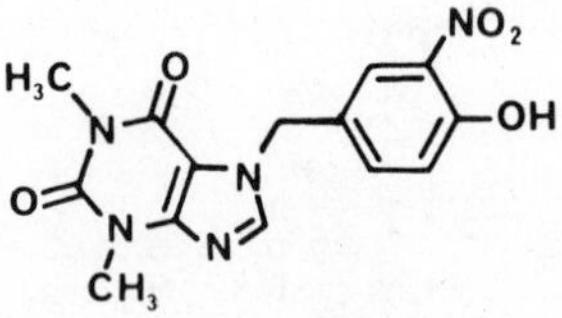

259

3.2.3. Tunicates (Urochordata).

The tunicates, also known as ascidians, are a unique class of marine invertebrates related in part to the vertebrate animals by the presence of a notochord in the larval stage. Alkaloidal compounds are found in this class, and the first secondary metabolites to be recognized were a series of substituted pyrrolidinones, the polyandrocarpidines A–D (260–263), isolated from a Pacific tunicate of the genus *Polyandrocarpa* [173,174]. The polyandrocarpidines possess both antimicrobial and cytotoxic properties. As in other classes of marine invertebrates, indole derivatives are common tunicate metabolites. The tunicate *Dendrodoa grossularia*, from Brittany, contains dendroine (264) and grossularine (265) [175,176]. Dendroine possesses a 1,2,4-thiadiazole functionality which is unprecedented in natural products.

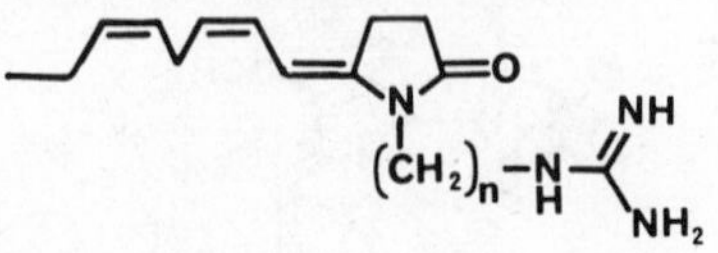

260 n=4
261 n=5

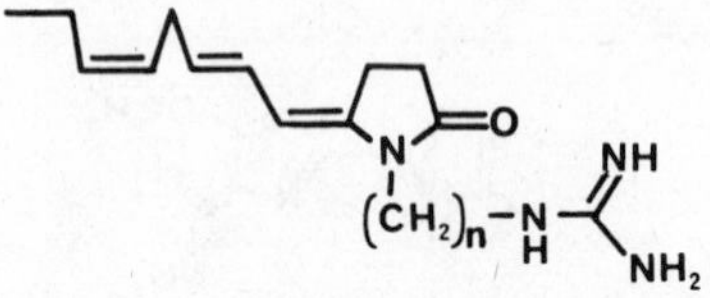

262 n=4
263 n=5

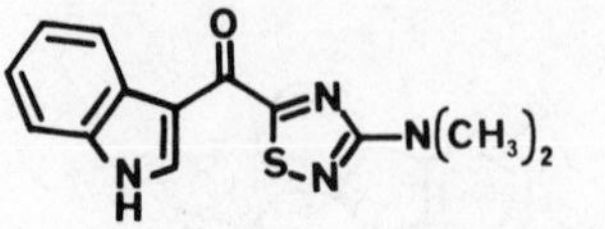

264

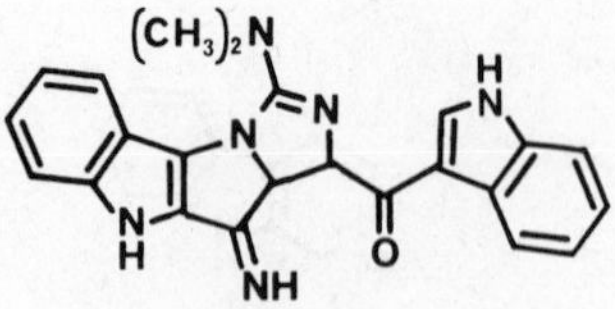

265

The Caribbean tunicate *Eudistoma olivaceum* has been the subject of considerable recent study, and over 15 interesting new indole alkaloids, the eudistomins, possessing antiviral and antimicrobial activities, have been isolated [177,178]. Eudistomins A and M (**266,267**) were shown to be pyrrolocarbolines, while eudistomins G, H, I. P, and Q (**272–276**) are the corresponding dihydro derivatives. Eudistomins D, J, N, and O (**268–271**) are various hydroxyl and bromine substitution isomers of the parent carboline nucleus, and the eudistomins C, E, K, and L (**277–280**) are extraordinary in possessing oxathiazepine rings.

An interesting class of alkaloidal compounds that has been observed in the tunicates are the thiazole-containing cyclic peptides. These compounds are found in a specific family of tunicates, the Didemnidae, which are frequently found in tropical marine environments. Ulicyclamide (**281**) and ulithiacyclamide (**282**) were the first examples of this class isolated from *Lissoclinum patella* [179,182]. Subsequent studies of this same organism resulted in the isolation of patellamides A–C (**283–285**) [180,181] and three other related peptides **286–288** [182]. A related peptide, ascidiacyclamide (**289**), was also reported from an unidentified Barrier Reef tunicate [183]. A

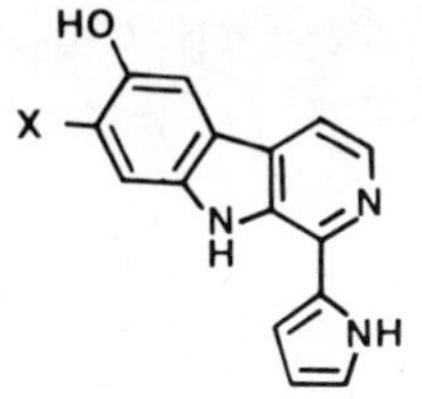

| | 266 | X = Br |
| | 267 | X = H |

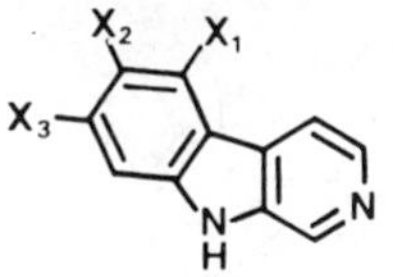

	X_1	X_2	X_3
268	Br	OH	H
269	H	OH	Br
270	H	Br	H
271	H	H	Br

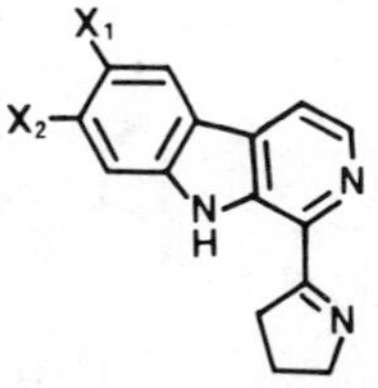

	X_1	X_2
272	H	Br
273	Br	H
274	H	H
275	OH	Br
276	OH	H

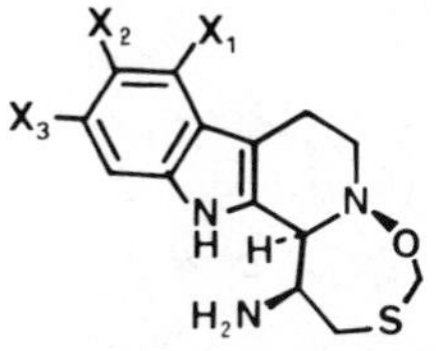

	X_1	X_2	X_3
277	H	OH	Br
278	Br	OH	H
279	H	H	Br
280	H	Br	H

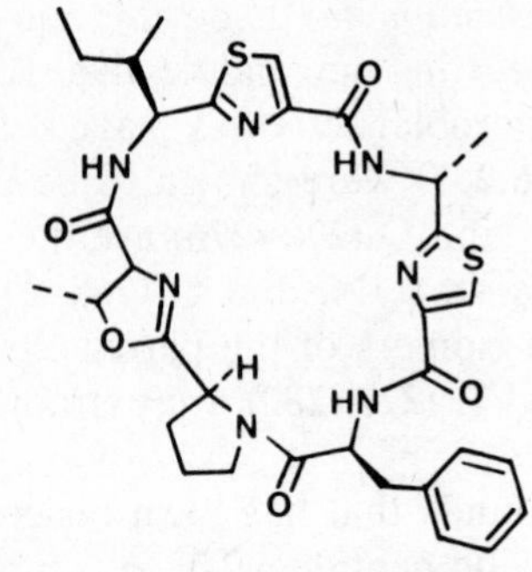

281

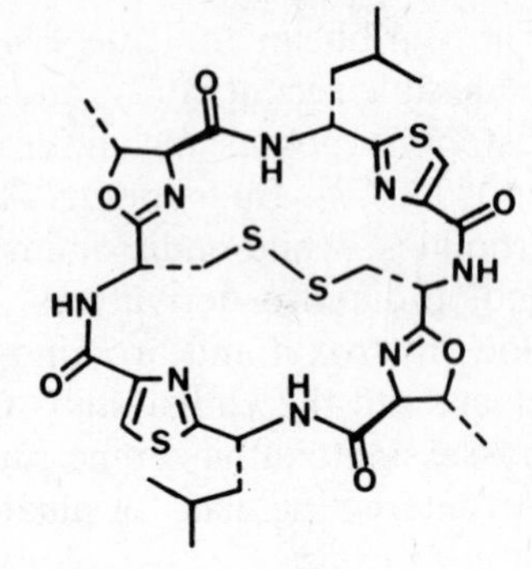

282

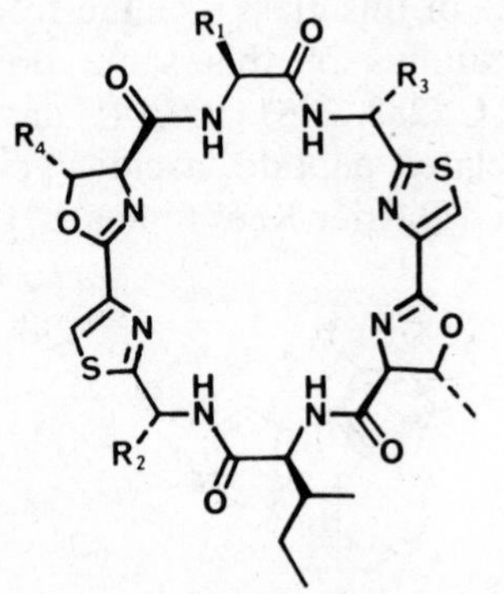

	R₁	R₂	R₃	R₄
283	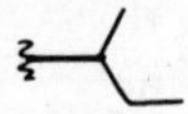	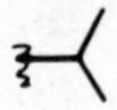		H
284		CH₃	Ph	CH₃
285		CH₃	Ph	CH₃

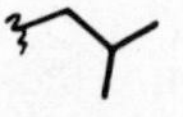

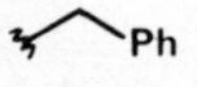

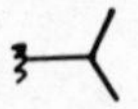

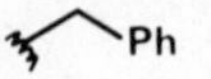

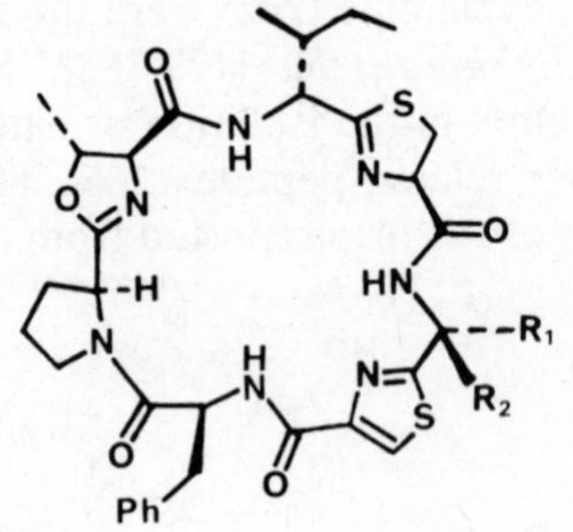

	R₁	R₂
286	CH₃	H
287	H	CH₃

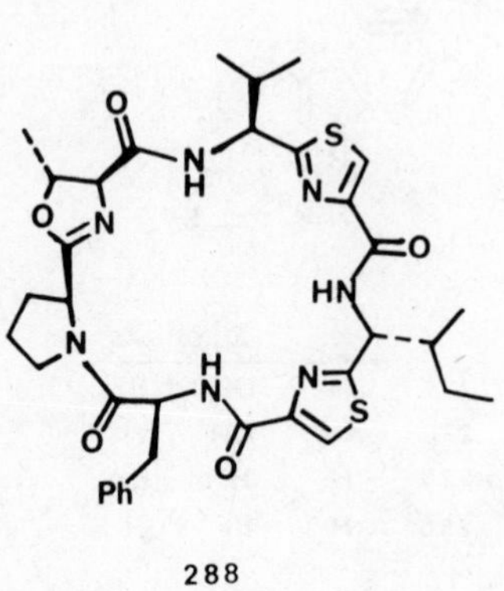

288

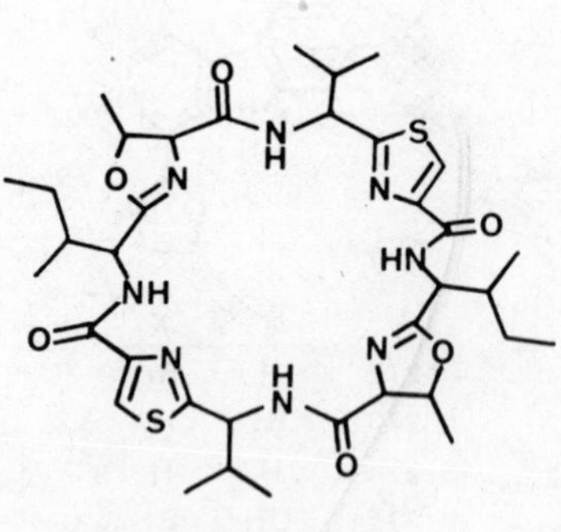

289

314

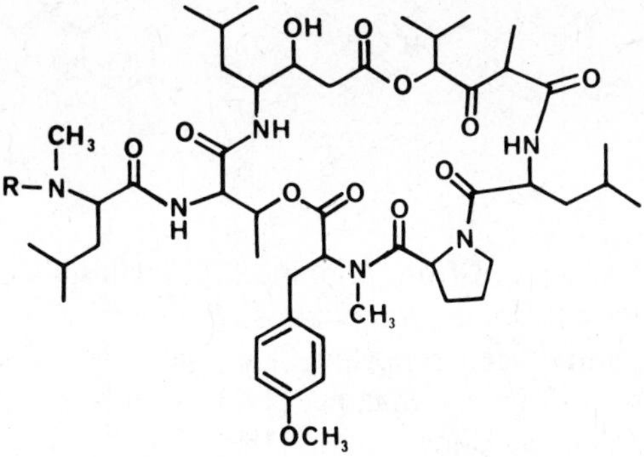

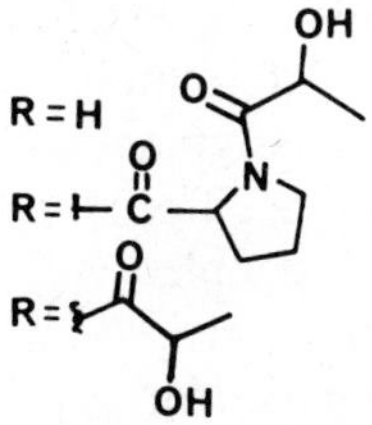

group of antiviral and cytotoxic cyclic peptides, the didemnins, have been isolated from a Caribbean tunicate of the genus *Trididemnum*. Didemnins A (**290**), B (**291**), and C (**292**) lack thiazole components and would appear to possess considerable potential in the therapeutic treatment of viral disease and some forms of human cancer [184,185].

3.2.4. Coelenterates (Cnidaria). Alkaloids are only rarely found in marine coelenterates and they also appear to be rather restricted to the animals within the order Zoanthidea. The bright yellow pigmentation of numerous zoanthids of the genus *Parazoanthus* has been shown to be due to the presence of a unique class of heteroaromatic amines known collectively as the zoanthoxanthins [5,7]. Zoanthozanthin (**293**), isolated from *P. axinellae*, was the first example of this unique class to be described [186,187]. Later, pseudozoanthoxanthin (**294**), which possesses an isomeric guanidine cyclization pattern, was isolated from *P. axinellae* [188]. Since these early reports over 20 variations (mainly *N*-methyl isomers) of these metabolites have been described. Although the biogenetic origins of these tetrazacyclopentazulene skeletons are not known, they are thought to arise via the dimerization of

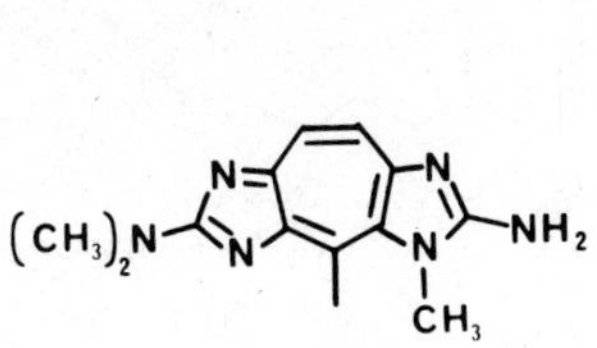

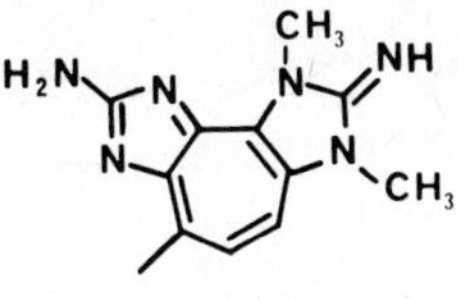

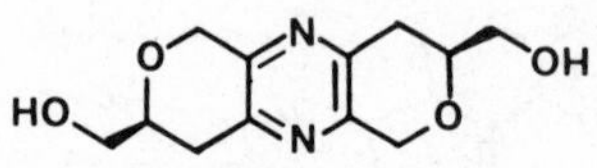

295

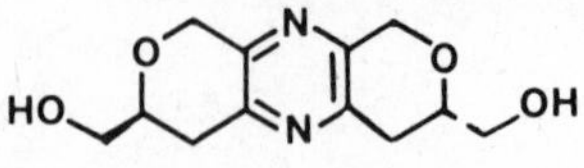

296

two C$_5$ metabolites derived from arginine [5]. Zoanthids of the genus *Palythoa* also produce alkaloidal compounds. *P. tuberculosa* from Japan, for example, has recently been reported to contain the pyrazine derivatives palythazine (**295**) and isopalythazine (**296**), which appear on structural grounds to be derived by dimerization [189]. The truly extraordinary molecule palytoxin (**297**) is a complex alkaloid found in several *Palythoa* species, but in particular in *P. toxicus*. Palytoxin is the most potent nonprotein toxin known, and its final structure assignment was the culmination of over 15 years of active research [190–192]. The structure of palytoxin was provided by interpretation of NMR data coupled with X-ray structure assignment of degradation products. The final stereochemistry of a significant portion of this complex molecule was provided largely on the basis of organic synthesis [193–195].

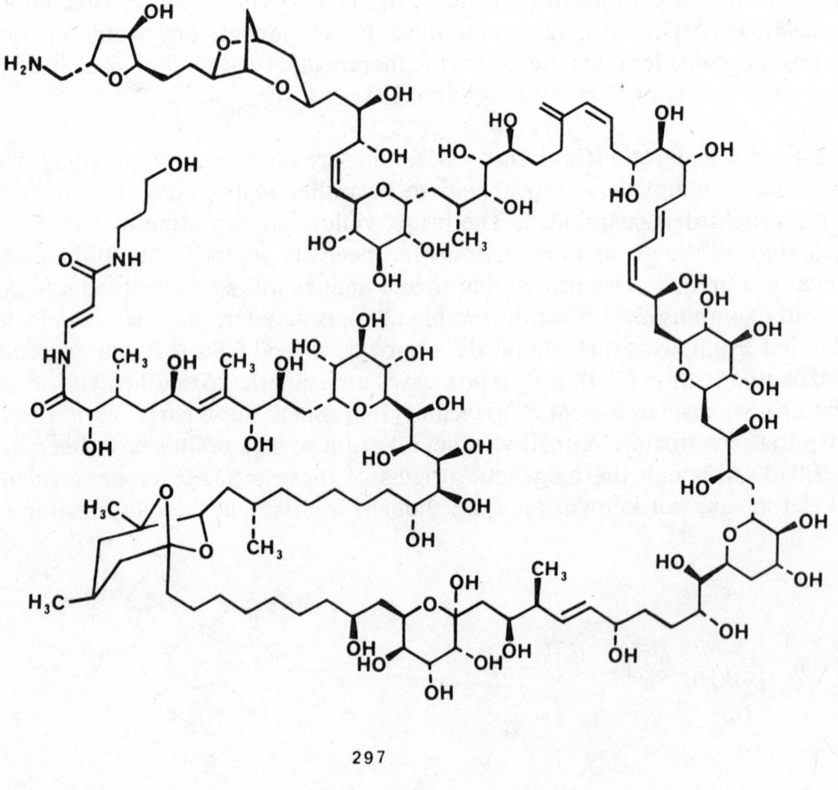

297

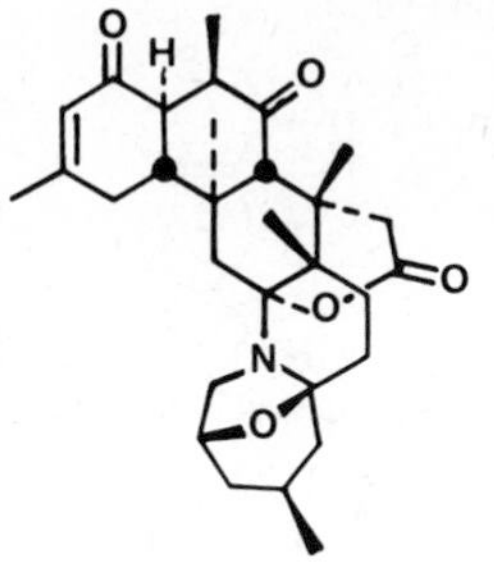

298

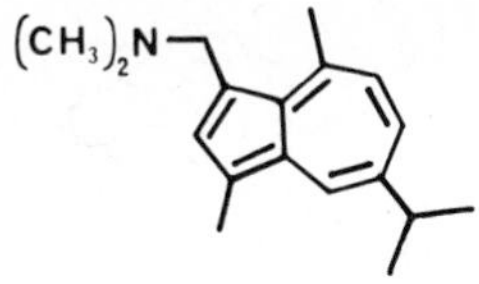

299

In a very recent report the unique structure of zoanthamine (**298**) was reported [196]. This very novel and unprecedented new compound was isolated from an unidentified *Zoanthus* species collected along the Indian coast. Although rare, a single example exists in which an alkaloidal metaboite has been isolated from a coelenterate of a different group. A deep sea gorgonian, or sea whip (Order Gorgonaceae) was shown to contain the amino guiazulene derivative **299** [197].

3.2.5. Molluscs (Mollusca). Several marine molluscs have been identified as containing nitrogenous compounds. The shell-less molluscs, or sea slugs, possess several unique compounds. The California sea slug *Navanax inermis*, for example, contains several polyunsaturated pyridine derivatives known as navenone A (**300**), 3-methyl navenone A (**301**), and 3,5-di-*cis*-navenone A (**302**) [198]. These derivatives, along with the corresponding phenyl and phenol analogs, are pheromones known to be involved in the alarm response behavior of *Navanax*. The sea slug *Bursatella leachii pleii* was reported to contain the aromatic bisnitrile, bursatellin (**303**) [199]. Bursatellin is among the only two or three nitriles isolated thus far from marine sources.

In recent studies of the cytotoxic metabolites of marine molluscs, the structure of the cyclic peptide dolastatin 3 (**304**) was suggested on spectral grounds [200]. Dolastatin 3 was isolated from the Indian Ocean sea hare *Dolabella auricularia* using bioassays to follow its potent antileukemia activity. A recent synthesis of the proposed structure of dolastatin 3 [201] has, however, shown the structure to be incorrect. An interesting similarity exists between dolastatin 3 and the ascidian-derived cyclic peptides discussed in Section 3.2.3.

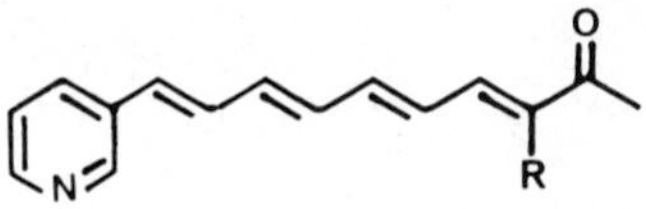

300   R = H
301   R = CH₃

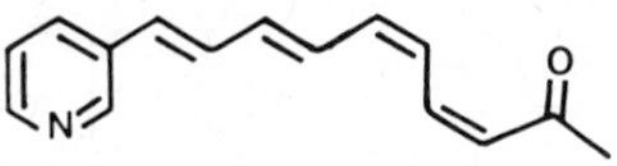

302

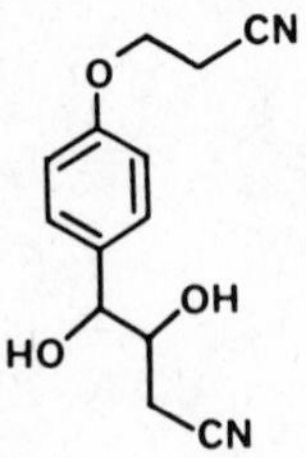

303

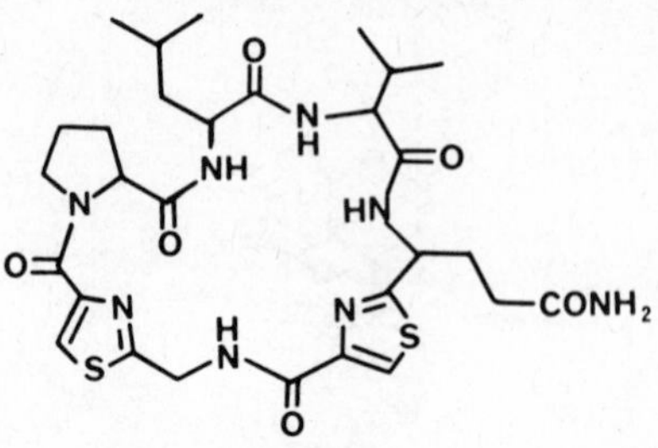

304

(to be revised)

Although I have generally attempted to avoid nitrogenous marine pigments, which have more widespread distribution, no discussion of the gastropod (shelled) molluscs could be complete without reference to the ancient indigo pigment Tyrian Purple (305). Tyrian Purple is a deep blue-purple compound that, since ancient times, has been obtained from the hypobranchial glands of numerous molluscs of the genera *Murex, Purpura,* and *Dicathais.* The pigment itself is produced photochemically from several colorless intermediates that are naturally occurring. The structure of Tyrian Purple as 6,6'-dibromoindigotin (305) was established in the mid 1970s [202] and more securely illustrated by a recent X-ray measurement [203,204].

Two complex compounds, surugatoxin (306) and neosurugatoxin (307), partially related to Tyrian Purple in that they also contain the 6-bromoindole nucleus, have been isolated from the edible Japanese ivory shell *Babylonia*

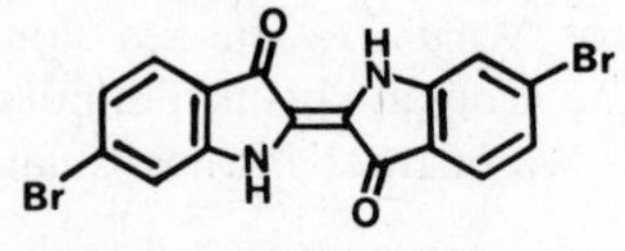

305

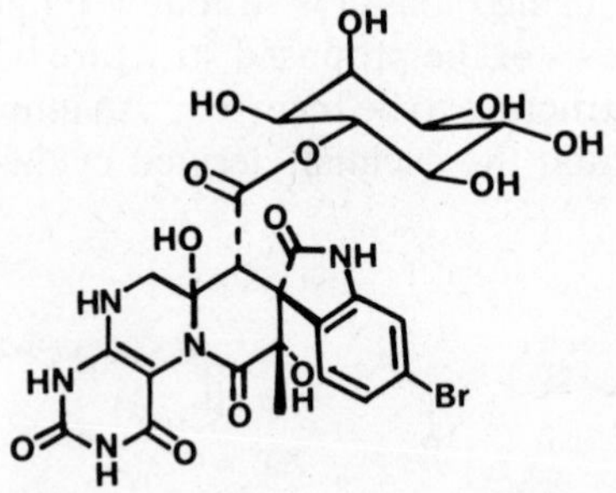

306

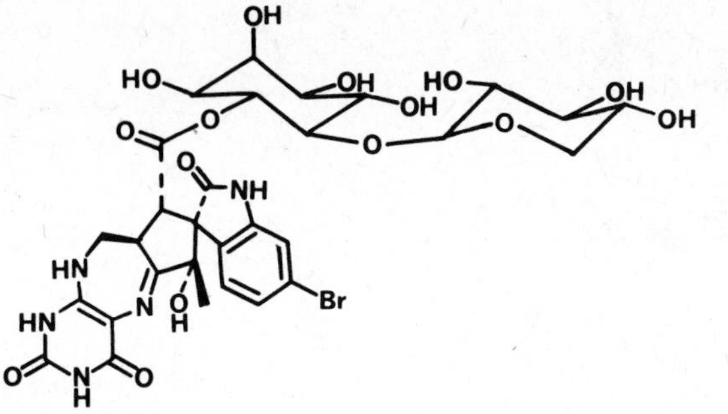

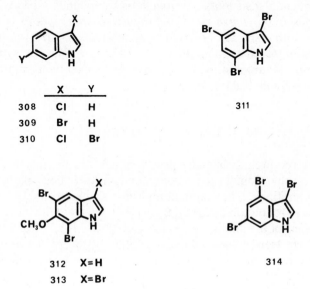

307

japonica [205,206]. These two metabolites are potent neurotoxins, and their subsequent isolation was a result of several human intoxications from consuming *B. japonica* in Suruga Bay, Japan. Most specimens of this gastropod did not contain the surugatoxins; hence it was concluded that they may be of microbial origin [207]. Although both toxins were defined by X-ray crystallography, Hashimoto suggests [207] that some uncertainty may exist in the position of the double bond in surugatoxin (**306**).

3.2.6. Miscellaneous Marine Invertebrates. The remaining nitrogenous metabolites from marine invertebrates consist mainly of indole and pyrrole derivatives found in numerous marine worms. The acorn worm *Ptychodera flava laysanica* (Hemichordata), for example, contains a complex mixture of the odorous halogenated indole derivatives **308–313** [208,209]. In a study of another worm, *Balanoglossus carnosus*, 3,4,6-tribromoindole (**314**) was

	X	Y
308	Cl	H
309	Br	H
310	Cl	Br

311

312 X=H
313 X=Br

314

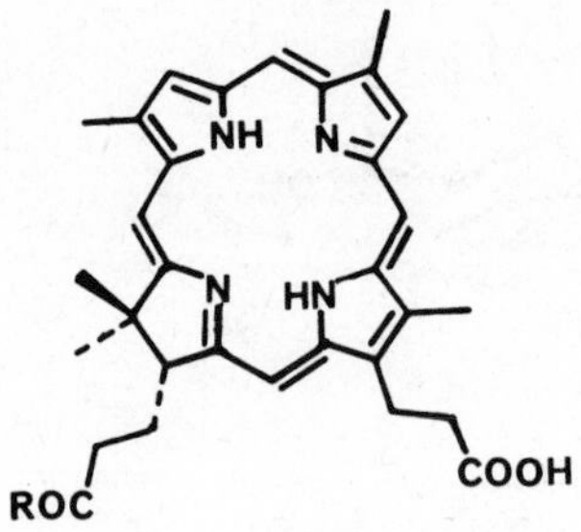

	R
315	OH
316	N-valyl
317	N-isoleucyl
318	N-leucyl
319	N-alloisoleucyl

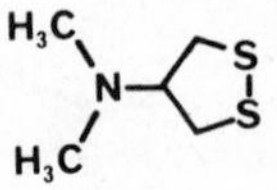

320

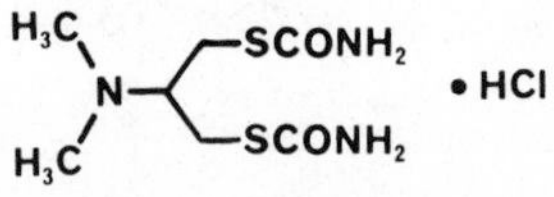

321

isolated [210]. These compounds closely resemble the halogenated indoles found in certain blue-green and red seaweeds (Sections 2.2.1. and 2.2.3.). A unique tetrapyrrole pigment, bonellin (315), has been the subject of much recent interest owing to its unique biological properties [211]. Bonellin is produced only by the female echurian worm *Bonellia viridis*, and this unique compound has the interesting property of inducing the development of male larvae from those which are sexually undifferentiated. In separate studies numerous bonellin–amino acid conjugates (316–319) were also described as components of the body wall of the animal [212,213]. The polychaete worms (Phylum Annelida) also contain nitrogenous compounds. The toxic principle of *Lumbrinereis brevicirra* (= *Lumbriconereis heteropoda*) is the simple dimethylaminothiolane 320, referred to as nereistoxin [207,214]. Nereistoxin shows unusual toxicity against insects, and in the 1960s the synthetic pesticide "Cartap Hydrochloride" (321) was developed under the trade name PADAN for argicultural use in Japan.

4. ALKALOIDS FROM MARINE VERTEBRATES

Alkaloidal structures from marine vertebrates are limited to several fish species that either contain toxic compounds in their body parts or possess defensive secretions. The classical and best known example is the potent neurotoxin, tetrodotoxin (322), originally isolated from the liver and ovaries of puffer fishes of the family Tetraodontidae [215–218]. Puffer fish poisoning is only known from Japan, where the fillets of the puffer fish, known as fugu, are carefully prepared for human consumption. Fugu has been con-

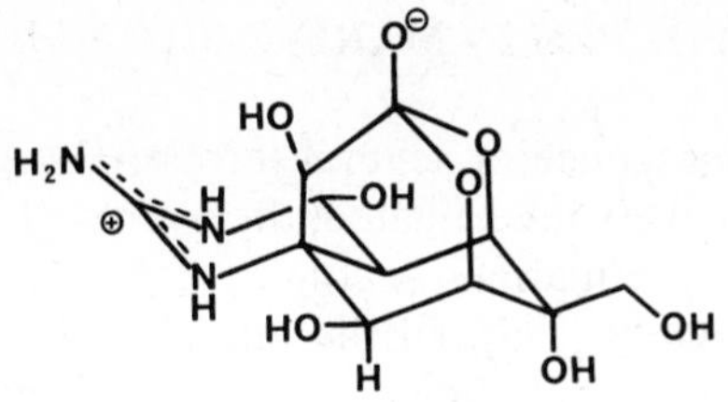

322

sidered a delicacy in Japan for over 1500 years despite the rather high incidence of its fatal effects. The unique structure of tetrodotoxin is suggestive of a origin in carbohydrate biosynthesis, but the precise biosynthesis of this potent toxin remains to be described. Curiously, tetrodotoxin is now recognized to be of a broader distribution than originally suspected. In subsequent investigations tetrodotoxin has been isolated from the Taiwan goby fish *Gobius cringer* [219], several newts of the genus *Taricha* [218], and from the skin of a Costa Rican frog of the genus *Atelopus* [220].

Certain classes of fishes, referred to as ichthyocrinotoxic, secrete toxins to repel potential predators. Such is the case with the flat fishes of the genus *Paradachirus*, and particularly with the Red Sea Moses Sole *P. marmoratus*, which secretes a toxin sufficiently potent to repel sharks. Although the active repellents from the secretion of *P. marmoratus* are at present unknown, a recent report has provided the chemical composition of the toxic secretion of the closely related Pacific Sole *P. pavoninus*. The toxic secretion contained six aminoglycoside saponins, the pavoninins 1–6 (**323–328**), which were identified on the basis of their spectral and chemical properties [221]. The pavoninins produce potent shark-repelling effects at levels of 5 mg/mL and are thought to act upon the olfactory rosette and bucal receptors of predatory animals.

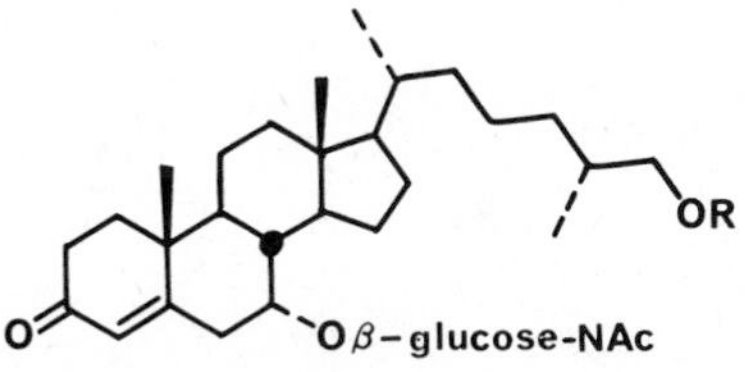

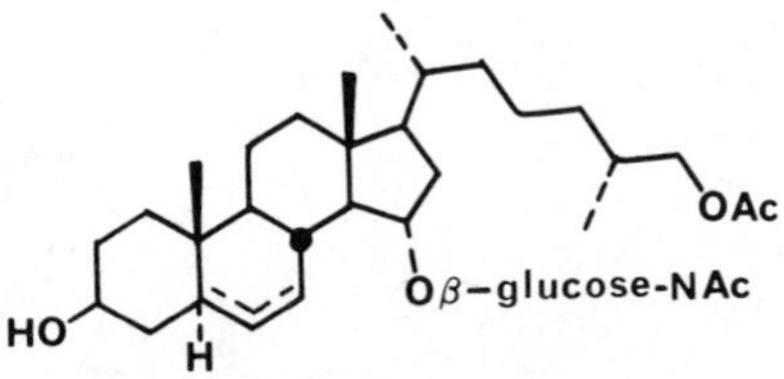

323 R = Ac
324 R = H

325 5en–3αOH
326 5,6H$_2$–3αOH
327 5en–3β OH
328 6en–3β OH

5. ALKALOIDS INVOLVED IN MARINE BIOLUMINESCENCE

Although the bioluminescence of several terrestrial organisms, such as the firefly, has been extensively investigated, the overwhelming majority of bioluminescent organisms are marine. A complete discussion of this topic was recently provided by Goto [222]. In the marine environment a taxonomic cross-section of organisms including bacteria, dinoflagellates, invertebrates, and vertebrate animals have evolved the biological production of light. Although not yet fully defined, most bioluminescent reactions appear to involve an enzyme-mediated oxygenation reaction of a small molecule, the "luciferin-luciferase system," or a "photoprotein" system, which appears to be less well understood. In the former luciferin-luciferase system, several "luciferins" have been isolated and shown to possess unique and specific alkaloidal structures. The molecule referred to as "Cypridina luciferin" was isolated and purified as the dihydrochloride salt first in 1966, and its structure was successfully defined as **329** [223,224]. In a similar study, the luciferin from the jellyfish *Aequorea aequorea* has been defined as the related alkaloid **330**, referred to as coelenterazine [225]. Subsequently coelenterazine was found in other bioluminescent organisms, such as the sea pansy *Renilla* and the squid *Watasenia* [222]. The oxygenation reactivity of both these luciferins has been shown to emanate from their respective dihydroimidazopyrazinone components.

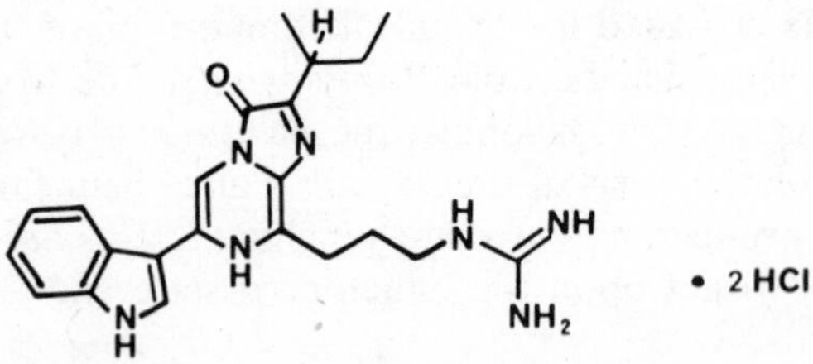

329

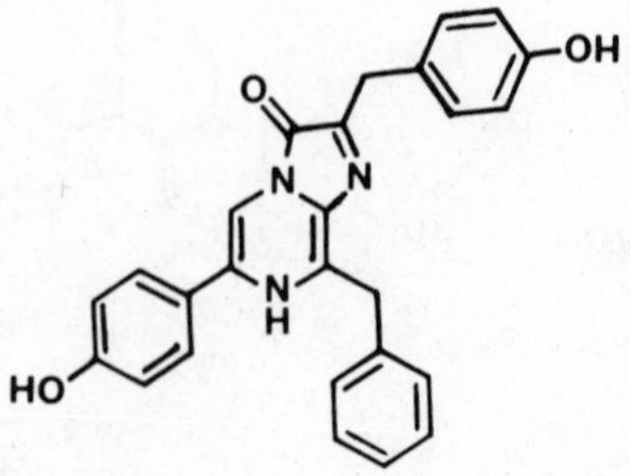

330

REFERENCES

1. S. W. Pelletier, "The Nature and Definition of an Alkaloid," in S. W. Pelletier, Ed., *Alkaloids: Chemical and Biological Perspectives*, Vol. 1, Wiley, New York, 1983, p. 26.

2. P. J. Scheuer, Ed., *Marine Natural Products: Chemical and Biological Perspectives*, Vols. I–V, Academic, New York, 1978–1982.

3. Y. Shimizu, "Dinoflagellate Toxins," in P. J. Scheuer, Ed., *Marine Natural Products: Chemical and Biological Perspectives*, Vol. 1, Academic, New York, 1978, p. 1.

4. E. Fattorusso and M. Piattelli "Amino Acids from Marine Algae," in P. J. Scheuer, Ed., *Marine Natural Products: Chemical and Biological Perspectives*, Vol. 3, Academic, New York, 1980, p. 95.

5. G. Prota, "Nitrogenous Pigments in Marine Invertebrates," in P. J. Scheuer, Ed., *Marine Natural Products: Chemical and Biological Perspectives*, Vol. 3, Academic, New York, 1980, p. 141.

6. R. E. Moore, "Constituents of Blue-Green Algae," in P. J. Scheuer, Ed., *Marine Natural Products: Chemical and Biological Perspectives*, Vol. 4, Academic, New York, 1981, p. 1.

7. L. Chevolot, "Guanidine Derivatives," in P. J. Scheuer, Ed., *Marine Natural Products: Chemical and Biological Perspectives*, Vol. 4, Academic, New York, 1981, p. 54.

8. C. Christophersen, "Marine Indoles," in P. J. Scheuer, Ed., *Marine Natural Products: Chemical and Biological Perspectives*, Vol. 5, Academic, 1983, p. 259.

9. D. J. Faulkner, *Nat. Products Rep.* **1** (3), 251 (1984).

10. D. J. Faulkner, *Nat. Products Rep.* **1** (6), 551 (1985).

11. E. J. Schantz, V. E. Ghazarossian, H. K. Schnoes, F. M. Strong, J. P. Springer, J. O. Pezzanite, and J. Clardy, *J. Am. Chem. Soc.* **97**, 1238 (1975).

12. J. Bordner, W. E. Thiessen, H. A. Bates, and H. Rapoport, *J. Am. Chem. Soc.* **97**, 6008 (1975).

13. C. F. Wichmann, W. P. Niemezura, H. K. Schnoes, S. Hall, P. D. Reichardt, and S. D. Darling, *J. Am. Chem. Soc.* **103**, 6977 (1981).

14. Y. Shimizu, C. Hsu, W. E. Fallon, Y. Oshima, I. Miura, and K. Nakanishi, *J. Am. Chem. Soc.* **100**, 6791 (1978).

15. G. L. Boyer, E. J. Schantz, and H. K. Schnoes, *J. Chem. Soc., Chem. Commun.* **1978**, 889.

16. C. F. Wichmann, G. L. Boyer, C. L. Divan, E. J. Schantz, and H. K. Schnoes, *Tetrahedron Lett.* **22**, 1941 (1981).

17. Y. Shimizu and C. P. Hsu, *J. Chem. Soc., Chem. Commun.* **1981**, 314.

18. M. Alam, Y. Oshima, and Y. Shimizu, *Tetrahedron Lett.* **23**, 321 (1982).

19. M. Kobayashi and Y. Shimizu, *J. Chem. Soc., Chem. Commun.* **1981**, 827.

20. S. Hall, P. B. Reichardt, and R. A. Neve, *Biochem. Biophys. Res. Commun.* **97**, 649 (1980).

21. M. Ikawa, K. Wagner, T. L. Foxall, and J. J. Sasner, Jr., *Toxicon.* **20**, 747 (1982).

22. Y. Shimizu, M. Norte, A. Hori, A. Genenah, and M. Kobayashi, *J. Am. Chem. Soc.* **106**, 6433 (1984).

23. M. Alam, R. Sanduja, M. B. Hossain, and D. van der Helm, *J. Am. Chem. Soc.* **104**, 5232 (1982).

24. M. DiNovi, D. A. Trainor, K. Nakanishi, R. Sanduja, and M. Alam, *Tetrahedron Lett.* **24**, 855 (1983).

25. F.-J. Marner and R. E. Moore, *Phytochemistry* **17**, 553 (1978).

26. C. J. Simmons, F.-J. Marner, J. H. Cardellina, II, R. E. Moore, and K. Seff, *Tetrahedron Lett.* **1979**, 2003.

27. J. H. Cardellina, II, and R. E. Moore, *Tetrahedron Lett.* **1979**, 2007.

28. C.-L. Yeh, W. T. Colwell, and J. I. DeGraw, *Tetrahedron Lett.* **1978**, 3987, and references cited therein.

29. J. H. Cardellina, II, F.-J. Marner, and R. E. Moore, *J. Am. Chem. Soc.* **101**, 240 (1979).

30. J. S. Mynderse and R. E. Moore, *J. Org. Chem.* **43**, 4359 (1978).

31. J. H. Cardellina, II, D. Dalietos, F.-J. Marner, J. S. Mynderse, and R. E. Moore, *Phytochemistry* **17**, 2091 (1978).

32. F.-J. Marner, R. E. Moore, K. Hirotsu, and J. Clardy, *J. Org. Chem.* **42**, 2815 (1977).

33. D. C. Carter, R. E. Moore, J. S. Mynderse, W. P. Niemezura, and J. S. Todd, *J. Org. Chem.* **49**, 236 (1984).

34. J. H. Cardellina, II, F.-J. Marner, and R. E. Moore, *Science* **204**, 193 (1979).

35. M. Takashima, H. Sakai, and K. Arima, *Agri. Biol. Chem.* **26**, 660 (1962).

36. J. H. Cardellina, II, M. P. Kirkup, R. E. Moore, J. S. Mynderse, K. Seff, and C. J. Simmons, *Tetrahedron Lett.* **1979**, 4915.

37. R. S. Norton and R. J. Wells, *J. Am. Chem. Soc.* **104**, 3628 (1982).

38. W. W. Carmichael, D. F. Biggs, and M. A. Peterson, *Toxicon* **17**, 229 (1979) and references cited therein.

39. R. E. Moore, C. Cheuk, and G. M. L. Patterson, *J. Am. Chem. Soc.* **106**, 6456 (1984).

40. S. E. Vest, C. J. Dawes, and J. T. Romero, *Botanica Marina* **26**, 313 (1983).

41. G. Aguilar-Santos and M. S. Doty, in H. D. Freudenthal Ed., *Drugs from the Sea*, Marine Technology Society, Washington, D. C., 1968, p. 173.

42. M. S. Doty and G. Aguilar-Santos, *Pacific Sci.* **24**, 351 (1970).

43. S. C. Maiti, R. H. Thompson, and M. Mahendran, *J. Chem. Res.* **1978**, 1683.

44. M. R. Brennan and K. L. Erickson, *Tetrahedron Lett.* **1978**, 1637.

45. K. L. Erickson, H. R. Brennan, and P. A. Namnum, *Synth. Comm.* **11**, 253 (1981).

46. G. T. Carter, K. L. Rinehart, Jr., L. H. Li, S. L. Kuentzel, and J. L. Connor, *Tetrahedron Lett.* **1978**, 4479.

47. M. P. Kirkup and R. E. Moore, *Tetrahedron Lett.* **23**, 2087 (1983).

48. R. Kazlauskas, P. T. Murphy, R. J. Wells, J. A. Baird-Lambert, and D. D. Jamieson, *Aust. J. Chem.* **36**, 165 (1983).

49. I. Tsujino, K. Yabe, I Sekikawa, and N. Hamanaka, *Tetrahedron Lett.* **1978**, 1401.

50. I. Tsujino, K. Yabe, and I. Sekikawa, *Botanica Marina* **23**, 65 (1980).

51. S. Takano, A. Nakanishi, D. Uemura, and Y. Hirata, *Chem. Lett.* **1979**, 419.

52. S. Murakami, T. Takemoto, and Z. Shimuzu, *J. Pharm. Soc. Jap.* **73**, 1026 (1953).

53. K. Daigo, *Yakagaku Zasshi* **79**, 350 (1959).

54. L. Minale, "Terpenoids from Marine Sponges," in P. J. Scheuer, Ed., *Marine Natural Products: Chemical and Biological Perspectives*, Vol. 1, Academic, New York, 1978, p. 175.

55. P. R. Burkholder, R. M. Pfister, and F. H. Leitz, *Appl. Microbiol.* **14**, 649 (1966).

56. R. J. Andersen, M. S. Wolfe, and D. J. Faulkner, *Mar. Biol.* **27**, 281 (1974).

57. S. J. Wratten, M. S. Wolfe, R. J. Andersen, and D. J. Faulkner, *Antimicrobial Agents Chemotherapy* **11**, 411 (1977).

58. N. N. Gerber, *Tetrahedron Lett.* **24**, 2797 (1983).

59. N. M. Gandhi, J. R. Patell, J. Gandhi, N. J. DeSouza, and H. Kohl, *Mar. Biol.* (Ber.) **34**(3), 223 (1976).

60. Y. Okami, K. Hotta, M. Yoshida, D. Ikeda, S. Kondo, and H. Umezawa, *J. Antibiotics* **32**, 964 (1979).

61. J. Vacelet, *J. Microscopie Biol. cell.* **23**, 271 (1975).

62. G. M. Sharma and P. R. Burkholder, *Tetrahedron Lett.* **1967**, 4147.

63. M. D'Ambrosio, A. Guerriero, R. DeClauser, G. DeStanchina, and F. Pietra, *Experientia* **39**, 1091 (1983).

64. E. Fattorusso, L. Minale, and G. Sodano, *J.C.S. Chem. Commun.* **1970**, 751.

65. W. Fulmor, G. E. Van Lear, G. O. Morton, and R. D. Mills, *Tetrahedron Lett.* **1970**, 4551.

66. G. E. Krejcarek, R. H. White, L. P. Hager, W. O. McClure, R. D. Johnson, K. L. Rinehart, Jr., J. A. McMillan, I. C. Paul, P. D. Shaw, and R. C. Brusca, *Tetrahedron Lett.* **1975**, 507.

67. D. B. Borders, G. O. Morton, and E. R. Wetzel. *Tetrahedron Lett.* **1974**, 2709.

68. K. H. Hollenbeak, F. J. Schmitz, P. N. Kaul, and S. K. Kulkarni, in P. N. Kaul and C. J. Sindermann, Eds., *Drugs and Food From the Sea, Myth or Reality?* University of Oklahoma Press, Norman, OK 1978, p.81.

69. K. Moody, R. H. Thomson, E. Fattorusso, L. Minale, and G. Sodano, *J. Chem. Soc. Perkin I,* **1972**, 18.

70. J. A. McMillan, I. C. Paul, Y. M. Goo, K. L. Rinehart, Jr., W. C. Krueger, and L. M. Pschigoda, *Tetrahedron Lett.* **22**, 39 (1981).

71. Y. Gopichand and F. J. Schmitz, *Tetrahedron Lett.* **1979**, 3921.

72. G. Cimino, S. De Rosa, S. De Stefano, R. Self, and G. Sodano, *Tetrahedron Lett.* **24**, 3029 (1983).

73. M. Rotem, S. Carmely, Y. Kashman, and Y. Loya, *Tetrahedron* **39**, 667 (1983).

74. D. M. Roll, C. W. J. Chang, P. J. Scheuer, G. A. Gray, J. N. Shoolery, G. K. Matsumoto, G. D. Van Duyne, and J. Clardy, *J. Am. Chem. Soc.*, in press, 1985.

75. R. Kazlauskas, R. O. Lidgard, P. T. Murphy, and R. J. Wells, *Tetrahedron Lett.* **21**, 2277 (1980).

76. R. Kazlauskas. R. O. Lidgard, P. T. Murphy, R. J. Wells, and J. F. Blount, *Aust. J. Chem.* **34**, 765 (1981).

77. S. Nishiyama and S. Yamamura, *Tetrahedron Lett.* **23**, 1281 (1982).

78. S. Nishiyama, T. Suzuki, and S. Yamamura, *Chem. Lett.* **1982**, 1851.

79. S. Nishiyama, T. Suzuki, and S. Yamamura, *Tetrahedron Lett.* **23**, 3699 (1982).

80. P. Djura, D. B. Stierle, B. Sullivan, D. J. Faulkner, E. Arnold, and J. Clardy, *J. Org. Chem.* **45**, 1435 (1980).

81. G. E. Van Lear, G. O. Morton, and W. Fulmor, *Tetrahedron Lett.* **1973**, 299.

82. G. Dellar, P. Djura, and M. V. Sargent, *J.C.S. Perkin I,* **1981**, 1679.

83. R. Kazlauskas, P. T. Murphy, R. J. Quinn, and R. J. Wells, *Tetrahedron Lett.* **1977**, 61.

84. K. Hollenbeak and F. J. Schmitz, *Lloydia*, **40**, 479 (1977).

85. P. R. Bergquist and R. J. Wells, "Chemotaxonomy of the Porifera: The Development and Current Status of the Field," in P. J. Scheuer, Ed., *Marine Natural Products: Chemical and Biological Perspectives*, Vol. 5, Academic, New York, 1983, p. 1.

86. P. Djura and D. J. Faulkner, *J. Org. Chem.* **45**, 735 (1980).

87. K. M. Taylor, J. A. Baird-Lambert, P. A. Davis, and I. Spence, *Fed. Proc. Am. Soc. Exp. Biol.* **40**(1), 15 (1981).

88. Y. Gopichand and F. J. Schmitz, *J. Org. Chem.* **44**, 4995 (1979).

89. R. J. Andersen, *Tetrahedron Lett.* **1978**, 2541.

90. R. J. Andersen and R. J. Stonard, *Can. J. Chem.* **57**, 2325 (1979).

91. R. J. Stonard and R. J. Andersen, *J. Org. Chem.* **45**, 3687 (1980).

92. R. J. Stonard and R. J. Andersen, *Can. J. Chem.* **58**, 2121 (1980).

93. G. Cimino, S. De Stefano, and L. Minale, *Experientia* **31**, 1387 (1975).

94. D. B. Stierle and D. J. Faulkner, *J. Org. Chem.* **45**, 4980 (1980).

95. S. Forenza, L. Minale, R. Riccio, and E. Fattorusso, *J. Chem. Soc. Chem. Commun.* **1971**, 1129; E. E. Garcia, L. E. Benjamin, and R. I. Fryer, *ibid.*, **1973**, 78.

96. L. Chevolot, S. Padua, B. N. Ravi, P. C. Blyth, and P. J. Scheuer, *Heterocycles* **7**, 891 (1977).

97. F. Cafieri, L. De Napoli, E. Fattorusso, C. Santacroce and D. Sica, *Tetrahedron Lett.* **1977**, 477.

98. F. Cafieri, L. De Napoli, E. Fattorusso and C. Santacroce, *Experientia* **33**, 994 (1977).

99. F. Cafieri, L. De Napoli, A. Iengo, and C. Santacroce, *Experientia* **34**, 300 (1978).

100. F. Cafieri, L. De Napoli, A. Iengo, and C. Santacroce, *Experientia* **35**, 157 (1979).

101. G. Cimino, S. De Stefano, and L. Minale, *Experientia* **30**, 846 (1974).

102. E. E. Garcia, L. E. Banjamin, and R. I. Fryer, *J. Chem. Soc. Chem. Commun.* **1973**, 78.

103. C. Cimino, S. De Stefano, L. Minale, and G. Sodano, *Comp. Biochem. Physiol.* **50B**, 279 (1975).

104. G. M. Sharma and P. R. Burkholder, *J. Chem. Soc. Chem. Commun.* **1971**, 151.

105. G. Sharma and B. Magdoff-Fairchild, *J. Org. Chem.* **42**, 4118 (1977).

106. H. Nakamura, Y. Ohizumi, and J. Kobayashi, *Tetrahedron Lett.* **25**, 2475 (1984).

107. Y. Ohizumi and T. Yasumoto, *J. Physiol.* **337**, 711 (1983).

108. R. P. Walker, D. J. Faulkner, D. Van Engen, and J. Clardy, *J. Am. Chem. Soc.* **103**, 6772 (1981).

109. G. M. Sharma, J. S. Buyer, and M. W. Pomerantz, *J. Chem. Soc. Chem. Commun.* **1980**, 435.

110. G. Cimino, S. De Rosa, S. De Stefano, L. Mazzarella, R. Puliti, and G. Sodano, *Tetrahedron Lett.* **23**, 767 (1982).

111. I. Kitagawa, M. Kobayashi, K. Kitanaka, M. Kido, and Y. Kyogoku, *Chem. Pharm. Bull.* **31**, 2321 (1983).

112. G. C. Harbour, A. A. Tymiak, K. L. Rinehart, Jr., P. D. Shaw, R. G. Hughes, Jr., S. A. Mizsak, J. H. Coats, G. E. Zurenko, L. H. Li, and S. L. Kuentzel, *J. Am. Chem. Soc.* **103**, 5604 (1981).

113. B. B. Snider and W. C. Faith, *J. Am. Chem. Soc.* **106**, 1443 (1984).

114. B. Sullivan, D. J. Faulkner, and L. Webb, *Science* **221**, 1175 (1983).

115. W. R. Kem, B. C. Abbott, and R. M. Coates, *Toxicon* **9**, 15 (1971).

116. F. J. Schmitz, K. H. Hollenbeak, and D. C. Campbell, *J. Org. Chem.* **43**, 3916 (1978).

117. E. Fattorusso, S. Forenza, L. Minale, and G. Sodano, *Gazz. Chim. Ital.* **101**, 104 (1971).

118. G. Cimino, S. De Rosa, S. De Stefano, A. Spinella, and G. Sodano, *Tetrahedron Lett.* **25**, 2925 (1984).

119. D. E. McIntyre, D. J. Faulkner, D. Van Engen, and J. Clardy, *Tetrahedron Lett.* **1979**, 4163.

120. J. M. Frincke and D. J. Faulkner, *J. Am. Chem. Soc.* **104**, 265 (1982).

121. H. Nakamura, J. Kobayashi, Y. Ohizumi, and Y. Hirata, *Tetrahedron Lett.* **23**, 5555 (1982).

122. F. J. Schmitz, S. K. Agarwal, S. P. Gunasekera, P. G. Schmidt, and J. N. Shoolery, *J. Am. Chem. Soc.* **105**, 4835 (1983).

123. W. Bergmann and D. C. Burke, *J. Org. Chem.* **20**, 1501 (1955).

124. W. Bergmann and D. C. Burke, *J. Org. Chem.* **21**, 226 (1956).

125. R. J. Quinn, R. P. Gregson, A. F. Cook, and R. T. Bartlett, *Tetrahedron Lett.* **21**, 567 (1980).

126. E. Cullen and J. P. Devlin, *Can. J. Chem.* **53**, 1690 (1975).

127. R. J. Capon and D. J. Faulkner, *J. Am. Chem. Soc. Chem.* **106**, 1819 (1984).

128. H. Nakamura, H. Wu, Y. Ohizumi, and Y. Hirata, *Tetrahedron Lett.* **25**, 2989 (1984).

129. H. Wu, H. Nakamura, J. Kobayashi, Y. Ohizumi, and Y. Hirata, *Tetrahedron Lett.* **25**, 3719 (1984).

130. A. Iengo, L. Mayol, and C. Santacroce, *Experientia* **33**, 11 (1977).

131. M. R. Hagadone, P. J. Scheuer, and A. Holm, *J. Am. Chem. Soc.* **106**, 2447 (1984).

132. F. Cafieri, E. Fattorusso, S. Magno, C. Santacroce, and D. Sica, *Tetrahedron* **29**, 4259 (1973).

133. M. Adinolfi, L. De Napoli, B. Di Blasio, A. Iengo, C. Pedone, and C. Santacroce, *Tetrahedron Lett.* **1977**, 2815.

134. E. Fattorusso, S. Magno, L. Mayol, C. Santacroce, and D. Sica, *Tetrahedron* **30**, 3911 (1974).

135. B. Di Blasio, E. Fattorusso, S. Magno, L. Mayol, C. Pedone, C. Santacroce, and D. Sica, *Tetrahedron* **32**, 473 (1976).

136. J. E. Thompson, R. P. Walker, S. J. Wratten, and D. J. Faulkner, *Tetrahedron* **38**, 1865 (1982).

137. L. Minale, R. Riccio, and G. Sodano, *Tetrahedron* **30**, 1341 (1974).

138. B. J. Burreson, C. Christophersen, and P. J. Scheuer, *Tetrahedron* **31**, 2015 (1975).

139. B. J. Burreson, P. J. Scheuer, J. S. Finer, and J. Clardy, *J. Am. Chem. Soc.* **97**, 4763 (1975).

140. M. R. Hagadone, B. J. Burreson, P. J. Scheuer, J. S. Finer, and J. Clardy, *Helv. Chim. Acta* **62**, 2484 (1979).

141. H. Nakamura, J. Kobayashi, Y. Ohizumi, and Y. Hirata, *Tetrahedron Lett.* **25**, 5401 (1984).

142. J. T. Baker, R. J. Wells, W. E. Oberhansli, and G. B. Hawes, *J. Am. Chem. Soc.* **98**, 4010 (1976).

143. R. Kazlauskas, P. T. Murphy, R. J. Wells, and J. F. Blount, *Tetrahedron Lett.* **21**, 315 (1980).

144. S. J. Wratten, D. J. Faulkner, K. Hirotsu, and J. Clardy, *Tetrahedron Lett.* **1978**, 4345.

145. C. W. J. Chang, A. Patra, D. M. Roll, P. J. Scheuer, G. K. Matsumoto, and J. Clardy, *J. Am. Chem. Soc.* **106**, 4644 (1984).

146. S. J. Wratten and D. J. Faulkner, *J. Am. Chem. Soc.* **99**, 7367 (1977).

147. S. J. Wratten, D. J. Faulkner, D. Van Engen, and J. Clardy, *Tetrahedron Lett.* **1978**, 1391.

148. S. J. Wratten and D. J. Faulkner, *Tetrahedron Lett.* **1978**, 1395.

149. von W. Hofheinz and W. E. Oberhansli, *Helv. Chim. Acta* **60**, 660 (1977).

150. R. Kazlauskas, R. O. Lidgard, R. J. Wells, and W. Vetter, *Tetrahedron Lett.* **1977**, 3183.

151. C. Charles, J. C. Braekman, D. Daloze, B. Tursch, and R. Karlsson, *Tetrahedron Lett.* **1978**, 1519.

152. C. Charles, J. C. Braekman, D. Daloze, and B. Tursch, *Tetrahedron* **36**, 2133 (1980).

153. J. E. Biskupiak and C. M. Ireland, *Tetrahedron Lett.* **25**, 2935 (1984).

154. K. L. Erickson and R. J. Wells, *Aust. J. Chem.* **35**, 31 (1982).

155. R. Kazlauskas, P. T. Murphy, and R. J. Wells, *Tetrahedron Lett.* **1978**, 4945.

156. J. C. Braekman, D. Daloze, P. Macedo de Abreu, C. Piccinni-Leopardi, G. Germain, and M. van Meerssche, *Tetrahedron Lett.* **23**, 4277 (1982).

157. M. Nakagawa, M. Endo, N. Tanaka, and L. Gen-Pei, *Tetrahedron Lett.* **25**, 3227 (1984).

158. Y. Kashman, A. Groweiss, and U. Shmueli, *Tetrahedron Lett.* **21**, 3629 (1980).

159. A. Groweiss, U. Shmueli, and Y. Kashman, *J. Org. Chem.* **48**, 3512 (1983).

160. I. Spector, N. R. Shochet, Y. Kashman, and A. Groweiss, *Science* **219**, 493 (1983).

161. R. M. Rosser and D. J. Faulkner, *J. Org. Chem.* **49**, 5157 (1984).

162. A. Sato and W. Fenical, *Tetrahedron Lett.* **24**, 481 (1983).

163. J. S. Carlé and C. Christophersen, *J. Am. Chem. Soc.* **101**, 4012 (1979).

164. J. S. Carlé and C. Christophersen, *J. Org. Chem.* **45**, 1586 (1980).

165. J. S. Carlé and C. Christophersen, *J. Org. Chem.* **46**, 3440 (1981).

166. P. Wulff, J. S. Carlé, and C. Christophersen, *J.C.S. Perkin I*, **1981**, 2895.

167. P. Wulff, J. S. Carlé, and C. Christophersen, *Comp. Biochem. Physiol.* **71B**, 523 (1982).

168. P. Wulff, J. S. Carlé, and C. Christophersen, *Comp. Biochem. Physiol.* **71B**, 525 (1982).

169. J. L. C. Wright, *J. Nat. Prods.* **47**, 893 (1984).

170. B. Carté and D. J. Faulkner, *J. Org. Chem.* **48**, 2314 (1983).

171. A. Blackman and D. Matthews, presented at the Royal Australian Chemical Society Meeting, Organic Division, Perth, Australia, May 1984.

172. S. W. Ayer, R. J. Andersen, H. Cun-heng, and J. Clardy, *J. Org. Chem.* **49**, 3869 (1984).

173. B. Carté and D. J. Faulkner, *Tetrahedron Lett.* **23**, 3863 (1982) and references cited therein.

174. K. L. Rinehart, Jr., G. C. Harbour, M. D. Graves, and M. T. Cheng, *Tetrahedron Lett.* **24**, 1593 (1983).

175. S. Heitz, M. Durgeat, M. Guyot, C. Brassy, and F. Beachet, *Tetrahedron Lett.* **21**, 1457 (1982).

176. C. Moquin and M. Guyot, *Tetrahedron Lett.* **25**, 5047 (1984).

177. K. L. Rinehart, Jr., J. Kobayashi, G. C. Harbour, R. G. Hughes, Jr., S. A. Mizsak, and T. A. Scahill, *J. Am. Chem. Soc.* **106**, 1524 (1984).

178. J. Kobayashi, G. C. Harbour, J. Gilmore, and K. L. Rinehart, Jr., *J. Am. Chem. Soc.* **106**, 1526 (1984).

179. C. Ireland and P. J. Scheuer, *J. Am. Chem. Soc.* **102**, 5688 (1980).

180. C. M. Ireland, A. R. Durso, Jr., R. A. Newman, and M. P. Hacker, *J. Org. Chem.* **47**, 1807 (1982).

181. J. E. Biskupiak and C. M. Ireland, *J. Org. Chem.* **48**, 2303 (1983).

182. J. M. Wasylyk, J. E. Biskupiak, C. E. Costello, and C. M. Ireland, *J. Org. Chem.* **48**, 4445 (1983).

183. Y. Hamamoto, M. Endo, M. Nakagawa, T. Nakanishi, and K. Mizukawa, *J. Chem. Soc. Chem. Commun.* **1983**, 323.

184. K. L. Rinehart, Jr., J. B. Gloer, R. G. Hughes, Jr., H. E. Renis, J. P. McGovren, E. B. Swynenberg, D. A. Stringfellow, S. L. Kuentzel, and L. H. Li, *Science* **212**, 933 (1981).

185. K. L. Rinehart, Jr., J. B. Gloer, J. C. Cook, Jr., S. A. Mizsak, and T. A. Scahill, *J. Am. Chem. Soc.* **103**, 1857 (1981).

186. L. Cariello, S. Crescenzi, G. Prota, S. Capasso, F. Giordano, and L. Mazzarella, *J. Chem. Soc. Chem. Commun.* **1973**, 99.

187. L. Cariello, S. Crescenzi, G. Prota, F. Giordano, and L. Mazzarella, *Tetrahedron* **30**, 3281 (1974).

188. L. Cariello, S. Crescenzi, G. Prota, and L. Zanetti, *Tetrahedron* **30**, 4191 (1974).

189. D. Uemura, Y. Toya, I. Watanabe, and Y. Hirata, *Chem. Lett.* **1979**, 1481.

190. R. E. Moore and G. Bartolini, *J. Am. Chem. Soc.* **103**, 2491 (1981).

191. D. Uemura, K. Ueda, Y. Hirata, H. Naoki, and T. Iwashita, *Tetrahedron Lett.* **22**, 2781 (1981).

192. Y. Shimizu, *Nature* **302**, 212 (1983).

193. L. L. Klein, W. W. McWhorter, Jr., S. S. Ko, K.-P. Pfaff, and Y. Kishi, *J. Am. Chem. Soc.* **104**, 7362 (1982).

194. S. S. Ko, J. M. Finan, M. Yonaga, Y. Kishi, D. Uemura, and Y. Hirata, *J. Am. Chem. Soc.* **104**, 7364 (1982).

195. H. Fujioka, W. J. Christ, J. K. Cha, J. Leder, Y. Kishi, D. Uemura, and Y. Hirata, *J. Am. Chem. Soc.* **104**, 7367 (1982).

196. C. B. Rao, A. S. R. Anjaneyula, N. S. Sarma, Y. Venkatateswarlu, R. M. Rosser, D. J. Faulkner, M. H. M. Chen, and J. Clardy, *J. Am. Chem. Soc.* **106**, 7983 (1984).

197. M. K. W. Li and P. J. Scheuer, *Tetrahedron Lett.* **25**, 4707 (1984).

198. H. L. Sleeper, V. J. Paul, and W. Fenical, *J. Chem. Ecol.* **6**, 57 (1980).

199. Y. Gopichand and F. J. Schmitz, *J. Org. Chem.* **45**, 5383 (1980).

200. G. R. Pettit, Y. Kamano, P. Brown, D. Gust, M. Inoue, and C. L. Herald, *J. Am. Chem. Soc.* **104**, 905 (1982).

201. Y. Hamada, K. Kohda, and T. Shioiri, *Tetrahedron Lett.* **25**, 5303 (1984).

202. J. T. Baker, *Endeavour* **33**, 11 (1974).

203. P. Susse and C. Krampe, *Naturwissenschaften* **66**, 110 (1979).

204. S. Larsen and F. Watjen, *Acta Chem. Scand.* **A34**, 171 (1980).

205. T. Kosuge, H. Zenda, A. Ochiai, N. Masaki, M. Noguchi, S. Kimura, and H. Narita, *Tetrahedron Lett.* **1972**, 2545.

206. T. Kosuge, K. Tsuji, K. Hirai, K. Yamaguchi, T. Okamoto, and Y. Iitaka, *Tetrahedron Lett.* **1981**, 3417.

207. Y. Hashimoto, *Marine Toxins and Other Bioactive Marine Metabolites*, Japan Scientific Societies Press, Tokyo, 1979.

208. T. Higa and P. J. Scheuer, *Heterocycles* **3**, 227 (1976).

209. T. Higa and P. J. Scheuer, in D. J. Faulkner and W. H. Fenical, Eds., *Marine Natural Products Chemistry*, Plenum, New York, 1977, pp. 35–43.

210. T. Higa, T. Fujiyama, and P. J. Scheuer, *Comp. Biochem. Physiol.* **65B**, 525 (1980).

211. A. Pelter, J. A. Ballantine, P. Murray-Rust, V. Ferrito, and A. F. Psaila, *Tetrahedron Lett.* **1978**, 1881.

212. A. Pelter, A. Abela-Medici, J. A. Ballantine, V. Ferrito, S. Ford, V. Jaccarini, and A. F. Psaili, *Tetrahedron Lett.* **1978**, 2017.

213. L. Cariello, M. De Nicola Guidici, L. Zanetti, and G. Prota, *Experientia* **34**, 1427 (1978).

214. T. Okaichi and Y. Hashimoto, *Bull. Jap. Soc. Sci. Fish.* **28**, 930 (1962).

215. K. Tsuda, S. Ikuma, M. Kawamura, R. Tachikawa, K. Sakai, C. Tamura, and O. Amakasu, *Chem. Pharm. Bull.* (Tokyo) **12**, 1357 (1964).

216. T. Goto, Y. Kishi, T. Takahashi, and Y. Hirata, *Tetrahedron* **21**, 2059 (1965).

217. R. B. Woodward, *Pure Appl. Chem.* **9**, 49 (1964).

218. H. S. Mosher, F. A. Fuhrman, H. D. Buchwald, and H. G. Fisher, *Science* **144**, 1100 (1964).

219. T. Noguchi and Y. Hashimoto, *Toxicon* **11**, 305 (1973).

220. Y. H. Kim, G. B. Brown, H. S. Mosher, and F. A. Fuhrman, *Science* **189**, 151 (1975).

221. K. Tachibana, M. Sakaitanai, and K. Nakanishi, *Science* **226**, 703 (1984).

222.   T. Goto, "Bioluminescence of Marine Organisms" in P. J. Scheuer, Ed., *Marine Natural Products: Chemical and Biological Perspectives*, Vol. 3, Academic, New York, 1980, p. 179.

223.   Y. Kishi, T. Goto, Y. Hirata, O. Shimomura, and F. H. Johnson, *Tetrahedron Lett.* **1966,** 3427.

224.   Y. Kishi, T. Goto, S. Eguchi, Y. Hirata, E. Watanabe, and T. Aoyama, *Tetrahedron Lett.* **1966,** 3437.

225.   S. Inoue, S. Sugiura, H. Kakoi, K. Hashizume, T. Goto, and H. Iio, *Chem. Lett.* **1975,** 141.

Chapter Three

The Dimeric Alkaloids of the Rutaceae Derived by Diels-Alder Addition

Peter G. Waterman
Phytochemistry Research Laboratories
University of Strathclyde
Glasgow, Scotland

CONTENTS

1. INTRODUCTION   331
2. STRUCTURE ELUCIDATION   335
    2.1. Tyrosine Dimers,   335
       2.1.1. Alfileramine,   335
       2.1.2. Culantraramine and the Culantraraminol Isomers,   338
    2.2. Tryptophan Dimers,   341
       2.2.1. Borreverine and its Derivatives,   342
       2.2.2. Isoborreverine and its Derivatives,   345
    2.3. Indole Dimers,   347
       2.3.1. Yuehchukene,   347
    2.4. 2-Quinolone and Mixed 2- and 4-Quinolone Dimers,   348
       2.4.1. The *Ptelea* Dimers,   348
       2.4.2. The Vepridimerines,   352
       2.4.3. The Paraensidimerines,   357
       2.4.4. Geijedimerine   363
       2.4.5. General Comments on NMR and Mass Spectra of Quinolone Dimers,   364
    2.5. Glycobismine A,   365
3. SYNTHESIS   367
4. BIOSYNTHESIS   376
    4.1. One-Bond Dimers,   377
    4.2. Two-Bond Dimers,   377

4.3.   Three-Bond Dimers,   378
    4.3.1.   Alfileramine,   378
    4.3.2.   Paraensidimerines B and D,   378
    4.3.3.   Borreverine and Isoborreverine,   380
    4.3.4.   Yuehchukene,   380
4.4.   Four-Bond Dimers,   381
    4.4.1.   The Vepridimerines, Paraensidimerines A, C, E, F,
         and Geijedimerine   381
5.   SUMMARY   385
ACKNOWLEDGMENTS   386
REFERENCES   386

1.   INTRODUCTION

The plant family Rutaceae is a rich source of prenylated alkaloids [1] and coumarins [2]. The form of the prenyl unit varies considerably, the majority containing one center of unsaturation, most commonly as a 3,3-dimethylallyl group. The transformations that occur in the proliferation of prenyl structures proceed initially through the epoxidation of that original double bond, followed by ring opening of the epoxide, and then in some cases by the formation of new centers of unsaturation by dehydration (Scheme 1a). By this process the formation of a prenyldiene can be envisaged and with that Diels-Alder type condensation reactions can occur between side chains.

The first records of dimers of this type in the Rutaceae were for coumarins, a typical example being cyclobisuberodiene (thamnosin) (1) [2], formation of which can be envisaged through the condensation of two molecules of the diene depicted in Scheme 1b [3]. In all examples of coumarin dimers from the family the mode of cyclization is the same, variation in structure arising only from differing substitution patterns on the coumarin nucleus [2].

The first report for a Diels-Alder derived alkaloid dimer in the Rutaceae occurred only in 1978. Since that time a total of about 20 compounds have been reported (Table 1), and the structures of most have been established. These alkaloids are based wholly on either tyrosine, tryptophan, indole, or 2-quinolone nuclei, or on one 2-quinolone coupled with one 4-quinolone. This review details evidence leading to the identification of these alkaloids and, where information is available, also discusses possible routes for their biogenesis and synthesis. A recently isolated dimeric acridone alkaloid, glycobismine A, while not directly comparable to the Diels-Alder adducts that are the theme of this chapter, has enough similarities to be worthy of mention and is discussed briefly in Section 2.5.

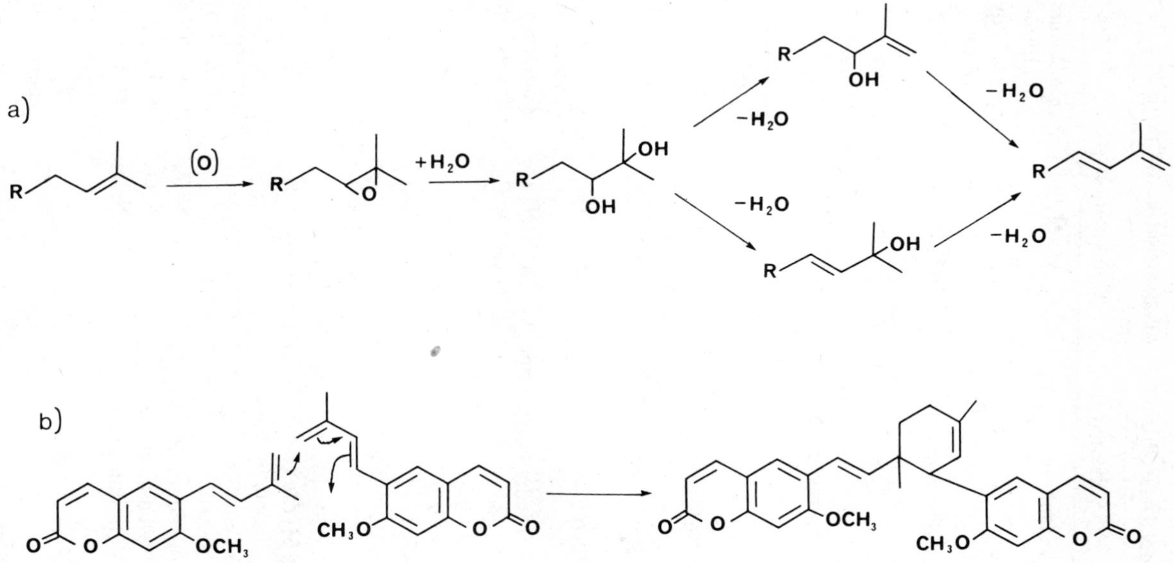

a)

b)

Scheme 1

Table 1. Diels-Alder Derived Dimeric Alkaloids of the Rutaceae and Their Sources

Type	Name	Composition	Source
Tyrosine	Alfileramine	$C_{30}H_{42}N_2O_2$	*Zanthoxylum punctatum* Vahl. [4] *Zanthoxylum coriaceum* A. Rich. [5]
	Culantraramine	$C_{32}H_{46}N_2O_2$	*Zanthoxylum culantrillo* H.B.K. [5] *Zanthoxylum procerum* Donn. Smith [6, 7]
	Culantraraminol	$C_{32}H_{48}N_2O_3$	*Zanthoxylum culantrillo* [5,7] *Zanthoxylum procerum* [7]
	Alloculantraraminol	$C_{32}H_{48}N_2O_3$	*Zanthoxylum procerum* [7]
	Epiculantraraminol	$C_{32}H_{48}N_2O_3$	*Zanthoxylum procerum* [7]
Tryptophan	Borreverine	$C_{32}H_{40}N_4$	*Flindersia fournieri* Planch. & Sieb. [8]
	N'-Methylborreverine	$C_{33}H_{42}N_4$	*Flindersia fournieri* [9]
	5-Hydroxy-4,5-dihydroborreverine	$C_{32}H_{42}N_4O$	*Flindersia fournieri* [10]
	Isoborreverine	$C_{32}H_{40}N_4$	*Flindersia fournieri* [8]
	N'-Methylisoborreverine	$C_{33}H_{42}N_4$	*Flindersia fournieri* [9]
	N,N'-dimethylisoborreverine	$C_{34}H_{44}N_4$	*Flindersia fournieri* [9]
	5-Hydroxy-4,5-isoborreverine	$C_{32}H_{42}N_4O$	*Flindersia fournieri* [10]
Indole	Yuehchukene	$C_{26}H_{26}N_2$	*Murraya paniculata* Jack. [11]
2-Quinolone	Pteledimeridine	$C_{30}H_{30}N_2O_4$	*Ptelea trifoliata* L [12]
	Pteledimericine	$C_{30}H_{30}N_2O_5$	*Ptelea trifoliata* [12]
	Vepridimerine A	$C_{34}H_{38}N_2O_8$	*Vepris louisii* Gilb. [14] *Araliopsis soyauxii* Engl. [J. F. Ayafor, pers. comm.]
	Vepridimerine B	$C_{34}H_{38}N_2O_8$	*Vepris louisii* [14] *Oricia renieri* Gilb. [14] *Araliopsis soyauxii* [J. F. Ayafor, pers. comm.]
	Paraensidimerine A	$C_{30}H_{30}N_2O_4$	*Euxylophora paraensis* Hub. [15]
	Paraensidimerine B	$C_{30}H_{32}N_2O_5$	*Euxylophora paraensis* [16, 17a]

334

Table 1. (*continued*)

Type	Name	Composition	Source
	Paraensidimerine C	$C_{30}H_{30}N_2O_4$	*Euxylophora paraensis* [15]
	Paraensidimerine D	$C_{30}H_{30}N_2O_4$	*Euxylophora paraensis* [17b]
	Paraensidimerine E	$C_{30}H_{30}N_2O_4$	*Euxylophora paraensis* [16, 17a]
	Paraensidimerine F	$C_{30}H_{30}N_2O_4$	*Euxylophora paraensis* [16, 17a]
	Paraensidimerine G	$C_{30}H_{30}N_2O_4$	*Euxylophora paraensis* [16, 17a]
2/4-Quinolone	Pteledimerine	$C_{30}H_{38}N_2O_2$	*Ptelea trifoliata* [13]
	Geijedimerine	$C_{28}H_{26}N_2O_4$	*Geijera balansae* Schintz and Guill [18]
	Vepridimerine C	$C_{34}H_{38}N_2O_8$	*Vepris louisii* [14] *Oricia renieri* [14] *Araliopsis soyauxii* [J. F. Ayafor, pers. comm.]
	Vepridimerine D	$C_{34}H_{38}N_2O_8$	*Oricia renieri* [14]

2. STRUCTURE ELUCIDATION

2.1. Tyrosine Dimers

2.1.1. Alfileramine.   The stem and root barks of about 70 species of the genus *Zanthoxylum* have been investigated [19,20] and a wide range of furoquinoline- and 1-benzyltetrahydroisoquinoline-derived alkaloids have been reported. By contrast, leaves of *Zanthoxylum* species have received little attention. Alfileramine was first isolated from a methanolic extract of the leaves of *Z. punctatum* as a white solid, $C_{30}H_{42}N_2O_2$, m.p. 185°–187°C, lacking optical activity [4]. No other alkaloids were isolated from the leaf extract, although several were detected. Chromatographic analysis of the crude extract confirmed the presence of alfileramine prior to acid/base treatment. Benzophenanthridine and aporphine alkaloids typical of the genus were obtained from the stems and branches [21], but alfileramine could not be detected.

The following spectral features of alfileramine were noteworthy:

1. A simple phenol (UV) with one replaceable proton (δ 9.30).
2. A base peak in the mass spectrum for m/z 58 ($CH_2{=}N^+Me_2$).
3. Two 6H signals at δ 2.33 and 2.38, indicating $2 \times NMe_2$.

4. Six aromatic protons that, with high-field NMR, were resolved into patterns for two 1,2,4-trisubstituted benzene rings.

5. Only 24 signals observed in the ^{13}C NMR spectrum (Table 2), demonstrating a degree of symmetry in the structure leading to a number of equivalent or near equivalent resonances. Immediately significant signals included triplets at δ_c 61.71, 61.59 and 33.11 (2C) for two CH_2CH_2N units and singlets at δ_c 150.86 and 152.98 p.p.m. for oxygen-bearing carbons in the aromatic nuclei; one of the oxygens must

Table 2. Characteristics of Tyrosine-Derived Dimers from ^1H and ^{13}C NMR Data

Position	^{1}H NMR					^{13}C NMR				
	4	4-(MeI)₂	3	7	8a	4	4-(MeI)₂	3	7[a]	8a[a]
C(1)						131.32	132.87	73.56[c]	131.1	134.5
C(2)						38.19	39.30	27.53[b]	39.1	36.7
C(3)			3.60			48.24	48.57[b]	50.94	36.7	33.9
C(4)			3.60			33.34	33.18	32.63	32.1	23.0
C(5)	4.44			4.25	4.20	48.24	49.04[b]	50.94	48.5	50.5
C(6)	5.53	5.31		5.45	5.53	132.30	132.87	37.03	123.3	124.5
C(7)	1.75	1.72	1.67	1.78	1.77	23.36	23.26	27.11[b]	23.1[b]	26.1
C(8)						77.35	77.88	75.60[c]	144.6	73.6
C(9)	1.84	1.52	1.47	4.25	0.68	20.04	20.00	19.16	110.4	30.4
C(10)	0.56	0.31	0.87	1.38	0.64	27.88	27.81	27.75[b]	23.3[b]	30.4
CH_2-α						33.11	28.68	32.34	30.1	30.2
						33.11	28.68	32.34	30.1	30.2
CH_2-β						61.59	67.40	60.70	58.7	58.8
						61.71	67.47	60.70	58.7	58.8
N-Me₂	2.32	3.26	2.33	2.26	2.28	45.05	53.49	44.31	42.5	42.5
	2.38			2.32	2.32	45.09	53.58	61.19	42.5	42.5
C(1)'/C(5)'-						125.82	125.04	120.97	126.1	127.2
C(2)'/C(6)'						125.82	126.28	123.87	126.7	127.2
						126.62	127.27	125.01	126.9	127.2
						127.01	127.80	126.20	127.8	128.7
						127.01	127.88	127.08	128.0	129.5
						128.74	128.76	128.95	130.5	131.4
						129.69	129.19	128.95	134.2	132.2
						130.84	129.19	130.11	134.2	133.1
C(3)'						115.20	115.99	114.34	109.9	110.6
						116.95	117.16	115.85	110.4	110.8
C(4)'						150.86	151.52	149.69	155.5	155.0
						152.98	153.47	153.12	156.9	159.9
OMe			3.68	3.79					54.7	55.0
			3.75	3.8					55.2	55.4

[a] As formate salts.
[b–c] Chemical shifts in any column with identical superscripts are interchangeable.

be the hydroxyl and the other must be involved in formation of a further ring system.

On this basis alfileramine was assigned the partial structure **2**. The linking $C_{10}H_{15}$ unit included three methyl groups (δ 0.56, 1.75, 1.84) and a vinylic proton (δ 5.53), which decoupling experiments showed to be part of a sequence C=CHCHCH—. From the ^{13}C NMR spectrum the presence of a double bond was confirmed, together with a vinylic methyl (δ_c 23.36), while a singlet at δ_c 77.35 indicated a fully substituted carbon linked to oxygen.

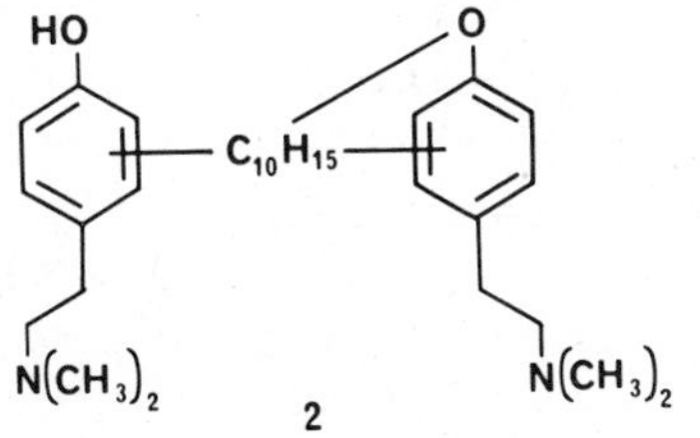

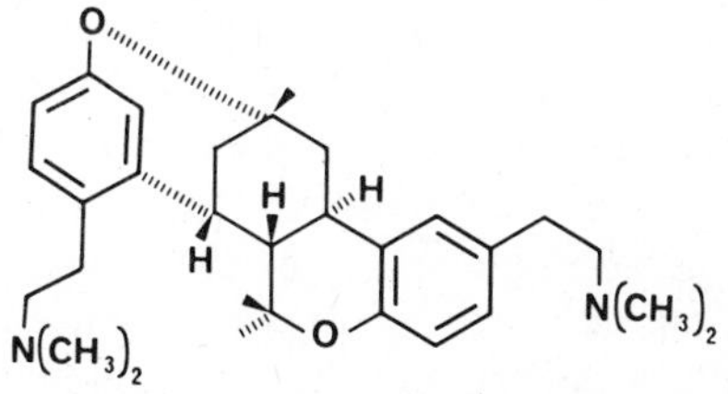

Treatment of alfileramine with methyl iodide gave the dimethiodide, m.p. 253°–255°C, which showed the expected deshielding of carbons adjacent to the quaternary centers (Table 2). Treating alfileramine with HBr/EtOH gave a crystalline solid with the same empirical formula as alfileramine but lacking both phenol and olefinic functions. This compound, named isoalfileramine (^{13}C NMR, Table 2), was shown by X-ray analysis to be structure **3** or its antipode [22].

Isoalfileramine was obviously formed by rearrangement of **4** through the free phenol of hordenine unit A and a double bond between C(1) and C(6) or alternatively between C(1) and C(2) of the cyclohexene ring. Caolo and Stermitz [4] argued for alfileramine having structure **4** by drawing a comparison with the analogous $\Delta_{1,6}$ and $\Delta_{1,2}$-tetrahydrocannabinols (**5a**) [23]. In

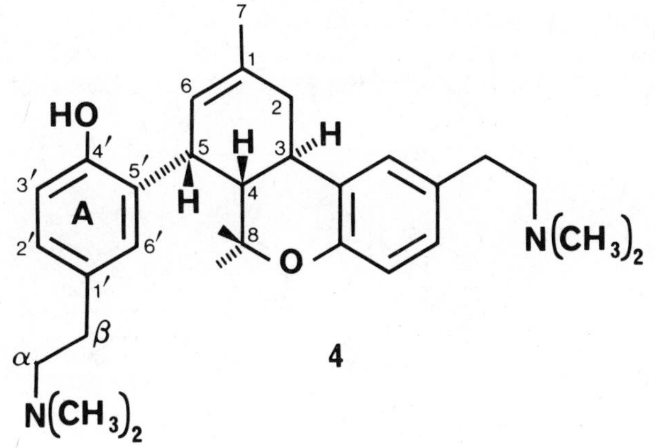

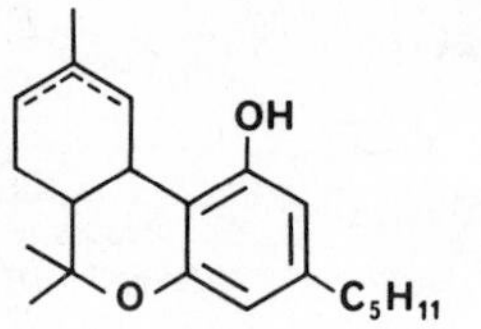

5a 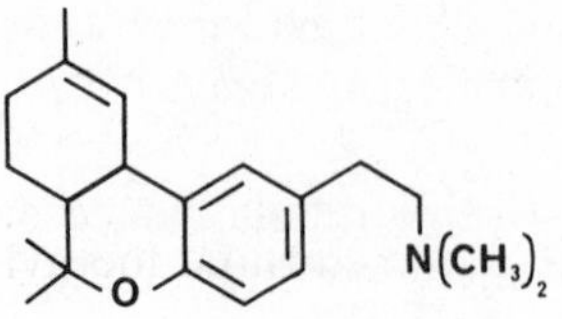   5b

the $\Delta_{1,6}$ compound the vinylic proton resonated at δ 5.45, while in the $\Delta_{1,2}$ compound it was deshielded to δ 6.42 because it then lies in the plane of the aromatic ring (cf. δ 5.53 in alfileramine).

Retrospective analysis of the spectral data obtained for alfileramine is in agreement with the proposed structure **4**. The highly shielded resonance for one methyl can be attributed to the axial substituent on C(8), which lies in the π-cloud generated by the aromatic nucleus of hordenine unit A. Finally, the mass spectrum of alfileramine can be interpreted in terms of a retro Diels-Alder (rDA) fragmentation leading to two ions at m/z 231 (Scheme 2).

One further interesting observation made by Caolo and Stermitz [4] concerned the possible structure of the unidentified alkaloid cannabamine B, reported from the leaves of *Cannabis sativa* [24]. Cannabamine B (mol. wt. 299) has a fragmentation pattern that suggests the structure **5b**, an obvious analogue of both the cannabinoids and alfileramine.

2.1.2. Culantraramine and the Culantraraminol Isomers.

A search for further sources of alfileramine by Stermitz and co-workers led to its isolation from the leaves of *Zanthoxylum coriaceum* [5]. A similar investigation of the leaves of another species, *Z. culantrillo*, yielded three lignans and a new alkaloid, culantraramine, $C_{32}H_{46}N_2O_2$ (alfileramine + C_2H_4) [5], and a compound tentatively identified as a naturally occurring salt of alfileramine. As in other analyses of *Zanthoxylum* species the dimer was found only in the leaves, whereas other parts of *Z. culantrillo* gave aporphines and furoquinolines [5].

Culantraramine differed from alfileramine in being nonphenolic, and in having two methoxy groups (responsible for the additional C_2H_4). The occurrence of 2 × OMe precluded the presence of a pyran ring as in alfiler-

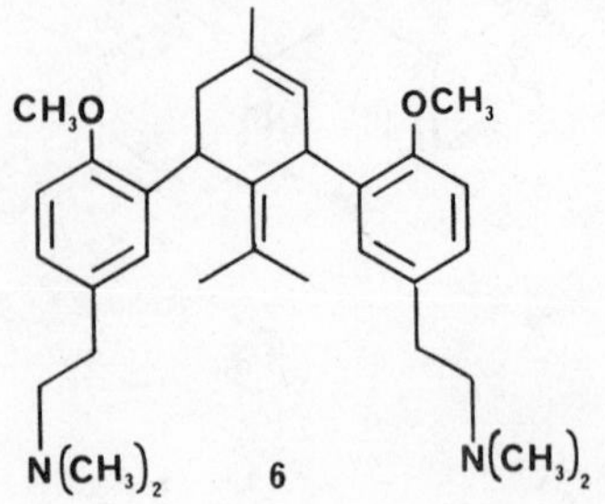

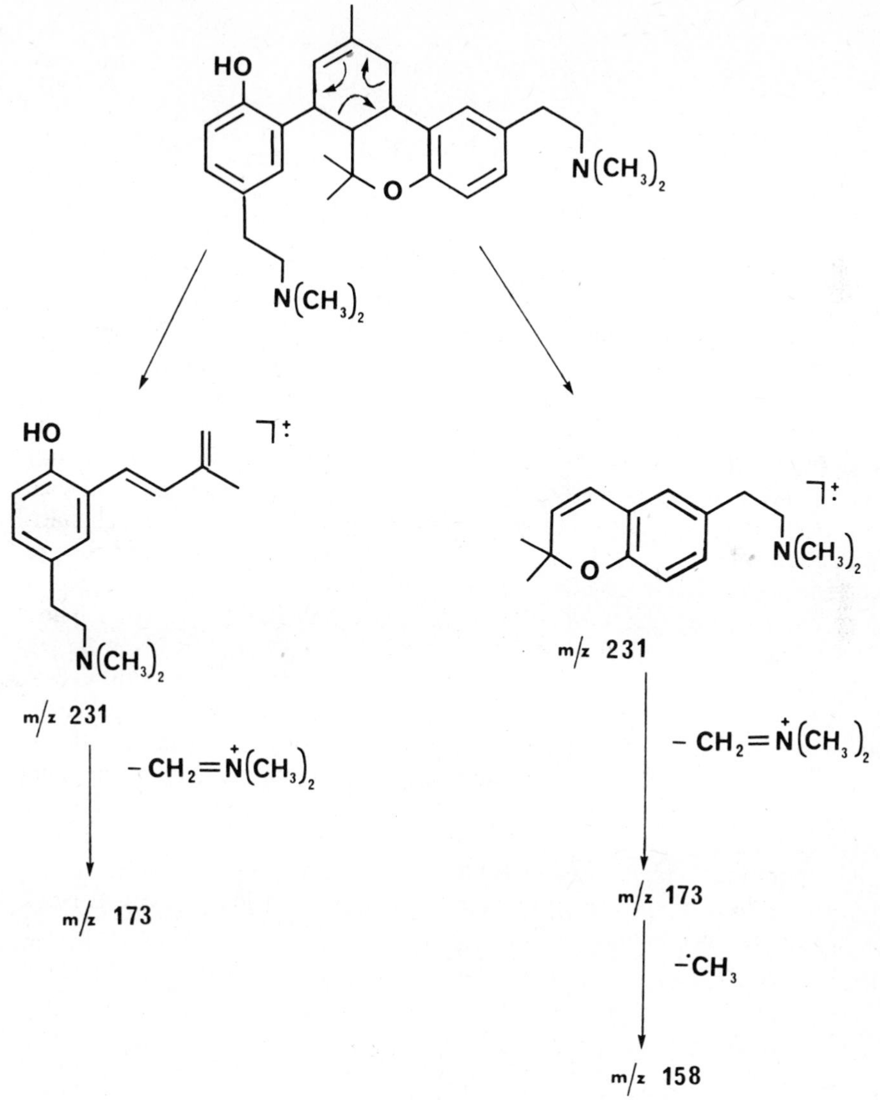

Scheme 2

amine, and this was confirmed by the absence of a resonance comparable to the δ_c 77.35 signal for C(8) in **4**. The ^{13}C NMR spectrum was useful in indicating the presence of two double bonds in the C_{10}-terpenoid portion of culantraramine. On this basis culantraramine was tentatively assigned structure **6**, with the relative stereochemistry at C(3) and C(5) undefined. Culantraramine was initially reported to be optically active ($+57°$) [5] but this has subsequently been shown not to be the case [7].

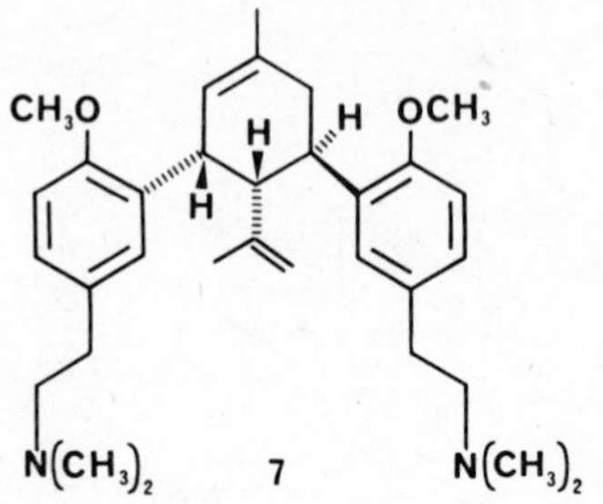

7

Culantraramine was later isolated in greater amounts from the leaves of *Zanthoxylum procerum* [7], and as a result of a more complete analysis its structure has undergone a minor revision to **7**; only two *C*-methyl resonances being observed, at δ 1.83 and 1.38. *Z. procerum* also yielded three related dimeric alkaloids, of which the major was culantraraminol (**8a**). Culantraraminol has an analysis corresponding to 18MU (H₂O) more than **7**; the electron impact mass spectrum showed facile loss of H_2O but the NH_3 chemical ionization mass spectrum confirmed the M^+. The ¹H NMR spectrum revealed three *C*-methyl signals, one vinylic (δ 1.82) and two at δ 0.72 p.p.m. A resonance at $δ_c$ 73.6 required a quaternary oxygen-bearing carbon. These data all suggested structure **8a** and the relationship between **7** and **8a** was established by conversion of **8a** to **7** in 80% yield by refluxing with $KHSO_4$ in CH_2Cl_2 for 5 hr. Culantraramine has, in turn, been converted to isoalfileramine (**3**) by refluxing in 48% HBr for 10 min followed by basification of the reaction mixture and extraction. Thus **7** and **8a** must have the same stereochemistry as alfileramine, with a *cis*-configuration between H(4) and H(5) and a *trans*-configuration between H(3) and H(4), the two hordeninyl units therefore being *trans*-substituted. An alkaloid isolated from *Z. culantrillo* only in a mixture with culantraramine [5], and initially called isoculantraramine, is culantraraminol.

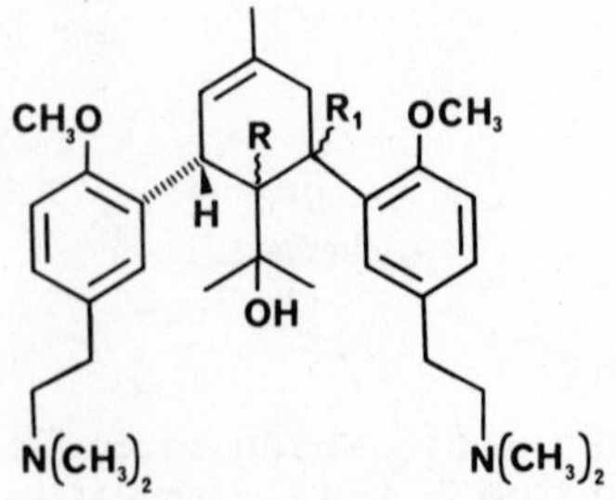

	R	R₁
8a	H-β	H-α
8b	H-α	H-β
8c	H-β	H-β

Two minor alkaloids isolated from *Zanthoxylum procerum* [7] were isomeric with **8a**. Decoupling experiments showed the sequence CH₂CHCHCH in each and, like **8a**, the absence of coupling between H(5) and H(6) suggested that H(5) was in the same configuration as in **7** and **8a**. One of the dimers was characterized by the relative deshielding of both Me(9) and Me(10), which resonated at δ 1.21 and 1.26 p.p.m. By contrast all of the other dimers of this type have a highly shielded methyl resonance that must lie within the shielding cone from one of the aromatic nuclei. The absence of this shielding indicated that the C(4)-isopropanol substituent must lie on the opposite side of the cyclohexene ring to the aromatic nuclei, which required structure **8b**. This conclusion was supported by a coupling constant of 11 Hz between H(4) and H(5) and an absence of coupling between H(3) and H(4), which would be predicted for **8b** if the cyclohexene ring occurred as a slightly twisted boat with H(3)—H(4) and H(4)—H(5) substitutions both *trans*. Compound **8b** has been given the trivial name alloculantraraminol.

The second minor product had one methyl shielded and varied from **8b** primarily in having a large coupling constant of 11.2 Hz between H(3) and H(4). It must therefore be assigned structure **8c** with H(3)—H(4) and H(4)—H(5) *cis* and has been given the trivial name of epiculantraraminol.

In a further examination of *Zanthoxylum culantrillo* leaves [7] rapid extraction and immediate TLC analysis confirmed the presence of **7** and **8a** in the extracts.

2.2. Tryptophan Dimers

The genus *Flindersia* occurs widely in Southeast Asia [25] and numerous investigations have shown it to be a good source of coumarins [2] and quinoline alkaloids [1]. Root bark of *F. fournieri*, from New Caledonia, yielded three of the most common furoquinoline alkaloids, dictamnine, γ-fagarine, and skimmianine [26]. The twigs and leaves were also a rich source of alkaloids (0.26%) [27] but, by contrast, none of the nine alkaloids obtained were quinolines. Two of them were identified as simple harmane derivatives, borrerine (**9**), an alkaloid previously isolated from *Borreria verticillata* (L.)Mey (Rubiaceae) [28], and its isomer isoborrerine (**10**) [29], which is known only from *F. fournieri*. Both borrerine and isoborrerine have been

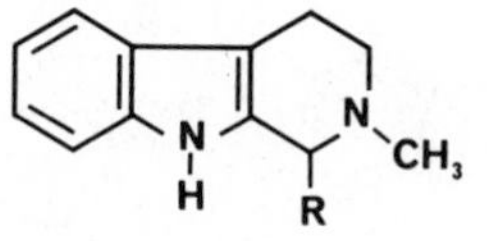

9   R = CH=C(CH₃)₂

10   R = CH₂CCH₃
           ‖
           CH₂

Table 3.   **Melting Points and UV Maxima of Borreverine, Isoborreverine, and Derivatives**

Alkaloid	Reference	m.p.	λ_{max}(log ϵ)			
11	[8]	137°C	230(4.80)	251(4.39)	285(4.39)	293(4.38)
13	[9]	—	226(4.53)	250(3.96)	286(4.01)	294(4.01)
14	[10]	—	226(4.52)	251(3.97)	285(4.07)	293(4.06)
15	[8]	—	226(4.72)		288(4.15)	294(4.15)
16	[9]	—	224(4.65)		286(4.11)	294(4.11)
18	[9]	209°C	225(4.79)		286(4.23)	294(4.23)
19	[10]	—	225(4.69)		286(4.14)	294(4.15)
21	[34]	—	225(4.17)		283(3.54)	292(3.49)

synthesized [29,30]. In addition to **9** and **10** seven dimeric alkaloids based on two tryptophan units were isolated after extensive column chromatography [27]. These alkaloids form two closely related groups based on the major compounds, borreverine and isoborreverine. Melting points (where available) and UV spectra are given in Table 3. All seven alkaloids are reported to be optically inactive.

2.2.1.   Borreverine and its Derivatives.   Borreverine (**11**) made up 22% of the total alkaloid extract of twigs and leaves [8]. It is the only dimeric alkaloid of the Rutaceae to have been recorded from another family, like borrerine having originally been isolated from *Borreria verticillata* [31]. The chemistry of borreverine has previously been reviewed [32]. The identity of **11** isolated from *Flindersia fournieri* was confirmed by direct comparison with material from the earlier study [31] in which structure had been established by X-ray crystallography. Table 4 shows ^{1}H and ^{13}C NMR data for **11**; these data have been discussed in a recent paper by Tillequin et al. [33]. Major spectral

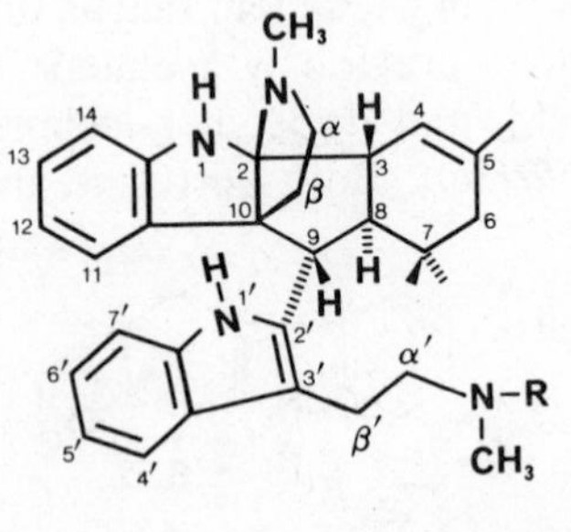

11   **R = H**

13   **R = CH₃**

Table 4. Characteristics of Tryptophan and Indole Dimers from NMR

	11	13	14	15	16	18	19	21
				^1H NMR Data				
H-4	5.60	5.63		5.40	5.48	5.41		5.70
N-Me	2.55	2.42	2.45	2.42	2.36	2.35	2.38	
	2.58	2.42	2.49	2.44	2.37	2.37	2.43	
		2.56			2.41	2.40		
					2.44			
7-Me	0.33	0.33	0.21	0.76	0.77	0.74	0.59	0.86
	0.92	0.91	1.10	1.06	1.10	1.08	1.36	1.10
5-Me	1.68	1.69	1.21	1.69	1.69	1.72	1.26	1.66
				^{13}C NMR Data				
C(2)	93.4			143.6			143.3	145.1
C(3)	47.4^a			38.1			52.9	38.3
C(4)	126.2			120.4			51.1^a	120.5
C(5)	138.1			133.6			66.2/71.3	140.2
C(6)	47.9			41.5			51.0^a	41.0
C(7)	32.5			32.7			32.5	33.4
C(8)	53.6			59.5			52.1	60.8
C(9)	47.6^a			55.4			54.3	37.5
C(10)	69.7			103.3			100.9	120.5^d
C(10a)	131.9			133.5			131.4	124.2
C(11)	118.4			119.5			117.9	118.2^b
C(12)	118.7			119.9			119.4	119.4^b
C(13)	128.1			122.8			122.4	122.9
C(14)	109.5			112.1			111.9	111.6^c
C(14a)	151.6			137.6			136.1	130.1
C(2')	133.9			133.9			132.3	122.2^a
C(3')	110.3			111.7			111.2	118.5^d
C(3a')	128.1			129.3			127.9	126.8
C(4')	118.1			119.1			117.7	119.2^b
C(5')	118.4			119.5			118.9	119.5^b
C(6')	120.9			121.1			121.1	122.1^a
C(7')	110.3			110.3			109.4	111.1^c
C(7a')	134.8			134.4			132.9	136.5
CH$_2$-β	39.3			25.2^a			24.1	
CH$_2$-β'	24.9			25.5^a			24.1	
CH$_2$-α	54.9			52.9^b			52.9	
CH$_2$-α'	53.1			53.2^b			52.1	
N-Me	36.3			35.9^c			35.6	
N-Me'	34.7			36.1^c			35.1	
7-Me$_{ax}$	23.5			28.1			29.1/29.4	28.8^f
7-Me$_{eq}$	28.5			29.1			30.8	29.0^f
5-Me	20.7			24.1			33.5/34.5	23.9

$^{a-f}$ Superscripts within the same column indicate interchangeable signals.

features relevant to the identification of **11** and pertinent to the structure elucidation of the other dimers are:

1. A shoulder at 251 nm in the UV spectrum and ^{13}C NMR resonances at δ_c 93.4, 69.7, and 151.6 p.p.m. (singlets for C(2), C(10) and C(14a), respectively), indicative of a dihydroindole.
2. The deshielding of a β-CH$_2$ resonance to δ_c 39.3 p.p.m. (ca. 24.9 for the acyclic β'-CH$_2$.
3. The base peak in the mass spectrum at m/z 172, which can be attributed to fragment **12**.

A minor alkaloid with a UV spectrum identical to **11** and base peak at m/z 172 was obtained in amorphous form and had an analysis corresponding to C$_{33}$H$_{42}$N$_4$ (borreverine + CH$_2$). The ^{1}H NMR spectrum (Table 4) differed from that of **11** only by the occurrence of a 6H singlet and a 3H singlet at δ 2.42 and 2.56 p.p.m., respectively, indicating the presence of an additional *N*-methyl group, and the absence of a signal for one of the nonindolic NH protons. The identity of this alkaloid as *N'*-methylborreverine (**13**) was confirmed by its synthesis from **11** [9], which was achieved by dissolving **11** in a mixture of 30% formaldehyde and glacial acetic acid (4–1), followed by the addition of boron cyanohydride and agitation at 0° for 40 min. Dilution of the reaction mixture, addition of ammonia, and extraction into chloroform gave crude **13**, which was purified by column chromatography.

A second minor alkaloid with the general characteristics of the borreverine group was also isolated as an amorphous base, C$_{32}$H$_{42}$N$_4$O [10]. Its identity as the 5-hydroxy derivative of 4,5-dihydroborreverine (**14**) followed from the IR spectrum (3400 cm^{-1}), loss of the H(4) olefinic signal, and shielding of the 5-methyl resonance to δ 1.21 from about 1.70 (Table 4). The direct relationship with **11** was confirmed by dehydration of **14** in benzene with trifluoracetic acid under reflux for 90 min, which yielded **11** [10]. Highfield (400 MHz) ^{1}H NMR of **14** revealed that the signal for the 5-methyl group was actually two singlets, which has led to the suggestion [10] that **14** exists as a mixture of the two configurations of C(5). This is considered further in the discussion on the corresponding isoborreverine derivative (see below).

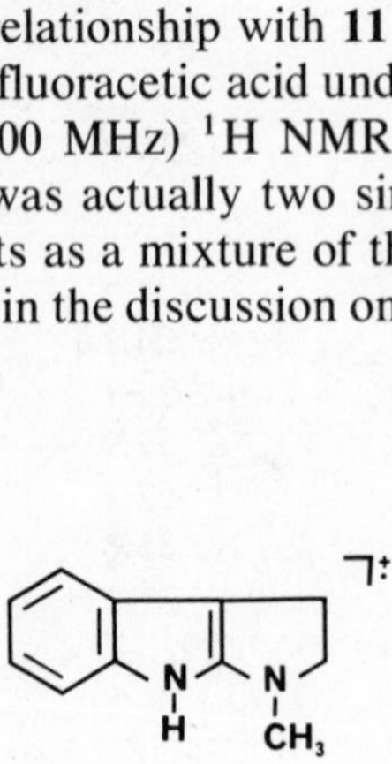

12

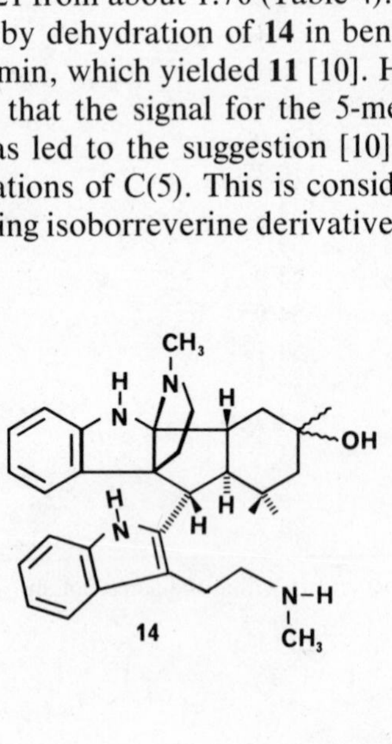

14

2.2.2. Isoborreverine and its Derivatives. The second major dimeric alkaloid (25% of the total), also had an analysis corresponding to $C_{32}H_{40}N_4$ and shared many of the spectral characteristics of borreverine [8]. Major differences were:

1. The UV spectrum lacked the 250 nm shoulder and was typical for indole nuclei only (Table 3).

2. In the ^{1}H NMR spectrum (Table 4) the resonances for the olefinic proton and N-methyl groups were slightly shielded in comparison with **11**, but there was less shielding of one of the geminal C-methyl groups. Three NH protons were observed, one indolic.

3. In the ^{13}C NMR spectrum (Table 4) resonances confirmed the absence of dihydroindole and its replacement by a second indole unit. Signals for the two β-CH$_2$ signals were now almost equivalent and similar to those for the free β'-CH$_2$ of **11**. Signals for the terpenoid moiety were at considerable variance with those for **11**, suggesting a marked change in this portion of the structure.

4. Acetylation at room temperature for 48 hr yielded a diacetate, confirming 2 × NHMe groups, whereas in **11** only a monoacetate was formed.

Acetylation was, in fact, the key to recognition of the structure of isoborreverine as **15**. The diacetate was identical with a rearrangement product

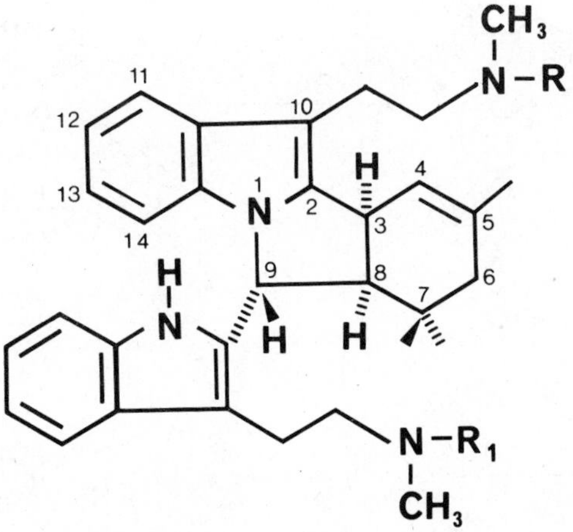

	R	R$_1$
15	H	H
16	CH$_3$	CH$_3$
18	H	CH$_3$

obtained during the acetylation of **11** at elevated temperatures, the structure **15** having been determined by X-ray analysis [31]. Comparative data on the NMR spectra of **11** and **15** are given in Table 4, and the more important features of the mass spectrum of **15** are shown in Scheme 3.

Reports of the isolation of three minor alkaloids in the isoborreverine series followed. The *N,N'*-dimethyl derivative (**16**) was obtained [9] and the structure confirmed by methylation of isoborreverine using the same method employed for borreverine. A significant feature of the mass spectrum of **16** was the presence of the *m/z* 199 fragment **17** (R = Me, Scheme 3). The mono-*N*-methyl derivative (**18**) was obtained from the same plant in crystalline form (Table 3) [9] and its identity denoted by the molecular ion and fragmentation and from the presence of three *N*-methyl resonances in the ¹H NMR spectrum (Table 4). The stereochemistry of **18** was confirmed by *N*-methylation to give **16**. Placement of the additional methyl group on the *N'*-nitrogen rather than on the *N*-nitrogen was confirmed by the absence of the *m/z* 199 ion, seen in the mass spectrum of **16**, and its replacement by an ion at *m/z* 185 (**17**, R = H). As this fragment can arise only from the larger tetracyclic part of the molecule, the monomethyl compound must be assigned structure **18**.

Finally, the 5-hydroxy-4,5-dihydro derivative of isoborreverine (**19**) was reported [10] and characterized using the same procedures previously outlined for the corresponding derivative of borreverine (**14**), including dehy-

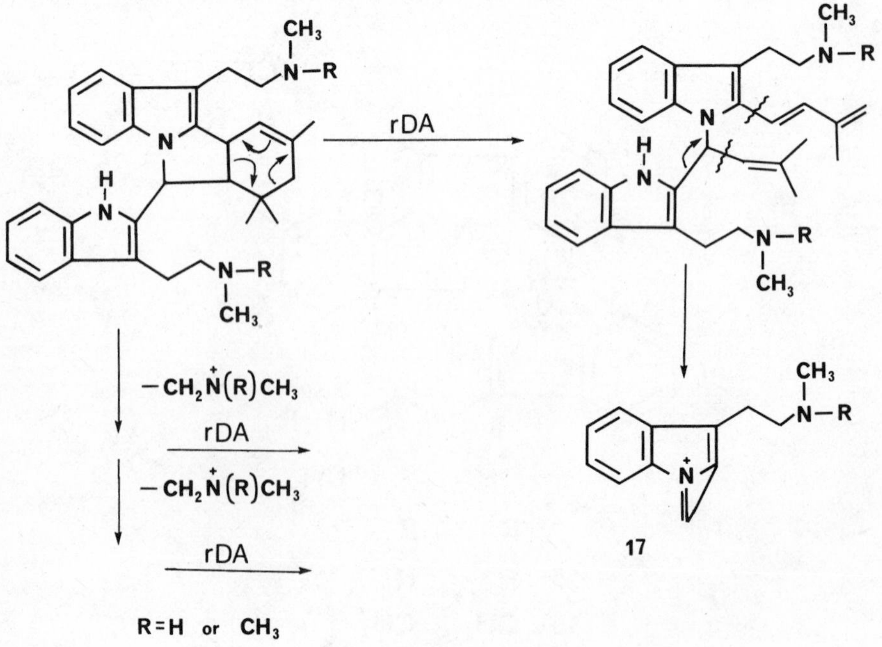

Scheme 3

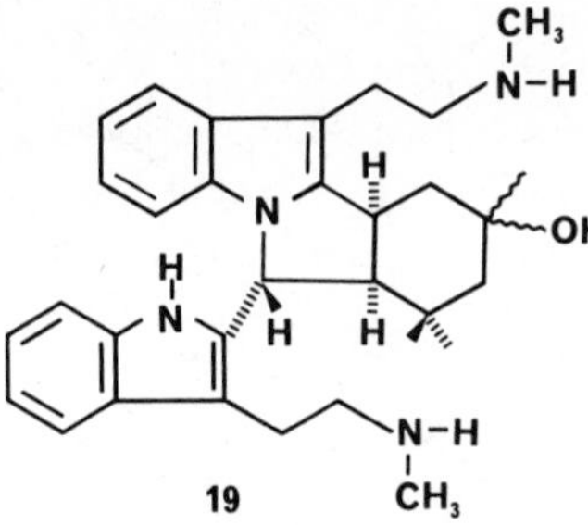

19

dration with TFA to establish comparable stereochemistry to **15**. Both ^{1}H and ^{13}C NMR spectra of **19** have been recorded [10] and are listed in Table 4. Again the resonance for the 5-methyl group appeared as two distinct signals in the ^{1}H spectrum, indicating the occurrence of the two configurations. Further support for this assertion came from the ^{13}C spectrum where the resonance for oxygen-bearing carbon C(5) was observed as two signals, at δ_c 71.3 and 66.2 p.p.m.

2.3. Indole Dimers

2.3.1. Yuehchukene. The genus *Murraya* belongs to the small subtribe Clauseninae, which is the major source of carbazole alkaloids [34]. *M. paniculata*, a small tree found in China and some neighboring countries, has yielded carbazoles, furoquinolines, acridones (all prenylated), and 3-formylindole [1] and is also the source of a large number of prenylated coumarins [2].

In a recent investigation of the coumarin-rich roots trace amounts of an alkaloid were detected, and after long and complex isolation procedures this was finally separated in a yield of between 20 and 50 p.p.m. The alkaloid, which has been given the trivial name yuehchukene (*M. paniculata* = *Yueh-Chu*), lacked optical activity. The mass spectrum displayed a molecular ion that corresponded to $C_{26}H_{26}N_2$, and that lost a methyl radical, and exhibited a major ion (**20**) at m/z 130. The presence of indole was also indicated by the UV spectrum, which agreed closely with data for the isoborreverine group (Table 3).

The ^{1}H NMR spectrum (360 MHz) revealed all 26 protons, major features being 3 × C—Me and an olefinic proton typical of the trimethylcyclohexene system of the *Flindersia* dimers (Table 4), 2 × indolic NH (δ 7.58, 8.00),

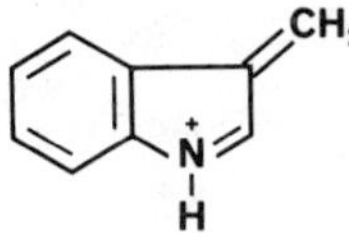

20

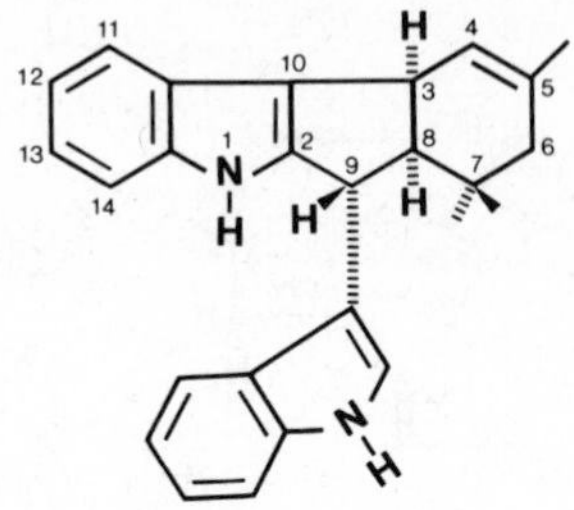

21

and nine protons in the aromatic zone for two unsubstituted indole aromatic rings and H(2) of one indole unit. The remaining six protons were made up of an AB quartet (J 15 Hz) at δ 1.63 and 2.28 for the isolated methylene (C(6)) and signals (1H each) at δ 3.15 (dd, J 8.5, 7), 4.02 (m), 4.56 (d, J 8.5) and 5.70 p.p.m. (br. s). A series of decoupling experiments revealed these protons to form a $=$CH(δ 5.70)—CH(4.02)—CH(3.15)—CH(4.56) system. The ^{13}C NMR spectrum is listed in Table 4 and can be seen to correlate closely with that of isoborreverine.

On this basis yuehchukene was tentatively assigned structure **21** with the C_{10}-unit linked to one indole unit through C(2) and C(3) (cf. N(1) and C(2) in isoborreverine) and to the other through C(3) only [11]. Acetylation of **21** gave a mono-N-acetyl derivative (δ 2.58), which was obtained in two crystalline forms. An X-ray study of the cubic crystalline form [35] agreed with the proposed structure, N-acetylation having occurred only at the less hindered N' position.

Yuehchukene has the same relative stereochemistry as isoborreverine with H(3) and H(8) *cis*, the J of 7.5 Hz permitting a dihedral angle approaching 0°. On the other hand H(8) and H(9) are *trans* with a large dihedral angle and J of 8.5 Hz. The dihedral angle between H(3) and H(4) is close to 90° and only a small coupling constant is seen.

2.4.   2-Quinolone and Mixed 2- and 4-Quinolone Dimers

2.4.1.   The *Ptelea* Dimers.   *Ptelea trifoliata*, a North American species, has been the subject of detailed phytochemical analysis that has revealed numerous quinolone and furoquinolone alkaloids exhibiting a wide range of modified isoprene units [1]. These include several, such as ptelefolidine (**22**),

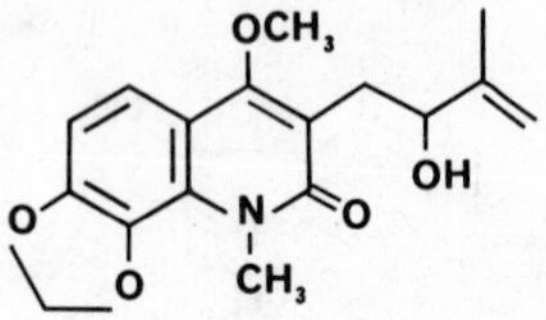

22

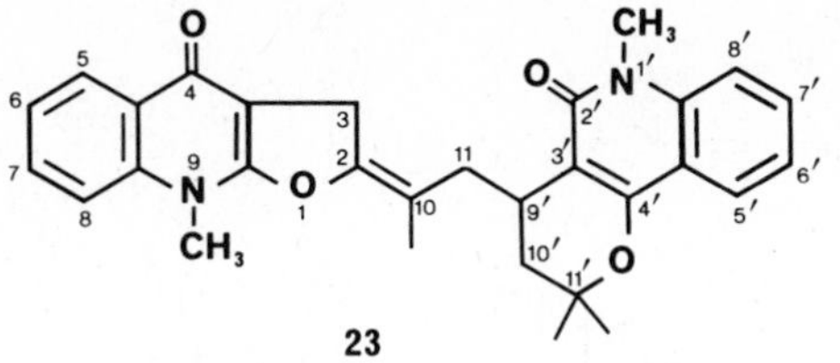

23

with 2-hydroxyisopentenyl units that, through dehydration, could give the 1,3-diene depicted in Scheme 1*a*.

Pteledimerine (**23**).   This compound was the first "Diels-Alder" dimeric alkaloid to be reported from the Rutaceae [13]. It was obtained from a root bark methanol extract by silica gel chromatography, had the composition calculated for $C_{30}H_{30}N_2O_4$, and gave a complex UV spectrum (Table 5) that was modified by addition of HCl. The IR spectrum exhibited carbonyl bands at 1645 and 1630 cm^{-1} together with a further strong band of 1550 cm^{-1}, which were taken [13], on comparison with published data [36], to indicate the presence of both 2- and 4-quinolone nuclei.

Deshielded aromatic protons at δ 8.32 and 8.07 p.p.m., each showing *ortho*- and *meta*-coupling, were typical of H(5) of 4-oxyquinolines, and together with a further six aromatic protons resonating between δ 7.17 and 7.67 p.p.m., required the presence of two quinolones unsubstituted in their benzenoid rings. Two *N*-methyl groups resonated at δ 3.64 and 3.67 p.p.m. and further singlets at δ 1.85 (3H) and 1.54 p.p.m. (6H) were assigned to an olefinic *C*-methyl and the geminal methyls of an isoprenyl-derived 2,2-dimethyldihydropyran system, respectively. The remaining seven protons took the form of an isolated methylene (δ 3.44, s), two further signals, each for 2H, at δ 2.08 (d, *J* 5.7 Hz) and δ 2.16 (d, *J* 5.4 Hz) and a multiplet at δ 3.21 p.p.m. (1H). Major features of the mass spectrum were the appearance of ions at *m/z* 241 and 188, which were assigned to **24** (this is equivalent to *N*-methylflindersine, which has been isolated from *Ptelea trifoliata* [1]) and **25**, which were envisaged [13] as deriving through initial fission of **23** between C(11) and C(9').

The structure given for pteledimerine rests solely on the above data. No information is available regarding optical activity. The ^{1}H NMR data out-

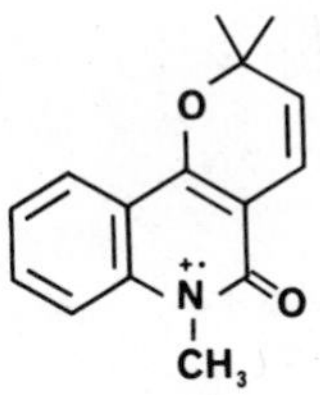

24 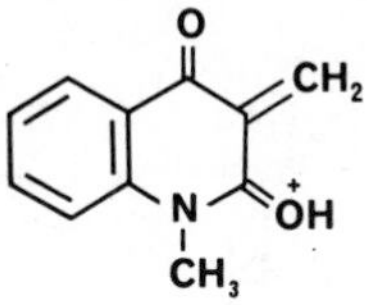   **25**

Table 5. Some Quinolone Dimers and Their UV Maxima

Alkaloid	Reference	λ_{max} (log ϵ)				
23	[13]	232.5(4.64)	252(4.20)	277.5(3.83)	287(3.89)	316(4.06)	328.5(3.98)
26	[12]	226.5(4.84)	266(4.53)	270(4.04)	286(4.05)	316.5(4.06)	329.5(3.91)
Pteledimericine	[12]	231(4.76)	252(4.29)	277(3.96)	286(3.99)	314(4.16)	327(4.09)
32	[38]	233	247/255		295	313	326
35	[38]	236	251		295	314	326

lined above is generally supportive of the proposed structure. The more highly deshielded quinolone H(5) signal (δ 8.32) is typical of 4-quinolones and related alkaloids such as acridones, whereas the resonance at δ 8.07 is more in line with H(5) when the C(4) oxygen is either hydroxy, methoxy, or another ether linkage. The similar chemical shifts of the C(3) and C(10') methylene protons were typical of comparable monomers [37] and restricted rotation between C(9') and C(11) seems unlikely, so that equivalence of the C(11) protons (either δ 2.08 or 2.16) is also acceptable. The stereochemistry of H(9') has not been assigned, but if the dihydropyran ring is in the antic-ipated half-chair, it must be equatorial in order to rationalize the coupling constant with the equivalent H(10') protons (either 5.4 or 5.7 Hz).

Pteledimeridine (**26**).   This dimer (m.p. 340°–342°C), isomeric with **23**, was isolated from the same source [12]. The UV spectrum was similar to pteledimerine (Table 5), but there were differences in the IR spectrum (1650, 1625, 1578 cm^{-1} but no 1550 cm^{-1} band), which led Mester et al. [12] to suggest a dimer based solely on 2-quinolone monomers. The mass spectrum was identical with that of pteledimerine so that pteledimeridine was assigned the same dihydropyrano-2-quinolone moiety as **23**, but in this case the second monomer was suggested to be an angular dihydrofuro-2-quinolone.

The ^{1}H NMR spectrum of **26** confirmed its relationship to **23**. Two N-methyl (δ 3.61, 3.67) and three C-methyl (δ 1.86, 1.55 × 2) singlets and a CH$_2$—CH—CH$_2$ sequence seen as 2H doublets at δ 2.23 and 2.15 (J 6 Hz for each) and a 1H multiplet at δ 3.14 p.p.m. were in close agreement with **23**. The 2H singlet for the isolated methylene was somewhat deshielded at δ 3.71 (cf. δ 3.44). Six aromatic protons found between δ 7.09 and 7.61 p.p.m. were also comparable with **23**, and valuable evidence supporting the pres-ence of two 2-quinolone monomers came from the resonances for the two H(5) protons, which were now found as double doublets at δ 8.06 and 7.96 p.p.m.

Further evidence in support of the proposed structures for pteledimerine and pteledimeridine would be welcome.

Pteledimericine.   This unidentified alkaloid (m.p. 275°–277°C) was re-ported together with pteledimeridine [12] and gave a UV spectrum almost

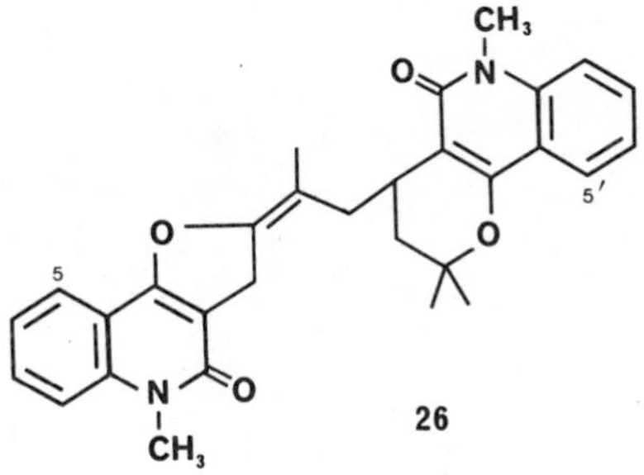

26

identical with that of pteledimerine (Table 5). Accurate mass measurement indicated $C_{30}H_{30}N_2O_5$, but no structure was proposed.

2.4.2. The Vepridimerines. Workers at the University of Yaounde, Cameroun, isolated three dimeric alkaloids from the stem bark of *Vepris louisii* and sent them to D. J. D. Connolly at the University of Glasgow for analysis. Independently, work in the reviewer's laboratory on the stem bark of *Oricia renieri*, collected in Rwanda, also yielded three dimeric alkaloids. These six isolates all gave analyses for the empirical formula $C_{34}H_{38}N_2O_2$ and had UV spectra indicative of quinolone dimers (Table 5). Fortunately the presence of identical or closely allied compounds in the two laboratories was recognized early, and the study was continued and published as a collaborative effort [14]. The ^1H NMR spectra of the dimers revealed that there were actually four separate structures among the six isolates. These were named vepridimerines A–D; *V. louisii* produced vepridimerines A–C and *O. renieri* vepridimerines B–D. The ^1H NMR spectra revealed the following common characteristics (Table 6):

1. Four methoxy substituents.
2. Two *N*-methyl substituents.
3. Three *C*-methyl substituents.
4. Four aromatic protons in the form of two AB quartets showing *ortho* coupling. In vepridimerines A and B the chemical shifts for both AB quartets were typical of H(5) and H(6) of 2-quinolones, while in vepridimerines C and D one of the AB quartets showed greater deshielding for a proton of one AB quartet typical of H(5) of a 4-quinolone. Thus vepridimerines A and B were considered to be dimers made up of two 7,8-dimethoxy-*N*-methyldihydroflindersine units (**27**) while in vepridimerines C and D one of these was replaced by the analogous 4-quinolone (**28**). In support of this the pyranoquinolone monomer veprisine (**29**) was identified as a major alkaloid in the extracts of both *V. louisii* [38] and *O. renieri* [39]. In all four dimers only three *C*-methyls were evident, indicating that linkage between the monomers must involve incorporation of a *C*-methyl group from one of the 2,2-dimethylpyran substituents.

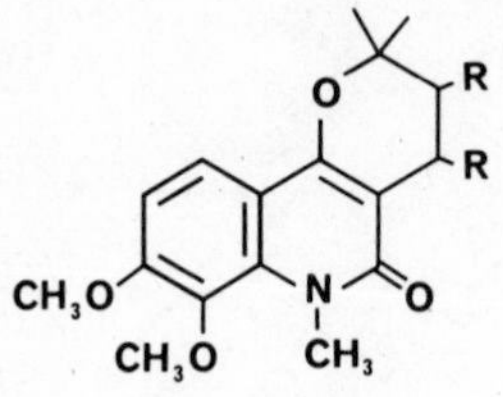

27 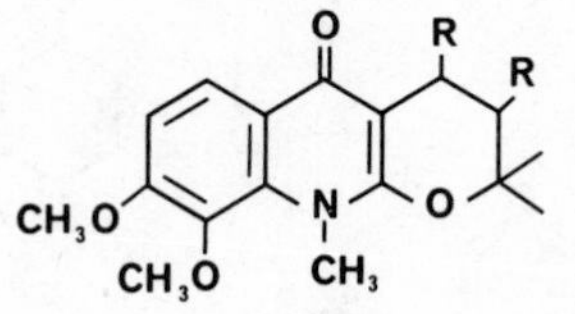   28 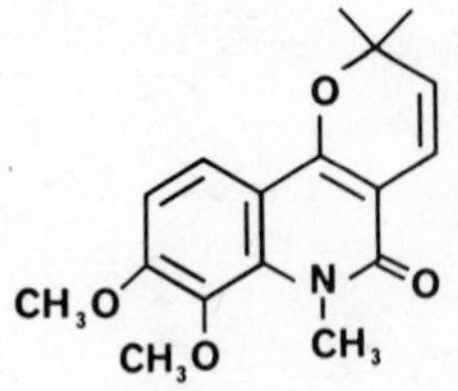   29

Table 6. Characteristics of Vepridimerines and Parensidimerines from ^{1}H NMR

Proton	Vepridimerines				Parensidimerines							
	A 30	B 32	C 34	D 35	A 41	B 48	C 42	D 47	E 44	F 43	G 45	isoG 46
N-Me	3.73, 3.73	3.72, 3.74	3.73, 3.75	3.72, 3.73	3.62, 3.68	3.75, 3.79	3.64, 3.69	3.65, 3.73	3.55, 3.55	3.63, 3.69	3.63, 3.63	3.64, 3.69
C-Me	1.49, 1.50, 1.81	1.30, 1.68, 1.87	1.49, 1.59, 1.82	1.37, 1.69, 1.88	1.55, 1.56, 1.86	1.22, 1.32, 1.39, 1.79	1.33, 1.72, 1.90	1.22, 1.70, 1.74, 1.94	1.34, 1.58, 1.61	0.99, 1.55, 1.96	1.38, 1.77, 2.02	1.26, 1.70, 2.11
O-Me	3.84, 3.89, 3.93, 3.95	3.83, 3.89, 3.91, 3.92	3.79, 3.89, 3.93, 3.95	3.73, 3.89, 3.92, 3.93								
OH						4.92						
H(4)/H(13)	7.69^a, 7.78	7.63^a, 7.66	8.05^a, 7.79	8.05^a, 7.67	8.08, 8.00	8.03, 8.03	7.96, 7.92	—, —	8.16, 8.05	8.08, 7.95	7.99, 7.99	8.00, 8.00
H-Ar^b	6.83^a, 6.87	6.79^a, 6.82	6.93^a, 6.92	6.93^a, 6.82	7.52–, 7.25	7.70–, 7.10	7.50–, 7.23	—, —	7.16–, 7.10	7.30–, 7.05	7.60–, 7.10	7.60–, 7.20
H(16)eq	3.10	3.90	3.26	3.93	3.20	—	3.89	—	2.07	3.84	7.60	4.12
H(16)ax	1.56	1.39	1.54	1.34	1.62	—	1.45	—	2.97	1.29	—	2.79
H(16a)	2.96	2.59	3.13	2.71	3.03	5.54	2.66	5.56	3.85	2.79	—	—
H(6a)	2.16	1.55	2.15	1.57	2.22	2.45	1.63	2.20	3.56	2.35	2.65	—
H(7)	3.61	3.20	3.58	3.16	3.69	3.35	3.27	3.92	3.90	3.77	3.63	4.10
H(19)eq	1.70	1.45	1.71	1.44	1.75	1.82^c	1.48	5.30	1.63	1.81	1.80	1.86
H(19)ax	2.14	2.12	2.10	2.13	2.21	1.97^c	2.17	—	2.27	2.35	1.99	2.08

353

Table 6. *(continued)*

	Vepridimerines				Parensidimerines							
Proton	A 30	B 32	C 34	D 35	A 41	B 48	C 42	D 47	E 44	F 43	G 45	isoG 46
	Coupling Constants Observed Between Protons H(A) and H(G), Values in Hz											
16$_{eq}$/16$_{ax}$	14	15	14	14	14	—	13	—	15	14	—	20
16$_{eq}$/16a	5	4	5	4	6	—	4	—	5	4	—	—
16$_{eq}$/6a	0	0	0	0	0	—	0	—	0	0	2	—
16$_{eq}$/19$_{eq}$	2	0	2	0	2	—	0	—	0	2	1	—
16$_{ax}$/16a	13	13	13	13	14	4	13	3	14	12	—	—
16a/6a	6	13	7	13	8	0	12	0	11	12	0	0
6a/7	1	3	1	4	1	3	4	9	0	2	4	3
7/19$_{eq}$	3	2	2	2	3	6	4	—	5	3	4	3
7/19$_{ax}$	3	3	2	3	3	3	5	—	2	2	3	3
19$_{eq}$/19$_{ax}$	14	14	14	13	13	13	13	—	14	13	14	12

[a] In the vepridimerines there are 2 × ABq systems. This superscript denotes the set of signals for 1 × ABq system.
[b] Signals for six protons in the parensidimerines.
[c] Interchangeable.

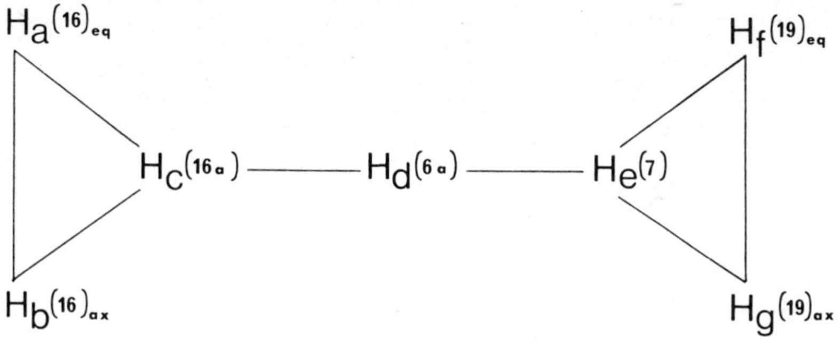

H_a (16)$_{eq}$ H_f (19)$_{eq}$

H_c (16 a) ——— H_d (6 a) ——— H_e (7)

H_b (16)$_{ax}$ H_g (19)$_{ax}$

Figure 1

5. The remaining seven protons in each vepridimerine were defined by high-field (360 MHz) ^{1}H NMR, which showed them to occur in the sequence $CH_2CHCHCHCH_2$ (Fig. 1). The chemical shifts and coupling constants for these protons in each dimer are given in Table 6. Coupling patterns noted for vepridimerines A and C were more or less identical but differed from those for vepridimerines B and D in respect of the coupling constant between the protons designated H_c and H_d in Fig. 1 (J of 6.1–6.8 Hz in A and C and 12.5–12.6 Hz in B and D). These differences were taken to indicate a *cis*-ring junction for vepridimerines A and C and a *trans*-ring junction in B and D.

Using these arguments it appeared that the four dimers were differentiated as follows:

Vepridimerine A. 2-Quinolone dimer with *cis*-ring junction.
Vepridimerine B. 2-Quinolone dimer with *trans*-ring junction.
Vepridimerine C. Mixed 2/4-quinolone dimer with *cis*-ring junction.
Vepridimerine D. Mixed 2/4-quinolone dimer with *trans*-ring junction.

Vepridimerine A, m.p. 343°–345°C, was assigned structure **30** on the basis of the above arguments (the numbering system of Jurd et al. [15] has been

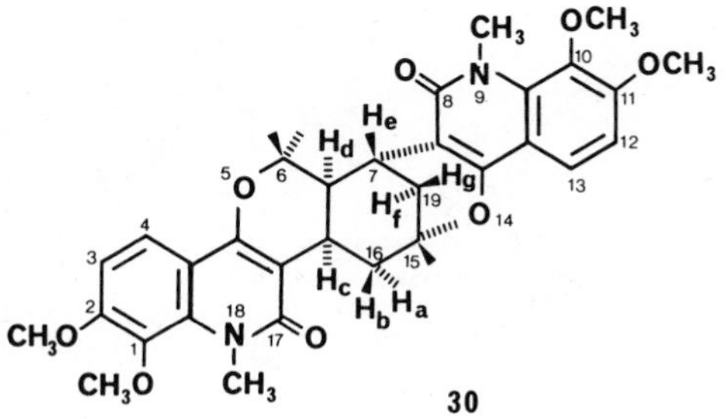

30

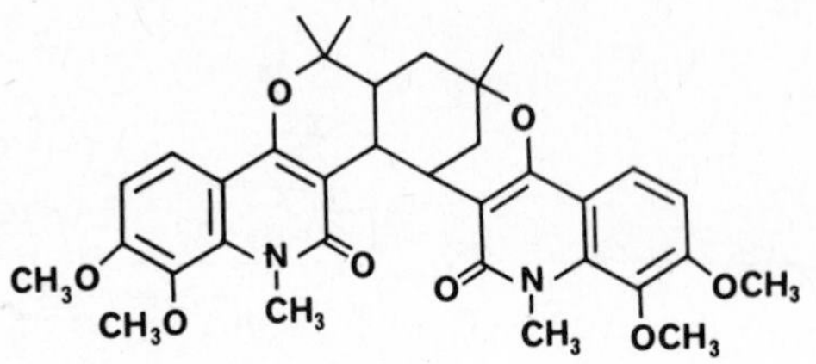

31

adopted here for ease of comparison with the paraensidimerines). The alternative structure **31** was discarded because it could not account for the strong deshielding seen in protons $H(16)_{eq}$, $H(6a)$, and $H(7)$, which must be caused by their proximity to the deshielding cones of the two quinoline carbonyls. In **31** deshielding of that order would only be seen in proton $H(7)$. A further interesting point regarding the ^1H NMR spectrum of **30** was the observation of W-bond coupling between protons $H(16)_{eq}$ and $H(19)_{eq}$, which required the cyclohexane ring to be in the chair conformation.

The structure of vepridimerine B (**32**), m.p. 278°–279°C, followed from that of vepridimerine A, there being only two major features distinguishing their ^1H NMR spectra (Table 6). These were (1) a large coupling constant (J 12.5 Hz) between $H(16a)$ and $H(6a)$ and (2) the absence of the W-bond coupling seen in **30**. The latter observation was important in that it indicated that in **32** the cyclohexane ring had adopted the boat conformation that molecular models show minimizes interaction between the quinolone nucleus of the right-hand monomer and the geminal methyl groups of the left-hand monomer (as drawn). In **32** there was again pronounced deshielding of $H(16)_{eq}$, $H(16a)$, and $H(7)$, that for $H(16)_{eq}$ being greater than in **30**, that for the others somewhat less.

Vepridimerine C, m.p. 272°C, had the stereochemistry of vepridimerine A and the ^1H NMR differences already discussed in relation to the presence of a 4-quinolone nucleus. Two types of structure were possible: **33** with the 4-quinolone on the right and **34** where it is on the left (as drawn). Structure **34** was preferred because of the greater deshielding of $H(16)_{eq}$ and $H(16a)$ compared with **30**, which indicated change of structure in relation to them while chemical shifts for the other cyclohexane protons remained virtually

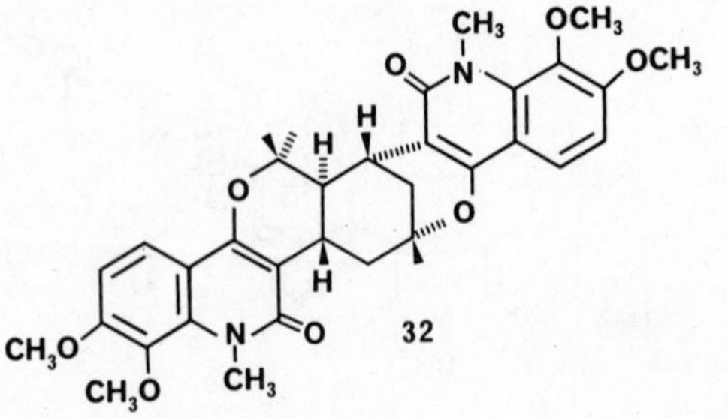

32

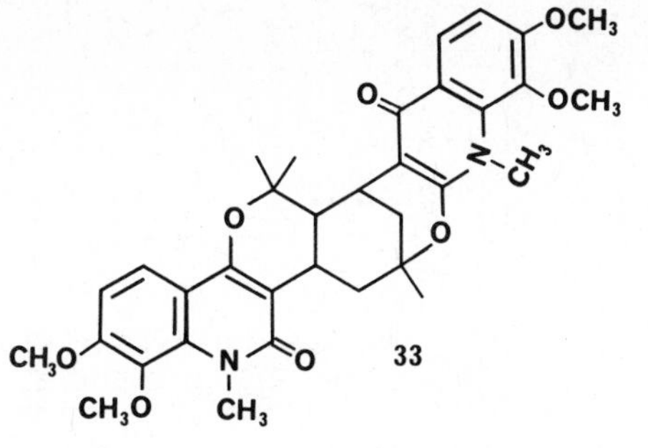

33

unchanged in comparison with **30**. The W-bond coupling previously seen in **30** was also observed here.

Vepridimerine D, m.p. 269–270°C [40], was assigned structure **35** by analogy with **32**, applying the arguments used above for vepridimerines A and C.

^{13}C NMR spectra were obtained for vepridimerines A C and are listed in Table 7. Mass spectral data were obtained for all four alkaloids. The important features of NMR spectra and mass fragmentation patterns are discussed together with those for the paraensidimerines in Sections 2.4.3 and 2.4.5.

2.4.3. The Paraensidimerines. *Euxylophora paraensis*, a tree indigenous to the tropical forests of South America, has been the subject of a considerable number of reports, and the bark is well known as a major source of indoloquinazoline alkaloids (general structure **36**). Examination of the heart-

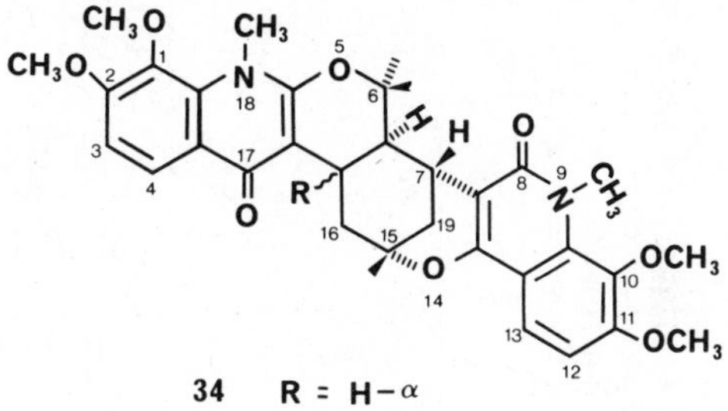

34 R = H–α

35 R = H–β

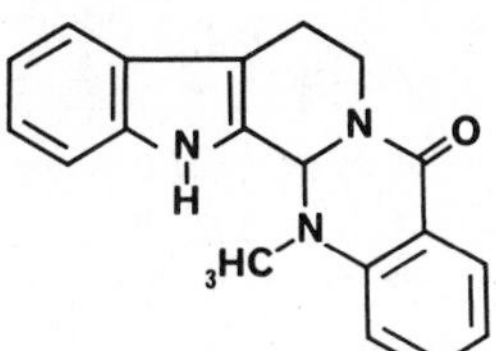

36

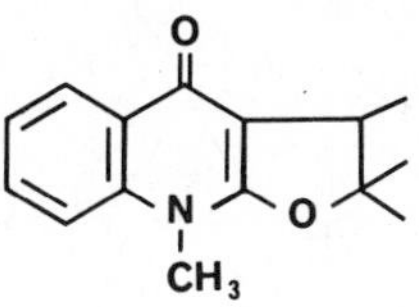

37

Table 7. Characteristics of Vepridimerines and Parensidimerines from ^{13}C NMR Data

Signal	Vepridimerines			Parensidimerines							
	A 30	B 32	C 34	A 41	B 48	C 42	D 47	E 44	F 43	G 45	isoG 46
C(1)/	136.4/	136.7/	137.6/	113.5/	113.8/	113.5/	113.5/	115.7/	113.5/	113.6/	113.7
C(10)/	136.8	136.8	136.8	113.6	114.0	113.6	113.9	115.4	113.6	113.6	113.7
C(2)/	155.0/	155.0/	155.0/	130.2/	130.6/	130.2/	130.4/	131.4/	130.1/	130.2/	130.3/
C(11)/	154.7[a]	154.5[a]	155.5[a]	130.3	131.8	130.2	131.6	131.5	130.5	131.0	130.8
C(3)/	106.9/	107.1/	108.4/	121.4/	121.8/	121.4/	121.6/	122.4/	121.3/	121.4/	121.5/
C(12)/	107.0	107.1	107.2	121.5	122.0	121.4	121.6	122.9	121.5	121.5	121.6
C(4)/	119.0/	118.6/	122.1/	123.3/	123.9/	122.9/	123.9/	124.2/	123.1/	123.4/	123.3/
C(13)/	119.2	118.9	119.3	123.5	124.0	123.3	123.9	124.2	123.7	123.9	123.5
C(4a)/	111.8/	112.8/	120.5/	115.8/	115.6/	114.5/	115.7/	114.5/	115.8/	115.9/	116.2/
C(13a)	112.5	112.9	112.0	116.5	115.8	114.5	115.8	118.3	116.6	116.4	116.4
C(4b)/	154.8/	154.9/	157.9/	155.0/	156.1/	154.9/	155.9/	155.9/	155.6/	155.5/	155.6/
C(13b)	157.7[a]	155.7[a]	156.8[a]	158.0	157.3	155.9	157.1	158.3	159.2	155.9	156.6
C(7a)/	105.9/	105.6/	107.3/	108.0/	103.6/	107.6/	103.7/	118.4/	106.6/	104.9/	108.0/
C(16b)	107.5	112.4	100.6	109.6	109.6	107.6	108.1	119.1	108.3	110.4	109.9
C(8)/	163.4/	163.3/	163.5/	161.9/	162.5/	161.9/	162.5/	164.6/	162.4/	161.3/	161.4/
C(17)/	164.3	164.2	176.3	162.8	163.1	162.8	162.6	164.8	163.0	162.4	162.5
C(9a)/	133.8/	134.1/	134.9/	138.6/	138.3/	138.8/	138.6/	140.3/	138.7/	138.6/	138.6/

358

C(18a)	134.0	134.2	134.0	138.7	140.1	138.8	140.1	140.7	138.7	138.7	139.3
C(6)	79.9	81.6	83.1	79.9	81.1	81.5	80.3	79.2	80.4	82.0	81.4
C(15)	76.8	78.3	76.9	77.0	64.1	78.5	64.2	74.9	78.0	73.0	76.4
C(6a)	43.6	52.4	43.4	43.6	29.4^d	52.2	29.4	42.3	53.6	48.0	136.1
C(7)/	26.2/	25.6/	25.5/	26.5/	44.7	25.7/	45.1	26.6/	26.9/	26.4	26.1
C(16a)	27.6	26.3	27.5	27.7	70.0	26.5	132.5	28.7	28.0	130.4	123.5
C(16)	39.5	39.8	39.8	39.5	52.5	39.7	127.6	51.2	42.8	128.1	42.3
C(19)	32.2	31.0	32.2	32.2	—	31.1	—	32.7	39.1	31.2	35.9
N-Me	32.9	33.1	33.3	28.9^b	29.1^d	29.0^b	29.3	30.2^c	28.4^b	29.0	28.7^e
	33.3	33.6	35.2	29.1^b	29.6^d	29.4^b	29.3	30.5^c	29.4	29.1	29.1^e
C-Me	24.8	20.7	25.2	24.9	21.3	20.7	18.2	32.5^c	19.0	22.8	24.6
	28.6	28.5	28.6	28.7^b	26.1	28.5^b	21.1	—	28.1^b	27.8	26.6
	29.1	29.2	29.0	29.2^b	29.6^d	29.3^b	25.9	—	28.6^b	27.8	29.4^e
					31.4		29.3				
OMe	56.1	56.2	56.1								
	56.2	56.2	56.2								
	61.4	61.4	61.2								
	61.5	61.4	61.5								

^{a,b} Interchangeable.

^{c} Resonances for five methyl groups in these three signals.

^{d} Three signals for 2 × N-Me 1 × C-Me, and one tertiary carbon.

^{e} 2 × N-Me and 1 × C-Me, unassigned.

/ links equivalent carbons from both parts of the dimer and chemical shift data, which are interchangeable between them.

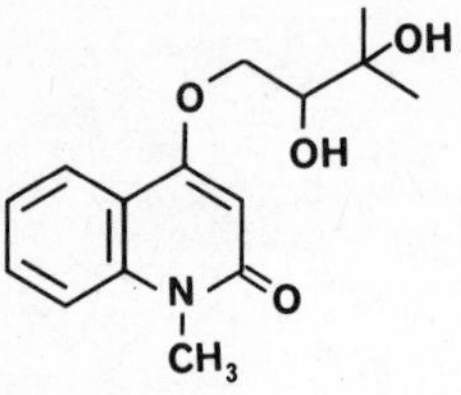

38 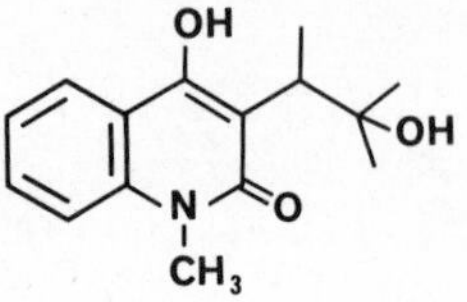   39 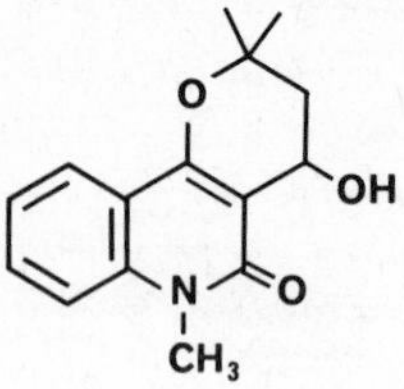   40

wood by Jurd et al. [15–17] gave no indoloquinazolines but yielded a number of monomeric quinolones, notably *N*-methylflindersine (**24**), lemnobiline (**37**), two novel 2-quinolones (**38**, **39**), and a dihydropyranoquinol-2-one (**40**) [17], as well as seven optically inactive dimeric 2-quinolones. These dimers, the paraensidimerines, are, as a group, related to the vepridimerines but exhibit greater structural and stereochemical diversity. Data on ^{1}H and ^{13}C NMR are given in Tables 6 and 7. The following discussion concerning their structure elucidation does not follow the chronological order of isolation and publication but deals first with those most similar to the vepridimerines.

Paraensidimerines A (**41**) *and C* (**42**).   Paraensidimerine A (m.p. 311°–312°C) and paraensidimerine C (m.p. 210°C as a hydrate) were both obtained in about 0.7% yield [17b]. Each had an analysis consistent with $C_{30}H_{30}N_2O_4$, and NMR spectra revealed two *N*-methyl and three *C*-methyl substituents, together with signals for two 2-quinolone nuclei with unsubstituted aromatic rings [15]. Resonances for the isoprene unit were in close agreement with those for vepridimerines A and B (Tables 6 and 7).

The two dimers were characterized by X-ray analysis of their dinitro derivatives, which were prepared by heating in concentrated nitric acid on

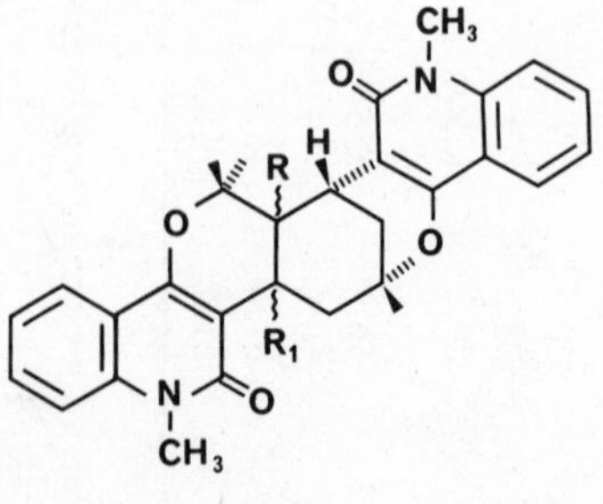

	R	R_1
41	H-α	H-α
42	H-α	H-β
43	H-β	H-α
44	H-β	H-β

a steam bath for 3 min [15]. This established that paraensidimerine A (**41**) differed from vepridimerine A (**30**) only in the absence of methoxy substituents. Paraensidimerine C (**42**) was likewise linked to vepridimerine B (**32**). The X-ray studies confirmed the inferences made concerning the stereochemistry of the cyclohexane system of the vepridimerines by Ngadjui et al. [14]. The only difference between the two dimers was at C(16a), with H(16a) axial and *cis* to H(6a) in paraensidimerine A, and axial and *trans* to H(6a) in paraensidimerine C. In both the relationship between H(7) and the C(16) methyl was *cis* and equatorial, in agreement with observations by Barnes et al. [41], and H(7) was *trans* to H(6a). As predicted from the vepridimerines [14], paraensidimerine A (*cis/trans* between C(16a)–C(6a)–C(7)) had the cyclohexane in a chair conformation (cf. vepridimerines A and C), and showed W-bond coupling between the equatorial protons of C(16) and C(19) methylenes. In paraensidimerine C, on the other hand, the *trans/trans* system leads to a boat conformation (cf. vepridimerines B and D).

Paraensidimerines E (**44**) *and F* (**43**).   The minor dimers, paraensidimerines E (m.p. 289°–290°C) and F (m.p. 310°C) [17a], were isomeric with A and C. Paraensidimerine F showed a coupling pattern for the cyclohexane protons that was similar to paraensidimerine C (Table 6), with a large coupling between H(16a) and H(6a) denoting their placement *trans* to one another. However, the occurrence of W-bond coupling between H(16)$_{eq}$ and H(19)$_{eq}$ required a chair conformation rather than the boat found in dimer C. Furthermore H(6a) and H(7) were appreciably deshielded by comparison with those protons in dimer C. These data require the alternative *trans/cis* configuration in F (**43**), with H(6a) axial and *beta* and H(16a) axial and *alpha*, which is opposite to paraensidimerine C (**42**).

Paraensidimerine E, like F, exhibited a large coupling between H(6a) and H(16a), but in this case no coupling at all could be detected between H(6a) and H(7). Bearing in mind that dimers A, C, and F represent *cis/trans*, *trans/trans*, and *trans/cis* isomers, then paraensidimerine E must be the *cis/cis* isomer (**44**). Jurd et al. [17a] pointed out that, if the *cis/cis* isomer were to exist in the chair conformation, then the two quinolone nuclei would lie parallel to one another and in close proximity. That this was not the case in dimer E was indicated by the absence of W-bond coupling between equatorial H(16) and H(19) protons. On the other hand if dimer E were in the boat conformation, the dihedral angles between H(6a) and H(16a) and between H(6a) and H(7) would be 60° and 0°, respectively. Clearly this also could not be the case as it would not permit the observed coupling constant of 11 Hz between H(6a) and H(16a) and the absence of coupling between H(6a) and H(7). Jurd et al. [17a] have suggested that paraensidimerine E actually exists with the cyclohexane ring twisted out of an original chair conformation by repulsion between the quinolone centers so that the dihedral angles are approximately: H(6a)–H(7), 90°; H(6a)–H(16a), 0°; H(16a)–H(16)$_{ax}$, 0°; H(16a)–H(16)$_{eq}$, 120°.

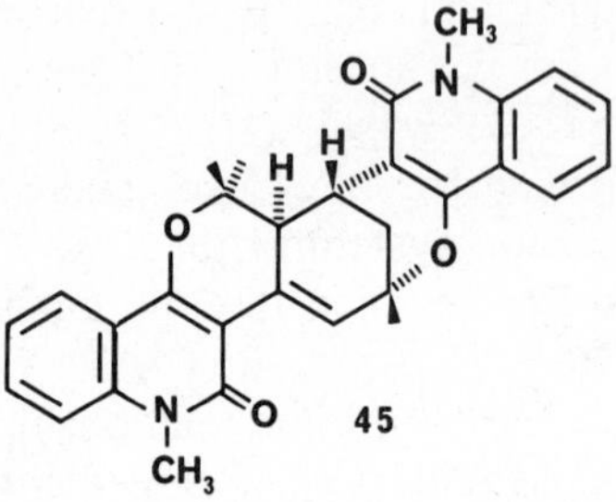

Paraensidimerine G (**45**). This dimer was found as a minor contaminant of C and was purified by chromatography over silica gel [17a]. It gave a melting point of 286°–287°C and had an analysis agreeing with $C_{30}H_{28}N_2O_4$, 2H less than the dimers already discussed. The NMR spectra revealed the close structural similarity of dimer G to A, C, E, and F, the only changes being in relation to the cyclohexane nucleus where one methine and one methylene signal were lost and were replaced by resonances for a trisubstituted double bond (δ_c 130.4 d, 128.1 s) and a methine proton (δ 7.60), thus requiring the presence of a cyclohexene unit rather than the original cyclohexane.

As the generation of the double bond had to remove one of the methylene groups, there were two theoretical possibilities for its location, C(7)–C(19) or C(16)–C(16a). It is not possible to make the C(7)–C(15) ring junction with a C(7)–C(19) double bond, thereby leaving C(16)–C(16a) as the only tenable site and leading to the assignment of structure **45**. The olefinic proton exhibited coupling (Table 6) to both H(6a) and to H(19)$_{eq}$, but there was no measurable coupling between H(6a) and H(7). These data require the C(6a)–C(7) junction to be *trans*, as in dimers A and C, and the existence of the cyclohexene in a half-chair allowing the observed long range couplings of H(16).

The identity of dimer G was confirmed by its formation, in quantitative yield, from dimer C by oxidation using DDQ in benzene under reflux for 2 hr [17a]. Similar treatment of dimer A yielded **45**, but in a yield of 80%, with a minor product that was characterized as **46** (δ_c 123.5 s, 136.1 s; two methylene resonances—see Tables 6 and 7).

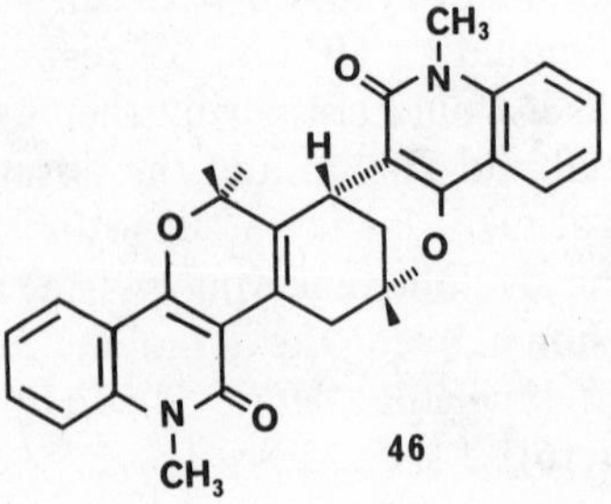

Paraensidimerines B (**48**) *and D* (**47**).   These dimers represent an alternate mode of cyclization but a numbering system analogous to that used for the other dimers is employed for ease of relating NMR data.

Paraensidimerine D (m.p. 259°C, $C_{30}H_{30}N_2O_4$) was the first dimer to be isolated from *Euxylophora paraensis* [17b]. It was identified as **47** by X-ray analysis and differs from the previously discussed dimers in that, while the monomers are similarly linked twice, no methyl group is involved in the dimerization. Instead one linkage involves oxygen and the resulting C_{10}-unit includes four *C*-methyl groups, *cis*-fused pyran rings, and a 2-methylpropenyl side chain substituent at C(7). To minimize interaction between methyl groups of the two pyrans, the bispyran system appears to be twisted out of a chair conformation. Paraensidimerine D was initially reported to be optically active [17b], but this was subsequently corrected [15].

In **47** the C(6a)–C(7) protons are *trans*, the dihedral angle between H(6a) and H(7) was measured as 86.9°, and no coupling was visible. This meant that the resonances for the pyran system were divided into two pairs of doublets, H(6a)–H(15) (3 Hz) and H(7)–H(16) (9 Hz) (Table 6). The ^{13}C NMR spectrum (Table 7) was reported separately [15], the major feature being the presence of an oxymethine carbon (C(15)) resonating at δ_C 64.2.

Paraensidimerine B (m.p. 286°–287°C) had a molecular formula $C_{30}H_{32}N_2O_5$, and the IR spectrum indicated the occurrence of a hydroxyl (3450 cm^{-1}) [17a]. It differed from D only in the loss of the side chain double bond and its replacement by resonances for a methylene and a carbinol carbon (see Tables 6 and 7). On this basis paraensidimerine B can be characterized as **48**.

2.4.4.   Geijedimerine.   (Note added in proof.) A further quinolone dimer, geijedimerine (m.p. 205–207°C, $C_{28}H_{26}N_2O_4$, $[\alpha]_D$ 0°), has recently been reported from *Geijera balansae* collected in New Caledonia [18]. Its UV spectrum was indicative of a mixed 2-/4-quinolone dimer (λ_{max} 253 (3.59), 277 (3.44), 288 (3.47), 313 (3.53); + HCl λ_{max} 245, 277, 286, 304). High-field (270 MHz) ^{1}H NMR revealed a cyclohexane system with the same stereochem-

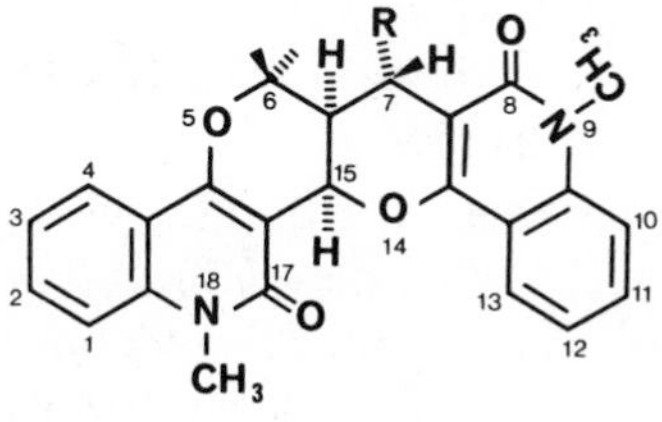

47   R = $\overset{16}{C}H = C(CH_3)_2$

48   R = $CH_2\ \underset{\underset{OH}{|}}{C}(CH_3)_2$

istry as paraensidimerine F (**43**), that is, with H(16a) *alpha* and *trans* to H(6a), which is, in turn, *cis* to H(7). Chemical shifts and coupling constants for the cyclohexane protons are reported as follows: δ 3.76 (ddd, *J* 13, 4, 2 Hz, H(16)$_{eq}$), 3.46 (ddd, *J* 4, 3, 2 Hz, H(7)), 2.57 (td, *J* 13, 4 Hz, H(16a)), 2.15 (dd, *J* 13, 2 Hz, H(19)$_{ax}$), 2.06 (dd, *J* 13, 3 Hz, H(6a)), 1.71 (ddd, *J* 13, 4, 2, Hz, H(19)$_{eq}$), 1.30 (t, *J* 13 Hz, H(16)$_{ax}$), 1.00, 1.46, 1.81 (3 × CH$_3$). Like paraensidimerine F, geijedimerine lacks substitution in the benzenoid rings; it is also the first of the quinolone dimers to lack *N*-methylation. The 4-quinolone system has been assigned to the left-hand side (identical to vepridimerines C (**34**) and D (**35**)) on the basis of a comparison of the ^{1}H NMR spectra of paraensidimerine F and *N*-methyl geijedimerine [18]. This is substantiated by the recent synthesis of vepridimerine E [47b], which has the same H(16a)–H(6a)–H(7) stereochemistry as geijedimerine and paraensidimerine F and which gives a ^{1}H NMR spectrum very similar to that of geijedimerine.

2.4.5. General Comments on NMR and Mass Spectra of Quinolone Dimers.

^{1}H and ^{13}C NMR data for the vepridimerines and paraensidimerines are listed in Tables 6 and 7 and features of importance to the characterization of individual compounds have been discussed in Sections 2.4.2 and 2.4.3. The following are some more general points of interest.

1. For the quinolone systems both ^{1}H [37] and ^{13}C data [42] were in good agreement with previously published information and proved valuable in distinguishing between 2- and 4-quinolones.

2. Dimers with a C(16a)–C(6a) *trans*-system had the C(6a) resonance at about δ$_C$ 53, whereas in the *cis*-isomers it was δ$_C$ 42–44. This is due to the *gamma*-gauche effect of the axial C(6)-methyl in the *trans*-isomers (β and axial in **32, 34, 42**; α and axial in **43**).

3. As a corollary to point 2, the nonequivalence in the resonances of the C(6)-methyls is greater in the C(16a)–C(6a) *trans*-dimers than in the *cis*-dimers (0.32–0.56 p.p.m. against 0.01–0.24 p.p.m. apart). The particularly strong deshielding of the axial C(6)-methyl in **43** is due to it lying close to the C(8)-carbonyl.

4. Both oxygen-bonded carbons (C(6) and C(15)) were slightly more deshielded in the *trans*-isomers.

5. The appreciable nonequivalence in the C(16) protons is attributed to their positions relative to the C(17) carbonyl. In the C(6)–C(16a) *trans*-isomers the H(16)$_{eq}$ proton lies within 2 Å of the C(17) carbonyl and resonates between δ 3.8–3.9, as compared with 2.5–2.8 Å for the *cis*-isomers, where it resonates at δ 3.1–3.2 p.p.m.

Table 8 lists the five major fragments recorded in the mass spectra of each of the vepridimerines and paraensidimerines (except for B and D). In most

Table 8.   Five Most Abundant Fragments in the Mass Spectra of Vepridimerines and Paraensidimerines

Fragment	Vepridimerines				Parensidimerines				
	A [40]	B [38]	C [38]	D^a [38]	A [15]	C [15]	E [17a]	F [17a]	G [17a]
M^+	x	x	x	x	x	x	x	x	x
M^+—CH_3	x				x	x			x^b
M^+—C_3H_7							x		x
49 m/z 306–308	x	x	x	x	x	x	x	x	x
50 m/z 292–294	x	x	x	x	x	x		x	x
51 m/z 256							x	x	
52 m/z 242		x	x	x	x				
53 m/z 226	x	x	x	x	x			x	x

^{a} Only four fragments above 10%.
^{b} m/z 463 [M^+—CH_3—2H].

cases the molecular ion was also the base peak. In all spectra there was a significant ion for M^+—CH_3 because of loss of one of the C(6)-methyl groups. Protonation of O(5) and fission of O(5)—C(6) seems likely to be the first step to loss of C_3H_7 (Scheme 4). Most dimers had major ions in the clusters m/z 306–308 and 292–294 (60 MU higher in the vepridimerines). These ions are tentatively attributed to fission of the O(15)—C(16) and C(7)—C(7a) bonds, probably through the same protonation sequence invoked for the loss of C_3H_7 (Scheme 4). This would generate **49** (m/z 308) and then **50**. Other important ions (**51–53**) can be derived by fragmentation through the cyclohexane ring.

2.5.   Glycobismine A

Recent studies on *Glycosmis citrifolia* Lindl. growing in Taiwan have led to the isolation of a large number of alkaloids, one of which was been identified [43] as the dimeric acridone glycobismine A (**54**). Glycobismine A, $C_{37}H_{34}N_2O_6$, m.p. 256°–258°C, was optically inactive, and its UV spectrum showed the anticipated maxima and log ϵ values for a dimeric acridone (λ_{max}

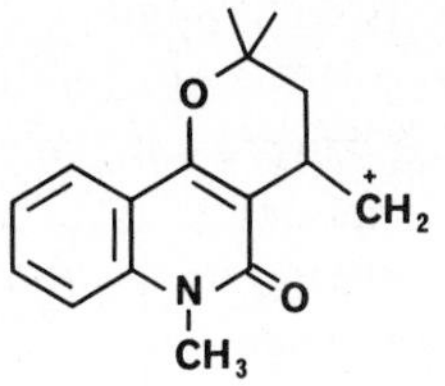

51

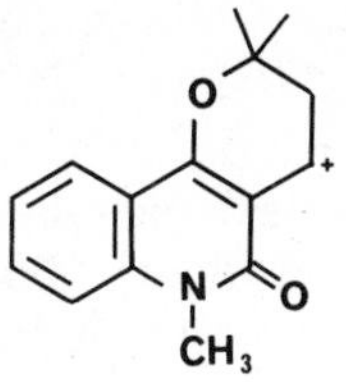

52

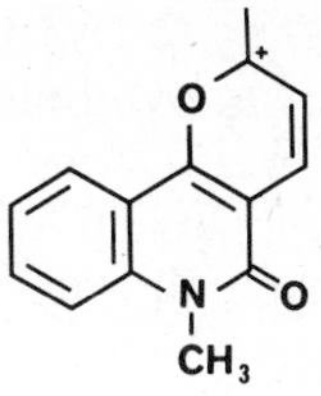

53

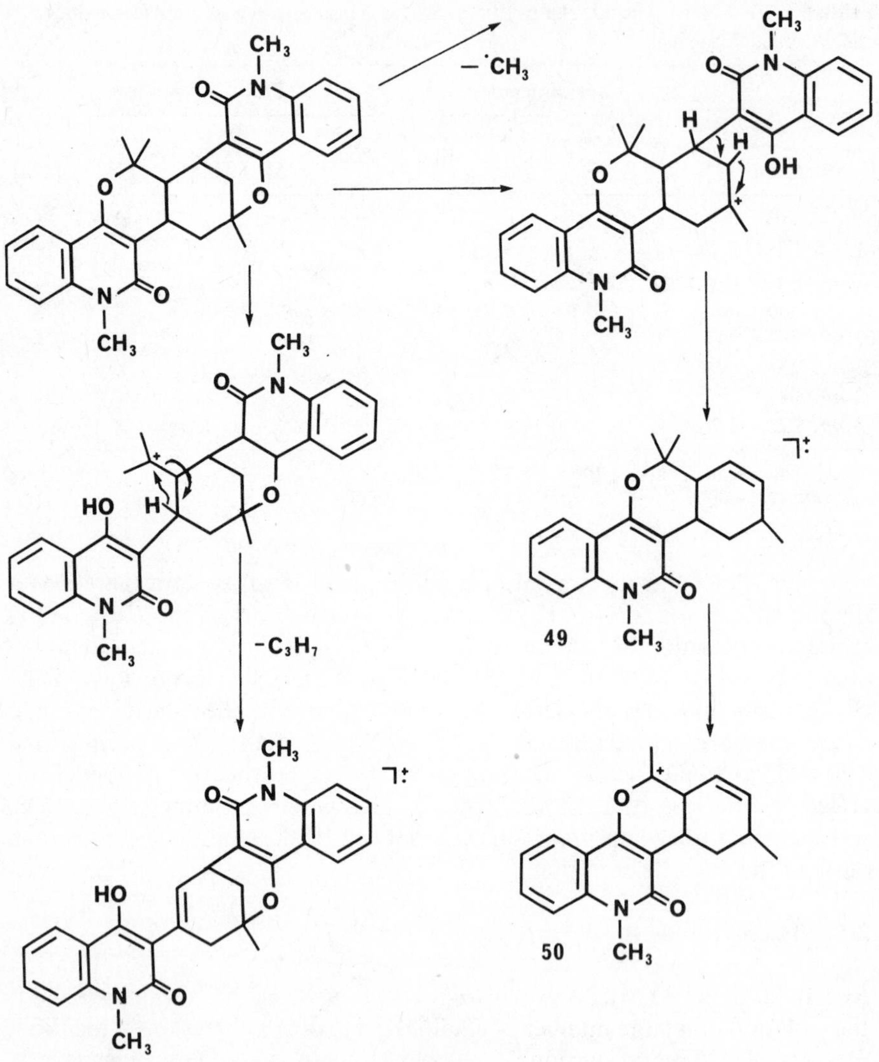

Scheme 4

235 (4.72), 246 (4.75), 282 (3.73), 300 (4.62), 336 (4.13), 372 (4.20), 423 (4.02)). The ^{1}H NMR spectrum revealed resonances for the two H-bonded C(1) hydroxy group protons at δ 16.51 and 14.72. Resonances assigned from the ^{1}H NMR and ^{13}C NMR spectra and shown on structure **54** agree with those anticipated for a dimer of the type depicted in which one monomer is linked through a nucleophilic center on the 1,3-dioxygenated A-ring to the dihydropyran ring of a second monomer. Placement of the linkage on C(4′) of the dihydropyran ring follows from the ^{13}C NMR resonances for C(3′) and C(4′), which were found at $δ_c$ 38.3 and 23.6 p.p.m., respectively.

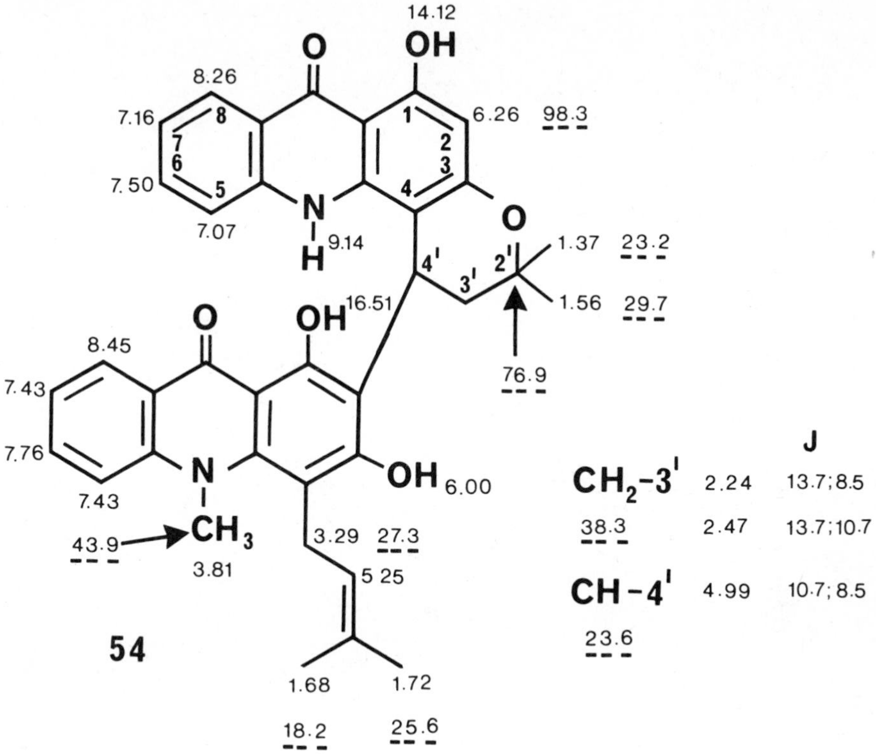

14.12
OH

O

8.26

7.16 8

7 6

7.50 5

7.07

N H 9.14

1 6.26 98.3

2

3

4

O

4' 3' 2'

1.37 23.2

1.56 29.7

76.9

O

8.45

7.43

7.76

7.43

OH 16.51

N

CH₃

43.9

3.81

OH 6.00

3.29 27.3

5 25

54

1.68 1.72

18.2 25.6

J

CH₂-3' 2.24 13.7;8.5

38.3 2.47 13.7;10.7

CH-4' 4.99 10.7;8.5

23.6

```
NB  -  ¹³C NMR chemical shift values underlined.
```

The angular situation of the pyran ring was established from the ^{13}C resonances for the two N-Me groups in the N,O,O,O-tetramethyl derivative. These were found at δ_c 43.2 and 43.9 p.p.m., indicative [44] of one substituent *peri* to the N-methyl group on the acridone nucleus in each case. Likewise the resonance for the free A-ring position (δ_c 98.3) was typical of C(2), the resonance for a C(4) tertiary carbon usually being found near δ_c 90.

3. SYNTHESIS

Successful syntheses have been achieved for culantraramine and culantraraminol, borreverine, isoborreverine, yuehchukene, and vepridimerines A–D.

For the attempted synthesis of culantraramine (**7**) Schroeder and Stermitz [7] prepared the diene **55** from *p*-methoxyhordenine (**56**) by a three-step process (Scheme 5). Attempts to form culantraramine dimers by allowing **55** to stand in xylene were unsuccessful, giving instead a 3–2 mixture of *endo*- and *exo*-forms of the dimer **57** in a yield of 60%. While not giving the

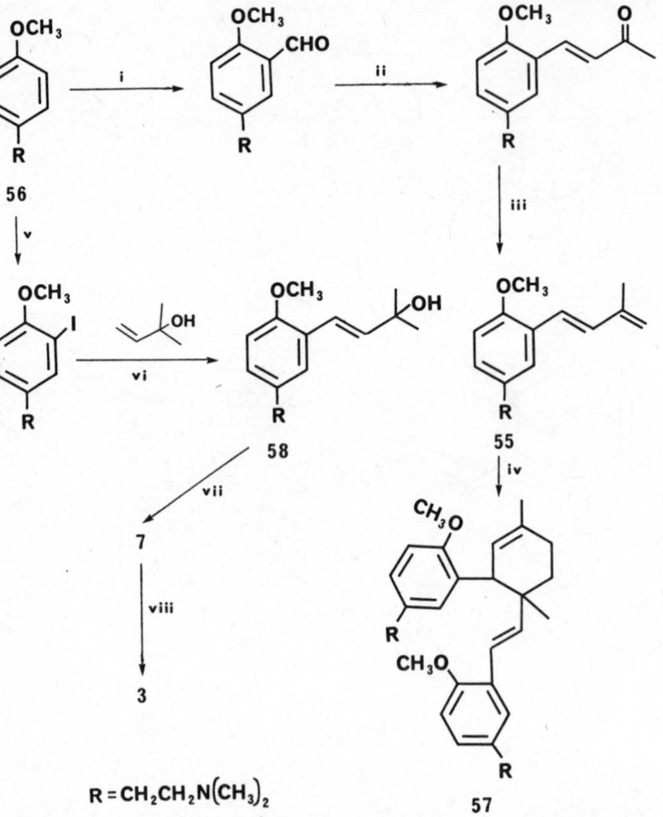

R = CH₂CH₂N(CH₃)₂

Scheme 5. (*i*) BuLi in DMF (69% yield); (*ii*) OH⁻/(CH₃)₂CO (81% yield); (*iii*) Ph₃PCH₃I/BuLi (65% yield); (*iv*) xylene for 10 days; (*v*) I₂/BuLi (81% yield); (*vi*) Pd(OAc)₂ (63% yield); (*vii*) AcCl, 1 hr; (*viii*) 48% HBr.

required product, this is an interesting route that should be applied to the synthesis of the coumarin dimers such as **1** and that may have implications for the biosynthesis of Diels-Alder adducts in the Rutaceae (see Section 4). As an alternative to **55** the corresponding hydroxy derivative **58** was obtained by treatment of the 3-iodo derivative of **56** with 2-methyl-3-buten-2-ol (Scheme 5). Addition of acetyl chloride to **58** gave, on standing for 1 hr, **7** in a yield of 80% [7], while treatment with 48% HBr in EtOH at 25°C for 15 min also gave 80% **7**, this time in company with other minor alkaloidal products. As previously noted more vigorous treatment of **7** with 48% HBr led to its conversion to isoalfileramine (**3**) in a 70% yield. The use of 1*M* HCl with **58** also led to a high yield of alkaloids, including **7**, **8a**, and **8c** in relative yields of about 55–25–5. The mechanism of this reaction is discussed in Section 4.2.

A biomimetic route to isoborreverine (**15**) and borreverine (**11**) has been reported by Tillequin et al. [45]. The monomeric precursor borrerine (**9**), dissolved in benzene and treated with trifluoroacetic acid at 65°C for 30 min, gave approximately equal quantities of **11** and **15** in an overall yield of 80%. The route is presumed to involve an initial protonation of **9** leading to its ring opening to give **59** and **60** with dimerization then following the pathways outlined in Scheme 6. Cyclization between the side chains of **59** and **60**, with the latter acting as dienophile, would lead to the *cis*-fused intermediate **61** or the *trans*-intermediate **62**, depending on the stereochemistry of the internal double bond of the diene **59**. Removal of the quaternary charge on the nitrogen in **61** could act as the driving force for the second cyclization required to yield **15**. A similar process in **62** but with a second cyclization through C(3) of the nonquaternary indole (rather than N(1)) would generate a new quaternary center, removal of which would lead to the further cyclization involved in formation of the additional heterocyclic ring of borreverine **11** (Scheme 6). The resulting stereochemistry between C(3) and C(8) of the initial cyclization differs, being *cis* in **15** and *trans* in **11**. Following classical Diels-Alder stereochemistry it would be anticipated that the *trans*-stereochemistry would derive from a *cis*-internal double bond on the diene **59**; this is not the favored configuration for the double bond but would lead to a fast reaction rate. By contrast the *cis*-stereochemistry in **15** would arise from the favored *trans*-configuration of the internal double bond but with a

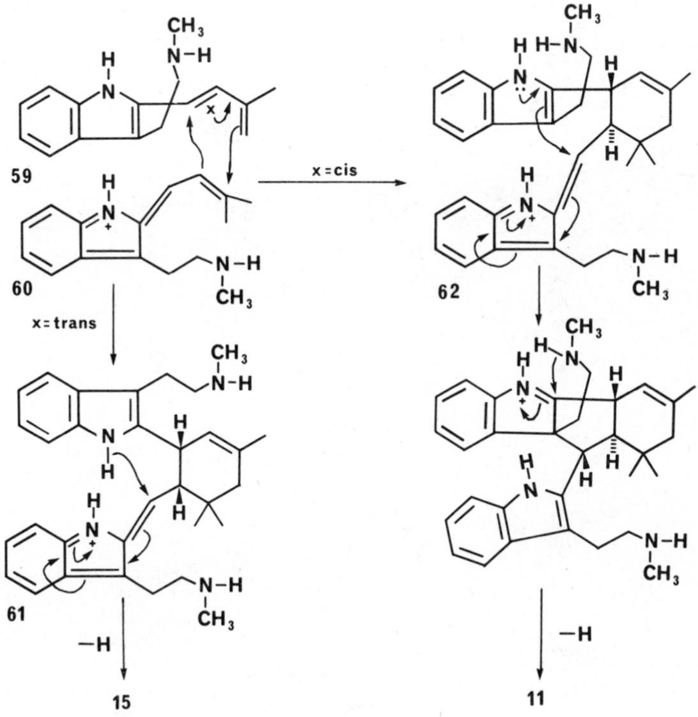

Scheme 6

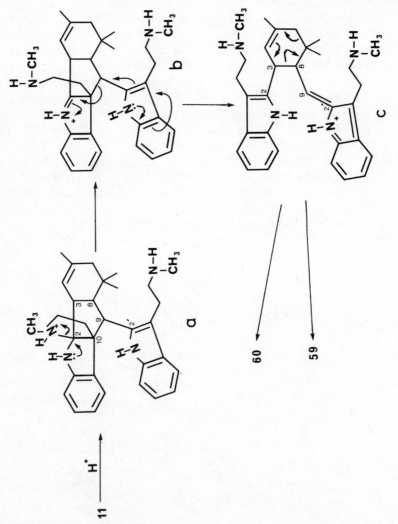

Scheme 7

reduced reactivity. Other acids were capable of initiating the reaction, but trifluoroacetic acid appeared to be the most efficient.

Under these reaction conditions Tillequin et al. [45] found that over a period of 12 hr **11** was completely converted into **15**. This is presumed to occur through protonation of **11** followed by a reversal of the pathway in Scheme 6 to return to the monomers **59** and **60**, or at least to the intermediate prior to this in which isomerization of the C(9)/C(2′) double bond to C(8)/C(9) in "c" could lead to the required change in stereochemistry of H(8) and hence to **15** (Scheme 7).

The synthesis of yuehchukene (**21**) originates in N-tosylated 3-formylindole (**63**) [46], which on treatment with a Grignard derivative of 3-chloro-2-methyl-l-propene gives **64** in an 80% yield (Scheme 8). Removal of the tosyl

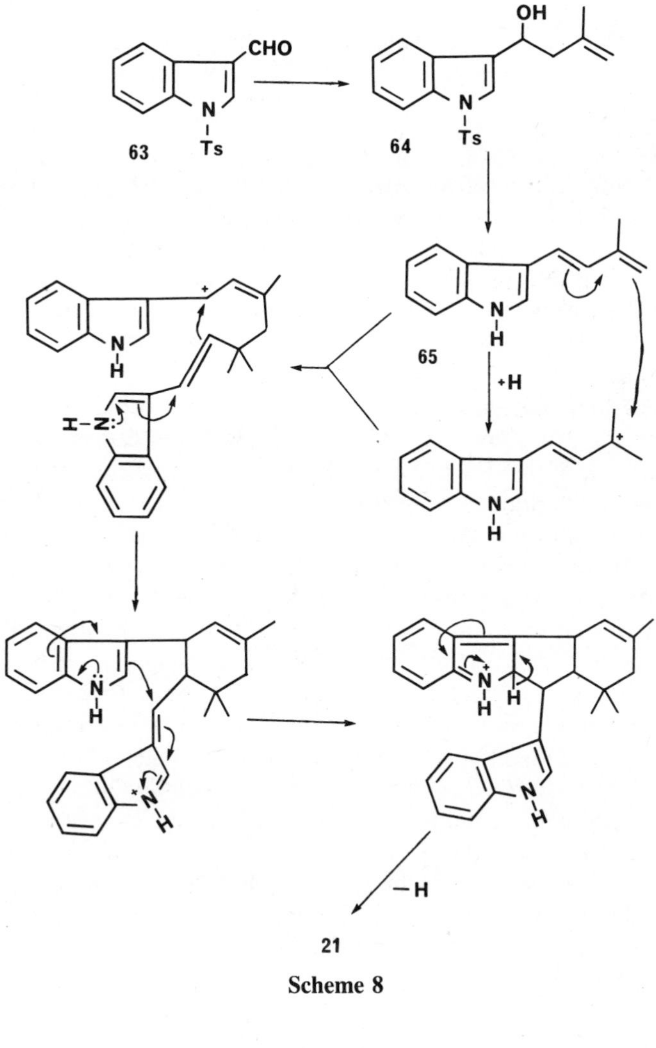

Scheme 8

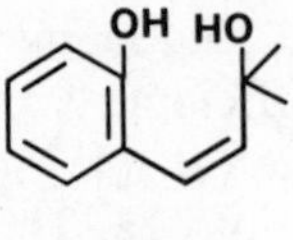

66

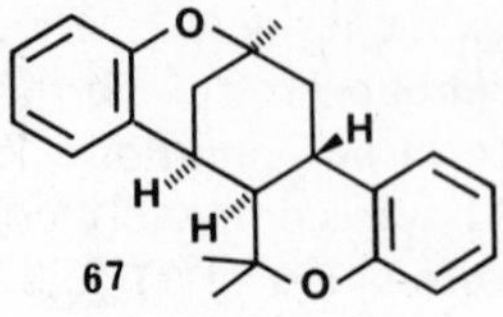

67

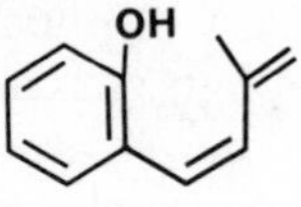

68

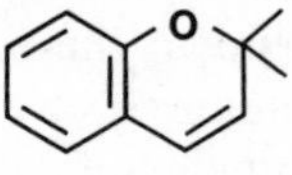

69

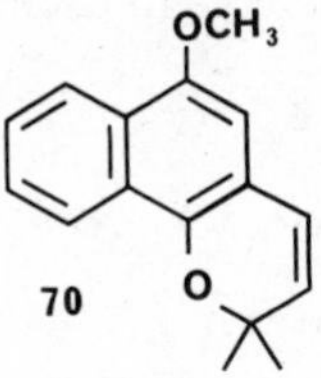

70

group and dehydration of **64** with 3*N* alcoholic NaOH gave the diene (**65**), which when treated with trichloroacetic acid in benzene was converted into **21** in a yield of 10%. The proposed route [46] is shown in Scheme 8. Examination of an alternative route wherein 3-formylindole underwent aldol condensation with acetone to give the 3-oxobutylidene product, which was then treated with methylenetriphenylphosphorane to give the diene (cf. Scheme 5) achieved only a very low yield.

A Diels-Alder dimerization closely allied to that required for formation of the vepridimerines and paraensidimerines was noted as early as 1963 [41] when treatment of the diol **66** with sulfuric acid under reflux in glacial acetic acid generated the dimer **67**. It is possible to envisage this occurring through three different intermediates, the diene **68**, the chromone **69**, or a cation arising through either loss of hydroxyl from **66** or protonation of the diene or chromone. It is noteworthy that the stereochemistry of the dimer is the alternative to that found in vepridimerines B and D and paraensidimerine C but identical with that for the minor dimer paraensidimerine F. This result has been confirmed by Ayafor et al. [47a], who have recently synthesised **67** and observed the W-bond coupling between the equatorial protons of the two methylenes indicative of the chair stereochemistry of the cyclohexane ring. A dimer reported to be derived from lapachenole (**70**) is presumed [41] to have an analogous structure to **67** but was originally misidentified [48].

Grundon et al. [49] have pioneered a possible route to paraensidimerines B and D. Conversion of 1,3-dimethyl-4-hydroxy-2-quinolone to the quinolone methide (**71**) by DDQ in benzene and in the presence of *N*-methylflindersine (**24**) led to a clean dimerization to **72** with the anticipated *cis*-ring junction (Scheme 9).

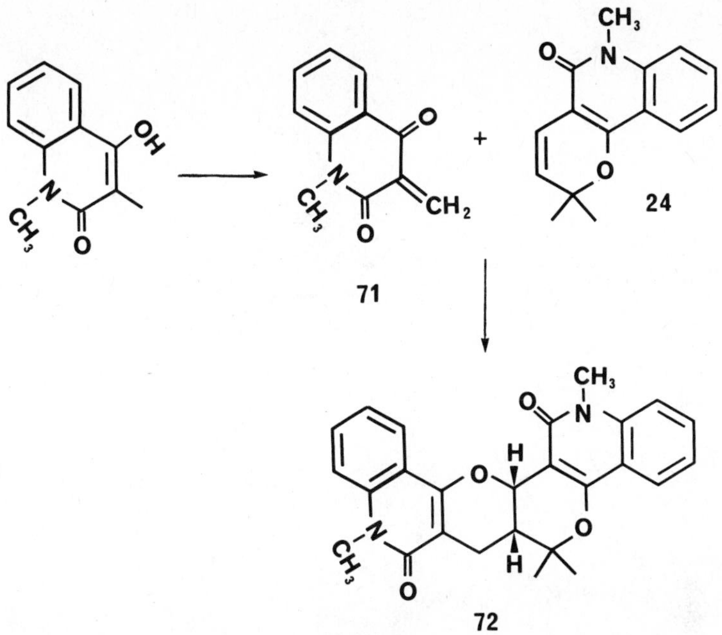

Scheme 9

In an exploration of routes to the quinolone dimers both Grundon and Rutherford [50] and Ayafor et al. [47a] have examined the behavior of pyranoquinolones under acid conditions. Treatment of N-methylflindersine (24) with hot acetic acid failed to cause dimerization but refluxing with either formic or trifluoroacetic acids gave a dimer, $C_{30}H_{30}N_2O_4$, in 57–68% yield [50]. The mass spectrum clearly indicated that this dimer was made up of dihydrodimethylpyranoquinolone and dimethylpyranoquinolone units, which left four possible structures through linkage between C(3)′ and C(4)′ of one pyran unit to C(3)′ or C(4)′ of the other.

The identity of the dimer as 73 (Scheme 10) was resolved by examination of the ^{1}H NMR spectrum. The ABX system due to the protons of the dihydropyran were observed as double doublets centred at δ 3.58 (J = 10, 7 Hz), 2.24, and 2.02. The deshielded position of the methine proton indicated its benzylic nature and thereby required linkage to the dihydropyran to be through C(4)′. The resonance positions of the methylene protons were in good agreement with published data [13] for the comparable protons in pteledimerine (23). The isolated olefinic proton of the pyran ring resonated at δ 6.35, typical of H(4)′ but not H(3)′, which is normally found between δ 5.5 and 6.0, and thereby required linkage to be through C(3)′ of the pyran ring. Treatment of 73 with DDQ gave the corresponding dehydrodimer (74) in which olefinic protons resonated at δ 6.61 and 5.50 ($δ_c$ 131.2 and 123.1), confirming the asymmetric nature of 73.

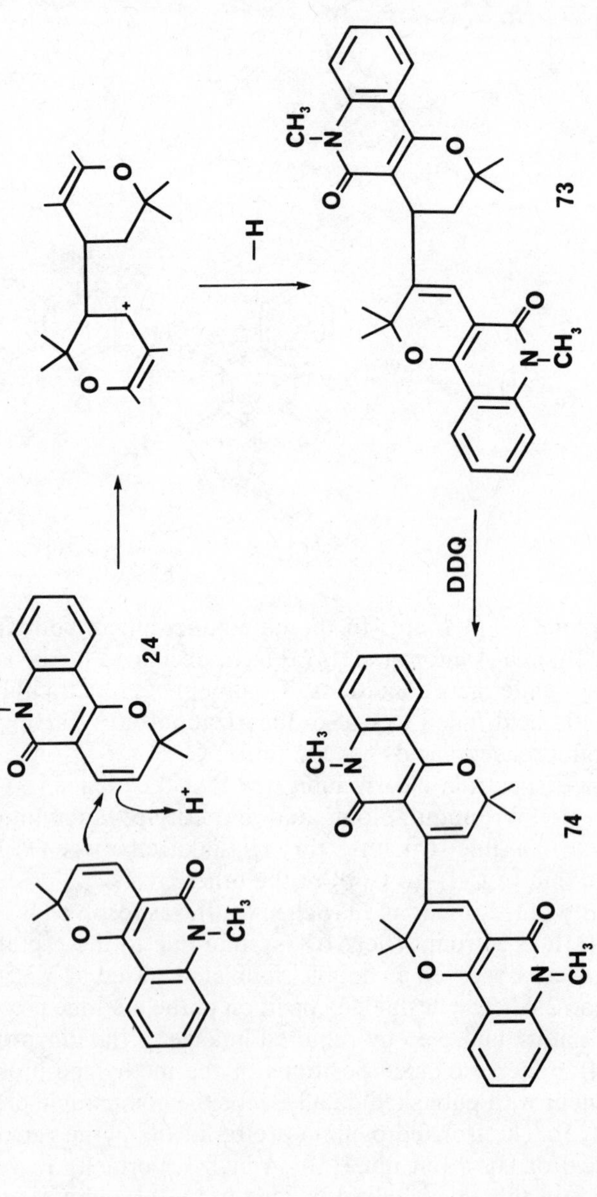

Scheme 10

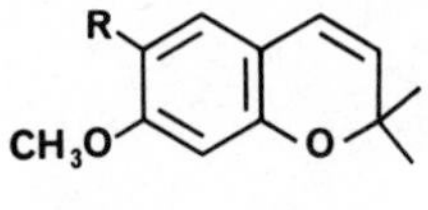

75 R = OCH$_3$

76 R = H

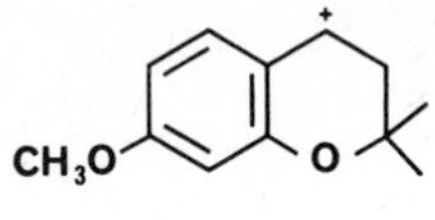

78

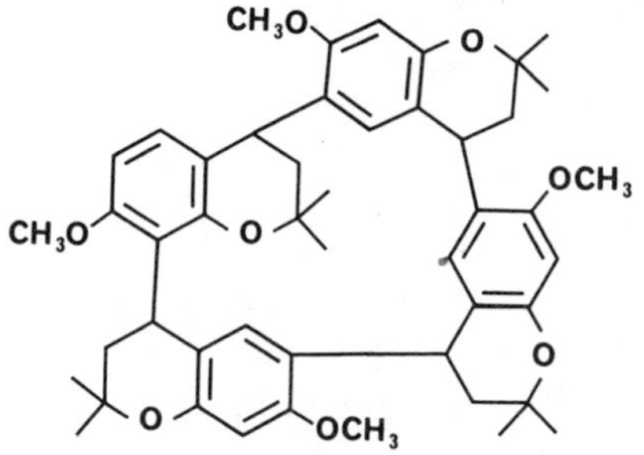

77

Dimer **73** is presumably formed by protonation of **24** followed by loss of a proton (Scheme 10). In an attempt to form the vepridimerines Ayafor et al. [47a] used veprisine (**29**) rather than **24** and obtained the comparable monomer to **73** substituted on the benzene rings. They also studied the acid-catalyzed reactions of the simple chromone **75**, which behaved similarly, but interestingly the monomethoxy derivative **76** gave the tetramer **77**, which must be formed via the cation **78** through its interaction with the electron-rich positions *ortho* to the methoxy substituent on the aromatic ring of **76**.

Very recently Ayafor et al. [47b] reported the successful biogenetic-type synthesis of vepridimerines A–D via the thermolysis of veprisine (**29**). Veprisine was sealed in a pyrex tube under reduced pressure and heated to 200°–220°C for 15 hr. This yielded seven products, of which five were isolated in an overall yield of 42%. Yields of the four known dimers were vepridimerine A (34%), B (38%), C(8%), and D (12%). It was suggested that the dimerization was initiated by conversion of veprisine to the quinoline quinone methide, which then gave the diene (**87**) via a sigmatropic shift. The route of cyclization could then follow the path depicted in Scheme 16 (see later). In addition to the four known dimers a further 2-/4-quinoline compound, vepridimerine E, was also obtained. Vepridimerine E has the same gross structure as vepridimerines C (**34**) and D (**35**) but with the cyclohexane stereochemistry of paraensidimerine F (**43**) and geijedimerine. ^{1}H NMR chemical shifts for the cyclohexane system were reported as follows: δ 3.80 (ddd, J 13.8, 4.3, 3 Hz, H(16)$_{eq}$), 3.65 (dt, J 2.9, 2.4 Hz, H(7)), 2.84 (td, J 12, 4.3 Hz, H(16a)), 1.97 (dd, J 12, 2.9 Hz, H(6a)), 1.93 (dd, J 13, 2.4 Hz, H(19)$_{ax}$), 1.77 (ddd, J 13, 3, 2.4 Hz, H(19)$_{eq}$), 1.17 (dd, J 13.8, 12 Hz, H(16a)$_{ax}$), 1.01, 1.48, 1.91 (3 × CH$_3$).

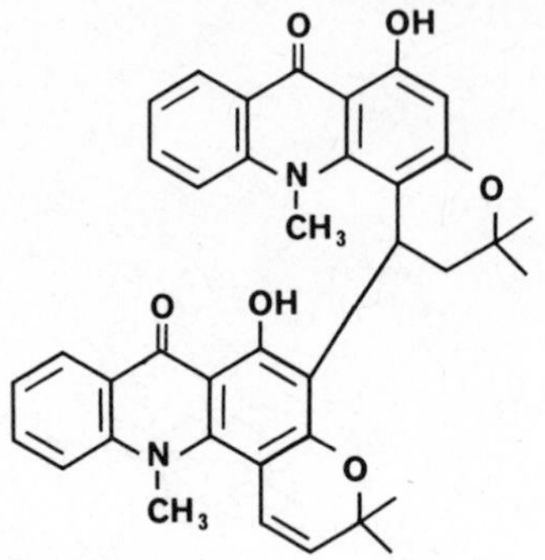

79

In the past year Cordell and his co-workers [51–53] have reported a series of dimers, trimers, and tetramers of the alkaloid noracronycine (see, for example, **79**) that would appear to be formed in a manner analogous to **73** and **77**. The acridone dimers prepared synthetically by Cordell et al. have recently been joined by the acridone dimer glycobismine A isolated from *Glycosmis citrifolia* (Section 2.5) and the synthesis by Cordell et al. appears to offer a simple route to glycobismine A.

4. BIOSYNTHESIS

Two related facts have permeated most thinking regarding the biogenesis of these dimers. First, all have been presumed to be derived from monomeric precursors by a Diels-Alder type cycloaddition mechanism. Second, all dimers so far reported appear to be optically inactive, suggesting that formation is not under enzymatic control, and so the normal rules for Diels-Alder type reactions should apply. Given that the latter is true and that the precursors from which they derive are reactive compounds, it is valid to speculate that the dimers are actually extraction artifacts and not present in the plant at all. However, in several cases, their presence has been detected in plant extracts obtained solely by extraction with nonpolar organic solvents, so that there seems little doubt that they are present in the dried plant material. Whether this is true for living plant material remains unknown.

The following is a review of the speculations made by various authors concerning the mode of formation of these dimers. It is noteworthy that Grundon [3] proposes that the pyranoquinolines that are the precursors of many of these dimers may themselves be formed via a prenyl diene (quinone methide **80**). The route via a 2,4-dioxygenatedquinolone with a 2′,3′-dioxygenated-3,3-dimethylallyl substituent at C(3) (**81**), which undergoes subsequent ring closure and oxidation to the pyran by dehydration, remains a plausible alternative, although it should favor formation of linear rather than angular pyranoquinolines (Scheme 11). In this section possible routes to the

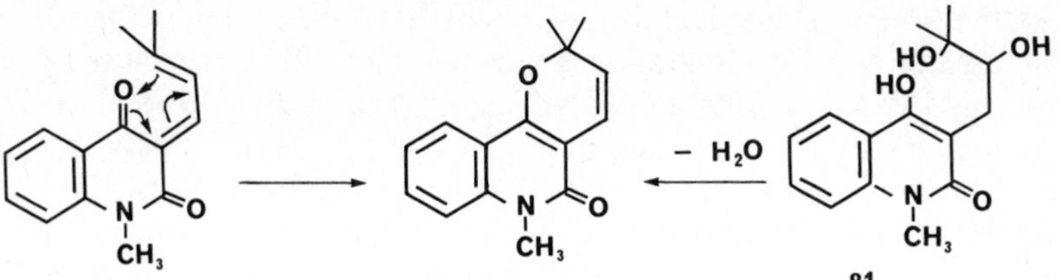

Scheme 11

different dimers are followed in terms of increasing complexity, that is in the number of bonds that are formed during the dimerization process.

4.1. One-Bond Dimers

Only the *Ptelea* dimers fall into this category. Grundon [3] suggests that pteledimerine (**23**) is formed by the acid-catalyzed reaction of a 2-isopropenyldihydrofuroquinol-4-one intermediate (**82**), a type of alkaloid known to occur in *Ptelea*, with N-methylflindersine (**24**) (Scheme 12). Pteledimeridine (**26**) would originate from the corresponding angular 2-isopropenyldihydrofuroquinol-2-one. An alternative proposal has been made by Jurd et al. [15], who suggest a more complex route to the *Ptelea* dimers as an offshoot of the pathway leading to the paraensidimerines. This is discussed in Section 4.4.

4.2. Two-Bond Dimers

Culantraramine (**7**) and the culantrariminols (**8a–8c**) are in this category. Their formation could be envisaged as a classic Diels-Alder addition of two

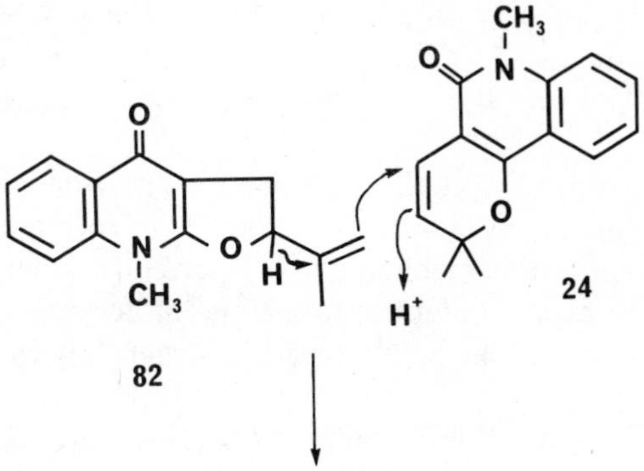

Scheme 12

units of the 3-methylbutadiene derivative of hordenine methylether (**83**, R = Me) to yield **7** and via hydration **8** (Scheme 13*A*). However, attempts to confirm this [7] by allowing a synthetically prepared diene to react (**84**, X = C_6H_5) gave only dimers linked in the manner of cyclobisuberodiene (**1**). Successful formation of **7** was achieved using a combination of the diene and the cation (**84a**) derived from either the diene or the corresponding 3-hydroxy-3-methylbut-1-ene (the latter had previously been employed by Barnes et al. to synthesize analogous dimers [41]). Stermitz suggests that this reaction represents a cation-diene rather than a classic Diels-Alder cycloaddition (Scheme 13*B*) and results in cation **85**, which can then be converted into **7**, **8**, or into alfileramine (**4**). The resulting stereochemistry of the culantraramine adducts has been discussed by Schroeder and Stermitz [7]. The cation **84a** will arise from **58** in which the double bond is *trans*. The major natural derivatives **7** and **8a** retain this *trans*-relationship in the corresponding H(3) and H(4) of the cyclohexene, while the second linkage leads to the expected *cis*-relationship between H(4) and H(5). Thus in the major products the cycloaddition must take place with retention of configuration in the cationic intermediates, while in the minor products **8b** and **8c** loss of configuration has occurred. Schroeder and Stermitz [7] liken this very plausible route for the "natural" formation of the culantraramine group to a classic nonsynchronous [4 + 2] cycloaddition reaction [54].

Another possible, although unlikely, alternative is that **7** and **8** might be formed by protonation of alfileramine (**4**) followed by loss of a proton and subsequent methylation (Scheme 13*C*).

4.3. Three-Bond Dimers

4.3.1. Alfilerame (4).

A possible route to **4** through the addition of a hordenine butadiene unit (**83**, R = H) to a preformed pyranohordenine (Scheme 13*C*) would be expected to yield *cis*-stereochemistry between C(3) and C(4). The most plausible pathway would appear to be through **85**, the secondary reaction between the cation and the hydroxy group to form the pyran system also being proposed for the vepridimerines and paraensidimerines (see below).

Caolo and Stermitz [4] initially proposed a different biosynthetic route involving the linkage of two hordenine units to a preformed C_{10}-precursor, dehydrocitral. This route is mechanistically perfectly satisfactory, but there seems no reason to expect culantraramine and alfileramine to be formed by a route that differs from that involved in the other dimers.

4.3.2. Paraensidimerines B (48) and D (47).

Jurd and Wong [17b] suggest that **47** could be formed by oxidative coupling of *N*-methylflindersine and the diene **86** (Scheme 14). Subsequent hydration of the butene side chain would give **48**. The involvement of precyclized *N*-methylflindersine receives

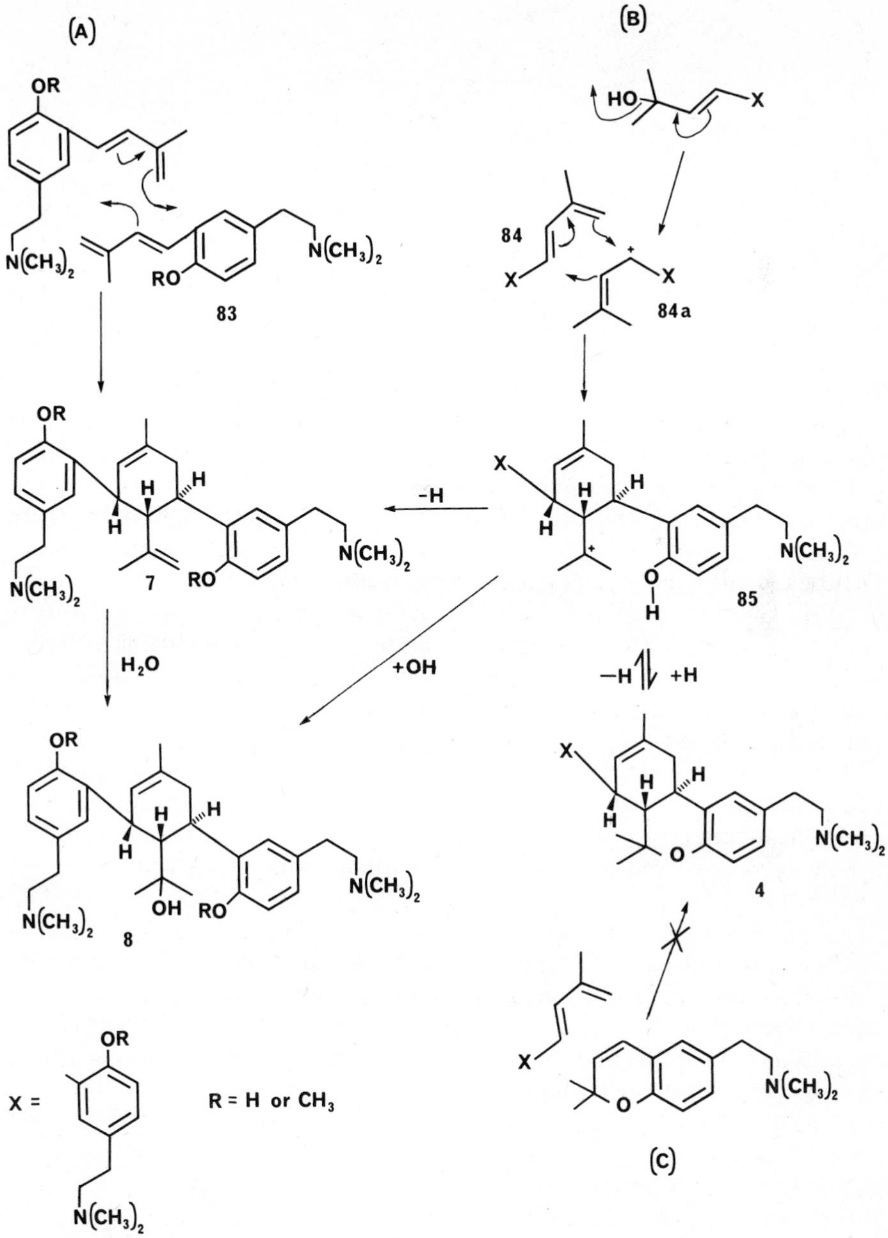

Scheme 13

379

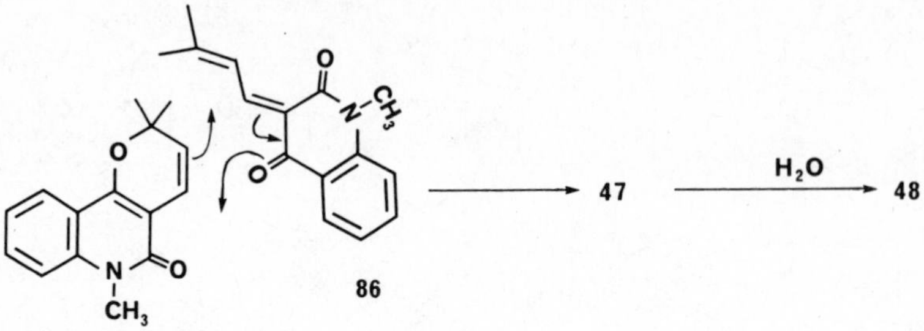

Scheme 14

support from the *cis*-stereochemistry between C(6a) and C(16a). Grundon [3] proposed that an alternative route via Diels-Alder cyclization takes place between **86** and the 4-hydroxyquinol-2-one prenyldiene with subsequent ring closure to give the dimethylpyran system. He concedes [3], however, that the route suggested by Jurd and Wong is probable in the light of synthetic studies that have shown *N*-methylflindersine to react in a comparable situation to yield a dimer analogous to paraensidimerine D (see Section 3, [49]).

4.3.3.   Borreverine (11) and Isoborreverine (15).   The biomimetic synthesis of both dimers from the monomer borrerine [45] has been discussed in Section 3 and could represent closely the natural route to these alkaloids. Likely precursors would be 2-prenyltryptamine, in the form of the tertiary diene **59** and the quaternary diene **60** (see Scheme 6). It is not known whether both **11** and **15** are being formed in *Flindersia fournieri* or whether **15** and its related dimers are being derived by conversion of **11** to **15**. Interestingly, despite its ready conversion to **15**, **11** is the major dimeric alkaloid of *Borreria verticillata* [45].

4.3.4.   Yuechukene.   The route to yuehchukene could be analogous to that suggested for isoborreverine. The precursors would be tertiary and quaternary 3-prenylindole dienes. The *cis*-stereochemistry of initial cycloaddition (cf. **15**) is followed by a further cyclization during discharge of the quaternary nitrogen, this time involving C(2) of the nonquaternary indole, rather than

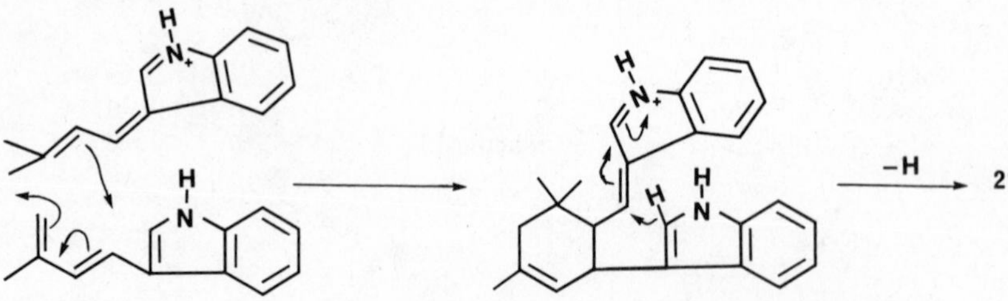

Scheme 15

N(1) (Scheme 15). The alternative *trans*-isomer (cf. **11**) has not been found but it must be remembered that **21** is known only as a trace constituent.

4.4. Four-Bond Dimers

4.4.1. The Vepridimerines, Paraensidimerines A, C, E, F, and G and Geijedimerine.

The most interesting biogenetic feature concerning this group of dimers is the stereochemistry between C(6a) and C(16a) and between C(6a) and C(7) (Table 9). Using classical Diels-Alder rules the four dimers *cis* between C(6a) and C(16a) would require a *cis*-dienophile precursor, while the five with a *trans*-configuration would need the *trans*-dienophile. For C(6a) and C(7) seven of the ten compounds have a *trans*-configuration indicative of a *cis*-internal double bond on the diene, while only three, the relatively minor paraensidimerines E and F and geijedimerine, have *cis*-stereochemistry and hence require a *trans*-internal double bond. This could be expected in as much as the rate of reaction of *cis*-dienes is fast, while that of *trans*-dienes is slow.

Formation of the C(6a)/C(16a) *cis*-compounds (vepridimerines A and C, paraensidimerines A and E—and also B and D) can be readily rationalized by the normal Diels-Alder condensation process. Routes have been proposed by both Grundon [3] and by Jurd et al. [15], the only difference being that the former commences with two 4-hydroxy-2-quinolone-3-prenyldienes (**87**) and the latter with one of these units and one of the corresponding pyranoquinolone (see Scheme 16). For Grundon's route a pentacyclic dimer (**88**) is required, which must then undergo two interactions between olefinic bonds and hydroxyl functions to give the two pyran rings, while in Jurd's route the hexacyclic intermediate (**89**) requires only one such interaction. For the mixed 2/4-quinolone dimers the only difference required would be

Table 9. Relative Stereochemistry of Diels-Alder Condensation Derived Ring Junctions in the Vepridimerines, Paraensidimerines and Geijedimerine

Compound	C(16a)—C(6a)	C(6a)—C(7)
Vepridimerine A (**30**)	*cis*	*trans*
Vepridimerine B (**32**)	*trans*	*trans*
Vepridimerine C (**34**)	*cis*	*trans*
Vepridimerine D (**35**)	*trans*	*trans*
Paraensidimerine A (**41**)	*cis*	*trans*
Paraensidimerine C (**42**)	*trans*	*trans*
Paraensidimerine E (**44**)	*cis*	*cis*
Paraensidimerine F (**43**)	*trans*	*cis*
Paraensidimerine F (**45**)	—	*trans*
Geijedimerine	*trans*	*cis*

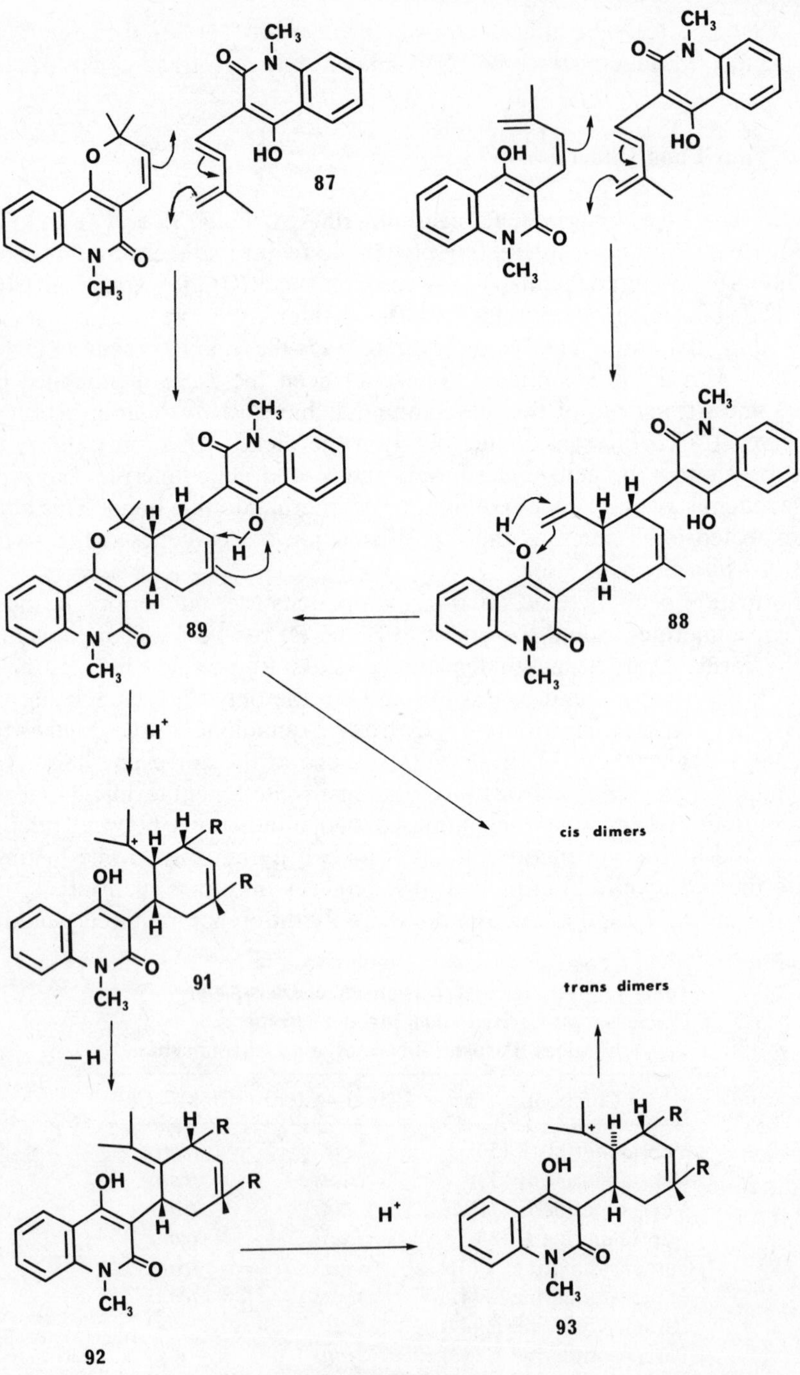

Scheme 16

90

replacement of the 4-hydroxy-2-quinolone-3-prenyldiene by its 2-hydroxy-4-quinolone isomer, which would yield the intermediate (**90**).

The route to the *trans*-dimers has been the subject of considerable discussion. Both Grundon [3] and Jurd et al. [15] suggest that isomerization of a *cis*-fused intermediate occurs. Grundon invokes this at the stage of the intermediate **88** (Scheme 16), while Jurd et al. suggest protonation of O(5) in intermediate **89** to give the carbonium ion **91**, which can isomerize through **92** to yield the *trans*-isomer **93**, which may then recyclize. It is noteworthy that a protonation of this type is also presumed to be the driving force behind the conversion of borreverine into isoborreverine (Section 3) and of interconversions in the isoalfileramine, alfileramine, cultantraramine group. In their successful synthesis of vepridimerines A–E Ayafor et al. [47b] obtained slightly higher yields of the *trans* dimers (B and D) than of the *cis* analogues (A and C), while the additional dimer (E) was also *trans*, although in this last case it differed by being *cis* between C(6a) and C(7).

Neither Grundon nor Jurd et al. appear to consider it possible that the C(6a)/C(16a) *trans*-stereochemistry could arise directly. Possible direct dienophile precursors for the *trans*-dimers would include the *trans*-form of the 4-hydroxyquinolone (**87**) or the *o*-hydroxy dienophile (**94**), analogues of which have been reported to dimerize to give *trans*-products [41] and which are implicated [7] in the generation of the culantraramine-type compounds (Section 4.2 and Scheme 13). In fact, in view of the observations made by Schroeder and Stermitz [7] it is now an interesting possibility that formation of the quinolone dimers could involve the cation (**95**) derived from **94**. The latter could react with **87** to give the intermediate **91**, but with *trans*-stereochemistry (Scheme 17).

Finally it should be noted that Jurd et al. [15] propose that the cation **91** could also act as an intermediate in the formation of the *Ptelea* dimers. Scheme 17 illustrates the proposed route to pteledimeridine (**26**); the route to pteledimerine differs only in **96** generating a 2-hydroxy-4-quinolone rather than a 4-hydroxy-2-quinolone.

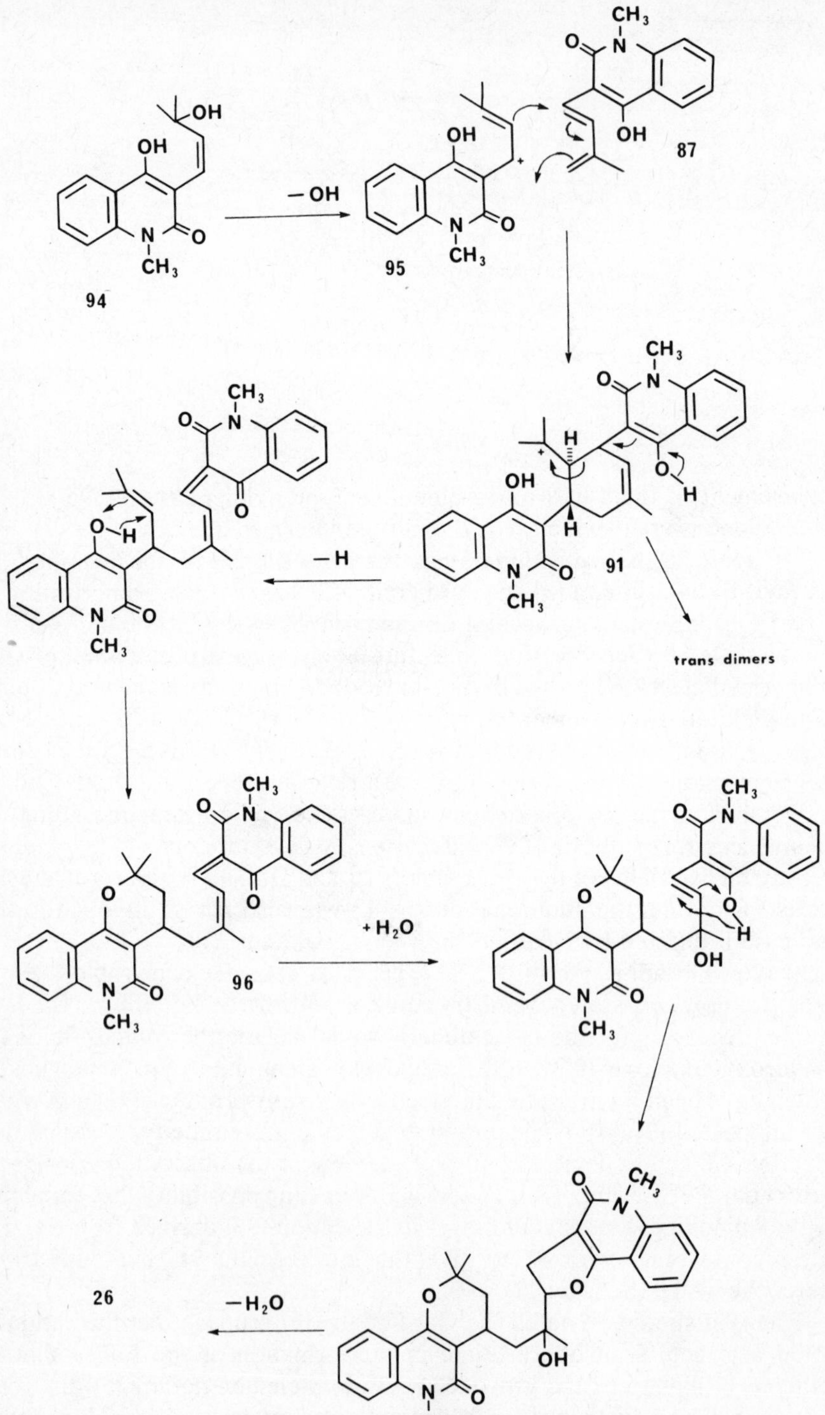

Scheme 17

5. SUMMARY

The Diels-Alder type dimeric alkaloids of the Rutaceae, whether true natural products or artifacts, are an interesting group of compounds that have come to light only in the past six years. Their characterization illustrates very well the power of high-field NMR to resolve problems of stereochemistry. The successful synthesis and X-ray analysis of several of the alkaloids is reassuring with regard to the integrity of structures proposed entirely on the basis of spectroscopic data. We can expect more dimers of this type to be found in the next few years.

One interesting point that needs to be further explored concerns the apparent difference in dimerization processes leading to the dimeric alkaloids and coumarins in the Rutaceae. Schroeder and Stermitz [7] have recently addressed this problem, and their comments offer an interesting and very plausible explanation for the observed dichotomy. First, they note that dimers of the coumarin type, in which the terminal double bond of a diene acts as the dienophile (Scheme 1), never co-occur with dimers in which the cycloaddition appears to involve the internal bond of a diene acting as the dienophile, which is the case in most of the alkaloids discussed here. They argue that where the terminal double bond is involved the dimerization process represents a true Diels-Alder reaction. This most often leads to dimers of the cyclobisuberodiene-type (1) but can also generate dimers such as the diclausenan diasterioisomers (97) recently reported from *Clausena wildenowii* Wight & Arn. (Rutaceae) [55]. By contrast where the dimerization apparently involved the internal double bond the precursor is in reality not the diene but the cation derived from the corresponding 3-hydroxy-3-methylbut-2-ene (e.g., 84a, Scheme 13). Certainly much of what data is available on facile, biomimetic, syntheses of these alkaloids and their congeners (reviewed in Section 3) does not contradict this convincing hypothesis.

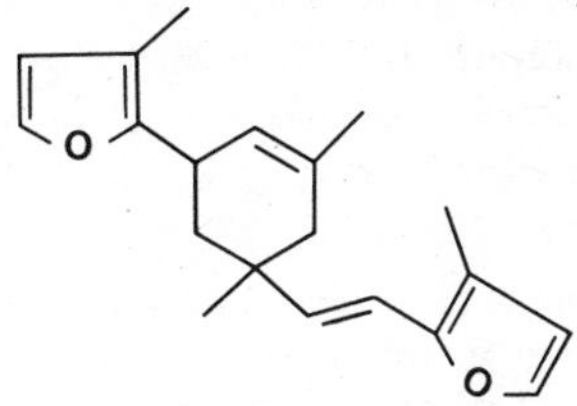

97

Another area that needs exploration is biological activity. Yuehchukene has been examined, and is reported to be a highly active anti-implantation compound at very low dosage in rats; it apparently lacks the estrogenic activity that is a problematical side effect of many anti-implanation compounds [11]. Nothing at all is known regarding the biological activity of any of the other dimers.

ACKNOWLEDGMENTS

The following are thanked for their generous supply of unpublished data: Dr. J. Ayafor, Department of Chemistry, University of Yaounde; Dr. J. D. Connolly, Department of Chemistry, University of Glasgow, Dr. G. E. Cordell, College of Pharmacy, University of Chicago, Prof. F. R. Stermitz, Department of Chemistry, Colorado State University, and Dr. F. Tillequin, Faculty of Pharmacy, Université Rene Descartes, Paris. Both Dr. A. I. Gray, Trinity College, Dublin, and Prof. F. R. Stermitz are warmly acknowledged for comments on various drafts of the manuscript.

REFERENCES

1. I. Mester, in P. G. Waterman and M. F. Grundon, Eds., *Chemistry and Chemical Taxonomy of the Rutales*, Academic, London, 1983, p. 31.
2. A. I. Gray, in P. G. Waterman and M. F. Grundon, Eds., *Chemistry and Chemical Taxonomy of the Rutales*, Academic, London, 1983, p. 97.
3. M. F. Grundon, in P. G. Waterman and M. F. Grundon, Eds., *Chemistry and Chemical Taxonomy of the Rutales*, Academic, London, 1983, p. 9.
4. M. A. Caolo and F. R. Stermitz, *Tetrahedron* **35**, 1487 (1979).
5. J. A. Swinehart and F. R. Stermitz, *Phytochemistry* **19**, 1219 (1980).
6. R. T. Boulware and F. R. Stermitz, *J. Nat. Prod.* **44**, 200 (1981).
7. D. A. Schroeder and F. R. Stermitz, *Tetrahedron,* (in press).
8. F. Tillequin, M. Koch, M. Bert, and T. Sevenet, *J. Nat. Prod.* **42**, 92 (1979).
9. F. Tillequin and M. Koch, *Phytochemistry* **18**, 1559 (1979).
10. F. Tillequin and M. Koch, *Phytochemistry* **18**, 2066 (1979).
11. Y-C. Kong, K. F. Cheng, R. C. Cambie, and P. G. Waterman, *J. Chem. Soc. Chem. Commun.* **1985**, 47.
12. I. Mester, J. Reisch, K. Szendrei, and J. Korosi, *Ann. Chem.* **1979**, 1785.
13. J. Reisch, I. Mester, J. Korosi, and K. Szendrei, *Tetrahedron Lett.* **1978**, 3681.
14. T. Ngadjui, J. F. Ayafor, B. L. Sondengam, J. D. Connolly, D. S. Rycroft, S. A. Khalid, P. G. Waterman, N. M. D. Brown, M. F. Grundon, and V. N. Ramachandran, *Tetrahedron Lett.* **23**, 2041 (1982).
15. L. Jurd, R. Y. Wong, and M. Benson, *Aust. J. Chem.* **35**, 2505 (1982).
16. L. Jurd and M. Benson, *J. Chem. Soc. Chem. Commun.* **1983**, 92.
17. (a) L. Jurd, M. Benson, and R. Y. Wong, *Aust. J. Chem.* **36**, 759 (1983); (b) L. Jurd and R. Y. Wong, *Aust. J. Chem.* **34**, 1625 (1981).
18. S. Mitaku, A.-L. Skaltsounis, F. Tillequin, M. Koch, J. Pusset, and G. Chauviere, *J. Nat. Prod.* **48**, 772 (1985).
19. F. Fish and P. G. Waterman, *Taxon* **22**, 177 (1973).
20. P. G. Waterman, *Biochem. Syst. Ecol.* **3**, 149 (1975).
21. F. R. Stermitz and I. Sharifi, *Phytochemistry* **16**, 2003 (1977).
22. M. A. Caolo, O. P. Anderson, and F. R. Stermitz, *Tetrahedron* **35**, 1493 (1979).
23. R. A. Archer, D. B. Boyd, P. V. Demarco, I. J. Tyminski, and N. L. Allinger, *J. Am. Chem. Soc.* **92**, 5200 (1970).

24. F. K. Klein, H. Rappoport, and H. W. Elliott, *Nature* **232,** 258 (1971).

25. T. G. Hartley, *J. Arnold Arboretum* **50,** 481 (1969).

26. F. Tillequin, M. Koch, and T. Sevenet, *Plantes medicinales phytother* **14,** 4 (1980).

27. F. Tillequin, R. Rousselet, M. Koch, M. Bert, and T. Sevenet, *Ann. Pharm. Franc.* **37,** 543 (1979).

28. J. L. Pousset, J. Kerharo, G. Maynart, X. Moneur, A. Cave, and R. Gouterel, *Phytochemistry* **12,** 2308 (1973).

29. F. Tillequin and M. Koch, *Phytochemistry* **19,** 1282 (1980).

30. M. D. de Maindreville, J. Levy, F. Tillequin, and M. Koch, *J. Nat. Prod.* **46,** 310 (1983).

31. J. L. Pousset, A. Cave, A. Chiaroni, and C. Riche, *J. Chem. Soc. Chem. Commun.* **1977,** 261.

32. G. A. Cordell and J. E. Saxton, in R. G. A. Rodrigo, Ed., *The Alkaloids*, Vol. 20, Academic, New York, 1981, p. 25.

33. F. Tillequin, M. Koch, and A. Rabaron, *J. Nat. Prod.* **48,** 120 (1985).

34. P. G. Waterman, in P. G. Waterman and M. F. Grundon, Eds., *Chemistry and Chemical Taxonomy of the Rutales*, Academic, London, 1983, p. 377.

35. Y.-C. Kong, C.-N. Lam, K.-F. Cheng, and T. C. W. Mak, *Tetrahedron Lett.*, submitted.

36. H. Rapoport and K. G. Holden, *J. Amer. Chem. Soc.* **82,** 4395 (1960).

37. A. A. Robertson, *Aust. J. Chem.* **16,** 451 (1963).

38. T. B. Ngadjui, J. F. Ayafor, and B. L. Sondengam, *Tetrahedron Lett.* **1980,** 3293.

39. S. A. Khalid and P. G. Waterman, *Phytochemistry* **20,** 2761 (1981).

40. S. A. Khalid, Ph.D. Thesis, University of Strathclyde (1982).

41. C. S. Barnes, M. I. Strong, and J. L. Occolowitz, *Tetrahedron* **19,** 839 (1963).

42. N. M. D. Brown, M. F. Grundon, D. M. Harrison, and S. A. Surgenor, *Tetrahedron* **36,** 3579 (1980).

43. H. Furukawa, T.-S. Wu, C.-S. Kuoh, T. Sato, Y. Nagai, and K. Kagei, *Chem. Pharm. Bull.* **32,** 1647 (1984).

44. H. Furukawa, M. Yogo, and T.-S. Wu, *Chem. Pharm. Bull.* **31,** 3084 (1983).

45. F. Tillequin, M. Koch, J. L. Pousset, and A. Cave, *J. Chem. Soc. Chem. Commun.* **1978,** 826.

46. K.-F. Cheng, Y.-C. Kong, and T.-Y. Chan, *J. Chem. Soc. Chem. Commun.* **1985,** 48.

47. (a) J. F. Ayafor (unpublished); (b) J. F. Ayafor, B. L. Sondengam, J. D. Connolly, and D. S. Rycroft, *Tetrahedron Lett.* **37,** 4529 (1985).

48. R. Livingstone and M. C. Whiting, *J. Chem. Soc.* **1955,** 3631.

49. M. F. Grundon, V. N. Ramachandran, and B. M. Sloan, *Tetrahedron Lett.* **1981,** 3105.

50. M. F. Grundon and M. J. Rutherford, *J. Chem. Soc. Perkin Trans.* **I,** 197 (1985).

51. S. Funayama, G. A. Cordell, H. Wagner, and H. L. Lotter, *J. Nat. Prod.* **47,** 143 (1984).

52. S. Funayama and G. A. Cordell, *Planta Med.* **48,** 263 (1983).

53. S. Funayama and G. A. Cordell, *Heterocycles* **20,** 2379 (1983).

54. H. M. R. Hofmann, *Angew. Chem. Int. Ed.* **23,** 1 (1984).

55. G. S. R. Subba Row, B. Ravindranath, and V. P. Sashi Kumar, *Phytochemistry* **23,** 399 (1984).

Chapter Four

Teratology of Steroidal Alkaloids

Richard F. Keeler
USDA ARS Poisonous Plant Research Laboratory
Logan, Utah

CONTENTS

1. INTRODUCTION 389
2. REVIEW OF TERATOLOGY PRINCIPLES 390
3. SPONTANEOUS CYCLOPIA IN SHEEP AND IDENTIFICATION OF THE PLANT *VERATRUM CALIFORNICUM* AS THE RESPONSIBLE AGENT 392
4. INVESTIGATIONS IDENTIFYING THE TERATOGENS AS STEROIDAL ALKALOIDS 394
5. VARIATION OF CONGENITAL DEFORMITIES INDUCED AND THE SPECIES SPECIFICITY 396
6. BIOLOGICAL SITE OF ACTION OF THE TERATOGENS 399
7. STRUCTURAL AND CONFIGURATIONAL SPECIFICITY FOR TERATOGENICITY AMONG STEROIDAL ALKALOIDS 400
8. STUDIES ON THE BIOCHEMICAL MECHANISM OF ACTION OF STEROIDAL ALKALOID TERATOGENS 409
9. NATURAL OCCURRENCE OF KNOWN AND SUSPECTED STEROIDAL ALKALOID TERATOGENS IN FOODS AND FEEDS 414
REFERENCES 422

1. INTRODUCTION

Induction of congenital defects by constituents of the diet of humans or animals is more common than once believed. But until nearly 1940 the consensus among scientists and laypersons was that the developing mammalian embryo was quite protected *in utero* from placental passage of hazardous compounds. Compounds inducing congenital defects (teratogens) were un-

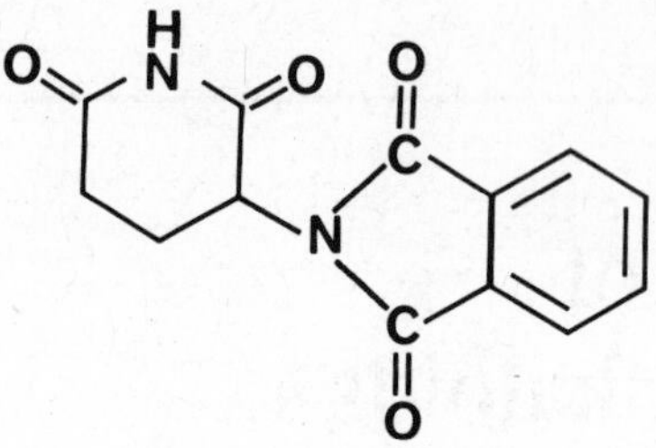

1

known, and such defects were attributed to other causes. Most believed congenital defects to be of genetic origin. But a considerable body of information has been accumulated during the past several decades in a field now called teratology, which has established the teratogenic hazard of many compounds—principally certain drugs, industrial compounds, and compounds in foods and feeds [1,2].

The "at risk" status of the developing embryo, and the *in utero* absorption from maternal exposure to many of these teratogens have accounted for the occurrence of some of the spontaneous incidents of congenital defects in humans and animals. These exposures have occurred under varied circumstances and have induced occasionally as many as a few thousand deformed offspring. The most striking example in humans was the unfortunate incident with the sedative drug thalidomide (**1**), which had induced limb and other deformities in about 7000 children before its teratogenicity became apparent in about 1960 [3,4]. Fortunately epidemics of congenital defects are rare in humans, but not so in domestic livestock, where the incidence of congenital defects under some circumstances has approached 20–30% of offspring.

It was the study of the natural occurrence of a condition called monkey face lamb disease, affecting several thousand lambs yearly, that provided the earliest knowledge of the teratogenicity of steroidal alkaloids. Today much information exists on that subject. In this chapter I describe the historical development of our understanding of both the biological and chemical aspects of the teratology of steroidal alkaloids.

2. REVIEW OF TERATOLOGY PRINCIPLES

The "Principles of Teratology" enunciated by Wilson [2] describe a number of important factors governing induction of congenital deformities by teratogens. Those principles in paraphrased form are now listed and discussed to aid the reader in understanding the material that follows.

Principle 1. Susceptibility to a teratogen depends upon conceptus genotype. Genetic inheritance is not solely responsible for congenital defects as was once believed. Even so, teratogen susceptibility does vary among genotypes. There can be very marked variation in susceptibility among species, strains, or breeds.

Principle 2.   *The conceptus must be exposed at the susceptible developmental stage.*   During development of a conceptus, events that give rise to body form and function are rigidly controlled in time and sequence. Formation of organs and tissues begins at specific times. Cells divide in programmed sequence to produce anatomical features. For a teratogen to produce a specific deformity, it must exert its influence at exactly the right time in gestation. This time, which can be called the insult time for a given deformity, varies among species partly because of the great differences in total gestation time. For example, the primitive streak/neural plate stage of embryonic development, at which time embryos may develop cyclopia if insulted by a teratogen, occurs at about day 7 in rabbits, but not until day 14 in sheep.

Principle 3.   *Teratogens exert their effects by specific mechanisms.* Virtually any biochemical reaction can be altered in rate or inhibited totally *in vivo* by extraneous compounds. If a reaction so altered is of sufficient importance in embryonic development, a congenital defect may result. But for a conceptus to be adversely affected by a specific teratogen, events must be occurring in the embryo at the time of insult that can be altered by the inhibition capabilities of that teratogen. Because lengthy reaction sequences are involved in the formation of any morphologic feature, it is not surprising that structurally dissimilar teratogens may give rise to similar deformities by altering different sequential reactions.

Principle 4.   *Among effects produced, teratogens can cause death as well as deformity.*   At doses higher than required to induce deformity, a teratogen may kill either the conceptus or the dam. Death of the conceptus results in either a resorbed fetus or an aborted fetus, with the former and possibly the latter going unnoticed in livestock or laboratory animals. Because of the necessity of regulatory screening of compounds required by the Toxic Substances Control Act, many teratologists [see, e.g., 5,6] hope to reduce the screening burden by restricting the use of the term teratogen to those compounds that induce a congenital deformity without causing maternal death or toxicity of any kind. But this arbitrary restriction is not applicable to teratogens derived from natural food or feed sources since levels ingested are not governed by physician prescription.

Principle 5.   *A teratogen must reach the conceptus.*   "Virtually all unbound chemicals in maternal plasma have access to the conceptus across the placenta" [2]. However, since most offspring are normal, it follows that the important consideration is whether the teratogen reaches the conceptus at a time and at a dose that will give rise to the effect.

Principle 6.   Incidence and severity of deformity is dose-dependent.
As with effects of drugs, the level of teratogen reaching the site of
action influences the degree of effect. Factors that influence incidence
and severity by way of dose include the amount ingested, the amount
surviving degradation in the maternal gut, the amount absorbed into
maternal circulation, the amount surviving metabolism in the liver
of the dam, and the amount passing the placenta and reaching the active
site in the conceptus.

3.   SPONTANEOUS CYCLOPIA IN SHEEP AND IDENTIFICATION OF THE PLANT *VERATRUM CALIFORNICUM* AS THE RESPONSIBLE AGENT

A very serious congenital deformity problem existed in certain highly in-
tensified lambing operations in Idaho many years ago [7–9]. Lambs were
born in some flocks with deformities of the head wherein there was usually
a cyclopic-type condition with a single centrally located eye (Fig. 1). Inci-

Figure 1.   Single globe cyclopic lamb induced by maternal ingestion of the *Veratrum
californicum* teratogen during pregnancy.

Figure 2. The range plant *Veratrum californicum*.

dence was often very high (up to 25% of the lamb crop), but the condition was found only on certain ranches and only in certain geographic locations. Owners whose flocks were affected reported that the condition was restricted to animals that had grazed only certain forest areas in the summer months. All range areas where the flocks had been grazing were in one of three forest areas—the Boise, Sawtooth, or Challis National Forests of Idaho.

At that time deformities in livestock were attributed almost universally to genetic causes. But a genetic cause did not seem likely for the cyclopic condition because considerable genetic variation existed among affected animals from ranch to ranch and from year to year, and there was no apparent correlation with incidence. Outbreaks of the condition seemed more related to area grazed within each ranching operation than to identity of rams or ewes [7].

Investigators in our laboratory began experimental work on that problem in about 1956. That work now provides a rather complete understanding of the disease. The original research group, led by W. Binns, was able to rule as unlikely a genetic basis for the disease by a breeding experiment involving ewes that had given birth to deformed lambs on a previous occasion bred to rams that were sons of these ewes [7]. Investigation also suggested there was no unusual elemental composition in flora, soil, or water from affected areas that might be incriminating [7].

According to the research group's observations, certain plants were common to all of the ranges where pregnant ewes that later gave birth to the malformed lambs had grazed. The researchers decided to test some of these plants for teratogenicity, particularly *Veratrum californicum* (Fig. 2). Pregnant ewes were gavaged that plant for periods of up to the first 30 days of gestation. Deformed lambs resulted [10]. Those were the first experimental trials incriminating a range plant as being teratogenic. I began cooperative work with that group in about 1962. By 1964–1965 experimental work had shown the precise insult period [11] to be the 14th day of gestation. This information, together with the fact that the plants are not widely distributed on the range, provided ranchers with a practical solution to the problem of cyclopic lambs. Ranchers now keep pregnant ewes away from the plant until two weeks have elapsed after rams are removed. Where incidence of the condition once approached 25–30%, with thousands of lambs deformed yearly, present incidence is less than 1% and could be zero if ranchers did not occasionally become careless.

4. INVESTIGATIONS IDENTIFYING THE TERATOGENS AS STEROIDAL ALKALOIDS

We continued our study of this problem even though the practical solution for the sheep disease was established. We felt that identification of the teratogen, elucidation of species specificity, analysis of plant content of the teratogen, study of the structure-activity relation, and an examination of its mechanism of action might lead to useful information applicable to prevention of related congenital defects in humans.

We speculated in 1964 [12] that the *Veratrum* teratogen would prove to be one of the many steroidal alkaloids known to be present in other members of the *Veratrum* genus [13]. The cell division and growth inhibitory properties of certain *Veratrum* alkaloids as reported earlier by others [14–16] made this possibility an attractive one. Pregnant sheep were therefore gavaged a variety of extracted *Veratrum* alkaloid-rich preparations to determine whether the alkaloid class might be responsible. Deformed offspring were induced by some of the preparations [12], suggesting the teratogen was readily extractable. The polarity of extractants used suggested that the teratogen, if alkaloidal, was probably a glycoside or parent alkamine—not an ester form of one or more of the 50–60 then known steroidal alkaloids of the *Veratrum* genus [13]. Of over a dozen *Veratrum* alkaloids tested, three proved capable of inducing the deformity condition. They were the known steroidal alkaloids jervine (2), another that we named cyclopamine (3), and a third that we named cycloposine (4). The latter two we originally designated alkaloids V and X [17–19]. Cyclopamine and cycloposine were present at high concentrations in *Veratrum californicum* and were therefore of principal concern to us. We assumed cyclopamine to be the most important, since its

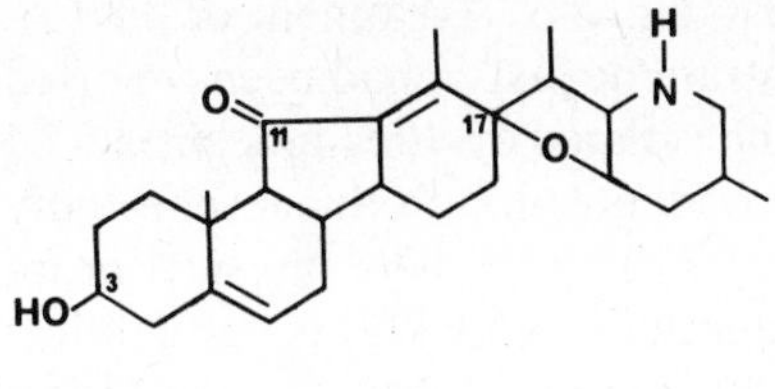

2

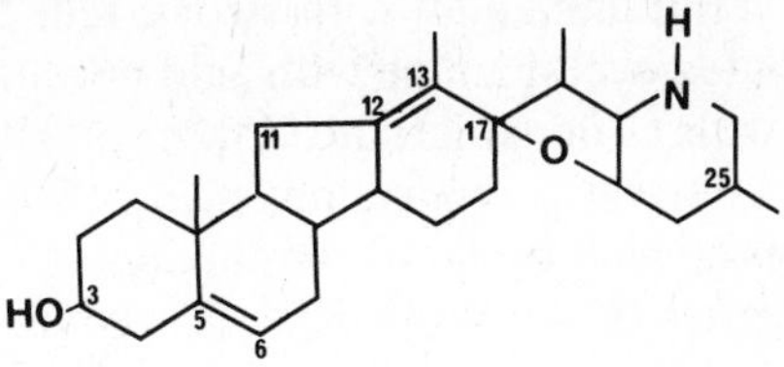

3

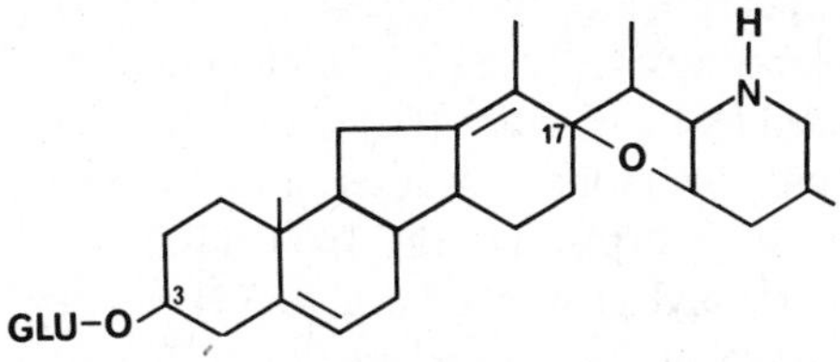

4

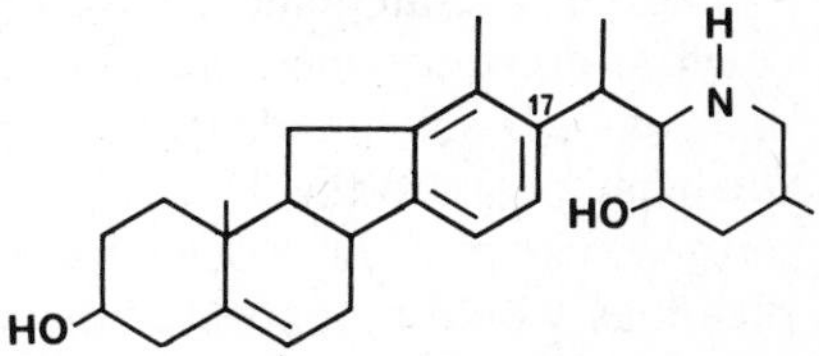

5

concentration was highest and the potency of each was roughly equivalent [17,18].

Structural analysis suggested close similarity among all three teratogens. Jervine was a compound of known structure [13], and cyclopamine was similar to jervine and unlike veratramine (5) according to our analysis in 1967–1968 by NMR, IR, MS, and other methods [20]. The properties of cyclopamine [20] appeared to be essentially identical to those reported for 11-deoxojervine by Masamune and co-workers [21], who had isolated and characterized 11-deoxojervine in 1964 from *Veratrum grandiflorum* in Japan shortly after we had begun isolation and feeding trial work.

The IR spectrum of our isolated cyclopamine [20] suggested a $\Delta^5,3\beta$-ol steroidal system (1050–1057 cm^{-1}), about the same 3500 cm^{-1} hydroxyl absorption as jervine, and only about half that of veratramine, which indicated the likelihood of a single rather than two OH groups. Peaks at 927, 984, and 1118 cm^{-1} suggested a ring oxygen system (the ether bridge) as with jervine. Cyclopamine had no aromatic ring absorption in the UV region and was therefore like jervine and unlike veratramine [20]. The probable empirical formula from elemental analysis and molecular weight was $C_{27}H_{41}NO_2$ [20]. Proton NMR suggested 41 or 42 protons, and assignments could be made consistent with those made for jervine and related compounds [22,23] with: nine protons on three methyl groups at C(19), C(21), and C(27) from two overlapping doublets and a singlet at 0.88 to 0.96 δ; three protons on another methyl group at C(18) from a singlet at 1.63 δ; one proton as the C(6) olefinic proton from a multiplet at 5.35 δ; and one proton on the C(3) as a multiplet at 3.4 δ [20]. MS analysis [20] showed an *m/z* parent peak at 411 with a fragmentation pattern related to that of jervine and unlike that of veratramine [24] with strong 125, 124, and 110 peaks. Treatment of the isolated cyclopamine with acid produced veratramine just as had been reported with 11-deoxojervine [25]. Acetylation of the isolated cyclopamine produced a derivative whose physical constants were identical with those reportedly produced by similar treatment of 11-deoxojervine [25]. The physical properties of the Wolff-Kishner reduction product of jervine [25] were identical with those of our isolated cyclopamine. Thus the properties of our isolated cyclopamine were essentially identical with properties of 11-deoxojervine, and so we assigned it that structure.

The third teratogenic compound, cycloposine, we believed to be 3-glucosyl-11-deoxojervine based on its properties [26]. It had an empirical formula of $C_{33}H_{51}NO_7$ and had the expected IR similarities [26] to cyclopamine. Each had peaks in the 927, 984, 1118 cm^{-1} regions characteristic of the ring oxygen system. Cycloposine had intense absorption in the 1000–1100 cm^{-1} region as would be expected of the hydroxyl groups of a glycosyl moiety. Contrary to cyclopamine, it did not have a sharp band in the 1057 cm^{-1} region characteristic of the Δ^5,3β-ol system suggesting, therefore, that the suspected glycosyl moiety was at C(3). There was no aromatic ring absorption in the UV spectrum, and thus the aglycone could not have been veratramine [26]. Comparison of the proton NMR spectra of cycloposine with that of cyclopamine showed similarities except for a broad region of peaks in the spectrum of cycloposine centered at 3.1 δ and assignable to the protons on the carbon atoms of the sugar residue [26]. The MS fragmentation analysis of cycloposine and cyclopamine differed as expected if the former were the glycoside of the latter. Cycloposine had a molecular ion peak at m/z 573 and the expected intense 125, 124, and 110 fragments characteristic of cyclopamine [26]. Physical properties were consistent with identity of the aglycone as cyclopamine. Acid hydrolysis of cycloposine to recover the parent alkamine and the glycosyl residue produced a sugar identified as D-glucose and an alkamine identified as veratramine rather than the expected cyclopamine [26]. But cyclopamine is very readily aromatized in ring D under even mild acid conditions to veratramine. Evidently the rate of aromatization was greater than was the rate of hydrolysis, which prevented accumulation of the expected hydrolytic product, cyclopamine. The evidence accumulated was consistent with assignment of the structure of cycloposine as 3-glucosyl-11-deoxojervine.

5. VARIATION OF CONGENITAL DEFORMITIES INDUCED AND THE SPECIES SPECIFICITY

Most of the deformed lambs observed [7–9] were single or double globe cyclopics. The appearance of the double globe cyclopics (Fig. 3) influenced Basque sheepherders to adopt the term "monkey face lamb disease" [7]. But for unknown reasons there was considerable variability in expression

Figure 3.   Double globe cyclopic lamb induced by maternal ingestion of the *Veratrum californicum* teratogen during pregnancy.

of the disease [7–9]. Some cyclopics had a proboscis above the single or double globe and some did not. Some lambs were microphthalmic or anophthalmic with variation in proboscis size or presence. Those with a severely manifested expression of the condition generally had a pronounced mandible that curved sharply upward (Fig. 4) around a shortened maxilla/premaxilla. Severely affected lambs died shortly after birth or were born dead. Surviving lambs with less severe expression of the condition had a short snout or cebocephalic appearance. Brains in affected lambs varied from slight fusion of cerebral hemispheres to a striking hydrocephalus in which only a rind of cerebral tissue remained. A single optic nerve was present in true cyclopics. The pituitary gland was usually absent in severe cases, which gave rise to an indefinite prolongation of gestation with gravid ewes becoming larger and larger as the fetus grew in size *in utero*. Prolonged gestation apparently occurred because of the absence of a hormonal message from the aplastic fetal pituitary gland required to trigger parturition. Twins with but one fetus

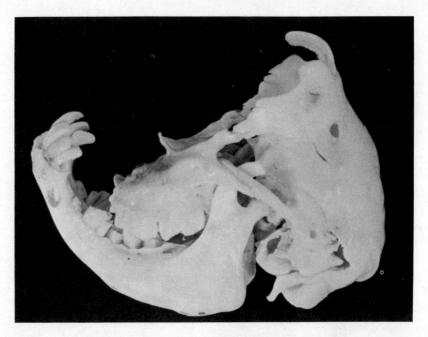

Figure 4. Mandibular curvature in a lamb induced by maternal ingestion of the *Veratrum californicum* teratogen during pregnancy.

affected were delivered at term because of the influence of the normal pituitary in the unaffected fetus [7–9].

About 30 mg/kg of cyclopamine or jervine [19] administered orally on the 14th day of gestation is required to induce terata in sheep. The primitive streak/neural plate stage of embryonic development coincides with the 14th day of gestation in sheep. But even with single day dosing in sheep there are serious logistical problems with sheep as an assay animal. Nonetheless we continued with sheep as the animal of choice in isolation and characterization of cyclopamine and the other teratogens despite logistical problems because of the possibility that there might be very narrow species specificity, as is often the case with teratogens [2]. Were that so for the cyclopic condition, we were concerned that we might be misled by another assay. Even though we dosed for only one day, the gestation period of sheep is 145 days, which does make a lengthy overall assay period. That lengthy period, coupled with the need for multigram quantities of pure compounds were the main reasons for logistical problems. Killing each treated ewe on about the 50th day of gestation, when deformities can be recognized in the fetus, shortened the assay considerably. When our experiments disclosed teratogen identity, we studied laboratory animal susceptibility and susceptibility of other livestock.

In addition to sheep, other domestic ruminants, specifically cattle and goats, proved to be susceptible [27]. Ruminants could have been affected as a result of microbial conversion in the rumen of a proteratogen to an active form. One could rule out such rumenal conversion if nonruminant laboratory animals were susceptible.

Rabbits were the first nonruminants tested, and they proved susceptible to the teratogenic action of cyclopamine or jervine [28] when these compounds were administered orally at 33–45 mg/kg. The primitive streak/neural plate stage in the rabbit is day 7, and that day proved to be the insult day for cephalic defects [29]. Double globe cyclopia and cebocephalia with a single or closely spaced double nostril were the common defects [28,29].

Cyclopamine or jervine also induced teratogenic effects in rats and hamsters and marginal effects in mice when gavaged during the primitive streak/neural plate stage of development at as low as 240 mg/kg, 170 mg/kg, and 180 mg/kg, respectively [30]. Microphthalmia and cebocephaly were induced at low incidence in Sprague-Dawley albino rats, a very low incidence of exencephalics were observed in Swiss Webster mice, while in Golden hamsters cebocephaly, encephalocele, exencephaly, and harelip were induced [30].

Sheep and nonruminant rabbits were about equally susceptible on a per kilogram basis, while other nonruminants—hamsters, mice, and rats—were less susceptible. Clearly conversion of cyclopamine by rumen microorganisms was not essential.

6. BIOLOGICAL SITE OF ACTION OF THE TERATOGENS

Without experimental investigations, the possibility could not be excluded that a metabolic conversion product derived from maternal hepatic microsomal enzyme action might be the teratogen. Furthermore the possibility existed that the compound expressed its activity only indirectly on the embryo by affecting the dam. These two possibilities now seem remote because of results from cooperative investigations of three different types where maternal hepatic enzymes play no role, and where the teratogen exerted its affect directly on the embryo. Those experiments studied the effect of the teratogens on chick embryos, the effect on sheep embryos by intrauterine injection, and the effect on rat embryo cultures.

Congenital deformities in hatched chicks resulted from direct application of 1–5 mg cyclopamine or jervine to the embryonic shield of windowed chicken eggs in work in Australia led by Bryden [31]. This showed that maternal metabolic alteration was not necessary for that species. The deformities were expressed mainly as cyclopia, cebocephaly, and microphthalmia. In experiments in sheep, Bryden found that intrauterine injection of as little as 1–2 mg of cyclopamine induced congenital deformities showing, that the dam could be completely bypassed in that species without loss of teratogenicity of cyclopamine [32]. Rat embryo culture experiments at the National Institute of Health using CD rat embryos provided the third line of evidence for a direct effect of the unaltered *Veratrum* alkaloid teratogen on the embryo. Sim and others [33] found that jervine altered the morphology of the rat embryo forebrain when the embryos were incubated with jervine; the jervine produced the appearance of an oblong head with the optic system closer to the ventral surface of the telencephalon than in controls, and with delayed evagination of the mesencephalon and rhombencephalon.

The experimental work consequently suggested that the *Veratrum* steroidal alkaloid teratogens exerted their action as a result of direct effect upon the embryo by the teratogen itself, unaltered by maternal rumen or hepatic microsomal enzyme metabolism.

7. STRUCTURAL AND CONFIGURATIONAL SPECIFICITY FOR TERATOGENICITY AMONG STEROIDAL ALKALOIDS

Our interest in the structural requirements for teratogenicity among steroidal alkaloids was stimulated in the late 1960s when it became evident that in sheep only certain *Veratrum* alkaloids were teratogenic [34]. The teratogenic compounds cyclopamine (3), cycloposine (4), and jervine (2) were mentioned above. They are C-nor-D-homo steroidal alkaloids with a fused furanopiperidine function attached spiro at carbon-17 of the steroid. Veratramine (5), also mentioned above, did not produce cyclopia or related cephalic deformities in sheep, is devoid of the ether bridge, has a hydroxyl group on the piperidine ring, and has a fully aromatic ring D. Muldamine, another compound we isolated from the plant as a principal alkaloid, likewise was not teratogenic in sheep. At that time [35] we assigned a provisional structure for muldamine (6) believing it to be like veratramine, a C-nor-D-homo steroid, devoid of the furan ring. The structure-teratogenic relationship of these five compounds (2–6), as we understood the structures at that time, suggested to us that either the furan ring itself or the configurational aspects of the molecule conferred by the ring were essential for teratogenicity [34].

Our provisional structure of muldamine proved to be in error, as demonstrated in recent studies led by Gaffield [36]. Examination of 100 MHz proton NMR and ^{13}C NMR spectra of muldamine in those recent studies showed the spectra to be inconsistent with the structure originally proposed for muldamine, as well as that proposed for its deacetyl derivative (7) because of the appearance of two C-methyl doublets in the proton spectrum and three quaternary carbon atoms in the carbon spectrum. The structure originally proposed for muldamine required three secondary C-methyl groups and two quaternary carbon atoms. Comparison of the carbon NMR data of muldamine with data from model compounds permitted structural revision of this steroidal alkaloid to (22S,25S)-22,26-epiminocholest-5-ene-3β-16α-diol-16-acetate (8) [36]. Therefore, deacetylmuldamine (9) appears

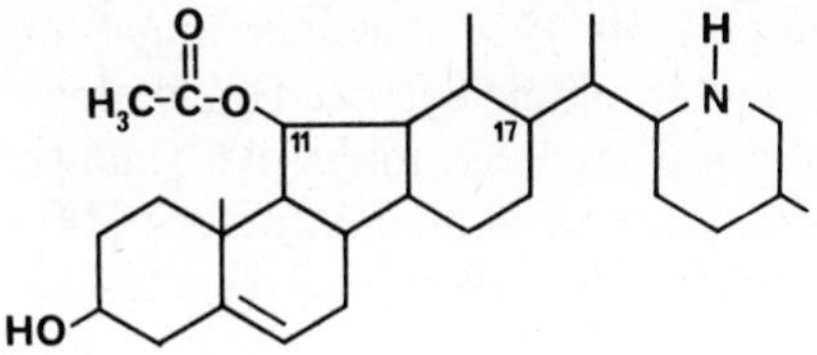

6

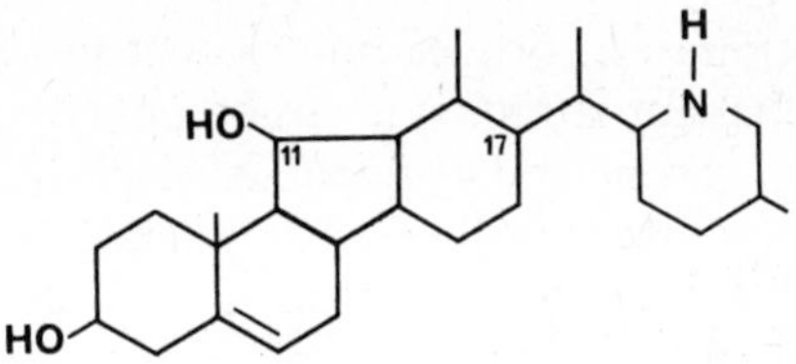

7

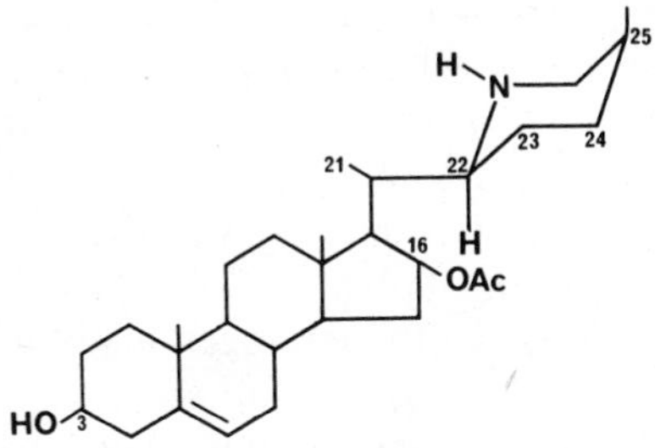

8

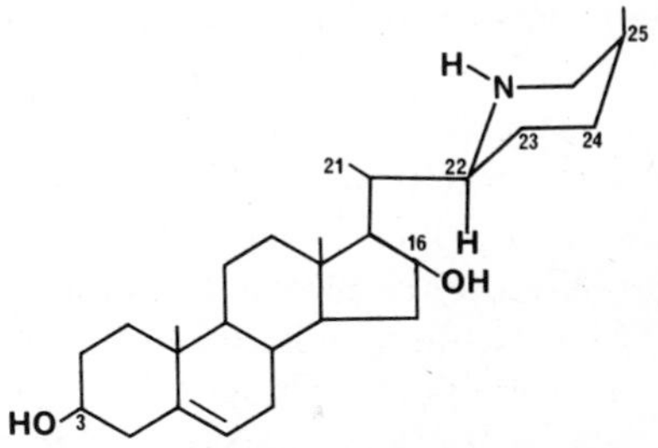

9

to be identical with an alkaloid from *Veratrum grandiflorum* named teinemine [37]. But, according to Gaffield et al. [36], the configuration of teinemine at C(22) is **S** and not **R** as originally reported by Kaneko and coworkers [37].

Gaffield [38] also examined the ^{13}C NMR spectral assignments for jervine and veratramine reported in the literature by Sprague and co-workers [39] and found them to be in error. Careful experiments allowed reassignment [38]. Using the correctly assigned ^{13}C NMR resonances for jervine and veratramine, it was possible to confirm that the structure we originally assigned cyclopamine (**3**) [20] was correct [38]. Further confirmation was obtained using a decoupled proton NMR spectrum of cyclopamine. Thus cyclopamine is indeed 11-deoxojervine. Examination of the ^{13}C NMR and proton NMR spectra of cycloposine, originally assumed to be 3-glucosyl-11-deoxojervine (**4**) [26], confirmed that original assignment [38].

These contemporary studies of the structures of cyclopamine, cycloposine, and muldamine [38], despite the correction in muldamine's structure, have not significantly altered those early conclusions [34] that the configuration imposed on the teratogen by the furan ring (the spiro attachment at carbon-17) played a large role in teratogenicity. According to configuration studies by other workers [40], that spiro attachment placed the nitrogen of the piperidine ring α to the plane of the steroid in both jervine and 11-deoxojervine (cyclopamine).

In our studies steroidal compounds (some alkaloidal, some hormonal, some sapogenins) had, with the exception of jervine, cyclopamine, and cycloposine, proved nonteratogenic in sheep [34]. By 1969 it seemed apparent that our attention in structure-activity studies should be centered on analogs of jervine and cyclopamine and alkaloids with a degree of molecular rigidity conferred by the spiro attached furanopiperidine or similar structure [34]. Preparation and testing of a variety of known analogs of jervine seemed like a logical approach [34].

Our attention was also drawn to an interesting paper by J. H. Renwick in 1972 [41], who hypothesized that spina bifida and anencephaly (ASB) in humans were the result of a teratogen in blighted potatoes. In that hypothesis he proposed that there was an epidemiologic correlation between incidence of these two conditions in humans and the yearly severity of potato blight. He suggested further that elimination of potatoes from the diet of pregnant women would reduce the incidence of ASB by 95%. We were attracted to Renwick's hypothesis because of the possible relationship between the *Veratrum* steroidal alkaloid teratogens and the steroidal alkaloids of potatoes. Perhaps if there were any validity to his hypothesis, the teratogen might be one of the potato alkaloids.

We set out therefore in 1973 to examine a number of questions. Were any of the potato alkaloids teratogenic? Could that question be resolved by testing teratogenicity of various blighted/nonblighted potato preparations in lab animals? What were the essential structural and configurational features required of steroidal alkaloid teratogens?

Among naturally occurring compounds, the best early approach on the potato alkaloid question seemed to be to test solanum alkaloids with the closest structural and configurational resemblance to cyclopamine. There were, of course, no solanum alkaloids with a fused furanopiperidine group attached spiro at carbon atom 17 of a C-nor-D-homo steroid. But spirosolane solanum alkaloids were close to meeting the requirement with their spiro connected furanpiperidine ring juncture. Sprague-Dawley rats were gavaged in 1973 with 240 mg/kg of the *Veratrum* teratogen, cyclopamine (**10**), or with either of the two spirosolanes, solasodine (**11**) or tomatidine (**12**), at the primitive streak/neural plate stage of embryonic development. Solasodine or tomatidine were not teratogenic even though cyclopamine was [42]. However, later experiments in hamsters [43,45] showed that solasodine (**11**) was

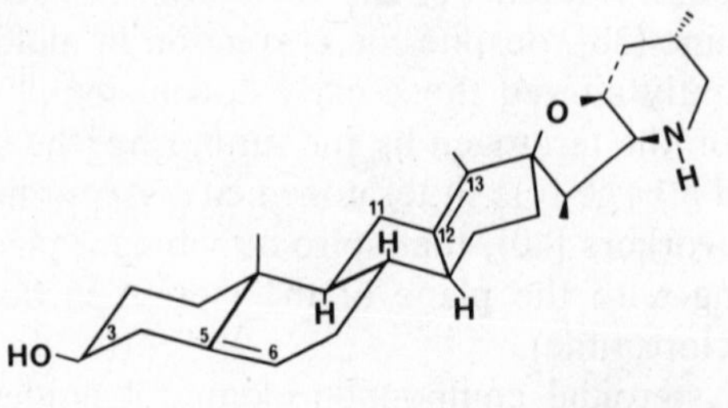

10

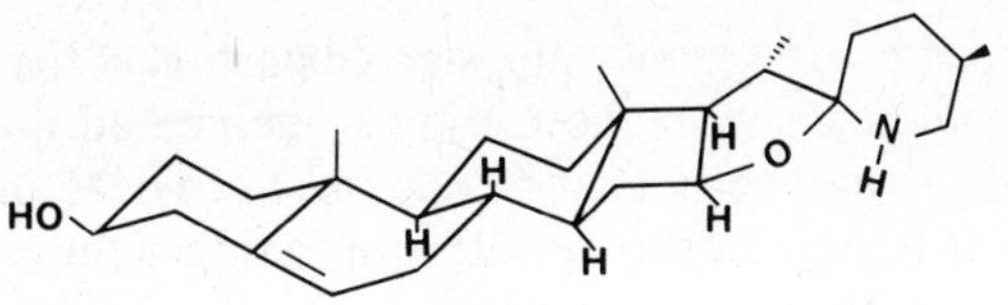

11

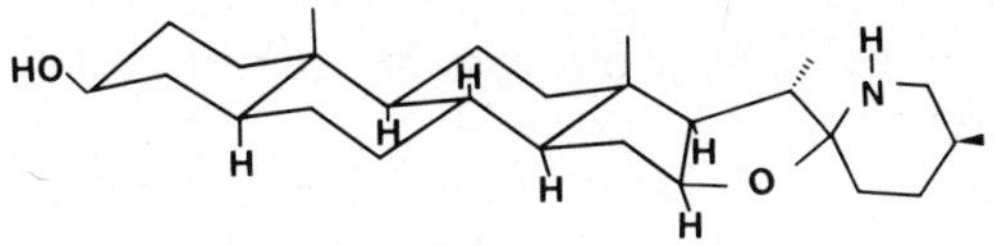

12

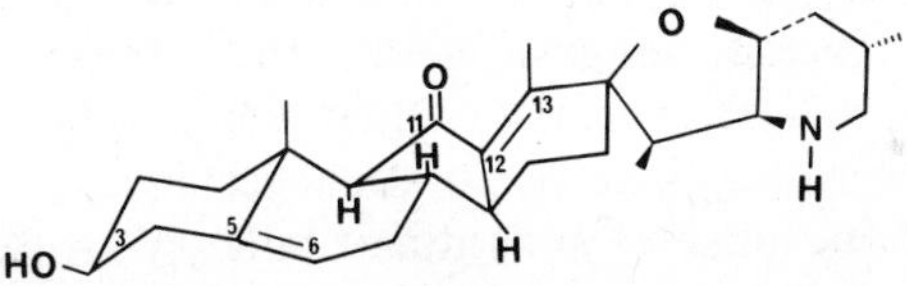

13

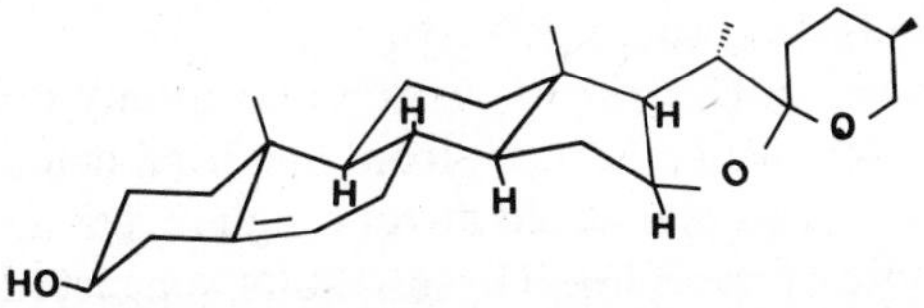

14

teratogenic at high dose as were both cyclopamine (**10**) and jervine (**13**). But the same experiments also showed no teratogenicity for tomatidine (**12**) or diosgenin (**14**), the nonnitrogen-containing analog of solasodine. We subsequently learned from a 1975 *Chemical Abstracts* entry that Russian workers had found solasodine to be teratogenic in rats [45], suggesting to us that in our earlier rat experiments we may have used too low a dose of solasodine.

The experiments suggested that the nitrogen in the piperidine ring was essential for teratogenicity and that its configurational position was critical [44], since there was a lack of teratogenicity with diosgenin and tomatidine despite the positive response from cyclopamine and solasodine. With a spiro connection between rings D and E or E and F the nitrogen of the piperidine ring projects either in the α- or β-position with respect to the steroid plane. The compounds with the α projecting nitrogen atom (jervine, cyclopamine, and solasodine) were teratogenic. With the nitrogen held β, as in tomatidine, the compound was not teratogenic. The steroid portion of the molecule could be the conventional form with the C ring 6 membered and the D ring 5 membered and need not be a C-nor-D-homo steroid as in the *Veratrum* teratogens. The apparent importance of the furan ring in the early compounds tested in sheep could have been due to the positioning of the nitrogen α with respect to the steroid ring system. Possibly the difference in activity between cyclopamine and solasodine might be accounted for by a greater extent of α nitrogen projection in cyclopamine than in solasodine, or perhaps might be accounted for by the greater nitrogen basicity of cyclopamine compared with solasodine [44].

In experiments in our laboratory Brown [46,47] prepared a series of jervine (13) and cyclopamine (10) analogs for testing in hamsters. Analogs without the 5-6 or 12-13 double bonds were teratogenic. Apparently neither double bond played an essential role in teratogenicity. 3-0-acetyljervine was active suggesting the hydroxyl at the 3-position was not essential. *N*-Methyljervine was teratogenic, suggesting that the nitrogen could be tertiary rather than secondary; however, the *N*-methylmethiodide, with no free electrons, on the nitrogen, was inactive. *N*-Butyljervine was not as teratogenic as jervine probably because of substituent bulk. *N*-Acetyljervine had very low activity, perhaps because of reduction in basicity, but *N*-formyljervine was of unexpectedly high activity. Thus the nitrogen with a free electron pair, of all the functionalities of the jervine molecule, appeared to be the most critical in conferring teratogenicity.

Hamster offspring, unlike sheep, were congenitally deformed at borderline statistical significance ($P < 0.05$) from maternal dosing with muldamine (8) [46] provided it was administered at very high levels. Deacetylmuldamine (9) was slightly more active [46]. These results suggested that a furan ring, though helpful, was possibly not absolutely essential in conferring teratogenicity. Subsequent experiments, as reported by Gaffield and Keeler [48], have helped explain why muldamine without the furan ring is of borderline teratogenicity in hamsters, and why deacetylmuldamine has a stronger activity than the parent compound as originally reported [46]. The ^{13}C NMR solution data [36] of deacetylmuldamine (9), as well as X-ray crystallographic studies [R. Y. Wong, unplublished data], according to Gaffield [38,48], indicated that the piperidine ring is oriented in a manner that allows intramolecular hydrogen bonding of the amino group to the C(16) hydroxy group (15). Conversely, downfield shifts of C(23) and C(24) signals in the ^{13}C NMR spectrum [38,48] of muldamine (8) indicated a piperidine ring orientation in which the amino group is directed away from the C(16) acetyl group (16). The intramolecularly hydrogen-bonded amino group of deacetylmuldamine (15) is therefore positioned below the α-face of the steroid, whereas the amino nitrogen of the weaker teratogen muldamine (16) is positioned on or above the β-face of the molecule, according to Gaffield [38,48]. Results are consistent with configuration-activity data for the spirosolanes [44], which

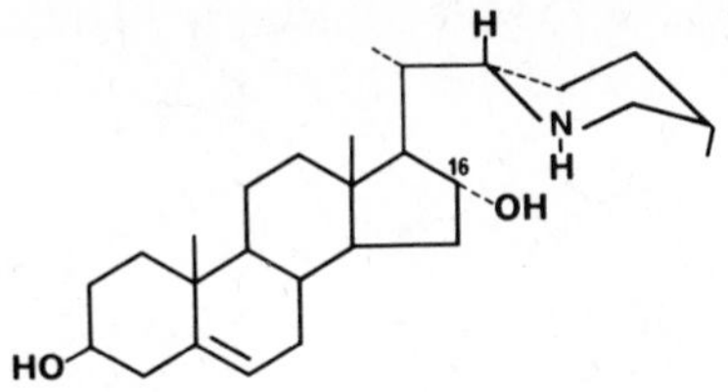

15

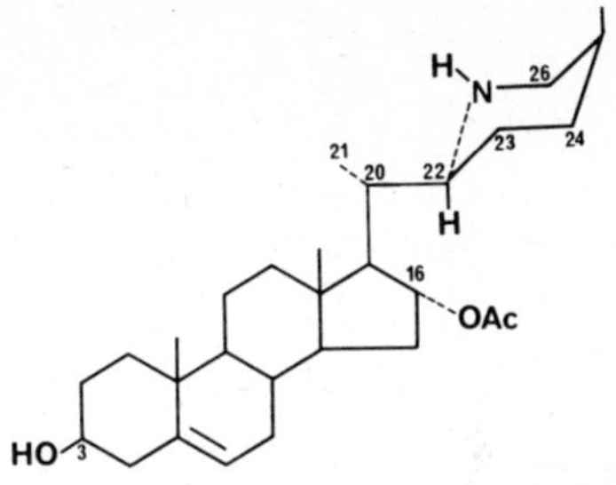

16

showed originally the importance of α-face presentation of the nitrogen atom for apparent bonding purposes.

Solanidan alkaloids, more typical of the alkaloids found in potatoes, are based on the C_{27}-carbon steroid skeleton of cholestane with a tertiary nitrogen shared by rings E and F, the nonsteroidal rings of the molecule. Important examples include solanidine (17), demissidine (18), and their glycosides. Because of Renwick's hypothesis, many workers investigated whether solanidan alkaloids might be teratogenic. By 1976 some investigators had reported teratogenicity of solanidine (17) or the glycoside solanine (19) in chick embryos [49,50], but most workers found no terata in lab mammals such as rats, rabbits, and mice [51–54].

Despite the reports of nonteratogenicity of the tertiary nitrogen solanidans in lab mammals, Brown, in our laboratory, believed that the teratogenicity

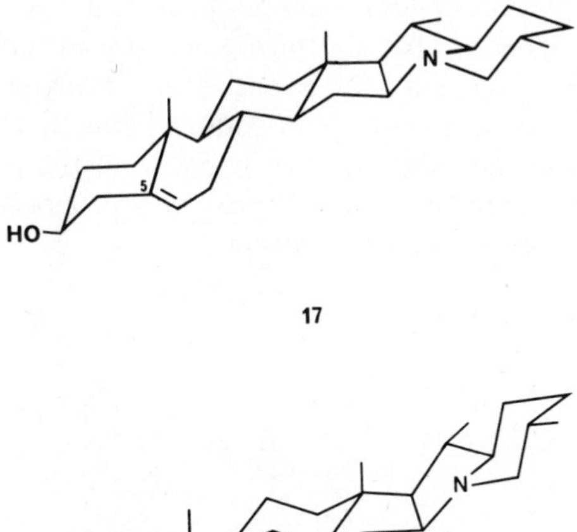

17

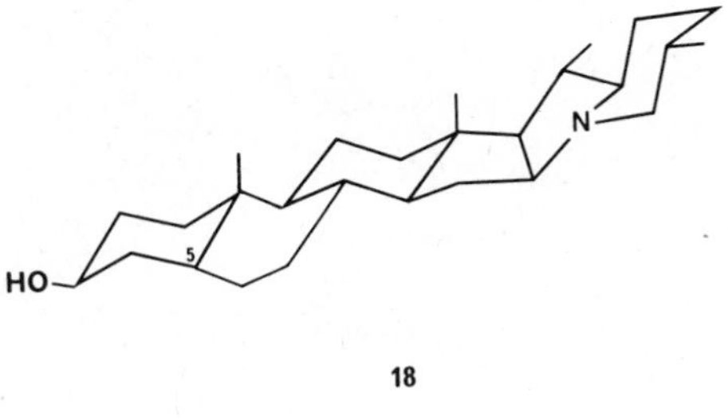

18

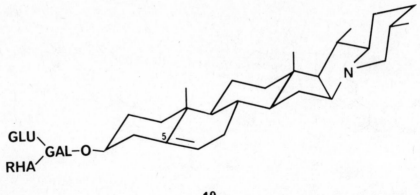

19

of *N*-methyljervine with its tertiary nitrogen suggested that tertiary nitrogen solanidans with nitrogen bonding capabilities α to the plane of the steroid should be active. Certain solanidan epimers of solanidine and demissidine, the **R-S** epimers at C(22) and C(25), were synthesized in 1961 by Sato and Ikekawa [55] and others. They provide a series of conformational epimers with the lone electron pair on the nitrogen projecting in either the α or β direction with respect to the plane of the steroid. Consequently Brown prepared some of these compounds for testing in hamsters, two 22S,22R epimers (**20**) having the nitrogen lone electron pair projecting α, and one 22R,25S epimer (**21**) having the pair projeting β. The two 22S,25R epimers with α projection were teratogenic. The 22R,25S epimer, which is in fact the naturally occurring compound demissidine, was not teratogenic. The other naturally occurring 22R,25S epimer, solanidine, was not tested [56].

Brown further believed that the nitrogen atom in the steroidal alkaloid teratogens may not be intrinsically essential for teratogenicity, but may only serve as an anionic bonding center that could be equally well served by some other atom. This idea seemed possible because of the relatively high teratogenicity of *N*-formyljervine. That activity was proposed to result, as one possibility, from a binding site different from the nitrogen atom, one in which

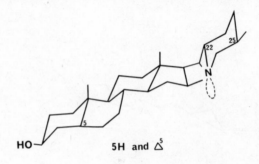

5H and Δ⁵

20

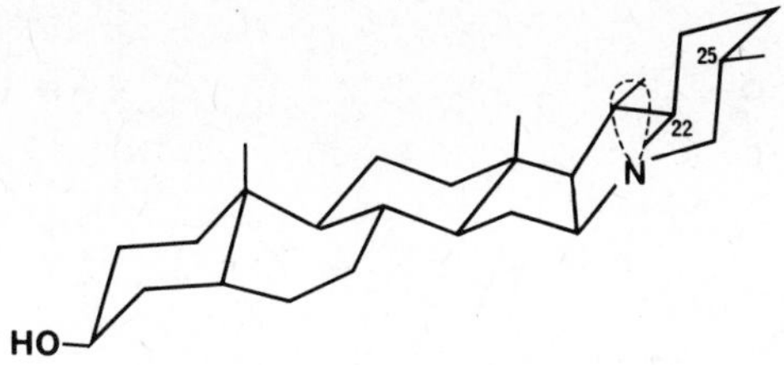

21

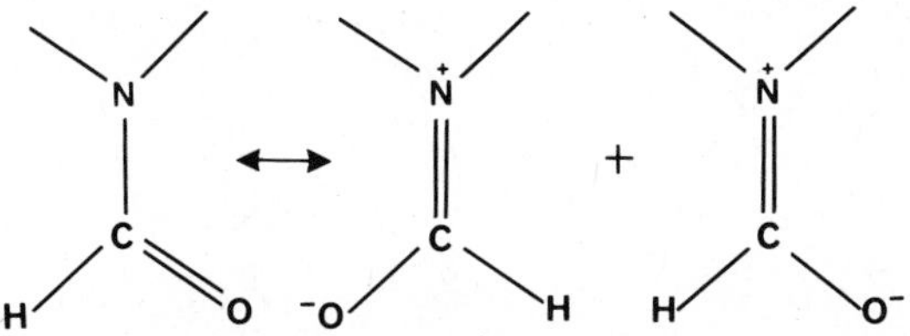

22

the oxygen could function as the anionic center (22) [47]. If there were no increased absorption from the gut with the N-formyljervine (and subsequent hydrolysis) compared with unsubstituted jervine, then that proposition might be a plausible explanation of the teratogenicity differences. This would be of some theoretical interest, although of less significance among naturally occurring alkaloids devoid of an N-formyl group, since in the latter the bonding appears to be directly to the nitrogen atom.

Brown also observed a weak teratogenicity in hamsters [57] of a mixture of isomeric N-formylsolasodines prepared by the method of Toldy and Radics [58,59], who reported that formylation of solasodine yields two isomers. The possibility that the activity resided in only one rather than both of the isomers prompted a reinvestigation by Gaffield and co-workers [60] of the structure of those compounds. Those isomers had been assigned structures by Toldy and Radics [58,59] based on an assumed restricted rotation about the C=N partial double bond (23). But Kusano et al. [61] proposed structures

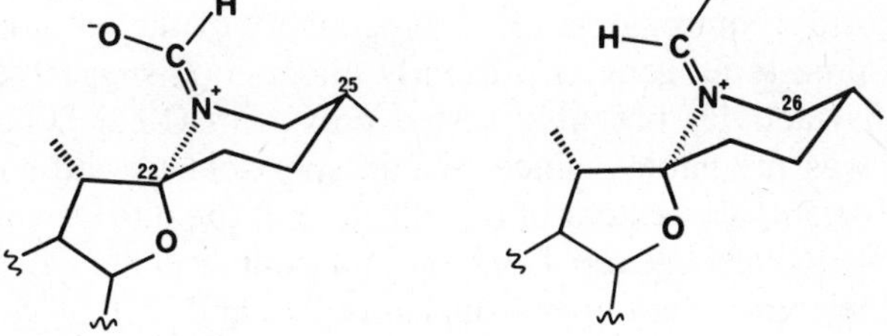

23

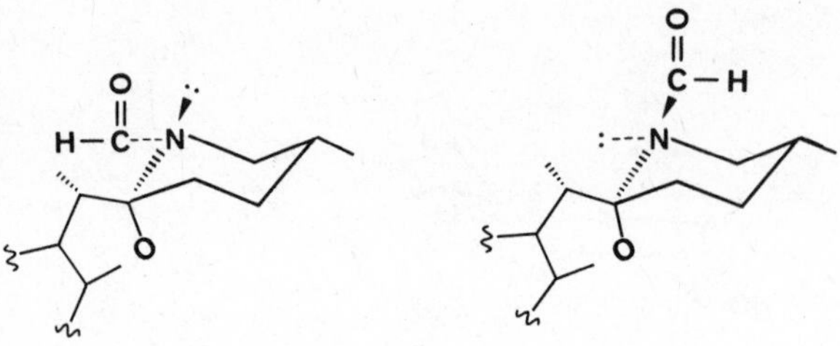

24

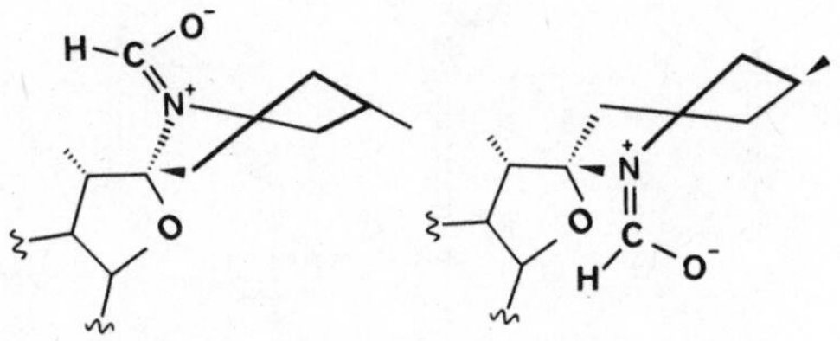

25

due to inversion about the nitrogen atom (**24**). Mass spectra as well as ^{1}H and ^{13}C NMR spectra have now shown, according to Gaffield and colleagues [60], that the isomerism is due rather to a difference in stereochemistry at C(22). The *N*-formylsolasodine isomers possess 22**R**,25**R** and 22**S**,25**R** stereochemistry and have nonchair ring-F conformations (**25**).

Based on these studies, Gaffield has suggested [38,48] that a reexamination of the biological activity of the isomeric *N*-formylsolasodines might elucidate which of the alternative mechanisms proposed [47] is correct for explaining the unexpectedly high teratogenicity of *N*-formyljervine. Oral administration of each isomeric derivative to hamsters and subsequent *in vivo* deformylation of the isomeric derivatives, after absorption by the animal, would produce steroidal amines having the amino group on opposite sides of the steroid plane. Then, assuming no further metabolism, each should exhibit differing degrees of teratogenicity. If, on the other hand, the alkaloid bonds to the active site in the embryo through the formyl group, the biological activity of the two isomers could conceivably be the same.

The structural requirements for teratogenicity of the naturally occurring steroidal alkaloid teratogens as presently understood suggest that our original belief that activity probably rested only with C-nor-D-homo type alkaloids [34] was premature, since certain spirosolanes and solanidans are teratogenic [44,56]. The extent of α projection of the nitrogen atom may not be as critical as we believed from the solasodine data [44], since 12,13-dihydrojervine (**26**) is teratogenic in hamsters, and in that compound the nitrogen, though α, is inclined away from the steroid α-face [46]. But at the moment it appears that steroidal alkaloids with a basic nitrogen in the ter-

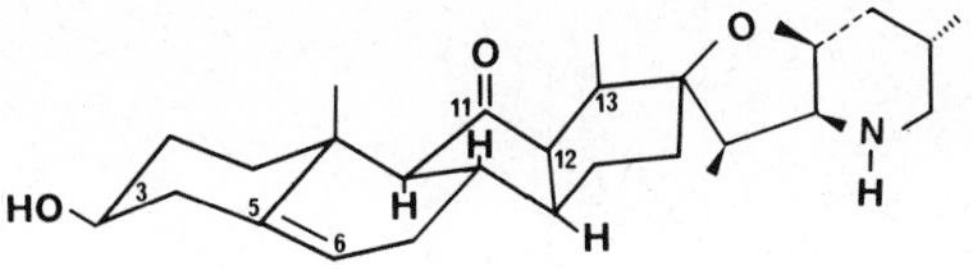

26

minal nonsteroidal F ring, shared or unshared with ring E, with nitrogen bonding capabilities α to the plane of the steroid, would be suspect as teratogenic.

8. STUDIES ON THE BIOCHEMICAL MECHANISM OF ACTION OF STEROIDAL ALKALOID TERATOGENS

By what mechanism do steroidal alkaloid teratogens exert their effect, and do all of them function by the same mechanism? A number of experiments have been performed to explore these questions. Competition with steroidal hormones that are essential to embryonic development is one possible mechanism because of the steroidal character of these teratogens. We have considered various possibilities that were essentially competitive inhibition between teratogen and hormone at whatever level of function contemporary knowledge allowed at the time—translation, transcription, or receptor binding [17,34,46].

Difference in pattern of terata response between the *Veratrum*, the spirosolane, and the solanidan teratogens suggests they may not be bonding to precisely the same receptor site. Pattern of terata response for these three compounds is compared in Table 1. Jervine induced a high incidence of cebocephalics, the 22S,25R solanidans a high incidence of harelip, and solasodine virtually all blebs and exencephalics. Variation in absorption or clearance rates could effectively produce a shift in insult timing to cause pattern variation, and so we must be cautious in drawing interpretations of that type. Consequently, we are not yet justified in assuming that all three teratogen types bond to the same receptor.

Experiments at the National Institutes of Health have been conducted to compare the pattern of terata response from jervine in various inbred mouse strains [62,63]. The data also allow comparison of incidence of terata induced by jervine and that reported for corticosteroids [64] in those same mouse strains. Table 2 shows that the incidence of cleft palate induced by jervine or by cortisone differs markedly among the mouse strains. Glucocorticoids such as cortisone at physiologic levels can function in fetal tissue by controlling morphologic and biochemical differentiation [64,65]. Binding of hormone with specific intracellular cytosol receptor proteins in target cells is the means by which these effects are mediated [65,66]. High doses of the

Table 1. Congenital Deformities Induced by Various Steroidal Alkaloid Teratogens in Simonsen Hamsters

Teratogen	Day of Dosing	Cebocephalic	Cranial Bleb (encephalocele)	Harelip	Exencephalic	Microphthalmic	Spina Bifida	Total Number of Deformities
Solasodine[a]	8		71%	2%	21%	15%	6%	34
22S,25R solanidans[b]	8		26%	48%	10%			20
Jervine[a]	8	52%	19%	22%	7%			27

[a] Data from [101].
[b] Data from [56].

Table 2. Comparison of Incidence in Inbred Mice of Cleft Palate Induced by Jervine or Cortisone Administration on Day 11 of Gestation

Strain of Mice	Jervine[a]	Cortisone[b]
Swiss Webster	Resistant	100%
A/J	100%	100%
C57BL/6J	100%	25%

[a] Data from [63].
[b] Summarized in [64].

glucocorticoid can disrupt development and cause various teratogenic effects. If the mechanism of action of jervine or cyclopamine teratogenicity was a competitive binding to the cortisone cytosol receptor, one might expect that the incidence and pattern of terata response in given mouse strains would be similar for cortisone and the *Veratrum* teratogens. The striking difference shown in Table 2 suggests dissimilar mechanisms of effect, or at least that different receptors might be involved.

In experiments in our laboratory using a different approach, if cyclopamine and jervine bound competitively to a cytosol receptor of an as yet unidentified but common hormone, then competitive inhibition studies might reveal which hormone. Consequently, we dosed Engle hamsters [67] with jervine in protocols that would produce terata. At that same time we dosed them with various hormones at a number of levels to see if terata response could be prevented or moderated. Table 3 shows compounds and levels used along with results. Simultaneous hormone administration did not prevent or moderate terata response to jervine, suggesting that jervine did not induce terata by competitive inhibition of these hormones. Thus neither the National Institutes of Health experiments nor those in our laboratory lent support to the notion of a competitive binding of the *Veratrum* teratogen to a steroidal hormone cytosol receptor.

Using still another approach, Hassell and others [68] incubated dissociated limb bud mesenchyme cells with cyclopamine. Those cells, which differentiate into chondrocytes to form cartilage, were inhibited by cyclopamine in cartilage proteoglycan synthesis and in DNA syynthesis. Cartilage proteoglycan synthesis as measured by uptake of ^{35}S-sulfate was inhibited by 54%, and DNA synthesis as measured by uptake of ^{3}H-thymidine was inhibited by 77%. By contrast, protein synthesis as measured by uptake of ^{35}S-methionine was only slightly reduced.

Further experiments by that group [69] showed that prechondrogenic mesenchyme, prior to condensation, is exquisitely sensitive to *Veratrum* teratogens. Prior to differentiation, exposure of limb bud mesenchyme cells to the *Veratrum* teratogen *in vitro* suppressed subsequent accumulation of cartilage proteoglycan, but treatment after differentiation had no such effect.

In experiments at our laboratory a very high proportion of lambs born to ewes gavaged on days 31, 32, and 33 with *Veratrum californicum* root, the crude drug from which cyclopamine is isolated, had severe stenosis of the trachea [70]. Preliminary experiments suggest cyclopamine is the cause. The defect was characterized by flattening of the trachea throughout the entire length (Fig. 5). Cartilaginous tracheal rings were reduced in number, non-uniform in size and shape, irregularly spaced, and of abnormal orientation. The smaller size and relatively noncurved shape of the cartilage rings of the stenotic trachea resulted in a nondistended lumen. Cartilage rings from normal neonatal lamb tracheas were prominently curved, had a relatively thicker superficial chondrogenic zone of proliferating cells on their outer than on their inner surface, and comprised a central plate of differentiated cartilage cells and matrix representing more than half of their cross-sectional substance. By contrast, cartilage rings from the stenotic trachea were thin-

Table 3. Hormons Tested and Found Inactive as Competitors of Jervine-Induced Malformations in Engle Hamsters[a,b]

Androgens
Epiandrosterone
Dehydroepiandrosterone[d]
Testosterone propionate[c,d]
Androsterone
Androstanedione

Corticosteroids
Hydrocortisone
Pregnenolone acetate[d]
Pregnenolone[d]
Corticosterone
Cortisone

Estrogens
Esterone[d]
Estriol[d]
17β-Estradiol (estradiol)[d]
17α-Estradiol[d]

Progestins
Ethisterone[d]
Progesterone

[a] 170 mg/kg jervine administered i.p. on day 7 with 85, 17, 8.5, or 1.7 mg/kg hormone all in 2% methyl cellulose in water (jervine dose exceeded ED_{50}).
[b] Data from [67].
[c] Hormone alone teratogenic at 85 mg/kg.
[d] Hormone alone produced resorptions at 85, 17, or 8.5 mg/kg.

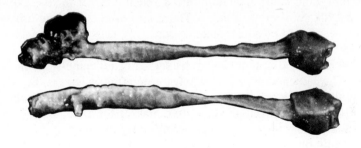

Figure 5.   Tracheal stenosis induced in twin lambs by maternal ingestion of *Veratrum californicum* during pregnancy. Note the flattening of both tracheas resulting in nondistended lumen.

ner, not curved, had chondrogenic cortical zones of similar width on their outer and inner surfaces, and comprised a thin central plate of differentiated cells and matrix representing little more than one-fifth of their cross-sectional substance.

Becauce the effect appeared to center on the cartilaginous rings of the trachea, the hypoplasia of the rings may represent an arrest of synthesis of that target tissue. If so, that lends support to the mechanism proposed by Hassell et al. [68]. A mechanism centered on a specific inhibition of proteoglycan synthesis seems logical, since the tracheal ring cartilage is derived from differentiation of endodermal trachea tube mesenchyme cells [71]. Thus a specific inhibition of proteoglycan synthesis either directly, or indirectly, through an effect on the genetic message, could be a mechanism by which cyclopamine exerts the various teratogenic effects expressed morphologically in such a variety of ways.

An alternative mechanism possibility is suggested by studies of Sim and others [72,73] in Australia. Those studies were stimulated by the reports that neurotransmitters play a critical role in early stages of embryonic development [74–76]. Catecholamines have been demonstrated throughout the lateral and dorsal cranial neuroepithelium during and after closure of the neural tube [77]. This distribution pattern coincides morphologically with the region of neuroepithelium disrupted by teratogenic *Veratrum* alkaloids. The pattern coincides in timing with the developmental stage at which cyclopia and related terata can be induced [33].

Consequently Sim et al. [72,73] proposed that catecholamine-secreting cells in embryonic neuroepithelium are a specific target for the expression of steroidal alkaloid teratogenicity. The effect of cyclopamine on the acetylcholine responses of catecholamine-secreting cells was examined using isolated chromaffin cells from bovine adrenal medulla [78,79]. These cells are considered to be a suitable model for catecholamine-secreting neural cells [80]. The catecholamine-containing chromaffin vesicles in the cells contain, in addition to catecholamines, large amounts of adenosine triphosphate (ATP), which may be used as a sensitive marker for exocytosis [81,82].

The initial time course for release of ATP by those cells was examined during stimulation by the antagonists acetylcholine or nicotine and the modification of ATP release by cyclopamine. Cyclopamine significantly decreased the acetylcholine- or nicotine-mediated ATP release [72,73], suggesting that cyclopamine may act upon the nicotine receptor, perhaps by competitive binding. The results were consistent with the reported effect of *Veratrum* teratogens on the embryonic cranial neuroepithelium [33]. Cells in the embryonic neuroepithelium contain catecholamines during a period coinciding with the insult timing for *Veratrum* teratogenicity. Inhibition of catecholamine release in embryonic neuroepithelium could induce deformities. Catecholamines are suggested as trophic factors at this stage of development [14,15]. It therefore seems reasonable that acetylcholine receptor–cyclopamine interaction could be a possible mechanism for *Veratrum* teratogenicity [73].

Further experimental work will be needed to determine which, if either, of these two likely mechanisms of action of *Veratrum* teratogens (proteoglycan inhibition or binding to the nicotinic receptor) is correct. It is not even possible to say at this juncture whether the varying categories of steroidal alkaloid teratogens (e.g., the furanopiperidine *Veratrum* alkaloids, the spirosolanes, and the solanidans) all function by the same mechanism if indeed either of these is correct.

9. NATURAL OCCURRENCE OF KNOWN AND SUSPECTED STEROIDAL ALKALOID TERATOGENS IN FOODS AND FEEDS

Veratrum steroidal alkaloids are found in a variety of nonfood plants. The genera include *Veratrum*, *Zygadenus*, and *Schoenocaulon* [83]. While they present no hazard to humans, they clearly present a potential teratogenic hazard to animals foraging upon them when they contain alkaloids of the teratogenic type such as jervine (13), cyclopamine (10), or cycloposine (4). But the potential depends upon many factors. The plant containing appropriate teratogens at sufficiently high concentration would have to be grazed at the susceptible gestational period in sufficient quantity by susceptible livestock. Not all sheep grazing *Veratrum californicum* give birth to cyclopic offspring. It seems doubtful that one could attribute human congenital deformities to *Veratrum* steroidal alkaloids. There is no reason to suppose humans would not be susceptible since there is wide species specificity, but the compounds have not been reported in plants consumed by humans [83].

Spirosolane alkaloids, on the other hand, such as the teratogen solasodine (11) and its glycosides, are found in plants used as food by humans in addition to those foraged by livestock. A tabular listing in 1968 [84] of plant sources of solasodine as aglycone or glycoside suggested it had been isolated even by then from about 100 different plants including various species in the *Solanum, Lycopersicon, Cestrum,* and *Cyphomandra* genera. Some are grazed by livestock and a few are common foods. *Solanum melongena* (eggplant), *S. quitoense* (the Andean Naranjilla), and *Cyphomandra betacea* (the tree tomato of Peru) are among those used as food. Spirosolanes other than solasodine that might be suspect as teratogenic if ingested by humans or animals include soladulcidine (27) (5,6-dihydrosolasodine) isolated from *S. dulcamera, S. megacarpa,* and *Lycopersicon pimpinellifolium* [84,85] and solanaviol (28), the C-12α-ol analog of solasodine, from *S. aviculare* [86].

Solanidan alkaloids are common to potatoes and tomatoes and questions of whether there are potential real world teratogenic hazards from solanidan alkaloids center largely on these two plants. We found no congenital defects in lambs whose dams were gavaged during pregnancy with high alkaloid plant and fruit material from both potato and tomato species [34]. But as mentioned above, Renwick [41] hypothesized that blighted potatoes were somehow responsible for human congenital spina bifida and anencephaly, suggesting that avoidance of potatoes by pregnant mothers would virtually eliminate those two conditions. That hypothesis [41] has been looked upon with skepticism by some, but not all, researchers in the intervening years since its enunciation.

Numerous reports [87–92] cast doubt on the validity of the epidemiologic relationships upon which Renwick's hypothesis was based, and particularly on the correlation between the level of blight induced by *Phytophthora infestans* and incidence of ASB. Furthermore, the birth of ASB children to mothers on potato-avoidance trials during pregnancy, as reported by Nevin and Merrett [93], seemed particularly strong evidence refuting Renwick's original hypothesis.

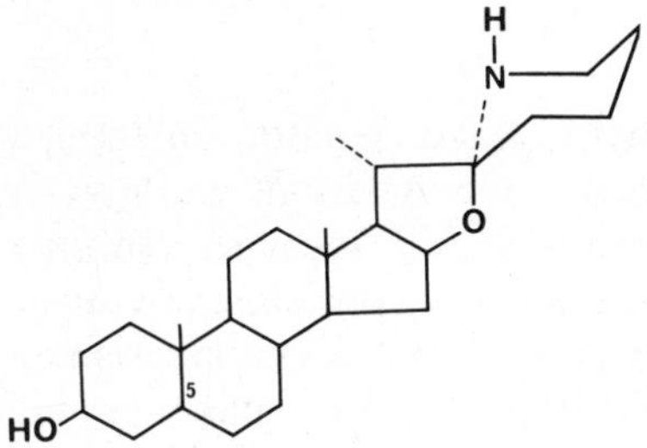

27 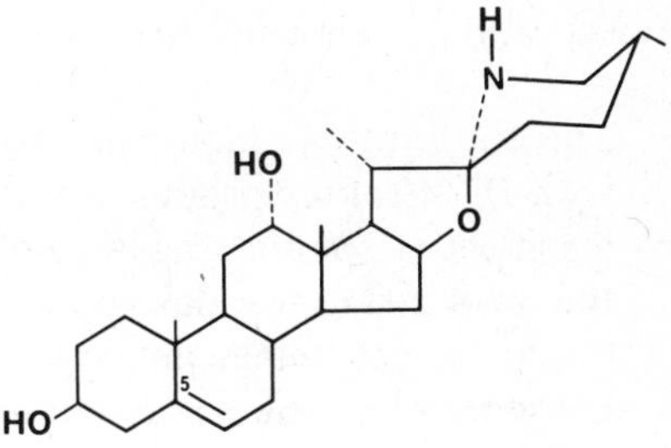   28

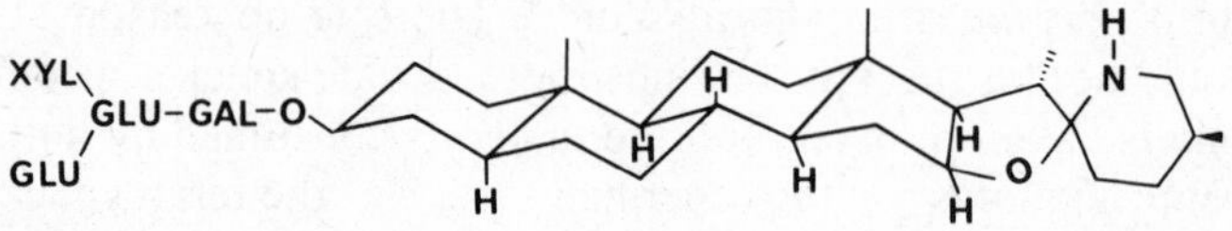

XYL\
　GLU–GAL–O
GLU/

H
N

H
H
H
H
O
H

29

RHA\
　GLU–O
RHA/

5

N

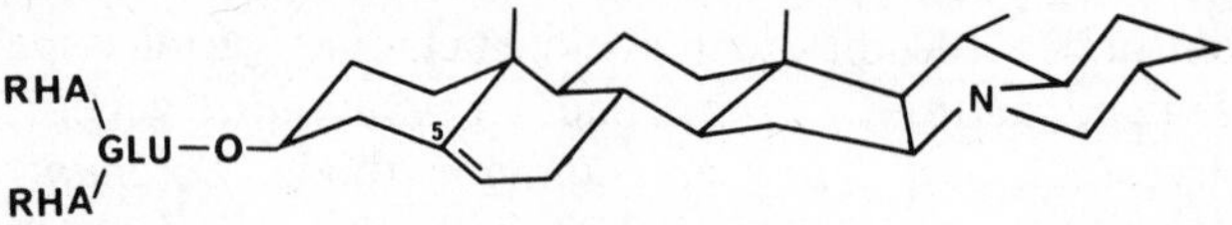

30

Other studies examined the question of teratogenicity for laboratory animals fed potato preparations or alkaloids derived therefrom. Poswillo et al. [94] found cranial osseous defects in neonatal marmosets from dams dosed with certain potato preparations, but these investigators were later unable to confirm the observation [95]. Swinyard and Chaube [96] reported that injection of extracted potato alkaloids or of pure solanine (**19**) in pregnant dams produced no neural tube defects in rats and rabbits. Ruddick et al. [97] reported that freeze-dried *Phytophthora*-blighted "Kennebec" potatoes were not teratogenic when fed to rats. Nishie et al. [52] reported that tomatine (**29**), α-chaconine (**30**), and α-solanine (**19**) failed to produce teratogenic effects in chicks. The intraperitoneal injection of solanine (**19**) to pregnant mice did not produce terata, according to Bell et al. [53]. In a further report, Chaube and Swinyard [54] reported no neural tube defects in offspring from rats subject to acute or chroniic i.p. injection of α-chaconine (**30**) or α-solanine (**19**), but some fetuses died. Pierro and co-workers [98], however, reported that α-chaconine (**30**) but not α-solanine (**19**) was teratogenic in inbred NAW/Pr mice when dams were dosed on days 8 or 9 of gestation. Midline facial defects were induced by day 8 dosing.

Mun and associates [49] showed that both pure solanine (**19**) and glycoalkaloids extracted from *Phytophthora*-blighted potatoes produced "rumplessness or trunklessness" in chick embryos when eggs were treated during early development. Jelinek et al. [50] verified and extended those observations, reporting that:

> The embryotoxic factor in ethanol extract of boiled potatoes infected with *P. infestans* produced the caudal regression syndrome and myeloschisis at somite stages of chicken embryos. After an injection on the third and fourth days a malformation syndrome consisting of cranioschisis, celosoma, and cardiac septal defects was a characteristic consequence. The same syndrome could be induced by injecting an equivalent amount of extract of healthy potatoes and by injecting solanine (**19**) in amounts corresponding to the solanine (**19**) concentration in the given extracts.

We tested teratogenicity of a variety of potato preparations. Neither blight-infested tuber material nor aged, high alkaloid tuber material was teratogenic in rats, mice, hamsters, or rabbits when gavaged at the primitive streak/neural plate stage of embryonic development [99,100]. None of the potato material we tested produced terata in any lab animals with but one exception. Sprouts that contained 200–400 times as much alkaloid as tubers were teratogenic in Simonsen hamsters with exencephaly and encephalocele the usual effects [101]. Sprouts from six different cultivars (Table 4) produced terata, but neither tubers nor peels from these same cultivars were teratogenic at doses up to four times as high as sprout dosages [101]. About 2500–3500 mg/kg dried sprout material was required to induce the effects.

We did not believe in 1978 that the teratogenicity of sprout material in this particular hamster strain was a reflection of a real world problem for humans [101]. For an average weight woman an equivalent dose would be about 1000 g wet weight of sprouts, a portion of the potato not eaten in any amount because of the bitter taste [101].

We did not establish whether the teratogenic effect of potato sprouts in Simonsen hamsters was due to alkaloids. It seemed likely, however, since the 3% acetic acid–extracted total glycoalkaloid fraction (TGA) from the sprouts produced the effect. Pregnant Simonsen hamsters were dosed with 370 mg/kg of a total glycoalkaloid preparation from Kennebec sprouts. Of 21 liters carried to term, 8 had a total of 23 deformed offspring [67].

Further experimentation will be needed to determine if that teratogen is an alkaloid. If it is, it will then have to be determined if it is a 22S,25R solanidan epimer with a α projecting electron pair on the nitrogen (which would certainly be unexpected since none have been isolated from common potatoes), or a spirosolane, or some other as yet unrecognized teratogen from common potatoes. Perhaps a 22S,25R-solanidan cannot be ruled out in view of reports [102,103] by Pakrashi and co-workers. They found, in *Solanum giganteum*, a naturally occurring solanidan, solanogantine (31), with the free electron pair of the ring nitrogen α to the plane of the steroid. The solanogantine occurred along with two other solanidans with β-nitrogen

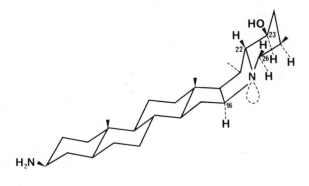

31

Table 4. Teratogenic Testing of Potato Preparations[a] in Simonsen Hamsters

Cultivar Treatment Group	Total Pregant Litters[b]	Total Litters with Malformed Fetuses	% Litters with Malformed Fetuses	Calculated Dose (as mg/kg of TGA)
H_2O control	522	7	1.3	
Kennebec				
Sprout	181	46	25.4 ($P < 0.0005$)	210
Peel	96	3	3.1 ($P > 0.1$)	20
Tuber	124	2	3.2 ($P > 0.1$)	0.5
Russet				
Sprout	70	18	25.7 ($P < 0.0005$)	140
Peel	31	0	0 ($P > 0.1$)	4
Tuber	23	1	4.3 ($P > 0.1$)	0.6
Targhee				
Sprout	58	6	10.4 ($P < 0.0005$)	160
Peel	38	1	2.6 ($P > 0.1$)	3
Tuber	31	1	3.2 ($P > 0.1$)	0.7
Sebago				
Sprout	52	7	13.5 ($P < 0.0005$)	150
Peel	16	0	0 ($P > 0.1$)	22
Tuber	32	2	6.3 ($P > 0.2$) [$P > 0.1$][c]	0.3
Nampa				
Sprout	55	10	18.2 ($P < 0.0005$)	130
Peel	34	2	5.9 ($P > 0.02$) [$P > 0.1$][c]	24
Tuber	38	1	2.6 ($P > 0.1$)	0.7
Norchip				
Sprout	46	4	8.7 ($P \cong 0.0005$) [$P < 0.005$][c]	170
Peel	34	1	2.9 ($P > 0.1$)	0.7
Tuber	35	0	0 ($P > 0.1$)	0.7

[a] Single average dose levels gavaged on the 8th day of gestation (3.7 g/kg sprouts, 14.8 g/kg tuber, 11.1 g/kg peel) [101,67].

[b] Carried to term or resorbed (10–40% overdose dam death).

[c] Yates correction applied (1 degree of freedom).

configuration [103]. But a spirosolane cannot be ruled out either, because when we consider the pattern of terata response described earlier in Table 1 by including sprouts and TGA from sprouts (see Table 5), an interesting pattern emerges. We see that the sprout and TGA patterns have greater similarity to the solasodine than to the 22S,25R-solanidan or the jervine patterns. Gregory and others [104] recently indicated that solasodine (**15**) was the aglycone of glycosides from *Solanum berthaultii*, a wild tuber-bearing solanum of potential usefulness in potato breeding.

The teratogenicity of sprouts that we demonstrated in Simonsen hamsters was investigated further by Renwick and colleagues, who sought to identify the teratogen and measure the *in vivo* retention of it in tissues of humans and hamsters. Renwick reported [105] that two potato alkaloid glycosides common to sprouts, α-solanine (**19**) and α-chaconine (**30**), induced a high incidence of exencephaly and encephalocele in offspring from pregnant hamsters gavaged the compounds on day 8 of gestation. With α-solanine, for example, 20 of 34 live litters had malformed individuals and about 24% of all offspring in these litters were malformed. These two solanidan glycosides are the principal alkaloidal glycosides in U.S. potato cultivars and the only alkaloidal glycosides in British cultivars [105]. More recently Renwick and colleagues [106] demonstrated that the British Potato Cultivar, Arran Pilot, is teratogenic in Simonsen hamsters. They showed through extraction and isolation experiments coupled with teratogenicity trials in that hamster strain that both α-solanine and α-chaconine isolated from that cultivar were teratogenic. Dosing with either compound induced defects in offspring in over half the litters. They concluded that the sprout teratogenicity could be accounted for, in the main, by those two compounds.

Of course, the naturally occurring solanidans α-chaconine (**30**) and α-solanine (**19**) are reported to be in the 22R,25S form (β-projection of nitrogen-free electrons). From our earlier data on teratogenicity of solanidan epimers, we would not have expected these compounds to be teratogenic unless enzyme-catalyzed inversion of the electron pair takes place through configurational alteration *in vivo*. Alternatively there may be some of the 22S,25R epimer naturally present in potatoes and in glycoside preparations as yet unrecognized by isolation. Another possibility is that the embryonic dose may be greatly enhanced when the glycosides are administered compared to the aglycone due to absorption differences. The resolution of this matter awaits further experimentation.

Renwick and co-workers [107] also reported on the kinetics of retention of tritiated solanidine (**17**) in human subjects. This investigation was prompted by the data of Nevin and Merrett [93], who had shown that potato avoidance by mothers during pregnancy did not prevent birth of ASB offspring among mothers who had previously delivered ASB children. The Renwick group reasoned that if potato alkaloids (solanidine or its glycosides) were responsible for induction of ASB, significant tissue storage would have to take place with a hypothesized release during pregnancy to maternal cir-

Table 5. Comparison of the Pattern of Terata Response in Simonsen Hamsters Induced by Various Steroidal Alkaloid Teratogens and by Potato Sprouts or the Total Glycoalkaloid Extract Therefrom.

Teratogen	Day of Dosing	Cebocephalic	Cranial Bleb (encephalocele)	Harelip	Exencephalic	Microphthalmic	Spina Bifida	Total Number of Deformities
Kennebec sprouts[a]	8	1%	63%		36%			83
TGA from sprouts[b]	8		43%		52%		5%	23
Solasodine[a]	8		71%	2%	21%		6%	34
22S,25R Solanidans[c]	8		26%	48%	10%	15%		20
Jervine[a]	8	52%	19%	22%	7%			27

[a] Data from [101].
[b] Data from [67].
[c] Data from [56].

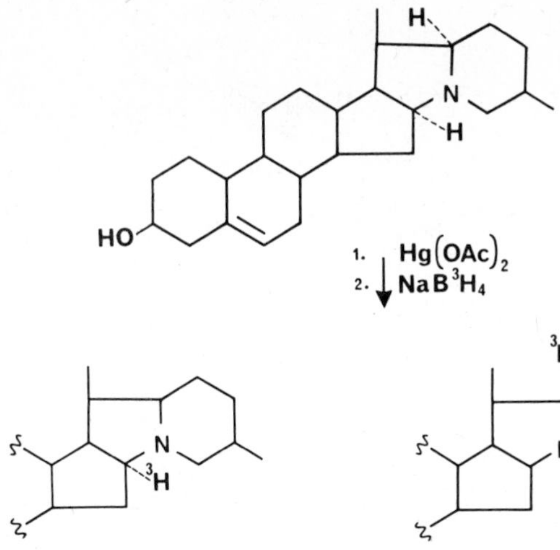

32

culation and thence to the embryo. In their experiments [107] solanidine (**17**) (identity confirmed by MS) was oxidized with mercuric acetate and subsequently reduced with tritated sodiumborohydride (**32**). The mixture of tritated solanidines (**32**) was administered to human subjects by I.V. injection. Urinary and fecal excretion rates of radiolabel were very slow with 90% or more of the label being sequestered somewhere in the body and excreted with a half life of from 34–68 days. Whether the sequestered label was still in the form of solanidine rather than some metabolite was not determined because of low concentrations; however, Renwick and his colleagues were able to isolate and identify solanidine from human postmortem liver samples from subjects that had died from cardiovascular disease. Assuming the retained label was still in the form of unchanged solanidine, their data allowed an approximation of 50 mg as the retained body burden of solanidine from daily intake and absorption of 1 mg/day. If mobilization of solanidine from innocuous loci occurred during the stress of pregnancy, those researchers reasoned, that mobilization might lead to sufficient embryonic insult to induce ASB [107].

Whether there is a real world teratogenic hazard to humans from consuming potato tubers remains to be clarified. If further research establishes that there is undue tissue storage of a potato teratogen not metabolized to an inactive form, then the possibility certainly exists.

One real world potential teratogenic hazard from potato material is in livestock fed potato waste products from processing plants. This material usually carries an extremely high content of sprouts. If species specificity to the sprout teratogen is wide enough to include livestock, perhaps terata in livestock could result. But the limited susceptibility among hamster strains to sprout-induced malformations argues against the possibility.

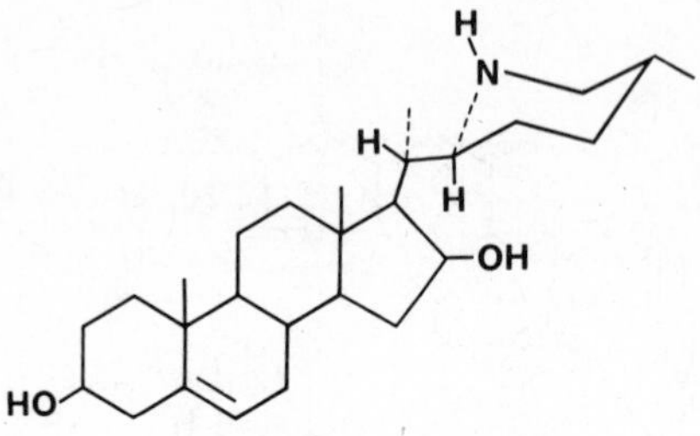

33

Whether there is any teratogenic potential from alkaloids of the final category, the 22,26-epiminocholestanes, rests largely, until we have further information, on the marginal teratogenicity in hamsters of muldamine (8) [46]. Further testing of muldamine and deacetylmuldamine are needed to provide statistical certainty concerning their teratogenicity. If they possess consistent, though reduced teratogenicity, then other 22,26-epiminocholestan alkaloids might be suspect—both early and more recently isolated examples [37,84,85,108–110] that differ mainly in piperidine ring or C(16) substituent configurations. One possibility is solaverbascine (33), 22S,25R-22,26-epiminocholest-5-ene-3β,16β-diol. That compound was recently isolated from *Solanum verbascifolium* by Adam and co-workers [111]. That plant has been used in folk medicine as an abortifacient [111], and often teratogens are abortifacient at higher doses.

REFERENCES

1. J. Warkany, *Congenital Malformations*, Year Book Medical Publishers, Chicago, 1971.

2. J. G. Wilson, *Environment and Birth Defects*, Academic, New York, 1973.

3. W. B. McBride, *Lancet* **2**, 1358 (1961).

4. W. Lenz, *Lancet* **1**, 271 (1962).

5. E. M. Johnson, *Teratology* **21**, 259 (1980).

6. R. M. Hoar, *Fund. Appl. Toxicol.* **4**, S335 (1984).

7. W. Binns, E. J. Thacker, L. F. James, and W. T. Huffman, *J. Am. Vet. Med. Assoc.* **134**, 180 (1959).

8. W. Binns, W. A. Anderson, and D. J. Sullivan, *J. Am. Vet. Med. Assoc.* **137**, 515 (1960).

9. W. Binns, L. F. James, J. L. Shupe, and E. J. Thacker, *Arch. Environ. Health* **5**, 106 (1962).

10. W. Binns, L. F. James, J. L. Shupe, and G. Everett, *Am. J. Vet. Res.* **24**, 1164 (1963).

11. W. Binns, J. L. Shupe, R. F. Keeler, and L. F. James, *J. Am. Vet. Med. Assoc.* **147**, 839 (1965).

12. R. F. Keeler and W. Binns, *Proc. Soc. Exp. Biol. Med.* **116**, 123 (1964).

13. S. M. Kupchan, J. H. Zimmerman, and A. A. Afonso, *Lloydia* **24**, 1 (1961).

14. M. Burroni, *Caryologia* **7**, 87 (1955).

15. D. L. Smith and L. D. Hiner, *J. Am. Pharm. Assoc.* **49,** 538 (1960).

16. H. Frank, *Zentralbl. Bakteriol. Parasitenkd. Infektionskr. Hyg. Abt.* 2, **113,** 128 (1959).

17. R. F. Keeler and W. Binns, *Can. J. Biochem.* **44,** 819 (1966).

18. R. F. Keeler and W. Binns, *Can. J. Biochem.* **44,** 829 (1966).

19. R. F. Keeler and W. Binns, *Teratology* **1,** 5 (1968).

20. R. F. Keeler, *Phytochem.* **8,** 223 (1969).

21. T. Masamune, Y. Mori, M. Takasugi, and A. Murai, *Tetrahedron Lett.* **16,** 913 (1964).

22. T. Masamune, M. Takasugi, M. Gohda, H. Suzuki, S. Kawahara, and T. Irie, *J. Org. Chem.* **29,** 2282 (1964).

23. T. Masamune, N. Sato, K. Kobayashi, I. Yamazaki, and Y. Mori, *Tetrahedron* **23,** 1591 (1967).

24. H. Budzikiewicz, C. Djerassi, and D. H. Williams, *Structural Elucidation of Natural Products by Mass Spectrometry*, Vol. 2, Holden Day, San Francisco, 1964, p. 21.

25. T. Masamune, Y. Mori, M. Takasugi, A. Murai, S. Ohuchi, N. Sato, and N. Katsui, *Bull Chem. Soc. Japan* **38,** 1374 (1965).

26. R. F. Keeler, *Steroids* **13,** 579 (1969).

27. W. Binns, R. F. Keeler, and L. D. Balls, *Clin. Toxicol.* **5,** 245 (1972).

28. R. F. Keeler, *Teratology* **3,** 175 (1970).

29. R. F. Keeler, *Proc. Soc. Exp. Biol. Med.* **136,** 1174 (1971).

30. R. F. Keeler, *Proc. Soc. Exp. Biol. Med.* **149,** 302 (1975).

31. M. M. Bryden, C. Perry, and R. F. Keeler, *Teratology* **8,** 19 (1973).

32. M. M. Bryden and R. F. Keeler, *J. Anat.* **116,** 464 (1974).

33. F. R. P. Sim, N. Matsumoto, E. H. Goulding, K. H. Denny, J. Lamb, R. F. Keeler, and R. M. Pratt, *Terat. Carc. Mut.* **3,** 111 (1983).

34. R. F. Keeler, *Teratology* **3,** 169 (1970).

35. R. F. Keeler, *Steroids* **18,** 741 (1971).

36. W. Gaffield, R. Y. Wong, R. E. Lundin, and R. F. Keeler, *Phytochemistry* **21,** 2397 (1982).

37. K. Kaneko, M. W. Tanaka, E. Takahashi, and H. Mitsuhashi, *Phytochemistry* **16,** 1620 (1977).

38. W. Gaffield, "Application of Nuclear Magnetic Resonance Spectroscopy to the Determination of Structure and Configuration of Steroidal Amine Teratogens," in A. A. Seawright, M. P. Hegarty, R. F. Keeler, and L. F. James, Eds., *Plant Toxicology*, Queensland Poisonous Plants Committee, Brisbane, Australia, in press.

39. P. W. Sprague, D. Doddrell, and J. D. Roberts, *Tetrahedron* **27,** 4857 (1971).

40. S. M. Kupchan and M. I. Suffness, *J. Am. Chem. Soc.* **90,** 2730 (1968).

41. J. H. Renwick, *J. Prev. Soc. Med.* **26,** 67 (1972).

42. R. F. Keeler, *Lancet* **1,** 1187 (1973).

43. R. F. Keeler, *Proc. Soc. Expt. Biol. Med.* **149,** 302 (1975).

44. R. F. Keeler, S. Young, and D. Brown, *Res. Comm. Chem. Path. Pharm.* **13,** 723 (1976).

45. Kh. I. Seifulla and K. E. Ryzhova, *Mater. Vses. Konf. Issled. Lek. Rast. Perspekt. Ikh. Ispolz Proizvod. Lek. Prep.* **1970,** 160 (1972); *Chem. Abstr.* **83,** 22383t (1975).

46. D. Brown and R. F. Keeler, *J. Agric. Food Chem.* **26,** 561 (1978).

47. D. Brown and R. F. Keeler, *J. Agric. Food Chem.* **26,** 564 (1978).

48. W. Gaffield and R. F. Keeler, "Structure and Stereochemistry of Steroidal Amine Teratogens" in M. Friedman, Ed., *Advances in Experimental Medicine and Biology*, Vol. 177, Plenum, New York, 1984, p. 241.

49. A. M. Mun, E. S. Barden, J. M. Wilson, and J. M. Hogan, *Teratology* **11**, 73 (1975).

50. R. Jelinek, V. Kyzlink, and C. Blattny, Jr., *Teratology* **14**, 335 (1976).

51. C. A. Swinyard and S. Chaube, *Teratology* **8**, 349 (1973).

52. K. Nishie, W. P. Norred, and A. P. Swain, *Res. Comm. Chem. Path. Pharm.* **12**, 657 (1975).

53. D. P. Bell, J. G. Gibson, A. M. McCarroll, G. A. McClean, and A. Geraldine, *J. Reprod. Fertil.* **46**, 257 (1976).

54. S. Chaube and C. A. Swinyard, *J. Toxicol. Appl. Pharmacol.* **36**, 227 (1976).

55. Y. Sato and N. Ikekawa, *J. Org. Chem.* **26**, 1945 (1961).

56. D. Brown and R. F. Keeler, *J. Agric. Food Chem.* **26**, 566 (1978).

57. D. Brown, personal communication to William Gaffield.

58. L. Toldy and L. Radics, *Tetrahedron Lett.* **1966**, 4753.

59. L. Toldy and L. Radics, *Kem. Kozlem.* **26**, 247 (1966).

60. W. Gaffield, M. Benson, W. F. Haddon, and R. E. Lundin, *Aust. J. Chem.* **36**, 325 (1983).

61. G. Kusano, T. Takemoto, N. Aimi, H. J. C. Yeh, and D. F. Johnson, *Heterocycles* **3**, 697 (1975).

62. K. S. Brown, F. R. Sim, A. Karen, R. F. Keeler, *Teratology* **21**, 30A (1980).

63. F. R. P. Sim, L. Omnell, R. F. Keeler, L. C. Harne, and K. S. Brown, manuscript in review.

64. R. M. Pratt and D. S. Salomon, "Biochemical Basis for the Teratogenic Effects of Glucocorticoids," in M. R. Jachau, Ed., *The Biochemical Basis of Chemical Teratogenesis*, Elsevier, New York, 1981, pp. 179–199.

65. B. W. O'Malley and A. B. Means, *Science* **183**, 610 (1974).

66. J. H. Clark, B. Markaverich, S. Upchurch, H. Eriksson, J. W. Hardin, and E. J. Peck Jr., *Recent Prog. Hormone. Res.* **36**, 89 (1980).

67. R. F. Keeler, "Mammalian Teratogenicity of Steroidal Alkaloids," in W. D. Nes, G. Fuller, and L.-S. Tsai, *Isopentenoids in Plants: Biochemistry and Function*, Marcel Dekker, New York, 1984, pp. 531–562.

68. J. R. Hassell, K. S. Brown, E. Horigan, I. Bleiberg, and R. F. Keeler, *Teratology* **27**, 48A (1983).

69. M. Campbell, K. S. Brown, B. Saunders, J. Hassell, E. Horigan, and R. F. Keeler, *Teratology* **29**, 22A (1984).

70. R. F. Keeler, S. Young, and R. Smart, *Teratology*, **31**, 83 (1985).

71. D. A. Green, *Arch. Otolaryngol.* **102**, 241 (1976).

72. F. R. P. Sim, B. G. Livett, C. A. Browne, S. E. Moore, R. F. Keeler and G. D. Thorburn, *Teratology* **30**, 50A (1984).

73. F. R. P. Sim, B. G. Livett, C. A. Browne, and R. F. Keeler, "Studies on the Mechanism of Veratrum Teratogenicity," in A. A. Seawright, M. P. Hegarty, R. F. Keeler, L. F. James, Eds., *Plant Toxicology*, Queensland Poisonous Plants Committee, Brisbane, Australia, in press.

74. D. F. Newgreen, I. J. Allen, H. M. Young, and B. R. Southwell, *Wilhelm Roux' Arch. Dev. Biol.* **190**, 320 (1981).

75. J. A. Wallace, *Am. J. Anat.* **165**, 261 (1982).

76. P. G. Layer, *Proc. Natl. Acad. Sci. U.S.A.* **80**, 6413 (1983).

77. I. E. Lawrence and H. W. Burden, *Am. J. Anat.* **137**, 199 (1973).

78. B. G. Livett, P. Boksa, D. M. Dean, F. Mizobe, and M. H. Lindenbaum, *J. Autonom. Nerv. Sys.* **7**, 59 (1983).

79. D. M. Dean and B. G. Livett, *Soc. Neurosci. Abstr.* **7**, 212 (1981).

80. E. W. Westhead and B. G. Livett, *Trends in Neurosciences* **6**, 254 (1983).

81. B. G. Livett and P. Boksa, *Canad. J. Physiol. Pharmacol.*, in press, (1984).

82. B. G. Livett, P. Boksa, and T. D. White, *Soc. Neurosci. Abstr.* **8**, 120 (1982).

83. S. M. Kupchan and A. W. By, "Steroid Alkaloids: The *Veratrum* Group," in R. H. F. Manske, Ed., *The Alkaloids*, Vol. 10, Academic, New York, 1968, pp. 193–285.

84. K. Schreiber, "Steroid Alkaloids: The *Solanum* Group," R. H. F. Manske, Ed., Vol. 10, Academic, New York, 1968, pp. 1–192.

85. K. Schreiber in J. G. Hawkes, R. N. Lester, and A. D. Skelding, Eds., *The Biology and Taxonomy of the Solanaceae*, Academic, New York, 1979, p. 193.

86. K. Kaneko, K. Niitsu, N. Yoshida, and H. Mitsuhashi, *Phytochem.* **19**, 299 (1980).

87. J. H. Elwood and G. MacKenzie, *Nature* **243**, 476 (1973).

88. I. S. Emanuel and L. E. Sever, *Teratology* **8**, 325 (1973).

89. B. Field and C. Kerr, *Lancet* **2**, 507 (1973).

90. L. Kinlen and A. Hewitt, *Brit. J. Prev. Soc. Med.* **27**, 208 (1973).

91. B. S. McMahon, S. Yen, and K. J. Rotham, *Lancet* **1**, 598 (1973).

92. P. S. Spiers, J. J. Pietrzyk, J. M. Piper, and D. M. Glebatis, *Teratology* **10**, 125 (1974).

93. N. C. Nevin and J. D. Merrett, *Brit. J. Prev. Soc. Med.* **29**, 111 (1975).

94. D. E. Poswillo, D. Sopher, and S. J. Mitchell, *Nature* **239**, 462 (1972).

95. D. E. Poswillo, D. Sopher, S. J. Mitchell, D. T. Coxon, R. F. Curtis, and K. R. Price, *Teratology* **8**, 339 (1973).

96. C. A. Swinyard and S. Chaube, *Teratology* **8**, 349 (1973).

97. J. A. Ruddick, J. Harwig, and P. M. Scott, *Teratology* **9**, 165 (1974).

98. L. J. Pierro, J. S. Haines, and S. F. Osman, *Teratology* **15**, 31A (1977).

99. R. F. Keeler, D. R. Douglas, and G. F. Stallknecht, *Proc. Soc. Expt. Biol. Med.* **146**, 284 (1974).

100. R. F. Keeler, D. R. Douglas, and G. F. Stallknecht, *Am. Potato J.* **52**, 125 (1975).

101. R. F. Keeler, S. Young, D. Brown, G. F. Stallknecht, and D. R. Douglas, *Teratology* **17**, 327 (1978).

102. S. C. Pakrashi, A. K. Chakravarty, and E. Ali, *Tetrahedron Lett.* **7**, 645 (1977).

103. S. C. Pakrashi, A. K. Chakravarty, T. K. Dhar, and S. Dan, *J. Indian Chem. Soc.* **LV**, 1109 (1978).

104. P. Gregory, S. L. Sinden, S. F. Osman, W. M. Tingey, and D. A. Chessin, *J. Agric. Food Chem.* **29**, 1212 (1981).

105. J. H. Renwick, *The Practitioner* **226**, 1947 (1982).

106. J. H. Renwick, W. D. B. Claringbold, M. E. Earthy, J. D. Few, and C. S. McLean, *Teratology* **30**, 371 (1984).

107. W. D. B. Claringbold, J. D. Few, and J. H. Renwick, *Xenobiotica* **12**, 293 (1982).

108. A. Usubillaga, C. Seelkopf, I. L. Karle, J. W. Daly, and B. Witkop, *J. Am. Chem. Soc.* **92**, 700 (1970).

109. G. J. Bird, D. J. Collins, F. W. Eastwood, B. M. K. C. Gatehouse, A. J. Jozra, and J. M. Swan, *Tetrahedron Lett.* **40**, 3653 (1976).

110. G. J. Bird, D. J. Collins, F. W. Eastwood, and J. M. Swan, *Aust. J. Chem.* **32**, 597 (1979).

111. G. Adam, H. Th. Huong, and N. H. Khoi, *Phytochem.* **19**, 1002 (1980).

Subject Index

Aaptamine, 300
Acanthellin-1, 303
Acetylated celenamides, 292
Acetylcholine receptor channel complex, 28
  histrionicotoxins, 84–88
N-Acetylhistamine, 206–207
3-O-Acetyljervine, 404
O-Acetylsamandarine, 167–168
Aconitine:
  polypeptide toxins allosteric
    interactions, 25–26
  radioactive analog binding, 30
  sodium channel activators, 24, 29
Acyclic aromatic hydroximines, 290
Acyliminium ion cyclization, 63, 65–68
  gephyrotoxin, 106–110
Adenosine triphosphate (ATP), 414
Aerophobins, 289–290
Aerothionin, 288
Agelasine, 301
Ageline, 301
Alfilerame, 378
Alfileramine, 335–338
  biosynthesis, 378
Alkaloid:
  amidine, 208–209
  amphibian, see Amphibian alkaloids
  background on research, 4–5
  defined, 4
  dimeric:
    background, 332–335
    Diels-Alder derived table, 334–335
    structure elucidation, 335–367
  imidazole, 206–208
  indole, 203–206
  lipophilic, 210–218
  macrocyclic, 307–309
  marine bioluminescence, 322
  marine invertebrates and
    microorganisms, 285–320
    bacteria, 285–286
    invertebrates, 287–320
  marine plants, 277–285
    phytoplankton, 277
    seaweeds, 278–285

marine vertebrates, 320–321
N-methylated or cyclized
  congeners, 210–218
nondendrobatid, 209–210
piperidine, 194–202
pyrrolidine, 202–203
steroidal, 307–309, 389–422
water-soluble, 210–218
see also specific alkaloids
Alkaloid 243A, 149
Alkaloid 251D, 126
Alkaloid 269AB, 149
Alkaloid 291A, 149
2-Alkyl-1-carboxypyrroles, 293–294
5-Alkylpyrrole-2-carboxaldehydes, 293–294
Alloculantraraminol, 341
Allodihydrohistrionicotoxin, 37, 45
Allopumiliotoxin, 128–130
  distribution, 225
  pharmacology, 144
  properties, 130, 132
    physical and spectral, 133–134
  structure-activity correlations, 144
  synthesis, 140–142
    enantiospecific, 140–141
Allopumiliotoxin 253A, 130
Allopumiliotoxin 267A,
  pharmacology, 143–144
Allopumiliotoxin 323B, 128, 170
Allopumiliotoxin 339B, 141–142
Allotetrahydrohistrionicotoxin, 37
Amathamides, 311
Amides, sponges, 306–307
Amidine alkaloids, 208–209
Amino acid:
  red algae, 284–285
  sponge derivatives, 306–307
Aminoimidazole derivatives, 295–297
Amphibian alkaloids:
  absence in captive-raised species, 222–223
  background, 4–5
  biological role, 210–221
  biosynthesis, 221–223
  distribution and taxonomic
    significance, 224–254

427

Amphibian alkaloids (*Continued*)
   batrachotoxins, 224
      other dendrobatids, 224–225, 254
      samandarines, 254
      tetrodotoxins, 254
   secondary metabolites, 210–218
   see also specific alkaloids
Amphimedine, 300
Anabaseine, 297–300
   pyridyl piperidines, 196
Anabasine, 195–196, 298
Anatoxin A, 281
Anencephaly, 402–415
Antibacterial activity of toxins, 220
Antipredator postures in amphibians,
     220–221
Antymycotic activity of toxins, 220
Aplysinopsin, 291
Aporphine, 335, 338
Arginine, 316
Ascidiacyclamide, 313
Ascidians, *see* Tunicates
Atelopidtoxin, *see* Zetekitoxin
ATPase (sarcoplasmic reticulum), 145
Axisonitrile-1, 303
Axonal transport blockade, batrachotoxin, 27
Azaspirane:
   intermediate 7 synthesis:
      (±)-depentylperhydrohistrionicotoxin,
       58
      iron-carbonyl complex, 55–56
      ozonolysis-condensation route, 55–57
      trimethylsilyl iodide, 55
   intramolecular Michael additions, 73–74
Azaspirocyclization, regiospecific, 76
Azaspiroketone, tetrahydroazepine ring
     contraction, 55–57
Azaspiro olefin syntheses, 53–58
Aziridine, 60

Bastadins, 290
Batrachotoxin:
   acetylcholine receptor channel
     complexes, 28
   axonal transport blockade, 27
   biological activity, 22–35
     pharmacological activity, 23–28
     radioactive analog binding, 30, 35
     structure, 28–30
     toxicity, 22–23
   calcium and potassium channels, 27–28
   cardiac ganglionic actions, 27
   chemical properties, 8
   Diels-Alder reaction, 35

distribution in *Phyllobates* genus, 224
early research, 6–7
effect on biological systems, 31–35
Ehrlich reaction, 8, 13
isolation procedure, 7
N-methylanthranilate analog, 29
properties, 12
   physical, 13–14
   spectral, 10–11, 13–14
   pyrrole moiety, 8, 28–29
sodium channel activation, 23–24
   altered properties, 26–27
   purification, 28
sodium channel blockers, 24–25
   allosteric interactions, 25–26
structure, 5–14
syntheses, 14–21
   ABC ring system, 17
toxicity, 13–14
Batrachotoxinin A, 7–9
   radioactive analog binding, 30, 35
   structure and activity, 28–29
   synthesis, 20
   toxicity, 22–23
   x-ray diffraction analysis, 10
Benzophenanthridine, 335
Biogenic amines, 203–204, 210–218
Biological systems, effects on:
   batrachatoxin, 31–35
   gephyrotoxin, 112
   histrionicotoxins, 92–95
Bioluminescence, marine, 322
Blow darts, 5–6
Blue-green algae:
   alkaloids, 278–282
     fresh-water, 281
   see also Organism index
Bonellin, 320
Borrerine, 341–342
Borreverine:
   derivatives, 342–344
   synthesis, 367–371
   three-bond dimers, 380
Botulinus toxin, 25
Bromogramine derivatives, 309–310
Bromoindoles, 291
   red algae, 283
Bromomethylindolex, red algae, 283
Bromopyrroles, 294
   marine bacteria, 285–286
Bromotyrosine derivatives, 298
   sponges, 287–290
Bryozoans (Ectoprocta), 309–311. *See also*
     Organism index

Bufodienolides, 186, 219
Bufothionine, 204
Bursatellin, 317
N-Butyljervine, 404
trans-2-n-Butyl-6-n-pentylpyrrolidine, 202
(5E, 9E)-3-n-Butyl-5-n-propylindolizidine,
     see Indolizidine 223AB
3-Butyl-5-propylindolizidine, stereoisomer
     synthesis, 119–121

Calcium channels:
  batrachotoxin, 27–28
  pumiliotoxin, 142–145
Calycanthine, 203, 210–218
Candicine, 204
Cannabamine B, 338
Captive-raised species, absence of toxic
     alkaloids, 222–223
Carbamylcholine, 89
  gephyrotoxin, 111
  histrionicotoxin, 89–90
Carbonimidic dichlorides, 306
Cartap hydrochloride, 320
Catecholamines, teratogens and, 413–414
Caulerpin, 282
Cebocephaly, 399
Celenamides, acetylated, 292
α-Chaconine, 416, 419
Chimonanthine, 203, 210–218
Chiriquitoxin, 186–187
  biological activity, 194
Chlorohyellazole, 281
Cleft palate incidence, 411
Clionamides, 292
Coelenterates (Cnidaria), 315–317. See also
     Organism index
Coelenterazine, 322
Congenital defects:
  diet and, 389–390
  potato blight, 402–409
  steroidal alkaloids, Simonsen hamsters,
     409–410
  variation and species specificity, 396–399
Convergence, alkaloid occurrence, 254
Corey-Kishi lactam 5:
  acliminiium ion pathway, 65
  ene reaction, 63–64, 67–69
  7,8-epi-isomer, 53–54
  synthesis, 51–54
Corey oxime 6 synthesis, 49–51
Coumarin, 332
  tryptophan dimers, 341
  yuechukene, 347–348
Crotamine, 25

Cryptobranchus alkaloid, 169
Culantraramine, 338–341
  biosynthesis, 377
  two-bond dimers, 377–378
  synthesis, 367–376
Culantraraminol:
  biosynthesis, two-bond dimers, 377–378
  isomers, 338–341
  synthesis, 367–376
Cyanohydrin, 294
Cyclic amides, sponges, 287–288
Cycloaddition reactions in histrionicotoxins,
     60–62
Cyclobisuberodiene, 332
Cycloneosamandaridine, 168
  iso designation, 176
  synthesis, 176–179
Cycloneosamandione, 166–168
  synthesis, 176–179
Cyclopamine, 394–396
  ATP release, 414
  biochemical mechanism, 411–414
  biological action site, 399–400
  dosage, 398
  natural occurrence, 414–422
  structural and configurational specificity,
     400–409
  tracheal stenosis, 412–413
Cyclopia, 399
Cycloposine, 394–396
  natural occurrence, 414–422
  structural and configurational specificity,
     400–409
Cypridina luciferin, 322

Deacetylmuldamine, 404
  teratogenicity, 422
Decahydroquinoline:
  biological activity, 162–163
  biomimetic route, 74–75
  compound I, 147–148
  compound II, 148
  compound III, 148
  distribution, 225
  properties, 149–150
    physical and spectral, 150
  structures, 148–151
  syntheses, 151–162
trans-Decahydroquinolines, 149
Dehydrobufotenine, 204–205
Demissidine, 406
Dendrobatid alkaloids:
  distribution and taxonomic significance,
     224–225, 254

Dendrobatid alkaloids (*Continued*)
  hypothetical biosynthetic pathways, 223
  occurrence in *Dendrobates* genus, 238–253
  occurrence in *Phyllobates* genus, 226–237
  see also specific alkaloids
Dendroine, 312
11-Deoxojervine, 395–396
Deoxyaminoglycosides, 286
Deoxyhistrionicotoxin:
  decahydroquinolines, 149
  natural occurrence, 83
2-Depentylperhydrohistrionicotoxin, 68
(±)-2-Depentylperhydrohistrionicotoxin,
      Michael addition, 74
2,5-Diallyl-*cis*-decahydroquinoline,
      148–149
Dibromophakellin, 295
Dibromophenethylammonium salt, 288
Dictamnine, 341
Didemnins, 315
Diels-Alder additions, gephyrotoxins,
      100–104
Diels-Alder dimeric alkaloids table, 334–335
Diels-Alder/Michael addition, gephyrotoxin,
      106–107
Diels-Alder reactions:
  intermolecular, pumiliotoxin C, 154–156
  intramolecular, pumiliotoxin C, 151–154
Dihydrobatrachotoxin, 15
7,8-Dihydrobatrachotoxinin A, synthesis, 19
  intermediate, 21
Dihydroflustramine C, 311
Dihydrogephyrotoxin, 97
  synthesis, 109–110
Dihydrohistrionicotoxin, 37
Dihydroindoles, 203
Dihydroxyquinoline, 298
  carboxylic acid, 298
Diketopiperazine derivatives, sponges, 307
Dimeric alkaloids:
  background, 332–335
  biological activity, 385
  biosynthesis, 376–384
    one-bond dimers, 377
    two-bond dimers, 377–378
    three-bond dimers, 378–381
    four-bond dimers, 381–384
  homobatrachotoxin, 12–13
  structure elucidation, 335–367
    glycobismine A, 365–367
    indole dimers, 347–348
    quinolone dimers, 348–365
    tryptophan dimers, 341–347

  tyrosine dimers, 335–341
  synthesis, 367–376
  trans-Dimer routes, 383
N,N-Dimethylhistamine, 206–207
Dimethyltryptamines, 291
Diosgenin, 403
2,5-Dipentyl-*cis*-decahydroquinoline, 148
2,6-Disubstituted (dehydro) piperidine,
      222–223
3,8-Disubstituted indolizidines, 223
2,5-Disubstituted pyrrolidines, 223
Diterpenoid isonitriles, 304–306
Dodecahydrohistrionicotoxin, 38
Dolastatin 3, 317
Domoid acid, 285
Dopamine, biological role, 210–221
Dysidin, 307

EGTA, pumiliotoxin pharmacology, 143
Embera Indians, 5–6
Embryo at risk status, 390
Enamides, 63, 65–66
Enamine, intramolecular Mannich
      reactions, 69
Enantioselective synthesis, indolizidine
      223AB, 123
Epiculantraraminol, 341
Epinephrine, biological role, 210–221
Epinine, 204
2-Epiperhydrohistrionicotoxin, 74–75
2,7-Epiperhydrohistrionicotoxin, 69, 71
Epoxides, stereospecific *cis*-enyne
      synthesis, 79–80
Eudistomins, 313

γ-Fagarine, 341
Fistularin, 288–289
Flustrabromine, 310
Flustramide, 310
Flustramines, 310–311
Foramides, 302–304
N-Formyljervine, 406–407
N-Formylsolasodines, 407–408
Fragilamide, 283–284
Frustraminols, 310–311
Furoquinolines, 338
  tryptophan dimers, 341

Geijedimerine, 363–364, 381
  four-bond dimers, 381–384
Gephran synthesis, 110–111
Gephyrotoxin, 36
  background, 95–96

bicyclic, *see* Indolizidine
biological activity, 110–112
  pharmacological activity, 111–112
  toxicity, 110–111
cis-enyne synthesis, 77–81
configurations, 95–96
defined, 95–96
effect on biological systems, 112
physical and spectral properties, 97
structure, 95–98
syntheses, 98–110
  acyliminium ion cyclization, 106–110
  Diels-Alder additions, 100–104
  Diels-Alder/Michael addition, 106–107
  gephran system, 110
  radical cyclization, 108, 110
  sigmatropic rearrangement, 104–106
  stereoselective hydrogenation, 98–100
tricyclic, 98
Gephyrotoxin 223AB, *see* Indolizidine
  223AB
Glycobismine A, 332, 365–367
  synthesis, 376
Goldman-Albright oxidation,
  batrachotoxin, 17
Gonyautoxin, 277
Granular cutaneous glands, 220
  function, 219–221
Grayanotoxin:
  radioactive analog binding, 30
  sodium channel activation, 24, 29
Green algae (Chorophyta), 282. *See also*
  Organism index
Grossularine, 312
Guanidine derivatives, 295–297
Guiazulene, 317

Halitoxin, 298
Hara-Oka alkaloid, 167, 169
  synthesis, 170–175
    proposed routes, 174
    regioselective D-ring lactone, 178–179
    samandarone conversion, 172–173
Harman, 311
Heteroaromatic bases, sponges, 300–302
Histamines:
  biological role, 210–221
  imidazole alkaloids, 206–207
Histrionicotoxin:
  analytical protocol, 38–39
  biological activity, 83–95
    acetylcholine receptor channel complex,
      84–88

binding radioactive
  perhydrohistrionicotoxin, 88–90
  potassium channel, 91–92
  sodium channels, 91
  toxicity and pharmacological activities,
    83–84
biomimetic route, 74–75
cis-enyne functionality, 77–81
distribution, 43, 45, 225
effects on biological systems, 92–95
gas chromatography, 38–41
grob cleavage, 81–82
hydrogen reduction, 82–83
mass spectra analysis, 38–43
  fragment ion, 41
N-methylation, 83
photoaffinity labeling, 86
properties, 41–43
  physical and chemical, 43–44
radioactive binding, 90
structure, 36–45
syntheses, 45–83
  acyliminium ion cyclization, 63, 65–68
  chemical modifications, 82–83
  cycloaddition reactions, 60–62
  intramolecular ene reaction, 62–63
  intramolecular Mannich reactions,
    69–70
  intramolecular Michael additions, 70–77
  spiro intermediates, 47–59
  stereoselective *cis*-enyne, 77–81
  trans-enyne, 77–81
thin-layer chromotography, 38–39
triisopropylsilyl (TIPS) groups, 78–79
Histrionicotoxin 235A, 38
Histrionicotoxin 259, 38
(±)-Histrionicotoxin, synthesis, 80–82
Δ-17-*trans*-Histrionicotoxin, 37
Homoaerothionin, 288
Homobatrachotoxin, 7
  spectral properties, 10–11
Homomorpholine:
  batrachotoxinin A analog, synthesis, 21
  propellane synthesis:
    A-B ring fusion, 15
    C-D ring fusion, 16
Homopumiliotoxin 223G, 133
Hormones, terata inhibition, 411–412
HTX-D compound, 37
Hydroxybatrachotoxin, 7
4β-Hydroxybatrachotoxin, 12
  toxicity, 23
4β-Hydroxyhomobatrachotoxin, 12

Hyellazone, 281

Imidazole alkaloids, 206–208
  spinaceamines, 208
Indole alkaloids, 203–206
  biogenic amines, 203–204
  blue-green algae, 281
  bryozoans, 310–311
  chlorinated, 281
  dehydrobufotenine, 204–205
  dihydroindoles, 203
  N-methylated or cyclized congeners,
    210–218
  miscellaneous marine invertebrates,
    319–320
  red algae, 283
  sponges, 290–292
  tetrahydrocarbolines, 205–206
Indole derivatives, 290–292
Indole dimers, 347–348
  characteristics, 348
  yuehchukene, 347–348
Indolizidine:
  biological activity, 123–124
    pharmacology, 124
    toxicity, 123
  distribution, 225
  properties, 115–116
  structure, 112–117
  syntheses, 117–123
    stereoisomers, 119–120
Indolizidine 167B, 114
Indolizidine 195B, 114–115
Indolizidine 205, 114–115
  properties, 117
Indolizidine 207A, 114–115
  properties, 117
Indolizidine 209D, 114–115
Indolizidine 223AB:
  origins, 112–113
  pharmacology, 124
  physical and chemical properties, 117
  structure, 113–114
  syntheses, 117–119
    enantioselective, 122–123
    stereoselective synthesis, 120–122
Indolizidine 225D, 115
Indolizidine 235B, 115
  properties, 117
Indolizidine 239AB:
  properties, 117
  structure, 113–114
Indolizidine 239CD:
  properties, 117

structure, 113–114
toxicity, 123
Indolizidine 239G, 115
Intramolecular ene reaction, 62–63
Intramolecular Mannich reactions, 69–70
Intramolecular Michael additions, 70–77
Iodotubercidin derivative 54, 284
Isoalfileramine, 337–338
Isobatrachotoxin, 9
Isoborrerine, 341–342
Isoborreverine, 342–344
  derivatives, 345–347
  synthesis, 367–371
  three-bond dimers, 380
Isocyanopupukeananes, 304
Isocycloneosamandaridine, 167–168,
    178–179
Isodihydrohistrionicotoxin, 36, 45
  acetylcholine receptor channel complex,
    87–88
Isodysidenin, 307
Isofistularin-3, 288–289
Isonitriles, 302–306
Isopalythazine, 316
Isoptilocaulin, 297
Isoquinoline alkaloids, 298–299
  sponges, 297–300
Isotetrahydrohistrionicotoxin, 37
  acetylcholine receptor channel complex, 87
  radioactive binding, 90
Isothiocyanates, 302–304
Istamycins, 286

Jervine, 394–396
  biochemical mechanism, 409–414
  dosage, 398
  natural occurrence, 414–422
  structural and configurational specificity,
    400–409

α-Kainic acid, 284
Kalihinol A, 305
Keramadine, 296

Lapachenole, 372
Latrunculins, 308–309
Leptodactyline, 204
Lipophilic alkaloids, 210–218
Luciferin-luciferase system, 322
Lyngbyatoxin A, 281

Macrocyclic alkaloids, 307–309
Magnesidins, 286
Majusculamides, 280

Malyngamide, 280
Malyngamide A, 280
Marine alkaloids, background, 274–275
Marine bacteria:
  alkaloids, 285–286
  see also Organism index
Marine bioluminescence, 322
Marine food web, 276
Marine invertebrates, 276
  alkaloids, 287–320
    bryozoans (Ectoprocta), 309–311
    coelenterates (Cnidaria), 315–317
    miscellaneous species, 319–320
    molluscs (Mollusca), 317–319
    sponges, 287–309
    tunicates (Urochordata), 312–315
Marine microorganisms:
  alkaloids, 285–320
  see also specific organisms
Marine phytoplankton, alkaloids, 277
Marine plants, defined, 276
Marine seaweeds, alkaloids, 278–285
  blue-green algae (cyanophta), 278–282
  green algae (chorophyta), 282
  red algae (Rhodophyta), 283–285
Marine vertebrates, alkaloids, 320–321
Mertensine, 283–284
Methanol extracts, 6
Methiodide, 164–166
Methylaplysinopsin, 291
3-O-Methylbatrachotoxin, 12
N'-Methylborreverine, 344
(5Z, 9Z) 3-Methyl-5-butylindolizidine, 124
3-O-Methyl-7,8-dihydro-20-
    epibatrachotoxinin A synthesis, 18
N-Methylflindersine, 373, 378–380
N-Methylhistamine, 206–207
N-Methyl geijedimerine, 364
3-O-Methylhomobatrachotoxin, 12
N-Methyljervine, 404–406
N-Methylmethiodide, 404
Microphthalmia, 399
Microsporine derivatives, red algae, 284
Midpacamide, 294
Millorins, 294
Mimosamycin, 299
Mimosin, 299
Molluscs (Mollusca), 317–319
Monkey face lamb disease, 396–397
Monophakellin, 295
Muldamine, 400, 404
  teratogenicity, 422

Navenone, 317

Nebenalkaloids, 168
Neodihydrohistrionicotoxin, 36
Neosamane, structure, 175–176
Neosaxitoxin, 277
Neosurugatoxin, 318–319
Nereistoxin, 320
Neurotoxins, 319
Neurotransmitters, teratogens and, 413
Nicotinic agonists, 88–89
Nicotinic antagonists, 88–89
Nitrendipine, pumiliotoxin B, 144
Nitrendipine, 90
Nitrile:
  dehydrobufotenine, 204–205
  marine sources, 317
Noanama Indians, 5–6
Noracronycine, 376
Noranabasamine, 195–196
Nucleosides:
  red algae, 284
  sponges, 300–302

Octahydrohistrionicotoxin, 37, 70, 72
  acetylcholine receptor channel complex,
    85, 87
  radioactive binding, 90
  synthesis, 82
(±)-Octahydrohistrionicotoxin, 72–73
Oenanthotoxin, 26
Oroidin, 295–296
Oxazolidine moiety in samandarine,
    164–165

Palythazine, 316
Palytoxin, 316
Paraensidimerines, 357–363, 381
  A and C, 360–361
  B and D, 362
    possible routes, 372
    three-bond dimers, 378–380
  C NMR characteristics, 358–359
  E and F, 361
    four-bond dimers, 381–384
    H NMR spectra, 364
  G, 362
  H NMR characteristics, 353–354
  mass spectra fragments, 365
  synthesis, 367, 372–376
Parallelism in alkaloid occurrence, 254
Patellamides, 313
Pavoninins, 321
Peptides:
  amphibian alkaloids, 219
  ascidian-derived cyclic, 317

Peptides (*Continued*)
 blue-green algae, 280
 thiazole-containing cyclic, 313–315
Perhistrionicotoxin, sodium channels,
  potency, 91
Perhydrogephyrotoxin, 98–99
 Diels Alder additions, 100–102
 Diels Alder/Michael addition, 106–107
 properties, 97–98
 sigmatropic rearrangement, 104–106
 synthesis:
  Diels-Alder additions, 100–102
  sigmatropic rearrangement, 104–106
  stereoselective hydrogenation, 98–99
Perhydrohistrionicotoxin, 45–48
 acetylcholine receptor channel complex,
  84–87
 [2+2] photocycloaddition, 62
 radioactive binding, 88–90
  noncompetitive blockers, 89–90
 racemic, 45–46
 synthesis:
  analog, 67
  Mannich intramolecular approach,
   69–70
  photoaffinity labeling analog, 67
 unsuccessful approach, 59
dl-Perhydrohistrionicotoxin, 85, 87–88
 acetylcholine receptor channel complex,
  85–87
(±)-Perhydrohistrionicotoxin, 60
Phakellins, 295–296
Phencyclidine, 90
[2+2] Photocycloaddition, 62–63
Photoprotein system, 322
Piperidine alkaloids, 194–202
 piperidines, 195
 pyridyl piperidines, 195–196
Piperidine-based alkaloids, 196–202
 base peak 70, 199
 other alkaloids, 200–202
  biological activity, 202
 tertiary amines:
  (base peak 140 or 126), 199–200
  (base peak 152), 196–197
  (base peak 166), 197–198
  odd integer base peak, 198–199
Plakinamines, 309
Polyandrocarpidines, 312
Polypeptide toxins, allosteric interactions,
  25–26
Potassium channel:
 batrachotoxin, 27–28
 histrionicotoxins, 91–92
 tetrodotoxin, 193

Potato sprouts:
 terata response, 420
 teratogenicity and, 402–409, 415–416
 teratogenic testing, 418
Prodigiosin, 286
2,5-di-*n*-Propyl-*cis*-decahydroquinoline,
  162
2-*n*-Propyl-*cis*-decahydroquinoline, 162
Pseudobatrachotoxin, 9
Pseudozoanthoxanthin, 315
Ptelea dimers, 348–352
 one-bond biosynthesis, 377
 pteledimericine, 351–352
 pteledimeridine, 351
 pteledimerine, 349, 351
Pteledimericine, 351–352
Pteledimeridine, 351
Pteledimerine, 349–350
 four-bond dimers, 383–384
Ptelefolidine, 348–349
Ptilocaulin, 297
Pukeleimides, 279–280
Pumiliotoxin:
 biological activity, 142–146
  pharmacology, 142–144
  toxicity, 142
 properties, physical and spectral,
  133–134
 structure-activity correlations, 144
 syntheses, 135–142
Pumiliotoxin 237A, 130
 enantiospecific synthesis, 135
 structure, 130
Pumiliotoxin 251D:
 configuration, 126
 pharmacology, 143–144
 spectra interpretation, 127–128
 structure, 126–127
 synthesis, enantioselective, 136–137
Pumiliotoxin 253A, 130
Pumiliotoxin 267C, 127–128
Pumiliotoxin 267D, 170
Pumiliotoxin 307B, 129–130
Pumiliotoxin 307F, 129–130
Pumiliotoxin 339A, 128
Pumiliotoxin A:
 distributions, 134–135
 intermediates, 223
 origins, 124–125
 pharmacology, 143–145
 properties, 130–131
 spectra interpretation, 127–128
 structure, 124–135
 syntheses, 135, 137–138
  enantioselective, 138

toxicity, 142
see also Allopumiliotoxin
Pumiliotoxin A (307A), 128–129
Pumiliotoxin A (307F), 129–130
Pumiliotoxin B:
 muscle contraction prolongations, 144–145
 origins, 124–125
 pharmacology, 142–144
 spectra interpretation, 127–128
 structure, 127
 synthesis, 136, 138–139
  enantioselective, 138–139
 toxicity, 142
Pumiliotoxin C:
 biological activity, 162–163
  pharmacology, 162
  toxicity, 162
 origins, 124–125, 148–149
 physical and spectral properties, 150–151
 syntheses, 151–162
  Diels-Alder/Michael addition, 155
  enantiospecific, 152
  indanone, 153
  intermolecular Diels-Alder reactions,
   154–156
  intramolecular Diels-Alder reactions,
   151–154
  intramolecular enamine cyclizations,
   157–160
  racemic and unnatural, 156, 158
  [3, 3]sigmatropic rearrangement,
   160–161
  tetrahydroindanones, 156–158
  unsuccessful routes, 161–162
 see also Decahydroquinoline; Piperidine-
   based alkaloids
(±)-Pumiliotoxin C, 154–156
Pyranoquinolones, 373–377
Pyrazine derivatives, 316
Pyridine:
 derivatives, 317
 sponges, 297–300
Pyrroles:
 derivatives, sponges, 292–295
 miscellaneous marine invertebrates,
  319–320
Pyrrolidine alkaloids, 202–203
Pyrrolidinones, 312

Quinazoline, conversion of tetrodotoxin,
 183–184
Quinoline alkaloid:
 bryozoans, 310–311
 sponges, 297–300
 tryptophan dimers, 341

Quinolizidine homopumiliotoxin 223G, 134
Quinolone dimers, 348–365
 geijedimerine, 363–364
 NMR and mass spectra, 364–365
 paraensidimerines, 357–363
 Ptelea dimers, 348–352
 synthesis, 373–375
 UV maxima, 350
 vepridimerines, 352–357

Red algae (Rhodophyta), 283–285. *See also*
  Organism index
Red tide phenomenon, 277
Renierone, 298
9-β-D-Ribofuranosyl-1-methylisoguanine,
 300–301

Saframycins, 299
Samandaridine, 167
Samandarine:
 biological activity, 182
 distribution, 254
 origin, 163–165
 physical and properties, 169–170
 selenium dioxide dehydrogenation,
  165–166
 structure, 163–170
 syntheses, 170–175
  apioxido bridged analog, 175
  interconversions, 180–182
  isomeric A-ring oxazolidine,
   173–174
  neosamane structure, 175–176
Samandarone:
 biological activity, 182
 configuration, 164
 structure, 167
 synthesis, 170–175
  Hara-Oka alkaloid, 172–173
  interconversions, 180–182
Samandenone, 167–168
Samandinine, 167–168
Samandiol, origin, 164–166
Samanine, 167, 169
 syntheses, 179–180
Saxitoxin, 277
 fresh-water, 281
 radioactive analog binding, 30
 sodium channel blocker, 24–25
Scalaradial, 295
Sceptrin, 296
Secondary metabolites:
 amphibians, 5
 antipredator postures, 221
 functions, 219–220

Secondary metabolites (*Continued*)
  occurrence in amphibian alkaloids,
    210–221
Serotonin, 203–204
  biological role, 210–221
Sesquiterpenoids, 306
Sigmatropic rearrangement, gephyrotoxin,
    104–106
Siphonodictidine, 297
Skimmianine, 341
Sodium channel activation:
  altered properties, 26–27
  bachtrachotoxin, 23–24
  histrionicotoxins, 91
  local anesthetic allosteric interactions, 26
  other alkaloids, 24
  polypeptide toxin allosteric interaction,
    25–26
  purification, 28
  selective blockers, 24–25
  tetrodotoxin, 193
Solanidan alkaloids, 405–406, 408
  biochemical mechanism, 409–414
  teratogenicity, 417
Solanidine, 406
  teratogenicity, 419, 421
Solanine, teratogenicity, 416
A-Solanine, 416, 419
Solanogantine, 417
Solanum alkaloids, 402
Solasodine, 402–409
  natural occurrence, 415
Solaverbascine, 422
Solidan alkaloids, natural occurrence, 415
Spina bifida, 402, 415
Spinaceamines, 208
Spiro intermediates:
  azaspiro olefin syntheses, 53–58
  spiroketolactam synthesis, 46–49
  spirolactam-olefin intermediate, 58–59
  spirooxime alcohol synthesis, 49–53
Spiroketolactam synthesis, 46–49
Spirolactam-olefin intermediate, 58–59
Spirooxime alcohol synthesis, 49–53
Spirosolane, 402–409
  biochemical mechanism, 409–414
  natural occurrence, 415
  teratogenicity, 417, 419
Sponges:
  amino acid derivatives, 306–307
  aminoimidazole and guanidine derivatives,
    295–297
  bromotyrosine derivatives, 287–290
  indole derivatives, 290–292

  isonitriles and related nitrogen-containing
    metabolites, 302–306
  macrocyclic and steroidal alkaloids,
    307–309
  pyridine, quinoline, and isoquinoline
    derivatives, 297–300
  pyrrole derivatives, 292–295
  uncommon nucleosides and related
    heteroaromatic bases, 300–302
Sponges (Porifera), 287–309. *See also*
    Organism index
Spongothymidine, 300
Spongouridine, 300
Spontaneous cyclopia, 392–394
  biological action site, 399–400
  variation and species specificity, 396–399
Stereoisomer synthesis, indolizidine 223AB,
    119–121
Stereoselective *cis*-enyne synthesis, 77–81
Stereoselective hydrogenation, gephyrotoxins,
    98–100
Stereoselective synthesis, indolizidine 223AB,
    120–122
Steroidal alkaloids, 307–309, 389–422
  background, 389–390
  biochemical mechanism, 409–414
  natural occurrence in foods and feeds,
    414–422
  potatoes, 402–409
  terata response *vs.* potato sprouts, 420
  teratogens as, 394–396
    structural and configurational specificity,
      400–409
  see also specific alkaloids
Surugatoxin, 318–319

Tambjamines, 311
Tarichotoxin, 182–184
Teinemine, 401
Teleocidin B, 281
Teratogens:
  hormone inhibition, 411–412
  steroidal alkaloids, 394–396
    biochemical mechanism, 409–414
    biological action site, 399–400
    occurrence in foods and feeds, 414–422
    structural and configurational specificity,
      400–409
Teratology:
  origins, 390
  principles, 390–392
Terpenoid amino acids, 284
Tetraacetyl clionamide, 292
Tetracyclic diisonitrile, 304–305

2,3,5,6-Tetrabromoindole, 283
7,8,16,17-Tetrahydrobatrachotoxinin A
  synthesis, 18
Tetrahydrocarbolines, 205–206
Tetrahydrohistrionicotoxin, 36
Tetrahydroindanones, 156–158
3,5,8-Trihydroxy-4-quinolinone, 298
Tetrodotoxin:
  background and origin, 183–186
  biological activity, 192–193
    chiriquitoxin, 194
    structural modifications, 193
    zetekitoxin, 194
  distribution, 254
  levels in atelopid frogs, 186
  marine vertebrates, 320–321
  pharmacological activity, 23
    blockade, 193
  quinazoline conversion, 183
  radioactive analog binding, 30
  sodium channel activation, altered
    properties, 26–27
  sodium channel blocker, 24–25
  structure, 182–188
    chiriquitoxin, 186–187
    tetrodotoxin, 182–186
    zetekitoxin, 187–188
  synthesis and chemistry, 188–192
    alternative route, 191–192
  toxicity, 192–193
Thalidomide, 390
[3+2] Thermal azide-olefin cyclization,
  60–61
[3+2] Thermal nitrone-olefin cyclization,
  61–62
Thiazole-containing cyclic peptides, 313
Tokuyama, Takashi (Dr.), 4
Tomatidine, 402–409
Tomatine teratogenicity, 416
N-Tosylated 3-formylindole, 371–372
Toxicity:
  amphibian alkaloids, 220–221
  see also specific alkaloids
Tracheal stenosis, cyclopamine, 412–413
Trypamine derivative, 310
Trypargine, 205–206
  synthesis, 209
Tryptamine, 204, 291
  biological role, 210–221
Tryptophan dimers, 338–341
  borreverine derivatives, 342–344

characteristics, 343
isoberreverine derivatives, 345–347
Tunicates (Urochordata), 312–315. See also
  Organism index
Tyramine, biological role, 210–221
Tyrian Purple, 318
Tyrosine derivatives, 298
Tyrosine dimers, 335–341
  alfileramine, 335–338
  characteristics, 336
  culantraramine and culantraraminol
    isomers, 338–341

Ulicyclamide, 313
Ulithiacyclamide, 313

Vepridimerines, 352–357
  C NMR characteristics, 358–359
  four-bond dimers, 381–384
  H NMR characteristics, 353–354
  mass spectra fragments, 365
  stereochemistry, 381
  synthesis, 367, 372–376
Vepridimerine A, 352–356
Vepridimerine B, 352–356
Vepridimerine C, 352–357
Vepridimerine D, 352–355, 357
Vepridimerine E, 375
Veratramine, 395–396
  structural and configurational specificity,
    400–409
Veratridine:
  polypeptide toxins allosteric interactions,
    25–26
  radioactive analog binding, 30
  sodium channel activation, 24, 29
Vinylharman, 311

Water-soluble alkaloids, 210–218

Xestospongins, 308

Yuechukene, 347–348
  biological activity, 385
  synthesis, 367, 371–372
  three-bond dimers, 380–381

Zetekitoxin, 187–188
  biological activity, 194
Zoanthamine, 317
Zoanthoxanthins, 315–317

Organism Index

Aaptos aaptos, 300
Ablystoma, 182
Acanthella genus, 305
Acanthella acuta, 303
Acanthella aurantiaca, 297
Adocia genus, 304–305
Aequorea aequorea, 322
Agelas genus, 296, 301
Agelas dispar, 301
Agelas nakamurai, 301
Agelas oroides, 294–295
Agelas sceptrum, 296
Amathia wilsoni, 311
Ambystoma, 193
Ambystoma tigrinum, 185
Ambystomidae family, 185
Amphanizomenon flosaquae, 277
Amphimedon genus, 300
Amphiuma means, 185
Amphiumidae family, 185
Amphizaminon flos-aquae, 281
Anabena flos-aquae, 281
Aneides luqubris, 185
Annelida phylum, 320
Anura amphibian order, 185
Aplysina genus, 287–309
   tyrosine derivatives, 298
Aplysia californica, 27
Ascidians, 312
Atelopus genus, 185
   tetrodotoxins, 254, 321
   tetrodotoxin table, 186
Atelopus chiriquiensis, 185–186
Atelopus cruciger, 185
Atelopus planispina, 185
Atelopus senex, 185
Atelopus varius, 185–187
Atelopus varius ambulatorius, 185–186
Atelopus varius varius, 185–186
Atelopus varius zeteki, 185
Atelopus zeteki, 185, 187
Axinella genus, 295
  unidentified species, 303
Axinella cannabina, 303
Axinella verrucosa, 296–297

Babylonia japonica, 318–319
Bacillus subtilis, 283–284
Balanoglossus carnosus, 319
Batrachoseps attenuatus, 185
Bombina genus, 220
Bonellia viridis, 320
Borreria verticillata:
  borrerine, 341
  borreverine, 342
   biosynthesis, 380
Brachycephalus genus, 254
Brachycephalus ephippium, 186
Bryozoans, 309–311
Bufo genus, 186
  alkaloid variations, 219
  gland secretion, 220
Bufo boreus, 185
Bufo marinus, 204
Bufonid genus:
  allopumiliotoxin, 134
  pumiliotoxins, 133
  tetrodotoxin, 185
Bufonidae family, 185
  dendrobatid alkaloids, 254
Bursatella leachii pleii, 317

Cacospongia mollior, 294
Cannabis sativa, 338
Caulerpa genus, 282
Caulerpa racemosa, 282
Centruroides genus, 25–26
Cestrum genus, 415
Chlorophyta, 282
Chondria species, 285
Chrombacterium species, 285
Clausena wildenowii, 385
Clinoa celata, 292
Cnidaria, 315–317
Coelenterates, 315–317
Colustethus genus, 219
Costaticella hastata, 311
Crypotethia crypta, 300
Cryptobranchidae family, 185
Cryptobranchus genus, 169
Cryptobranchus alleganiensis, 185

Cryptobranchus maximus, 169–170
Cyanophyta, 278–281
Cynops genus, 170
Cynops ensicauda, 185
Cynops pyrrhogaster, 182, 185
    tetrodotoxin, 193
Cyphomandra genus, 415
Cyphomandra betacea, 415

Dendrobates genus:
    alkaloid distribution, 224–225
    alkaloid occurrence, 238–253
        tables, 240–241, 244–245
    biologically active alkaloids, 219
    histrionicotoxin distribution, 43, 45
    indolizidine, 114
    piperidine alkaloids, 194–202
    pumiliotoxin, 134
    pyrolidiine alkaloids, 202–203
    summary of occurrence and properties,
        228–237
    toxicity, 221
    undescribed species, 225
Dendrobates abditus, alkaloid occurrence,
    238
Dendrobates altobueyensis, alkaloid
    occurrence, 238
Dendrobates arboreus, 225
    alkaloid occurrence, 238
Dendrobates auratus:
    alkaloid occurrence, 238–240
    allopumiliotoxin, 128
    histrionicotoxin, 38
    toxicity, 221
Dendrobates azureus, alkaloid occurrence,
    240
Dendrobates bombetes:
    alkaloid occurrence, 240
    pumiliotoxins, 133–134
    tertiary amines, 198
Dendrobates erythromos, alkaloid
    occurrence, 241
Dendrobates espinosai, 208
    alkaloid occurrence, 241
Dendrobates femoralis, alkaloid occurrence,
    241
Dendrobates fulguritus, alkaloid occurrence,
    241
Dendrobates granuliferus, 225
    alkaloid occurrence, 241
Dendrobates histrionicus:
    alkaloid distribution, 225
    alkaloid occurrence, 241–244
    alkaloid variation, 219

batrachotoxin sensitivity, 24
    biosynthesis, 222
    decahydroquinolines, 147–149
    hihydrogephyrotoxin, 97
    gas chromatographic profile, 40
    gephyrotoxins, 95–97
    histrionicotoxins, 36, 41, 43
    muscle contractions, 86
    piperidines, 195
    tertiary amines, 199
Dendrobates lehmanni:
    alkaloid distribution, 225
    alkaloid occurrence, 244
Dendrobates leucomelas, alkaloid
    occurrence, 244
Dendrobates minutus, alkaloid occurrence,
    245
Dendrobates myersi, alkaloid occurrence, 245
Dendrobates occultator:
    alkaloid distribution, 225
    alkaloid occurrence, 245
    tertiary amines, 199
Dendrobates opistomelas, 246
Dendrobates parvulus, alkaloid occurrence,
    246
Dendrobates petersi, alkaloid occurrence, 246
Dendrobates pictus, 208
    alkaloid occurrence, 246
Dendrobates pumilio:
    alkaloid distribution, 225
    alkaloid levels, 219
    alkaloid occurrence, 247–250
    allopumiliotoxin, 129
    amidine alkaloids, 208
    biosynthesis, 222
    decahydroquinolines, 146
    homopumiliotoxin 223G, 133
    indolizidine, 114
    pumiliotoxins, 124–125
Dendrobates quinquevittatus, alkaloid
    occurrence, 250
Dendrobates reticulatus, alkaloid occurrence,
    250
Dendrobates silverstonei, alkaloid
    occurrence, 250
Dendrobates speciosis (Panama),
    indolizidine, 114
Dendrobates speciosus:
    alkaloid distribution, 225
    alkaloid occurrence, 251
Dendrobates steyermarki, alkaloid
    occurrence, 251
Dendrobates tinctorius, alkaloid occurrence,
    251

Dendrobates tricolor:
  alkaloid occurrence, 252
  amidine alkaloids, 208
  pumiliotoxins, 126
Dendrobates trivittatus, 195
  alkaloid occurrence, 252
Dendrobates truncatus, alkaloid occurrence, 252
Dendrobates viridis, alkaloid occurrence, 253
Dendrodoa grosularia, 311
Dendrophryniscus stelzneri, 221
Dercitus genus, 291
Desmognathus quadramaculatus, 185
Dicathais genus, 318
Didemnidae family, 313
Digenia simplex, 285
Dinoflagellates, 277
Dinophyta, 277
Dolabella auricularia, 317
Dysidea genus, 306–307
Dysidea herbacea, 307

Echinodictyum genus, 302
Ectoprocta, 309–311
Ensatina eschscholtzi, 185
Eudistoma olivaceum, 311
Euxylophora paraensis, 357, 360, 363

Fascaplysinopsis reticulara, 291
Flindersia genus, 341–348
Flindersia fournieri, 341
  borreverine and isoborreverine biosynthesis, 380
Flustra foliacea, 310–311

Geijedimerine balansae, 363
Glycosmis citrifolia, 365–366, 376
Gobius cringer, 321
Gonyaulax genus, 277
Gonyaulax excavata, 277
Gonyaulax tamarensis, 277
Gorgonaceae order, 317
Gorgonians (sea whip), 317
Gymnodinium breve, 277

Halichondria genus, 303
Halichondria melanodocia, 292
Haliclona genus, 298
Haliclona erina, 293
Haliclona rubens, 298
Haliclona veridis, 298
Hapalosiphon fontinalis, 281
Hoplonemertinae, 297

Hyella caespitosa, 281
Hyla cinerea, 185
Hylambates maculatus, 206
Hylidae family, 185
Hymeniacidon genus, 304
Hymeniacidon aldis, 297
Hymeniacidon amphilecta, 305
Hypnea valendiae, 284

Ianthella genus, 287–309
Ianthella basta, 290
Ichthyocrinotoxic fish, 321
Iotrochota species, 291

Kassina senegalensis, 205–206

Latruncularia magnifica, 308–309
Laurencia brongniartii, 283
Laxosuberites, 293
Leimadophis genus, 221
Leiurus, toxins, 25
  radioactive batrachotoxin analog, 30
  structure and activity, 29
Leptodactylus, 220
Leptodactylus laticeps, 208
Leptodactylus pentadactylus, 208
Lissoclinum patella, 313
Litoria moorei, 208
Lumbriconereis heteropoda, 320
Lumbrinereis brevieirra, 320
Lycopersicon genus, 415
Lycopersicon pimpinellifolium, 415
Lyngbya majuscula, 279–281

Malapterurus electricus, 88
Mantella genus, dendrobatid alkaloids, 254
Mantella aurantiaca, 134–135, 209
Mantella madagascariensis, 45
  indolizidine, 117
  pumiliotoxin, 134, 209
  pumiliotoxin C, 149
Martensia fragilis, 283
Melanophryniscus genus, 254
Melanophryniscus moreirae, 134, 209
Mollusca order, 317–319
Monomarium, 202
Murex genus, 318
Murraya genus, 347–348
Murraya paniculata, 347
Mycobacterium smegmatis, 284
Myobatrachidae family, 135
  dendrobatid alkaloids, 254

Navanax inermis, 317

Necturus maculosus, 185
Nemertine worms, 297
Nictimystes disrupta, 208
Notophthalamus genus:
 tetrodotoxin, 170
 toxicity, 221
Notophthalamus viridescens, 182, 185, 193

Opossum, 221
Oricia renieri, 352
Oscarella lobalaris, 292–293

Palythoa genus, 316
Palythoa toxicus, 316
Palythoa tuberculosa, 316
Paradachirus genus, 321
Paradachirus marmoratus, 321
Paradachirus pavoninus, 321
Paramesotriton hongkongenisis, 185
Parazoanthus genus, 315
Parazoanthus axinellae, 315
Penicillium species, 302
Petrosia seriata, 307–308
Phakellia flabellata, 295–296
Pharaoh ants, indolizidines, 124
Phidolopora pacifica, 311
Phrynomerus bifasciatus, 220
Phyllidia varicosa, 304
Phyllobates genus:
 alkaloid distribution, 224
 batrachotoxin distribution, 7
 biologically active alkaloids, 219
 dendrobatid alkaloid occurrence, 226–227
 toxicity, 221
Phyllobates aurotaenia, 6–7, 11
 biosynthesis, 222
 dendrobatid alkaloid occurrence, 226
 histrionicotoxins, 36
 piperidine alkaloids, 196
Phyllobates bicolor, 6–7, 11
 dendrobatid alkaloid occurrence, 226
 piperidine alkaloids, 196
Phyllobates lugubris:
 alkaloid distribution, 224
 alkaloid occurrence, 226
Phyllobates terribilis, 6, 11–12, 203
 alkaloid distribution, 224
 alkaloid occurrence, 227
 biosynthesis, 222
 noranabasamine, 196
 pyridyl piperidines, 195
Phyllobates vittatus, 227
Phytophthora infestans, 280, 415–416
Phytoplankton, alkaloids, 227–278

Pipidae family, 185
Plakina genus, 309
Plasmopora viticola, 280
Plethodontidae family, 185
Polyandrocarpa, 311
Polyfibrospongia species, 291
Polyfibrospongia maynardii, 291
Porifera, 287–309
Potatoes:
 terata response pattern, 420
 teratogenic testing table, 418
Proteidae family, 185
Protogonyaulax species, 277
Psammaplysilla purpurea, 289
Pseudaxinyssa pitys, 306
Pseudomonas species, 285–286
Pseudomonas bromoutilis, 285
Pseudophryne genus, 210, 254
Pseudophryne corroboree, 170
 samandarine alkaloids, 254
Pseudophryne semimarimorata, 134, 170
 pumiliotoxins, 209
Ptelea genus, 377
Ptelea trifoliata, 348–349
Ptilocaulis cf. *spiculifera*, 297
Ptychodera falva laysanica, 319
Ptychodiscus brevis, 25, 277
Puffer fish, 320–321
Purpura genus, 318

Rana genus, 220
Rana catesbeiana, 221
Rana pipiens, 86, 185, 193
Ranidae family, 185
 dendrobatid alkaloids, 254
Reniera genus, 298
Renilla, 322
Rhadinaea genus, 221
Rhodophylis membranaceae, 283
Rhodophyta, 283–285
Rivularia firma, 281
Rutaceae family:
 background, 335
 biosynthesis, 368
 borreverine and derivatives, 342–344
 Diels-Alder derived dimeric alkaloids, 334–335
 dimerization process summarized, 385

Saccharomyces cerevisiae, 283
Salamander genus:
 alkaloid levels, 219
 samandarine alkaloids, 254
Salamandra atra, 163, 170

Salamandra maculosa, 220
Salamandra maculosa maculosa, 163
Salamandra maculosa taeniata:
  samandarine, 163, 168
  samandarone, 168
Salamandra salamandra, 163, 170, 185
Salamandridae family, 254
Schoenocaulon, 414
Sessibugula transluscens, 311
Siphonodictyon genus, 297
Sirenidae family, 185
Siren intermedia, 185
Siren lacertina, 185
Smenospongia genus, 291
Smenospongia aurea, 291–292
Smenospongia echina, 291
Solanum genus, 415
Solanum aviculare, 415
Solanum berthaultii, 419
Solanum dulcamera, 415
Solanum giganteum, 417
Solanum megacarpa, 415
Solanum melongena, 415
Solanum guitoense, 415
Solanum verbascifolium, 422
Spheroides vermicularis, 192–193
Sponges, 287–309
Staphyllococcus, 220
Staphylococcus aureus, indole alkaloids, 284
Streptococcus, 220
Streptomyces genus, 281
Streptomyces lavendulae, 299
Streptomyces tenjimariensis, 286

Tambje genus, 311
Taricha genus:
  tetrodotoxin, 170, 182, 193, 321
  toxin release, 221
Taricha granulosa, 184–185
  resistance to toxins, 193
  toxicity, 220–221
Taricha rivularis, 184–185
Taricha torosa, 182, 184–185
  toxicity, 220–221
Tedania digitata, 300
Tetradontidae family, 320–321
Tetraodontidae suborder, 182
Theonella cf. *swinhoei*, 304
Thorecta species, 291
Tityus genus sodium channels, 25–26
Torpedo ray, 84, 87–88

Trididemnum genus, 315
Triturus genus, 170
Triturus alpestris, 185
Triturus cristatus, 185
Triturus marmoratus, 185
Triturus torosa, 182
Triturus vulgaris, 185
Tunicates, 312–315

Urochordata, 312–315

Vepris louisii, 352
Veratrum genus:
  alkaloid properties, 394–396
  biochemical mechanism, 409–414
  structural and configurational alkaloid
    specificity, 400–407
  teratogenicity, 399–400
    timing, 413–414
Veratrum californicum:
  biochemical mechanism, 412
  causes spontaneous cyclopia, 392–394
  cyclopamine concentrations, 394–395
  cycloposine concentrations, 394–395
  illustration, 393
  occurrence of teratogens, 414–415
Veratrum grandiflorum, 395
  teinemine, 401
Verongia genus, 287–309
  tyrosine derivatives, 298
Verongia aerophoba, 288–289, 298
Verongia fistularis, 288–289
Verongia spengelii, 291
Verongia thiona, 288

Watasenia, 322

Xenopus laevis, 185
Xestospongia exigua, 308

Yueh-Chu, 347

Zanthoxylum genus, 335
Zanthoxylum coriaceum, 338
Zanthoxylum culantrillo, 338, 340–341
Zanthoxylum procerum, 340–341
Zanthoxylum punctatum, 335
Zoanthidea order, 315
Zoanthus species, 317
Zoobotryon verticillatum, 310
Zygadenus genus, 414